T0331661

MATHEMATICAL FEYNMAN PATH INTEGRALS AND THEIR APPLICATIONS

Second Edition

MATHEMATICAL FEYNMAN PATH INTEGRALS AND THEIR APPLICATIONS

Second Edition

Sonia Mazzucchi

University of Trento, Italy

 **World Scientific**

NEW JERSEY · LONDON · SINGAPORE · BEIJING · SHANGHAI · HONG KONG · TAIPEI · CHENNAI

Published by

World Scientific Publishing Co. Pte. Ltd.

5 Toh Tuck Link, Singapore 596224

USA office: 27 Warren Street, Suite 401-402, Hackensack, NJ 07601

UK office: 57 Shelton Street, Covent Garden, London WC2H 9HE

Library of Congress Cataloging-in-Publication Data
Names: Mazzucchi, Sonia, author.
Title: Mathematical feynman path integrals and their applications /
    Sonia Mazzucchi, University of Trento, Italy.
Description: Second edition. | New Jersey : World Scientific, [2022] |
    Includes bibliographical references and index.
Identifiers: LCCN 2021042126 | ISBN 9789811214783 (hardcover) |
    ISBN 9789811214790 (ebook) | ISBN 9789811214806 (ebook other)
Subjects: LCSH: Feynman integrals.
Classification: LCC QC174.17.F45 M39 2022 | DDC 530.12--dc23
LC record available at https://lccn.loc.gov/2021042126

British Library Cataloguing-in-Publication Data
A catalogue record for this book is available from the British Library.

For any available supplementary material, please visit
https://www.worldscientific.com/worldscibooks/10.1142/11679#t=suppl

Desk Editors: Balasubramanian Shanmugam/Lai Fun Kwong

Typeset by Stallion Press
Email: enquiries@stallionpress.com

Printed in Singapore

Preface to the first edition

Even if more then 60 years have passed since their first appearance in Feynman's PhD thesis, Feynman path integrals have not lost their fascination yet.

They give a suggestive description of quantum evolution, reintroducing in quantum mechanics the classical concept of trajectory, which had been banned from the traditional formulation of the theory. In fact they can be recognized as a bridge between the classical description of the physical world and the quantum one. Not only do they provide a quantization method allowing to associate, at least heuristically, a quantum evolution to any classical Lagrangian, but they also make very intuitive the study of the semiclassical limit of quantum mechanics, i.e. the study of the detailed behaviour of the wave function when the Planck constant is regarded as a small parameter converging to zero.

Nowadays the physical applications of Feynman's ideas go beyond non-relativistic quantum mechanics and include quantum fields, statistical mechanics, quantum gravity, polymer physics, geometry. Nevertheless in most cases Feynman path integrals remain a mathematical challenge as they are not well defined from a mathematical point of view.

Since 1960 a large amount of work has been devoted to the mathematical realization of Feynman path integrals in terms of a well defined functional integral. Despite the several interesting results that have been obtained in the last decades, the feeling that Feynman integrals are only an heuristic tool is still a widespread belief among mathematicians and physicists.

The present book provides a detailed and self-contained description of the rigorous mathematical realization of Feynman path integrals in terms of infinite dimensional oscillatory integrals, a particular kind of functional integrals that can be recognized as the direct generalization of classical oscillatory integrals to the case where the integration is performed on an infinite dimensional space, in particular on a space of paths.

The book describes the mathematical difficulties, the first results obtained in the 70's and the 80's, as well as the more recent development and applications. Special attention has been paid to enlightening the mathematical techniques, including infinite dimensional integration theory, asymptotic expansions and resummation techniques, without loosing the connection with the physical interpretation of the theory.

A large amount of references allows the reader to get a deeper knowledge of the most interesting mathematical results as well as of the modern physical applications.

I am grateful to many coworkers, friends and colleagues for fruitful discussions. Special thanks are due to S. Albeverio for his help and support, as well as for reading the manuscript and making lots of useful comments. I am also particularly grateful to G. Greco, V. Moretti, E. Pagani, M. Toller and L. Tubaro.

S. Mazzucchi

Preface to the second edition

Twelve years have passed since the publication of the first edition, during which new interesting results were obtained on the mathematical theory of Feynman path integrals. This second edition contains more than 100 pages of new material. Three completely new chapters have been added. They present a self-contained description of those elements of measure integration theory in finite and infinite dimension that are necessary for a deep understanding of the problems involved in the mathematical definition of Feynman path integrals. In particular, a unified view of infinite dimensional integration based on the concept of linear functional is presented in Chapter 4. Besides providing an analytical background for most existing approaches to Feynman integration, it has actually a wider scope and allows for the construction of functional integral representations of the solution of a particular class of evolution equations. In addition, the present edition contains a new section on a recent application of infinite dimensional oscillatory integrals to the Schrödinger equation for a charged particle in an external magnetic field. A careful reader will notice that even this seemingly simple model contains a number of complex problems and, even if only some of them have so far been solved, interesting ideas come into play. I hope that this book can be a source of inspiration for those young people who want to study the still challenging problem of Feynman path integration.

I am grateful to several colleagues and friends for useful discussions, in particular Sergio Albeverio, Nicolò Cangiotti, Nicolò Drago, Naoto Kumano-go, Valter Moretti, Ivan Remizov and Oleg Smolyanov.

Trento, Italy
May 2021

S. Mazzucchi

About the Author

Sonia Mazzucchi (Ph.D. University of Trento) is Associate Professor in Probability and Mathematical Statistics at the University of Trento. Her research interests include stochastic analysis and its applications to quantum physics. In 2007 and 2009 she was Alexander von Humboldt fellow at the Hausdorff Center for Mathematics of Bonn University, Germany. In 2007 she was awarded the Francesco Severi fellowship of INdAM (Italian National Institute of High Mathematics). She is author of several research papers and two monographs — *Mathematical Theory of Feynman Path Integrals: An Introduction* by S. Albeverio, R. Høegh-Krohn, S. Mazzucchi, Springer (2008), and *Mathematical Feynman Path Integrals and their Applications* (First edition) by S. Mazzucchi, World Scientific (2009) — on the mathematical theory of Feynman path integrals.

Contents

Preface to the first edition   v

Preface to the second edition   vii

About the Author   ix

1. Introduction   1

     1.1   Wiener's and Feynman's integration   6
     1.2   The Feynman functional   12
     1.3   Infinite dimensional oscillatory integrals   14

2. A primer in measure theory   19

     2.1   Measurable spaces and measures   19
     2.2   Borel measures   25
     2.3   Extension of a finitely-additive measure   26
     2.4   Measurable functions   30
     2.5   Image measure   32
     2.6   Integration of measurable functions   35
     2.7   Complex measures   38
     2.8   Fourier transform of complex measures on $\mathbb{R}^n$ 42

3. Measures on infinite dimensional spaces   45

     3.1   Non-existence of Lebesgue measure   46
     3.2   Kolmogorov theorem   50
     3.3   Wiener measure   59
     3.4   Projective limit spaces   68
     3.5   Non-existence of Feynman's measure   71

3.6 Complex Borel measures on Banach spaces and their
     Fourier transforms   74
3.7 Gaussian measures on Hilbert spaces and abstract Wiener
     spaces   79

4. Projective systems of functionals and the Fourier
    transform approach   87

4.1 Projective systems of functionals   88
4.2 Continuous extensions of projective systems of
     functionals   95
4.3 Projective systems of complex measure spaces and
     associated functionals   97
     4.3.1 Fresnel integrals   99
     4.3.2 Feynman functionals   101
     4.3.3 Complex Markov kernels, pseudoprocesses and
            high-order heat-type equations   102
4.4 The Fourier transform approach   106
4.5 Infinite dimensional Fresnel integrals   110
4.6 Infinite dimensional Fresnel integrals with
     polynomial phase   119

5. Infinite dimensional oscillatory integrals   127

5.1 Finite dimensional oscillatory integrals   128
5.2 The Parseval type equality   132
5.3 Generalized Fresnel integrals   135
5.4 Infinite dimensional oscillatory integrals   145
5.5 Polynomial phase functions   154

6. Feynman path integrals and the Schrödinger equation   169

6.1 The anharmonic oscillator with a bounded
     anharmonic potential   169
6.2 Time dependent potentials   179
6.3 Phase space Feynman path integrals   188
6.4 Quartic potential   198
6.5 Magnetic field   210

7. The stationary phase method and the semiclassical limit
   of quantum mechanics   231

    7.1   Asymptotic expansions 231
    7.2   The stationary phase method: Finite dimensional case . . 237
    7.3   The stationary phase method: Infinite dimensional case . 246
    7.4   The semiclassical limit of quantum mechanics 258
    7.5   The trace formula 269

8. Open quantum systems   277

    8.1   Feynman path integrals and open quantum systems . . . 277
    8.2   The Feynman-Vernon influence functional 285
    8.3   The stochastic Schrödinger equation 295

9. Alternative approaches to Feynman path integration   307

    9.1   Analytic continuation of Wiener integrals 307
    9.2   The sequential approach 311
    9.3   White noise calculus 314
    9.4   Poisson processes 318
    9.5   Further approaches and results 320

Bibliography   323

Index   343

Chapter 1

Introduction

One of the most challenging problems of modern physics is the connection between the macroscopic and the microscopic world, that is between classical and quantum mechanics. In principle a macroscopic system should be described as a collection of microscopic ones, so that classical mechanics should be deduced from quantum theory by means of suitable approximations. At a first glance the solution of the problem is not straightforward; indeed there are deep differences between the classical and the quantum description of the physical world.

In classical mechanics the state of an elementary physical system, for instance a point particle, is given by specifying its position q (a point in its configuration space) and its velocity $\dot{q}$; the time evolution in the time interval $[0, t]$ is described by a path $q(s)_{s \in [0,t]}$ in the configuration space. If the particle moves under the action of a force field described by the real-valued potential V, its dynamics is determined by the classical Lagrangian

$$\mathcal{L}(q, \dot{q}) := \frac{m}{2}\dot{q}^2 - V(q), \qquad (1.1)$$

where m is the mass of the particle. By the Hamilton's least action principle, the Euler-Lagrange equations of motion

$$\frac{d}{dt}\frac{\partial \mathcal{L}}{\partial \dot{q}} - \frac{\partial \mathcal{L}}{\partial q} = 0$$

follow by a variational argument. Indeed, the trajectory of the particle connecting a point x at time t_0 to a point y at time t is the path making stationary the action functional S:

$$\delta S_t(q) = 0, \quad S_t(q) = \int_0^t \mathcal{L}(q(s), \dot{q}(s))ds. \qquad (1.2)$$

The quantum description of a point particle appears at a first glance completely different. First of all the concept of trajectory is meaningless.

Indeed, Heisenberg's uncertainty principle states the existence of "incompatible observables"; the measurement of one of them destroys the information about the measurement of the other one. Position and velocity are the typical example; quantum mechanics forbids the knowledge of the couple $q(s), \dot{q}(s)$ for a time interval $[0, t]$ with arbitrary precision. In other words, from a quantum mechanical point of view the trajectory of a particle has no physical meaning as there is no way to measure it.

Contrary to classical mechanics, the state of a quantum particle moving in the d-dimensional Euclidean space is described by the so-called "wave function", i.e. a unitary vector ψ in the complex separable Hilbert space $L^2(\mathbb{R}^d)$. The physical meaning of the vector ψ is probabilistic; the probability that the result of the measurement of the position of the particle is contained in a Borel set $A \subset \mathbb{R}^d$, is given by the integral $\int_A |\psi(x)|^2 dx$.

The time evolution is determined by a strongly continuous one parameter group of unitary evolution operators $U(t) : L^2(\mathbb{R}^d) \to L^2(\mathbb{R}^d)$, whose infinitesimal generator is a densely defined self-adjoint operator $H : D(H) \subset L^2\mathbb{R}^d) \to L^2(\mathbb{R}^d)$ called the *quantum Hamiltonian*. By Stone's theorem the group $(U(t))_{t \in \mathbb{R}}$ is uniquely determined by H and one often writes $U(t) = e^{-\frac{i}{\hbar}Ht}$, where $\hbar$ is the reduced Planck constant. Under rather mild assumptions on the classical potential V, the domain $D(H)$ of Hamiltonian operator contains the set $C_0^\infty(\mathbb{R}^d)$ of compactly supported smooth functions and the action of H on vectors $\psi \in C_0^\infty(\mathbb{R}^d)$ is given by the differential operator

$$H\psi(x) = -\frac{\hbar^2}{2m}\Delta\psi(x) + V(x)\psi(x), \qquad x \in \mathbb{R}^d, \qquad (1.3)$$

where Δ is the Laplacian. Under suitable assumptions on the potential (see for instance [281]), H is essentially self-adjoint on $C_0^\infty(\mathbb{R}^d)$ and the group of evolution operators $U(t)$ is uniquely determined by Eq. (1.3). Further, the evolution of the state vector, i.e. the wave function at a given time t, can be described by the Schrödinger equation:

$$\begin{cases} i\hbar\frac{\partial}{\partial t}\psi(t,x) = -\frac{\hbar^2}{2m}\Delta\psi(t,x) + V(x)\psi(t,x) \\ \psi(0,x) = \psi_0(x) \end{cases} \qquad (1.4)$$

In 1948, following a suggestion by Dirac [120, 121], R. P. Feynman proposed a new suggestive description of quantum evolution. Feynman's aim was to provide a Lagrangian formulation of quantum mechanics by introducing the action functional as well as variational arguments in the theory, in analogy with classical mechanics. Feynman developed Dirac's ideas that in quantum dynamics the imaginary exponential of the action functional plays a fundamental role. According to Feynman's interpretation, the total transition amplitude $G_t(x, y)$ from the point x at time 0 to the point y at time t, i.e. the kernel of the evolution operator $U(t)$ evaluated at the points x, y, should be given by a sum over the contributions of all possible paths $\gamma : [0, t] \to \mathbb{R}^d$ such that $\gamma(0) = x$ and $\gamma(t) = y$:

$$G_t(x, y) = \int e^{\frac{i}{\hbar} S_t(\gamma)} D\gamma, \qquad (1.5)$$

where $D\gamma$ denotes a Lebesgue-type measure on the space of paths. Analogously, the solution of the Schrödinger equation, i.e. the wave function $\psi(t, x)$ evaluated at the time t in the point $x \in \mathbb{R}^d$, should be given by the integral over the space of paths $\gamma : [0, t] \to \mathbb{R}^d$ such that $\gamma(t) = x$:

$$\psi(t, x) = \int e^{\frac{i}{\hbar} S_t(\gamma)} \psi(0, \gamma(0)) D\gamma. \qquad (1.6)$$

In other words, according to Feynman's formulation, the time evolution of a quantum system should be given by a "sum over all possible histories".

Even if more than half a century has passed since Feynman's original paper [138], formulae (1.5) and (1.6) are still astonishing and have not lost their fascination yet. Indeed, Feynman's approach creates a bridge between the classical Lagrangian description of the physical world and the quantum one, reintroducing in quantum mechanics the classical concept of trajectory, which had been banned by the traditional formulation of the theory. It allows, at least heuristically, to associate a quantum evolution to each classical Lagrangian. Moreover it makes very intuitive the study of the semiclassical limit of quantum mechanics, i.e. the study of the behaviour of the wave function when the Planck constant $\hbar$ is regarded as a small parameter which is allowed to converge to 0. In fact, when $\hbar$ becomes small, the integrand $e^{\frac{i}{\hbar} S_t(\gamma)}$ behaves as a strongly oscillatory function and, according to an heuristic application of the stationary phase method (see Chapter 7), the main contribution to the integral should come from those paths which make stationary the phase functional S_t. These, by Hamilton's least action principle, are exactly the classical orbits of the system.

An intuitive justification of, at this stage still mysterious, Feynman's formula can be given by means of Trotter product formula [320, 91, 92]. Under suitable assumption on the potential V (see [91, 92, 320, 265] for more details) the evolution operator $U(t) = e^{-\frac{i}{\hbar} Ht}$ can be written in terms of a strong operator limit:

$$e^{-\frac{i}{\hbar} Ht} = \lim_{n \to \infty} \left(e^{-\frac{it}{\hbar n} H_0} e^{-\frac{it}{\hbar n} V} \right)^n, \qquad (1.7)$$

where $H_0 = -\frac{\hbar^2}{2m}\Delta$. In particular, by taking an initial datum $\psi_0 \in \mathcal{S}(\mathbb{R}^d)$, the solution of Eq. (1.4) with $V = 0$ can be written as

$$e^{-\frac{i}{\hbar}H_0 t}\psi_0(x) = \left(\frac{2\pi i\hbar t}{m}\right)^{-\frac{d}{2}} \int_{\mathbb{R}^d} e^{im\frac{|x-y|^2}{2\hbar t}}\psi_0(y)dy, \qquad (1.8)$$

and Eq. (1.7) gives

$$e^{-\frac{i}{\hbar}Ht}\psi_0(x) = \lim_{n\to\infty}\left(\frac{2\pi i\hbar t}{mn}\right)^{-\frac{nd}{2}} \int_{\mathbb{R}^{nd}} e^{\frac{i}{\hbar}\sum_{j=1}^n \left(\frac{m}{2}\frac{|x_j-x_{j-1}|^2}{(t/n)^2}-V(x_j)\right)\frac{t}{n}}$$

$$\psi_0(x_0)dx_0 \dots dx_{n-1}. \qquad (1.9)$$

If we now divide the time interval $[0, t]$ into n equal parts of amplitude t/n, and if for any path $\gamma : [0, t] \to \mathbb{R}^d$ we consider its approximation by means of a broken line path γ_n:

$$\gamma_n(s) := x_j + \frac{(x_{j+1}-x_j)}{t/n}(s-jt/n), \quad s \in [jt/n, (j+1)t/n], \quad j = 0\dots n-1,$$
$$(1.10)$$

where $x_j := \gamma(jt/n)$, then the exponent in the integrand of Eq. (1.9) can be regarded as the Riemann approximation of the action functional S_t evaluated along the path γ_n and the right hand side of the approximation formula (1.9) can be equivalently written as

$$\lim_{n\to\infty}\left(\frac{2\pi i\hbar t}{mn}\right)^{-\frac{nd}{2}} \int_{\mathbb{R}^{nd}} e^{\frac{i}{\hbar}\sum_{j=1}^n \left(\frac{m}{2}\frac{|\gamma(t_j)-\gamma(t_{j-1})|^2}{(t/n)^2}-V(\gamma(t_j))\right)\frac{t}{n}}$$

$$\psi_0(\gamma(t_0)\, d\gamma(t_0) \dots d\gamma(t_{n-1}). \qquad (1.11)$$

In other words, Eq. (1.6) can be regarded as an intuitive way to write the *time slicing approximation* (1.11).

These arguments present several issues, related for instance to the convergence of the approximation scheme provided by Trotter formula. Indeed, this gives the existence of the limit in (1.9) in the strong $L^2(\mathbb{R}^d)$ sense, while in most physical applications Feynman's formula is used for the construction of the kernel $G_t(x, y)$ in terms of the pointwise limit of approximations of the form

$$G_t(x, y) = \lim_{n\to\infty}\left(\frac{2\pi i\hbar t}{mn}\right)^{-\frac{nd}{2}} \int_{\mathbb{R}^{nd}} e^{\frac{i}{\hbar}\sum_{j=1}^n \left(\frac{m}{2}\frac{(x_j-x_{j-1})^2}{(t/n)^2}-V(x_j)\right)\frac{t}{n}}$$

$$dx_1 \dots dx_{n-1}.$$

with $x_0 = y$ and $x_n = x$ (see [268] for rigorous results on this problem). However, as we have already discussed above, the intuitive power of Feynman's formula goes beyond a simple mnemonic tool to write a limiting procedure. Indeed Feynman extended his approach to the description of the dynamics of more general quantum systems, including the case of quantum fields [139–141] and producing an heuristic calculus that, from a physical point of view, works even in cases where rigorous arguments fail.

Despite it successfully predicting power, Feynman path integrals lack of mathematical rigour. Feynman himself was conscious of this problem:

> [...] one feels like Cavalieri must have felt calculating the volume of a pyramid before the invention of the calculus. [138]

This sentence actually highlight what is actually the main problem of Feynman integration from a mathematical point of view. In fact, according to the approximation procedure explained above, we know how to compute the Feynman integral, but we do not actually know *what a Feynman path integral is*. This concept can be illustrated using an analogy with the elementary theory of the integration of functions $f : \mathbb{R} \to \mathbb{R}$. Some students confuse integration with the calculation of primitives and do not grasp that the real meaning of integration is the computation of *size of sets*, i.e. areas in the plane and volumes in the three-dimensional space. At this point in the discussion the mathematically minded reader may feel the need to define Feynman integration in the framework of traditional Lebesgue measure and integration theory. This could allow to interpret the heuristic Feynman's formula (1.6) as a "sum over all possible histories of the system". Besides this foundational motivation, there are strong reasons for not settling for the minimal approach in terms of piecewise linear approximations. First of all, Lebesgue integrals enjoy some important continuity properties that would play an important role in Feynman integration, in particular in those cases where approximations technique come into play, such as for instance in perturbation theory. Moreover it is important to stress that the traditional approximation procedure presents serious problem even in rather simple quantum system. An interesting example of this situation can be found in the Feynman path integral representation of the solution of the Schrödinger equation for a non-relativistic quantum particle moving under the influence of an external magnetic field $\mathbf{B}$ associated to a vector potential $\mathbf{a}$

$$i\hbar \frac{\partial}{\partial t} \psi(t, x) = \frac{1}{2} \left(-i\hbar \nabla - \mathbf{a}(x) \right)^2 \psi(t, x). \tag{1.12}$$

If we try to define the heuristic Feynman formula

$$\psi(t,x) = \int_{\gamma(t)=x} e^{\frac{i}{2\hbar}\int_0^t |\dot\gamma(s)|^2 ds + \frac{i}{\hbar}\int_0^t \lambda\mathbf{a}(\gamma(s))\cdot\dot\gamma(s)ds}\psi_0(\gamma(0))d\gamma, \qquad (1.13)$$

in terms of the traditional time-slicing approximation

$$\int e^{\frac{i}{2\hbar}\sum_i \frac{|\gamma(t_{i+1})-\gamma(t_i)|^2}{t_{i+1}-t_i} + \frac{i}{\hbar}\sum_i \lambda\mathbf{a}(\gamma(\tilde{t}_i))\cdot(\gamma(t_{i+1})-\gamma(t_i))}$$

$$\psi(\gamma(t_0))\frac{d\gamma(t_1)\cdots d\gamma(t_n)}{\prod_i (2\pi i\hbar(t_{i+1}-t_i))^{1/2}}, \qquad (1.14)$$

where $0 = t_0 < t_1 < t_2 < \cdots < t_n = t$ and $\tilde{t}_i \in [t_i, t_{i+1}]$, $i = 0, \ldots, n-1$, we discover that it present serious ambiguities. In particular different choices of the point $\tilde{t}_i \in [t_i, t_{i+1}]$ lead to different results and the only correct choice relies on the so-called "mid-point rule" which requires the vector potential **a** in (1.14) to be evaluated at the point $\tilde{t}_i \equiv \frac{\gamma(t_{i+1})+\gamma(t_i)}{2}$. This discussion will be resumed in Chapter 6. Here we just point out how this example highlights the need of a unifying theory of Feynman path integration able to go beyond the traditional approximation procedure and to provide the correct solution of Schrödinger equation without the need of introducing additional "ad hoc" rules.

1.1 Wiener's and Feynman's integration

When we try to interpret the heuristic integral (1.6) in the framework of Lebesgue theory we have to face mainly two mathematical difficulties.

First of all one has to implement a non-trivial integration theory on a space of paths, that is on an infinite dimensional space. It is reasonable to assume that the function space containing the "Feynman paths" has a metric or at least a topological structure. A possible candidate is the space of paths with "finite kinetic energy", that is the Hilbert space $\mathcal{H}_t$ of absolutely continuous paths $\gamma : [0,t] \to \mathbb{R}^d$ such that $\gamma(t) = 0$ and $\int_0^t |\dot\gamma(s)|^2 ds < \infty$ ($\dot\gamma$ denoting the distributional derivative of the path γ) endowed with the inner product

$$\langle \gamma_1, \gamma_2 \rangle = \int_0^t \dot\gamma_1(s)\dot\gamma_2(s)ds. \qquad (1.15)$$

The Hilbert space $\mathcal{H}_t$ is often called *Cameron-Martin space*.
Another possible choice is the Banach space $C([0,t], \mathbb{R}^d)$ of continuous paths $\omega : [0,t] \to \mathbb{R}^d$ such that $\omega(0) = 0$, endowed with the sup-norm $\|\omega\| = \sup_{s\in[0,t]} |\omega(s)|$. In the case where $d = 1$, this space will be denoted by C_t.

In both cases the expression $D\gamma$ in Eq. (1.6), denoting a Lebesgue "flat" measure, is meaningless. A rather simple argument shows that a Lebesgue-type measure cannot be defined on infinite dimensional Hilbert spaces. Indeed the assumption of the existence of a σ-additive measure μ which is invariant under rotations and translations and assigns a positive finite measure to all bounded open sets, leads to a contradiction. This can be easily proved by taking an orthonormal system $\{e_i\}_{i\in\mathbb{N}}$ in an infinite dimensional Hilbert space $\mathcal{H}$ and by considering the open balls $B_i = \{x \in \mathcal{H}, \|x - e_i\| < 1/2\}$; they are pairwise disjoint and their union is contained in the open ball $B(0, 2) = \{x \in \mathcal{H}, \|x\| < 2\}$. By the Euclidean invariance of the Lebesgue-type measure μ one can deduce that $\mu(B_i) = a$, for some constant $a \in (0, +\infty)$ and for all $i \in \mathbb{N}$. By the σ-additivity one has

$$\mu(B(0,2)) \geq \mu(\cup_i B_i) = \sum_i \mu(B_i) = \infty,$$

but on the other hand $\mu(B(0,2))$ should be finite as $B(0,2)$ is bounded. An analogous argument holds also for Banach spaces [158] and in particular for the space C_t and in fact it has a larger scope. This topic will be discussed in more detail in Section 3.1, showing that under rather general assumption it is impossible to define in an infinite dimensional linear space a measure that can be recognised as a reasonable and useful analogue of Lebesgue measure on $\mathbb{R}^n$. In other words, the expression $D\gamma$ in formulae (1.5) and (1.6) is not defined from a mathematical point of view and cannot be used as a reference measure, i.e. the measure with respect to which Feynman's measure has density $e^{i\frac{S_t}{\hbar}}$.

It is worthwhile to recall that integration theory on a space of continuous paths was already present at Feynman's time. In fact, the first non-trivial example of a probability measure on the space C_t had been already provided in 1923 by N. Wiener [332] in his work on Brownian motion, however there is no mention of Wiener integral in Feynman's paper.

The deep connection between Feynman's idea and Brownian motion was pointed out for the first time in 1949 by M. Kac [211, 212], who proved the celebrated Feynman-Kac formula (see Eq. (1.18) below and Section 3.3) after being inspired by a Feynman's lecture at Cornell University. In particular, Kac realized that the solution of the heat equation

$$\begin{cases} \frac{\partial}{\partial t}u(t,x) = \frac{1}{2}\Delta u(t,x) - Vu(t,x) \\ u(0,x) = u_0(x) \end{cases} \tag{1.16}$$

admits a mathematically rigorous path integral representation. Indeed, by replacing in Feynman's formula (in the simple case $V = 0$ and $\hbar = m = 1$)

the oscillatory term $e^{iS_t(\gamma)} = e^{i \int_0^t \frac{\dot{\gamma}(s)^2}{2} ds}$ with the not-oscillatory one $e^{-\int_0^t \frac{\dot{\gamma}(s)^2}{2} ds}$:

$$\frac{e^{iS_t(\gamma)} D\gamma}{\int e^{iS_t(\gamma)} D\gamma} \longrightarrow \frac{e^{-S_t(\gamma)} D\gamma}{\int e^{-S_t(\gamma)} D\gamma}$$

then the heuristic expression (1.6) can be interpreted in terms of a well defined integral on the space of continuous paths with respect to the Wiener measure W:

$$u(t, x) = \int_{C_t} u_0(\omega(t) + x) dW(\omega). \qquad (1.17)$$

In the case $V \neq 0$ Eq. (1.17) has to be replaced by

$$u(t, x) = \int_{C_t} u_0(\omega(t) + x) e^{-\int_0^t V(\omega(s) + x) ds} dW(\omega) \qquad (1.18)$$

.

and provides a probabilistic representation for the solution of the heat equation Eq. (3.20). Eq. (1.18) is now called *Feynman-Kac formula* and represents the first and most famous example of an extensively developed theory, which relates stochastic Markov processes and parabolic equations associated to 2nd order elliptic operators in such a way that the solution of the latter can be given as a functional of the former [122].

We give here only a brief description's of Wiener's measure and of Kac's result postponing the details to Chapter 3, in order to present the underlying ideas and show not only the similarities, but also the differences with Feynman's formulas.

Wiener measure is a σ-additive, Gaussian probability measure on the space $C([0, t], \mathbb{R}^d)$. Different construction procedures have been proposed (see for instance [219, 302, 265] and Section 3.5 of Chapter 3). We have chosen an intuitive approach that shows the analogies between Feynman's and Wiener's integration.

In the following for notational simplicity we shall assume that $d = 1$, but the whole discussion can be simply generalized to arbitrary dimension d. Let us consider the *cylinder sets*, i.e. the subsets of C_t of the form

$$C_{t_1,\dots,t_k;A} := \{\gamma \in C([0, t], \mathbb{R}) : (\gamma(t_1), \dots, \gamma(t_k)) \in A\}, \qquad (1.19)$$

where and A is a borel set in $\mathbb{R}^k$. In particular, if $A = I_1 \times \cdots \times I_k$ with $I_1, \dots, I_k$ intervals of $\mathbb{R}$, the set $C_{t_1,\dots,t_k;A}$ assumes the intuitive form

$$C_{t_1,\dots,t_k;I_1 \times \cdots \times I_k} := \{\gamma \in C([0, t], \mathbb{R}) : \gamma(t_1) \in I_1, \dots \gamma(t_k) \in I_k\}.$$

The Wiener measure of cylinder sets is given by the following formula:

$$W(C_{t_1,\ldots,t_k;I_1\times\cdots\times I_k}) = \left(\frac{1}{(2\pi)^k t_1(t_2-t_1)\ldots(t_k-t_{k-1})}\right)^{1/2}$$
$$\int_{I_k}\ldots\int_{I_1} e^{-\sum_{j=1}^k \frac{(x_j-x_{j-1})^2}{2(t_j-t_{j-1})}} dx_1\ldots dx_k, \quad (1.20)$$

where $t_0 \equiv 0$ and $x_0 \equiv 0$.

Eq. (1.20) admits a suggestive interpretation close to Feynman's own ideas. Let us introduce the collection of polygonal paths $\gamma \in C_t$ of the form

$$\gamma(s) = \sum_{j=0}^{k-1} \chi_{[t_j,t_{j+1}]}(s)\left(x_j + \frac{x_{j+1}-x_j}{t_{j+1}-t_j}(s-t_j)\right), \qquad s \in [0,t], \quad (1.21)$$

where $0 \leq t_1 \leq \ldots t_k \leq t$, $x_1,\ldots,x_k \in \mathbb{R}$, and the symbol $\chi_{[t_j,t_{j+1}]}$ denotes the characteristic function of the interval $[t_i,t_{i+1}]$. In fact, for any $j = 1,\ldots,k$ and for $s \in [t_j,t_{j+1}]$ the path γ coincides with the constant velocity path connecting x_j with x_{j+1}.

Let

$$S_t^\circ(\gamma) \equiv \frac{1}{2}\int_0^t |\dot\gamma(s)|^2 ds \qquad (1.22)$$

be the free action, i.e. the time integral of the kinetic energy of the path. In the case of a polygonal path γ of the form given in Eq. (1.21) it is easy to see that Eq. (1.22) reduces to

$$S_t^\circ(\gamma) \equiv \frac{1}{2}\sum_{j=0}^{k-1}\left|\frac{x_{j+1}-x_j}{t_{j+1}-t_j}\right|^2(t_{j+1}-t_j). \qquad (1.23)$$

If we define

$$D\gamma \equiv Z^{-1}\prod_{t\in\{t_1,\ldots,t_k\}} d\gamma(t), \qquad (1.24)$$

with

$$Z \equiv \left((2\pi)^k t_1(t_2-t_1)\ldots(t_k-t_{k-1})\right)^{\frac{1}{2}}, \qquad (1.25)$$

and $d\gamma(t) = dx_j$ per $t = t_j$, then Eq. (1.20) can be equivalently written as

$$\left(\frac{1}{(2\pi)^k t_1(t_2-t_1)\ldots(t_k-t_{k-1})}\right)^{1/2}\int_{I_k}\ldots\int_{I_1} e^{-\sum_{j=1}^k \frac{(x_j-x_{j-1})^2}{2(t_j-t_{j-1})}} dx_1\ldots dx_k,$$
$$= Z^{-1}\int_{I_k}\ldots\int_{I_1} e^{-S_t^\circ(\gamma)}\prod_{t\in\{t_1,\ldots,t_k\}} d\gamma(t) \quad (1.26)$$

In other words, the Wiener measure of cylinder sets can be heuristically written in terms of the following formula:

$$W(\gamma_{t_1} \in I_1, \dots, \gamma_{t_k} \in I_k) = \int_{\{\gamma(t_1) \in I_1, \dots, \gamma(t_k) \in I_k\}} e^{-S_t^\circ(\gamma)} D\gamma. \qquad (1.27)$$

The right hand side has a meaning as soon as we restrict ourselves to cylinder sets, as the infinite dimensional Wiener integration can be reduced to a (finite dimensional) integration on $\mathbb{R}^k$. In this case both the "normalized Lebesgue measure" $D\gamma$ and the factor $e^{-S_t^\circ(\gamma)}$ make sense. When the measure is extended on the whole Borel σ-algebra $\mathcal{B}(C_t)$, formula (1.27) has to been interpreted as a finite dimensional approximation: the key point is that, even if the single terms $D\gamma$ and $e^{-S_t^\circ(\gamma)}$ lose a well defined meaning, their combination is still meaningful and it gives exactly Wiener Gaussian measure.

If we now consider Eq. (3.20) and, analogously to the Schrödinger case, we write the corresponding heat semigroup in terms of the Trotter product formula [320, 91, 92], we obtain:

$$
\begin{aligned}
e^{-Ht} u_0(x) &= \lim_{n \to \infty} \left(e^{-\frac{t}{n} H_0} e^{-\frac{t}{n} V} \right)^n \\
&= \lim_{n \to \infty} \left(\frac{2\pi t}{nm} \right)^{-\frac{n}{2}} \int_{\mathbb{R}^n} e^{-\sum_{j=1}^n \left(\frac{(x_j - x_{j-1})^2}{(t/n)^2} - V(x_j) \right) \frac{t}{n}}
\end{aligned}
$$

$$u_0(x_0) dx_0 \dots dx_{n-1}. \qquad (1.28)$$

Now the latter line can be recognized as the finite dimensional approximation of the Wiener integral (1.18). In addition, according to the heuristic notation introduced above, the second line of Eq. (1.28) can be formally written

$$\lim_{n \to \infty} \int_{\gamma(t_1) \in \mathbb{R}, \dots, \gamma(t_n) \in \mathbb{R}} e^{\sum_{j=1}^n V(\gamma(t_j)) \frac{t}{n}} u_0(\gamma(0)) e^{-S_t^\circ(\gamma)} D\gamma,$$

where $t_j \doteq jt/n$, $j = 1, \dots, n$.

The analogies between Feynman's and Wiener's integral end at this stage because there is a deep difference between them, due to the presence of oscillations in the heuristic "Feynman measure" $e^{\frac{i}{\hbar} S_t(\gamma)} D\gamma$, as M. Kac himself writes:

> The occurrence of i (which is essential for Quantum Mechanics) makes manipulations with integrals like [formula (1.9)] extremely tricky. [213]

Wiener integration is a Lebesgue type integration, where the absolute convergence is fundamental. On the other hand, physical intuition leads us to stress the importance of the oscillatory behaviour of the integrand in Feynman's formula, which describes the concept of coherent superposition and of interference, which is typical of quantum phenomena. In principle the convergence of the integral should be given by the cancellations due to this oscillatory behavior and we should not expect to implement an integration theory in the Lebesgue's traditional way.

This fact was clearly explained by R.H. Cameron in 1960 [77], who proved that it is not possible to realize Feynman's measure as a infinite dimensional Gaussian measure with complex covariance (a complex version of Wiener measure) as it would have infinite total variation. We give here a sketch of Cameron's argument as it helps us to understand the core of the problem: the oscillations and the infinite dimensional setting. Cameron's starting point is Wiener measure on $C([0, t], \mathbb{R}^d)$ with a generic covariance $\sigma \in \mathbb{R}^+$. As we have seen above, it is completely determined by its value on cylinder sets (1.19). In the general case $\sigma \neq 1$ Eq. (1.20) becomes:

$$W_\sigma(A(t_1, \ldots t_k; I_1 \ldots I_k)) = \left(\frac{1}{(2\pi)^k t_1 (t_2 - t_1) \ldots (t_k - t_{k-1}) \sigma^k} \right)^{1/2}$$

$$\int_{I_1 \times \cdots \times I_k} e^{-\sum_{j=1}^k \frac{(x_j - x_{j-1})^2}{2\sigma(t_j - t_{j-1})}} dx_1 \ldots dx_k \quad (1.29)$$

Let us now assume that σ is a complex parameter and try to extend the definition of W_σ to this case. If we try to compute the total variation[1] $|W_{\sigma,k}|$ of the "complex measure" W_σ restricted to the collection of cylinder sets $C_{t_1,\ldots t_k; A}$ *with k fixed*, we find out that it depends on both σ and k in the following way:

$$|W_{\sigma,k}| = \left(|\sigma| \mathrm{Re}(\sigma^{-1}) \right)^{-k/2}.$$

By letting now $k \to \infty$ we can conclude that the total variation of the measure W_σ, with $\sigma \in \mathbb{C} \setminus \mathbb{R}^+$, cannot be finite. Furthermore, it is possible to see that the complex measure W_σ would have infinite total variation even on bounded sets[2]. In other words there is no σ-additive measure W_σ, with $\sigma \in \mathbb{C} \setminus \mathbb{R}^+$, on C_t such that its value on cylinder sets is given by

---

[1] We recall that the total variation of a complex Borel measure μ on a set A is given by

$$|\mu| = \sup \sum_i |\mu(A_i)|,$$

where the supremum is taken over all sequences $\{A_i\}$ of pairwise disjoint Borel subsets of A, such that $\cup_i A_i = A$.

[2] Even the Lebesgue measure on $\mathbb{R}^n$ has infinite total variation, but its total variation on bounded pluri-intervals is finite.

Eq. (1.29) and which allows the implementation of an integration theory in the Lebesgue's sense. We shall extensively discuss this issue in Chapter 3 in a wider context, showing that Cameron's no-go result is actually a particular case of a generalization of Kolmogorov existence theorem (see Theorems 3.4 and 3.7) to the case of complex measures on infinite dimensional spaces.

It is worthwhile to remark that, in the case $\sigma = i$, the Gaussian measure with covariance σ cannot have finite total variation, even when it is defined on a finite dimensional space. As an example we can consider the Fresnel integral

$$\int_{\mathbb{R}} e^{\frac{i}{2}x^2} dx. \tag{1.30}$$

Eq. (1.30) cannot be interpreted as an integral with respect to a complex measure $d\mu := e^{\frac{i}{2}x^2} dx$, as its total variation would be infinite:

$$|\mu| = \int |e^{\frac{i}{2}x^2}| dx = \int dx = \infty$$

The integral (1.30) has to be defined in an alternative way, for instance as an improper Riemann integral, and its convergence is given by the cancellations due to the oscillatory behaviour of the integrand, is such a way that:

$$\int_{\mathbb{R}} e^{\frac{i}{2}x^2} dx = \sqrt{2\pi i}.$$

We mention here that an alternative way for the definition of non-absolutely convergent integrals of the form (1.30) is the theory of Henstock-Kurzweil integral, which can also be applied to the theory of Feynman integration (see the monographs [163, 260] for a detailed account of this topic).

1.2   The Feynman functional

Cameron's result shows that it is impossible to realize Feynman integral as an absolutely convergent (Lebesgue) integral with respect to a "mysterious" σ-additive Feynman complex measure μ_F, heuristically written as

$$\mu_F(\gamma) = \frac{e^{\frac{i}{\hbar}S_t} d\gamma}{\int e^{\frac{i}{\hbar}S_t} d\gamma}, \tag{1.31}$$

as the latter cannot exist. Feynman integration requires an alternative approach.

Let us slightly change our point of view and recall that, by Riesz-Markov theorem (see Theorem 2.20), any complex regular Borel measure μ on a locally compact space X can be seen as an element of the dual of $C_0(X)$, the space of continuous complex valued functions vanishing at ∞, endowed with the sup-norm:

$$\|f\| = \sup_{x \in X} |f(x)|, \qquad f \in C_0(X).$$

In this way, the integral of a function $f \in C_0(X)$ with respect to a complex measure μ can be represented as the action on f of the functional $l_\mu \in C_0(X)^*$ associated to μ:

$$\int_X f(x) d\mu(x) \equiv l_\mu(f). \tag{1.32}$$

In Feynman's case, as we have seen so far, the left-hand side of Eq. (1.32) is not defined, as Feynman's measure does not exists, but one could try to make sense of the right-hand side of Eq. (1.32) by slightly changing the functional setting. In other words, one could try to realize the Feynman integral as a linear continuous functional on a sufficiently rich Banach algebra of functions, different from $C_0(X)$. This issue will be extensively discussed in Chapter 4. Here we limit ourselves to mention that such a functional should fulfill some conditions in order to mirror the features of the heuristic Feynman measure. In particular:

(1) It should behave in a simple way under "translations and rotations in path space", reflecting the fact that $D\gamma$ heuristically represents a "flat" measure.

(2) It should satisfy a Fubini-type theorem, concerning iterated integrations in path space (allowing the construction, in physical applications, of a one-parameter group of unitary operators).

(3) It should be approximated by finite dimensional oscillatory integrals, allowing a sequential approach, close to Feynman's original work.

(4) It should be related to probabilistic integrals with respect to the Wiener measure, allowing an "analytic continuation approach to Feynman path integrals from Wiener type integrals".

(5) It should be sufficiently flexible to allow a rigorous mathematical implementation of an infinite dimensional version of the stationary phase method and the corresponding study of the semiclassical limit of quantum mechanics.

Nowadays several implementation of this program can be found in the physical and in the mathematical literature. One of the first techniques introduced and widely developed, also in connection to quantum fields, is the analytic continuation of Wiener Gaussian integrals

[77, 265, 203, 323, 215, 123, 245, 260, 94, 317, 318]. The starting point of this approach is the transformation of variable formula for the Wiener integral with covariance σ:

$$\int f(\omega)dW_\sigma(\omega) = \int f(\sqrt{\sigma}\omega)dW(\omega). \tag{1.33}$$

As Cameron proved, the left-hand side of Eq. (1.33) is not defined when σ is complex. The leading idea of the analytic continuation approach is to give meaning to the right hand side of Eq. (1.33) in the case where $\sigma = i$ for a suitable class of analytic functions f.

Another alternative approach is the realization of Feynman measure as an infinite dimensional distribution. The idea was proposed by C. De Witt-Morette [114]. Its rigorous mathematical realization has been more recently undertaken in the framework of Hida calculus [183]. The latter approach has given particularly interesting results in the applications to Chern-Simons theory [16].

Other possible approaches involve "complex Poisson measures" [249, 67, 99, 232] and non-standard analysis [13]. All these approaches will be described in detail in Chapter 9.

1.3   Infinite dimensional oscillatory integrals

In this book we shall especially focus on the rigorous mathematical definition of Feynman path integrals in terms of the *infinite dimensional oscillatory integrals*.

The leading idea is the rigorous definition of an infinite dimensional analogue of the Fresnel integral (1.30), in particular of expressions as

$$\int_{\mathcal{H}} f(x)e^{\frac{i}{2\hbar}\|x\|^2}dx, \tag{1.34}$$

where $\mathcal{H}$ is a real separable infinite dimensional Hilbert space, $\hbar \in \mathbb{R}^+$ a positive parameter, $f : \mathcal{H} \to \mathbb{C}$ is a suitable function and $e^{\frac{i}{2\hbar}\|x\|^2}$ plays the role of the oscillatory factor, the density of an heuristic complex Gaussian measure.

The roots of this approach can be found in two papers by K. Ito appeared in the 60s [195, 196] (see also [33] for a discussion of Ito's work), but it was systematic developed by Albeverio and Høegh-Krohn in the 70s [17, 18] in terms of *infinite dimensional Fresnel integrals*.

In Albeverio and Høegh-Krohn's work, the integral (1.34) is defined by dualization. By taking a function f which is the Fourier transform of a complex bounded-variation measure μ_f on $\mathcal{H}$

$$f(x) = \int_{\mathcal{H}} e^{i\langle x,y\rangle} d\mu_f(y), \qquad (1.35)$$

the infinite dimensional Fresnel integral of f is defined as

$$\int_{\mathcal{H}} f(x)e^{\frac{i}{2\hbar}\|x\|^2} dx := \int_{\mathcal{H}} e^{-\frac{i\hbar}{2}\|x\|^2} d\mu_f(x). \qquad (1.36)$$

The integral on the right-hand side of (1.36) is absolutely convergent and well defined in Lebesgue's sense. The class of functions f of the form (1.35), endowed with a suitable norm, is a Banach algebra, the *Fresnel algebra*, denoted with $\mathcal{F}(\mathcal{H})$. The application

$$I_F : \mathcal{F}(\mathcal{H}) \to \mathbb{C}, \qquad f \mapsto \int_{\mathcal{H}} f(x)e^{\frac{i}{2\hbar}\|x\|^2} dx,$$

is a linear continuous functional.

In the application of this formalism to the Feynman path integral representation of the solution of the Schrödinger equation, by assuming that the potential V is the sum of an harmonic oscillator term and a bounded perturbation

$$V(x) = \frac{1}{2}x \cdot \Omega^2 x + V_1(x) \qquad (1.37)$$

(where Ω^2 is a positive definite symmetric $d \times d$ matrix and V_1 is the Fourier transform of a complex measure on $\mathbb{R}^d$), Albeverio and Høegh-Krohn prove that the solution can be represented as an infinite dimensional Fresnel integral on the Cameron-Martin space (see Eq. (1.15)).

In [17] an infinite dimensional version of the stationary phase method has been developed and applied to the semiclassical limit of quantum mechanics. Some of these results will be described in detail in Chapter 7.

The study of oscillatory integrals on infinite dimensional Hilbert spaces was further implemented by D. Elworthy and A. Truman [129], S. Albeverio and Z. Brzeźniak [7, 9], S. Albeverio and S. Mazzucchi [28, 37, 251, 39, 11]. In [129] a slightly different definition was proposed. The integral (1.34) is defined as the limit of a sequence of finite dimensional approximations. Each term of the sequence is a classical oscillatory integral on a finite dimensional space, which is defined as an improper Riemann integral, by modifying a definition proposed by Hörmander [186, 187]. This new definition of the integral (1.34) can be recognized as the infinite dimensional generalization of oscillatory integrals of the type (1.30).

The class of "integrable function" f is in principle different from the Fresnel class $\mathcal{F}(\mathcal{H})$ considered by Albeverio and Høegh-Krohn, as the definition of the infinite dimensional Fresnel integrals and of the infinite dimensional oscillatory integrals are different. However it is possible to prove that any function $f \in \mathcal{F}(\mathcal{H})$ is integrable and its infinite dimensional oscillatory integral, i.e. the limit of a sequence of finite dimensional approximations, exists and is given by Eq. (1.36), that in this new setting has to be interpreted as a theorem instead of a definition.

It is worthwhile to point out that the definition of infinite dimensional oscillatory integrals is rather flexible and allows not only to enlarge the class of integrable functions [28] to sets larger than $\mathcal{F}(\mathcal{H})$, but also to study several applications to quantum mechanics [27].

In this book theory and applications of infinite dimensional oscillatory integrals will be extensively presented.

- Chapter 2 provides some basic notions of measure and integration theory.

- Chapter 3 is a detailed presentation of those topics of measure and integration theory on infinite dimensional spaces that are of particular importance in the mathematical theory of Feynman path integrals.

- Chapter 4 describes a systematic implementation of a theory of infinite dimensional integration based on the concept of linear functional. It provides a unified view of probabilistic and oscillatory integration in infinite dimensions as well as the general framework for the theory of infinite dimensional Fresnel integrals described in Section 4.5.

- Chapter 5 contains the theory of finite and infinite dimensional oscillatory integrals.

- Chapter 6 describes the application of infinite dimensional oscillatory integrals to the rigorous mathematical realization of the Feynman path integral representation of the solution of the Schrödinger equation.

- Chapter 7 is devoted to the stationary phase method and its application to the study of the semiclassical limit of quantum mechanics.

- Chapter 8 shows that it is possible to generalize the definition of infinite dimensional oscillatory integrals in order to deal with complex-valued phase functions. Such a functional is applied to the solution of a stochastic Schrödinger equation appearing in the theory of continuous quantum measurement: the Schrödinger-Belavkin equation. A mathematical definition and construction of the Feynman-Vernon influence functional is also given.

- The last chapter is devoted to a brief description of some alternative approaches to the mathematical definition of Feynman path integrals and to their applications.

Chapter 2

A primer in measure theory

This chapter presents some basic notions of measure and integration theory. Besides the classical results, we present some notions that will play an important role in the development of Feynman integration theory such as, e.g., the theory of complex measures and their Fourier transform. For additional details we refer to, e.g., [57, 56, 70, 96, 293, 66].

2.1 Measurable spaces and measures

Definition 2.1. A *measurable space* is a couple $(\Omega, \mathcal{F})$ where Ω is a set and $\mathcal{F}$ is a σ- algebra of subsets of Ω, i.e. a family of sets satisfying the following conditions

- $\Omega \in \mathcal{F}$
- if $A \in \mathcal{F}$ then $A^c \in \mathcal{F}$
- if $\{A_n\} \subset \mathcal{F}$ is a sequence of subsets in $\mathcal{F}$, then $\cup_n A_n \in \mathcal{F}$.

The elements of a σ-algebra are often called *measurable sets*.
Trivial examples of σ-algebras are the power set $\mathcal{P}(\Omega)$ of Ω and the collection $\mathcal{F} = \{\emptyset, \Omega\}$. In non-trivial cases σ-algebras are constructed starting from rather small collections of sets according to the procedure described in the following definition.

Definition 2.2. Let $\mathcal{C} \subset \mathcal{P}(\Omega)$ be a collection of subsets of a set Ω. The intersection of all σ-algebras containing $\mathcal{C}$ is called σ-*algebra generated by* $\mathcal{C}$ and it is denoted with the symbol $\sigma(\mathcal{C})$:

$$\sigma(\mathcal{C}) := \bigcap_{\substack{\mathcal{C} \subset \mathcal{F} \\ \mathcal{F}\, \sigma-\text{algebra}}} \mathcal{F}$$

19

It is rather easy to prove that $\sigma(\mathcal{C})$ is a σ-algebra with the following properties:

$$\mathcal{C} \subset \sigma(\mathcal{C}) \tag{2.1}$$

$$\text{If } \mathcal{C} \subset \mathcal{F} \text{ and } \mathcal{F} \text{ is a } \sigma\text{-algebra, then } \sigma(\mathcal{C}) \subset \mathcal{F} \tag{2.2}$$

$$\text{If } \mathcal{C} \text{ is a } \sigma\text{-algebra then } \mathcal{C} = \sigma(\mathcal{C}) \tag{2.3}$$

$$\text{If } \mathcal{C} = \emptyset \text{ then } \sigma(\mathcal{C}) = \{\emptyset, \Omega\} \tag{2.4}$$

$$\text{If } \mathcal{C}_1 \subset \mathcal{C}_2 \text{ then } \sigma(\mathcal{C}_1) \subset \sigma(\mathcal{C}_2). \tag{2.5}$$

In particular, on the basis of property (2.2), $\sigma(\mathcal{C})$ is the smallest σ-algebra containing $\mathcal{C}$.

Example 2.1. If $\mathcal{C} = \{E\}$, with $E \subset \Omega$ and $S \neq \Omega, \emptyset$, then $\sigma(\mathcal{C}) = \{E, E^c, \Omega, \emptyset\}$.

Example 2.2. Let Ω be a topological space, and $\mathcal{C}$ is the corresponding topology, i.e. the collection of open sets, then the σ-algebra generated by $\mathcal{C}$ is called the *Borel σ-algebra*. If there is not ambiguity on the choice of the topology on Ω, then the Borel σ-algebra is denoted with the symbol $\mathcal{B}(\Omega)$.

Example 2.3. Let $\Omega = \mathbb{R}$ endowed with the euclidean topology and let us consider the following collections of subsets.

- $\mathcal{C}_1 = \{(-\infty, t], \, t \in \mathbb{R}\}$,
- $\mathcal{C}_2 = \{(a, b), \, a, b \in \mathbb{R}\}$,
- $\mathcal{C}_3 = \{(a, b], \, a, b \in \mathbb{R}\}$.

By applying property (2.5) above, it is rather simple to prove that all these families of sets generate the Borel σ-algebra of the real line

$$\mathcal{B}(\mathbb{R}) = \sigma(\mathcal{C}_i), \qquad i = 1, 2, 3.$$

Definition 2.3. Let $(X_i, \mathcal{F}_i)$, $i = 1, \ldots, n$, measurable spaces. The *product σ-algebra* $\times_{i=1}^n \mathcal{F}_i = \mathcal{F}_1 \times \cdots \times \mathcal{F}_n$ is the σ-algebra on the Cartesian product $\times_{i=1}^n X_i = X_1 \times \cdots \times X_n$ generated by the collection of product sets of the form $E_1 \times \cdots \times E_n$, where $E_i \in \mathcal{F}_i$, $i = 1, \ldots, n$.

For future use it is convenient to introduce the notion of λ-system (also called *Dynkin system*).

Definition 2.4. A class of sets $\mathcal{L} \subset \mathcal{P}(\Omega)$ is called λ-system if it satisfies the following conditions:

λ_1. $\Omega \in \mathcal{L}$;

λ_2. if $A \in \mathcal{L}$ then $A^c \in \mathcal{L}$.

λ_3. if $\{A_n\}_{n \in \mathbb{N}} \subset \mathcal{L}$ is a countable family of sets in $\mathcal{L}$ such that $A_n \cap A_m = \emptyset$ if $n \neq m$, then $\cup_n A_n \in \mathcal{L}$.

Remark 2.1. The set of properties $\lambda_1, \lambda_2, \lambda_3$ is equivalent to the set of properties $\lambda_1, \lambda_2', \lambda_3$ where:

λ_2'. if $A, B \in \mathcal{L}$, with $B \subset A$ then $A \setminus B \in \mathcal{L}$.

In fact the set of properties defining a λ-system is weaker than those appearing in the definition of a σ-algebra. In other words a σ-algebra is always a λ-system but the converse is not true. The following result shows which property needs to be added to the definition of λ-system to get a σ-algebra.

Definition 2.5. A class of sets $\mathcal{P} \subset \mathcal{P}(\Omega)$ is called π-system if it satisfies the following condition:

π. if $A, B \in \mathcal{P}$ then $A \cap B \in \mathcal{P}$.

In other words, a π-system is a class of sets closed under finite intersections. It is rather simple to prove the following result

Lemma 2.1. *If $\mathcal{C} \subset \mathcal{P}(\Omega)$ is both a π-system and a λ-system, then $\mathcal{C}$ is a σ-algebra.*

The following theorem finds several applications in different settings.

Lemma 2.2 (Dynkin). *Let $\mathcal{P}$ be a π-system and $\mathcal{L}$ a λ-system. If $\mathcal{P} \subset \mathcal{L}$ then $\sigma(\mathcal{P}) \subset \mathcal{L}$.*

Proof. Let us denote $\mathcal{L}_0 \subset \mathcal{P}(\Omega)$ the intersection of all λ-systems containing $\mathcal{P}$:

$$\mathcal{L}_0 := \bigcap_{\substack{\mathcal{P} \subset \mathcal{L} \\ \mathcal{L} \, \lambda-\text{system}}} \mathcal{L}.$$

The proof of the theorem is divided into three main steps:

(1) $\mathcal{L}_0$ is a λ-system containing $\mathcal{P}$ (this part of the proof is left as an exercise).

(2) $\mathcal{L}_0$ is a π-system. This point is rather complicated and is proved below.

(3) By 1 and 2 as well as Lemma 2.1, we can conclude that $\mathcal{P} \subset \mathcal{L}_0 \subset \mathcal{L}$ and that $\mathcal{L}_0$ is a σ-algebra containing $\mathcal{P}$, hence:

$$\mathcal{P} \subset \sigma(\mathcal{P}) \subset \mathcal{L}_0 \subset \mathcal{L}.$$

Let us now prove in some detail part 2. We have to show that if $A, B \in \mathcal{L}_0$ then $A \cap B \in \mathcal{L}_0$.

Let us consider a set $A \in \mathcal{L}_0$ and let us define the collection $\mathcal{L}_A \subset \mathcal{L}_0$ as:

$$\mathcal{L}_A := \{E \in \mathcal{L}_0 : A \cap E \in \mathcal{L}_0\}.$$

2.1. $\mathcal{L}_A$ is a λ-system. The proof of this part is left as an exercise.

2.2. $\mathcal{P} \subset \mathcal{L}_A$.

    2.2.a First of all, let us prove that is $A \in \mathcal{P}$ then $\mathcal{L}_0 \subset \mathcal{L}_A$. Indeed, if $A \in \mathcal{P}$ then $\mathcal{P} \in \mathcal{L}_A$ since $\mathcal{P}$ is a π-system and, by step 1., $\mathcal{P} \subset \mathcal{L}_0 \subset \mathcal{L}_A$.

    2.2.b By step 2.2.a we can deduce that $\forall A' \in \mathcal{P}$ and $\forall A \in \mathcal{L}_0$ one has that $A \cap A' \in \mathcal{L}_0$. This means that if $A \in \mathcal{L}_0$ then $\mathcal{P} \in \mathcal{L}_A$.

2.3 Eventually, by points 2.1 e 2.2 we obtain that $\mathcal{P} \subset \mathcal{L}_0 \subset \mathcal{L}_A$, which means that if $B \in \mathcal{L}_0$ then $B \cap A \in \mathcal{L}_0$. By the generality on $A \in \mathcal{L}_0$, we obtain that $\mathcal{L}_0$ is a π-system.

$\square$

Definition 2.6. A *measure* μ on a measurable space $(\Omega, \mathcal{F})$ is a positive map $\mu : \mathcal{F} \to [0, +\infty]$ such that

(1) $\mu(\emptyset) = 0$;
(2) $\mu(\cup_n E_n) = \sum \mu(E_n)$ for any countable family of sets $\{E_n\} \subset \mathcal{F}$ such that $E_n \cap E_m = \emptyset$ whenever $n \neq m$.

Property 2 is called σ-*additivity* of the set function μ. Condition 1 can be replaced by the, apparently weaker,

  1' $\mu(E) \neq +\infty$ for at least a set $E \in \mathcal{F}$.

If $\mu(\Omega) < +\infty$ then μ is said to be a *finite measure*, while if $\mu(\Omega) = 1$ then μ is a *probability measure*. A positive measure is said to be $\sigma-finite$ if there exists a countable family of measurable sets $\{E_n\} \subset \mathcal{F}$ such that $\Omega = \cup_n E_n$ and $\mu(E_n) < +\infty$ for all n.

Definition 2.7. A *measure space* is a triple $(\Omega, \mathcal{F}, \mu)$, where $(\Omega, \mathcal{F})$ is a measurable space and $\mu : \mathcal{F} \to [0, +\infty]$ is a measure on $\mathcal{F}$.

If μ is a probability measure then $(\Omega, \mathcal{F}, \mu)$ is called *probability space*.

Definition 2.8. Let $(\Omega, \mathcal{F}, \mu)$ be a measure space. A property P is said to hold *almost everywhere* if there exists a set $N \in \mathcal{F}$ such that $\mu(N) = 0$ and all points $\omega \in \Omega \setminus N$ have the property P.

Example 2.4. A trivial example of measure space can be obtained by taking $\mathcal{F} = \mathcal{P}(\Omega)$ and $\mu(E) = 0$ for all $E \in \mathcal{F}$.

Example 2.5. Let Ω be a non-empty set and fix an element $\omega \in \Omega$. The *Dirac measure* in Ω is the probability measure δ_ω on $(\Omega, \mathcal{P}(\Omega))$ defined as

$$\delta_\omega(E) := \begin{cases} 1 & \omega \in E \\ 0 & \omega \notin E. \end{cases} \tag{2.6}$$

Any measure μ on a measurable space enjoys the following properties:

i. Monotonicity. $\mu(E_1) \leq \mu(E_2)$ for all $E_1, E_2 \in \mathcal{F}$ such that $E_1 \subset E_2$;
ii. Finite additivity. $\mu(\cup_{i=1}^n E_i) = \sum_{i=i}^n \mu(E_i)$ for all $E_1, \ldots, E_n \in \mathcal{F}$ such that $E_i \cap E_j = \emptyset$ when $i \neq j$;
iii. Finite subadditivity. $\mu(\cup_{i=1}^n E_i) \leq \sum_{i=i}^n \mu(E_i)$ for all $E_1, \ldots, E_n \in \mathcal{F}$.

Additional continuity properties of measures are the following:

iv. Continuity from below. For all sequences $\{E_n\} \subset \mathcal{F}$ such that $E_n \subset E_{n+1}$, the following holds

$$\mu(\cup_n E_n) = \lim_{n \to \infty} \mathbb{P}(E_n), \tag{2.7}$$

v. Continuity from above. For all sequences $\{E_n\} \subset \mathcal{A}$ such that $E_{n+1} \subset E_n$ and $\mu(E_1) < \infty$, the following holds

$$\mu(\cap_n E_n) = \lim_{n \to \infty} \mu(E_n) \tag{2.8}$$

For a proof see Theorem 2.4 below, where the same result is obtained under milder assumptions.

By iii. and iv. one can prove the following property

vi. Countable subadditivity. $\mu(\cup_n E_n) \leq \sum_n \mu(E_n)$ for all $\{E_n\} \subset \mathcal{F}$.

We end this section by addressing the following problem: on how many sets is it necessary to define the value of a measure to define it completely? For finite or σ-finite measures the answer is provided by the following result, obtained as a simple consequence of Dynkin lemma.

Theorem 2.1. *Let $(\Omega, \mathcal{F})$ be a measurable space and let μ and ν be two measures on $(\Omega, \mathcal{F})$. Let $\mathcal{P} \subset \mathcal{F}$ be a π-system such that $\sigma(\mathcal{P}) = \mathcal{F}$. If the following conditions are satisfied*

- $\mu(E) = \nu(E)$ *for all $E \in \mathcal{P}$;*
- *there exists a family $\{E_n\}_{n \in \mathbb{N}} \subset \mathcal{P}$ of sets in $\mathcal{P}$ such that $\Omega = \cup_n E_n$ and $\mu(E_n) = \nu(E_n) < +\infty$ for all $n \in \mathbb{N}$;*

then μ and ν coincide on $\mathcal{F}$.

Proof.   Let $E \in \mathcal{P}$ such that $\mu(E) = \nu(E) < +\infty$ and define the family of sets $\mathcal{L}_E \subset \mathcal{F}$ as:

$$\mathcal{L}_E := \{A \in \mathcal{F} : \mu(A \cap E) = \nu(A \cap E)\}.$$

It is rather easy to prove that $\mathcal{L}$ is a λ-system, indeed:

- $\Omega \in \mathcal{L}$ (since $\mu(E) = \nu(E)$ by assumption).
- if $A \in \mathcal{L}_E$ then $A^c \in \mathcal{L}_E$. This follows by the equality

$$\mu(E \cap A^c) + \mu(E \cap A) = \mu(E) = \nu(E) = \nu(E \cap A^c) + \nu(E \cap A)$$

  and by the assumption $\mu(E) = \nu(E) < +\infty$.
- if $\{A_n\} \subset \mathcal{L}_E$ is a sequence of sets in $\mathcal{L}_E$ such that $A_n \cap A_m = \emptyset$ for $n \neq m$, then $\cup_n A_n \in \mathcal{L}_E$. This follows easily from the σ-additivity of the measures μ and ν.

Since, by assumption, $\mathcal{P} \subset \mathcal{L}_E$ and $\mathcal{P}$ is a π-system, by Dynkin lemma $\sigma(\mathcal{P}) \subset \mathcal{L}_E$, hence $\mathcal{L}_E = \mathcal{F}$. By the generality on of E we can conclude that for all $E \in \mathcal{P}$ such that $\mu(E) = \nu(E) < \infty$ and for all $A \in \mathcal{F}$ the following holds

$$\mu(A \cap E) = \nu(A \cap E). \tag{2.9}$$

Let us consider now the countable family of sets $\{E_n\}_{n \in \mathbb{N}} \subset \mathcal{P}$ such that $\Omega = \cup_n E_n$ and $\mu(E_n) = \nu(E_n) < +\infty$ for all $n \in \mathbb{N}$, and construct out of it a family $\{E'_n\}_{n \in \mathbb{N}} \subset \mathcal{F}$ of mutually disjoint subsets such that $\cup_n E_n = \cup_n E'_n$ and $E'_n \subset E_n$ by:

$$E'_1 := E_1, \qquad E'_n := E_n \cap (\cup_{k=1}^{n-1} E_k)^c, \quad k \geq 2.$$

Since $E'_n \in \mathcal{F}$ then $E'_n \cap A \in \mathcal{F}$ for all $A \in \mathcal{F}$ hence

$$\mu(E'_n \cap A) = \mu(E_n \cap E'_n \cap A) = \nu(E_n \cap E'_n \cap A) = \nu(E'_n \cap A).$$

The final result follows from the chain of equalities valid for any $A \in \mathcal{F}$:

$$\mu(A) = \sum_n \mu(A \cap E'_n) = \sum_n \nu(A \cap E'_n) = \nu(A).$$

$\square$

2.2 Borel measures

Definition 2.9. Let Ω be an Hausdorff topological space and $\mathcal{B}(\Omega)$ the Borel σ-algebra. A *Borel measure* on Ω is a measure $\mu : \mathcal{B}(\Omega) \to [0, +\infty]$ such that $\mu(K) < +\infty$ for all compact[1] sets $K \subset \mathcal{B}(\Omega)$. A measure $\mu : \mathcal{B}(\Omega) \to [0, +\infty]$ is said:

- *locally finite* if every point $\omega \in \Omega$ has a neighbourhood U such that $\mu(U) < +\infty$;
- *inner regular* if for every $A \in \mathcal{B}(\Omega)$

$$\mu(A) = \sup\{\mu(K), K \subset A \text{ and } K \text{ compact}\};$$

- *outer regular* if for every $A \in \mathcal{B}(\Omega)$

$$\mu(A) = \inf\{\mu(U), A \subset U \text{ and } U \text{ open}\};$$

- *regular* if it is both outer and inner regular;
- *Radon* if it is both locally finite and regular.

Remark 2.2. In an Hausdorff topological space every locally finite measure $\mu : \mathcal{B}(\Omega) \to [0, +\infty]$ is a Borel measure. Indeed, if K is a compact set, every point $\omega \in K$ admits a neighborhood U_ω with finite measure μ. Clearly the collection $\{U_\omega\}_{\omega \in K}$ is an open covering of K and, by compactness, there exists a finite covering $\{U_{\omega_j}\}_{j=1,\ldots,n}$ such that $K \subset \sqcup_{j=1}^n U_{\omega_j}$. Hence by monotonicity and subadditivity of μ we get $\mu(K) \leq \sum_{j=1}^n \mu(U_{\omega_j}) < +\infty$.

In fact, regular Borel measures naturally arise in a particular class of topological spaces.

Definition 2.10. A *Polish space* is a complete metric space with a countable dense subset.

Examples of Polish spaces are the Euclidean space $\mathbb{R}^d$ of every dimension $d \geq 1$ and, more generally, any real separable Banach space. The following result gives a simple sufficient condition for the regularity of a Borel measure on a Polish space. For a proof see, e.g., [57, 71].

Theorem 2.2. *Every finite Borel measure on a Polish space is regular.*

---

[1] Since by assumption Ω is an Hausdorff topological space, every compact set is closed, hence it is a Borel set.

Remark 2.3. For future use it is convenient to remark that finite Cartesian product of Polish spaces are Polish. More precisely, if S is a Polish space and J is a finite set, let S^J be the set of maps $f : J \to S$, which can be identified with the cartesian product $\times_{i=1}^n S = S \times \cdots \times S$ of n copies of S, n being the cardinality of J. If S^J is endowed with the product topology, then it is a Polish space and the corresponding Borel σ-algebra $\mathcal{B}(S^J)$ coincides with the product σ-algebra $\times_{i=1}^n \mathcal{B}(S)$. For the detailed proof see, e.g., [57].

2.3 Extension of a finitely-additive measure

The construction of measure spaces is a non-trivial task. Usually it is easier to consider a smaller class of sets $\mathcal{C} \in \mathcal{P}(\Omega)$ and define a set function μ on $\mathcal{C}$. Under suitable assumptions on $\mathcal{C}$ and on μ, this is sufficient for the construction of a unique measure on the σ-algebra generated by $\mathcal{C}$. In the most relevant cases this technique relies on a procedure that takes the name of C. Carathéodory.

Let us start by introducing the definition of *algebra* of sets.

Definition 2.11. A collection $\mathcal{A} \subset \mathcal{P}(\Omega)$ of a set Ω is called *algebra* of subsets of Ω if it satisfies the following properties

(1) $\Omega \in \mathcal{A}$
(2) if $E \in \mathcal{A}$ then $E^c \in \mathcal{A}$
(3) if $E_1, E_2 \in \mathcal{A}$ then $E_1 \cup E_2 \in \mathcal{A}$.

In fact, a σ-algebra is a family of sets $\mathcal{F} \subset \mathcal{P}(\Omega)$ (containing Ω) which is closed under the operation of countable union and/or intersection of its elements, while an algebra $\mathcal{A}$ is a family of sets $\mathcal{F} \subset \mathcal{P}(\Omega)$ (containing Ω) which is closed under the operation of finite union and/or intersection of its elements. Obviously a σ-algebra is also an algebra but, generally, the converse is not true, as the following example shows.

Example 2.6. Let $\Omega = \mathbb{R}$ and consider the collection $\mathcal{I}$ of intervals $(a, b] \subset \mathbb{R}$ with $-\infty \le a \le b \le +\infty$ (with the convention $(a, b] \equiv (a, b)$ if $b = +\infty$. Let $\mathcal{A}$ be the collection of subsets $E \subset \mathbb{R}$ obtained as finite disjoint union of intervals in $\mathcal{I}$. Specifically, a set $E \in \mathcal{A}$ is of the form

$$E = \cup_{i=1}^k (a_i, b_i]$$

for some $k \in \mathbb{N}$, $-\infty \le a_1 \le b_1 \le \cdots \le a_k \le b_k \le +\infty$.

It is rather simple to check that $\mathcal{A}$ is an algebra but it isn't a σ-algebra. For example the open interval $(0, 1)$ doesn't belong to $\mathcal{A}$ but it can be obtained as a countable union of sets in $\mathcal{A}$:

$$(0, 1) = \bigcup_{n \geq 1} \left(0, 1 - \frac{1}{n} \right].$$

Let us also introduce the notion of *finitely additive measure*.

Definition 2.12. A map $\mu : \mathcal{A} \to [0, +\infty]$ defined on an algebra $\mathcal{A} \subset \mathcal{P}(\Omega)$ is a *finitely additive measure* if $\mu(\emptyset) = 0$ and for any finite family $\{E_1, ..., E_N\} \subset \mathcal{A}$ such that $E_n \cap E_m = \emptyset$ if $n \neq m$, the following holds

$$\mu(\cup_{n=1}^{N} E_n) = \sum_{n=1}^{N} \mu(E_n) \qquad (2.10)$$

If a finitely additive measure is normalized, i.e. is $\mu(\Omega) = 1$ then it is called *finitely additive probability measure*

If μ is a finitely additive measure, it is rather simple to prove the following properties valid whenever $E_1, E_2, \ldots, E_n \in \mathcal{A}$:

- if $E_1 \subset E_2$ then $\mu(E_1) \leq \mu(E_2)$,
- $\mu(E_1 \cup E_2) + \mu(E_1 \cap E_2) = \mu(E_1) + \mu(E_2)$,
- if $E_1 \subset E_2$ and $\mu(E_1) < +\infty$ then $\mu(E_2 \setminus E_1) = \mu(E_2) - \mu(E_1)$,
- $\mu(\cup_{j=1}^{n} E_j) \leq \sum_{j=1}^{n} \mu(E_j)$

Moreover, if $\{E_n\} \subset \mathcal{A}$ is a countable family of mutually disjoint sets belonging to the algebra $\mathcal{A}$ such that $\cup_n E_n \in \mathcal{A}$ then $\sum_n \mu(E_n) \leq \mu(\cup_n E_n)$.

Example 2.7. Let $\Omega = \mathbb{R}$ and $\mathcal{A}$ be the algebra of sets defined in example 2.6. Let $\mu : \mathcal{A} \to [0, +\infty]$ be defined as $\mu(a, b] := |b - a|$ and, more generally, $\mu(\cup_{i=1}^{n}(a_i, b_i]) = \sum_{i=1}^{n} |b_i - a_i|$, where $-\infty \leq a_1 < b_1 < a_2 < b_2 < ... < a_n < b_n \leq +\infty$. It is simple to check that μ is a finite additive measure.

We say that a finitely additive measure μ on an algebra $\mathcal{A}$ is *σ-additive on $\mathcal{A}$* if for any countable family $\{E_n\} \subset \mathcal{A}$ such that $E_n \cap E_m = \emptyset$ for $m \neq n$ and $\cup_n E_n \in \mathcal{A}$, the following holds

$$\mu(\cup_n E_n) = \sum_n \mu(E_n)$$

Example 2.8. An example of finitely additive measure, which cannot be σ-additive, is provided by the map $\mu : \mathcal{P}(\Omega) \to [0, +\infty]$ defined on the power set of a infinite set Ω as $\mu(E) = 0$ if $\#E < \infty$ and $\mu(E) = \infty$ if $\#E = \infty$.

A useful tool for the proof of σ-additivity is the following result.

Theorem 2.3. *Let* $\mu : \mathcal{A} \to [0, +\infty]$ *be a finitely additive measure on an algebra* $\mathcal{A}$. *If* μ *is continuous at* $\emptyset$, *i.e. if for any sequence* $\{E_n\} \subset \mathcal{A}$ *such that* $E_{n+1} \subset E_n$ *and* $\cap_n E_n = \emptyset$ *the following holds*

$$\lim_{n \to \infty} \mathbb{P}(E_n) = 0,$$

then μ *is* σ-additive on $\mathcal{A}$.

Proof. Let $\{E_n\} \subset \mathcal{A}$ be a countable family in $\mathcal{A}$ of sets such that $E_n \cap E_m = \emptyset$ whenever $n \neq m$ and $\cup_n E_n \in \mathcal{A}$. Let us consider the sequence $\{\tilde{E}_n\}$ defined by $\tilde{E}_n := \cup_{n>N} E_n$. Since $\cup_n E_n = \cup_{n=1}^N E_n \cup \tilde{E}_N$, $\mathcal{A}$ is an algebra and, by assumption $E_n \in \mathcal{A}$ and $\cup_n E_n \in \mathcal{A}$, we can deduce that $\tilde{E}_N \in \mathcal{A} \; \forall N \in \mathbb{N}$. Furthermore $\tilde{E}_{N+1} \subset \tilde{E}_N$ and $\cap_n \tilde{E}_n = \emptyset$. We eventually obtain $\lim_{n \to \infty} \mu(\tilde{E}_n) = 0$, hence $\mu(\cup_n E_n) = \sum_n \mu(E_n)$. $\qquad \square$

The following result shows how σ-additivity is related to continuity properties of the measure.

Theorem 2.4. *Let* $\mu : \mathcal{A} \to [0, +\infty)$ *be a* σ-additive measure on an algebra $\mathcal{A}$. *Then*

  i. μ *is continuous from below, i.e. for any sequence* $\{E_n\} \subset \mathcal{A}$ *such that* $E_n \subset E_{n+1}$ *and* $\cup_n E_n \in \mathcal{A}$, *the following holds*

$$\mu(\cup_n E_n) = \lim_{n \to \infty} \mu(E_n),$$

  ii. μ *is continuous from above, i.e. for any sequence* $\{E_n\} \subset \mathcal{A}$ *such that* $E_{n+1} \subset E_n$, $\cap_n E_n \in \mathcal{A}$ *and* $\mu(E_1) < \infty$, *the following holds*

$$\mu(\cap_n E_n) = \lim_{n \to \infty} \mu(E_n).$$

Proof.

  i. Given a sequence $\{E_n\} \subset \mathcal{A}$ of sets in $\mathcal{A}$ such that $E_n \subset E_{n+1}$ and $\cup_n E_n \in \mathcal{A}$, let us construct the sequence $\{\tilde{E}_n\} \subset \mathcal{A}$ of disjoint sets of $\mathcal{A}$, defined as •

$$\tilde{E}_1 := E_1, \tilde{E}_n := E_n \setminus E_{n-1}.$$

It is easy to see that $\tilde{E}_n \in \mathcal{A} \; \forall n \in \mathbb{N}$, $\tilde{E}_n \cap \tilde{E}_m = \emptyset$ and $\cup_n E_n = \cup_n \tilde{E}_n$. Further $E_N = \cup_{n=1}^N \tilde{E}_n$, hence:

$$\mathbb{P}(\cup_n E_n) = \mathbb{P}(\cup_n \tilde{E}_n) = \lim_{N \to +\infty} \sum_{n=1}^N \mathbb{P}(\tilde{E}_n) = \lim_{N \to +\infty} \mathbb{P}(E_N).$$

ii. Let $\{E_n\} \subset \mathcal{A}$ be a sequence of sets in $\mathcal{A}$ such that $E_{n+1} \subset E_n$ and $\cap_n E_n \in \mathcal{A}$. Let us construct a sequence $\{\tilde{E}_n\} \subset \mathcal{A}$ of sets in $\mathcal{A}$ such that $\tilde{E}_n \subset \tilde{E}_{n+1}$ and $\cup_n \tilde{E}_n \in \mathcal{A}$, defined as $\tilde{E}_n = E_1 \setminus E_n$. Since $\mathcal{A}$ is an algebra, it is easy to see that $\tilde{E}_n \in \mathcal{A} \ \forall n \in \mathbb{N}$. Moreover $\cup_n \tilde{E}_n \in \mathcal{A}$ since

$$\cup_n \tilde{E}_n = E_1 \setminus \cap_n E_n.$$

Since $\mu(E_1) < +\infty$ by assumption, we have $\mu(\cup_n \tilde{E}_n) = \mu(E_1) - \mu(\cap_n E_n)$ and by point i. we eventually get

$$\lim_{n \to \infty} \mu(\tilde{E}_n) = \mu(E_1) - \lim_{n \to \infty} \mu(E_n) = \mu(\cup_n \tilde{E}_n)$$

hence $\mu(\cap_n E_n) = \lim_{n \to \infty} \mu(E_n)$.

$\square$

Let us assume a finitely additive measure $\mu_0 : \mathcal{A} \to [0, +\infty]$ on an algebra $\mathcal{A}$ of sets is given. Does it exist an extension μ of μ_0 to the σ-algebra $\sigma(\mathcal{A})$ generated by $\mathcal{A}$? In other words, does it exist a σ-additive measure $\mu : \sigma(\mathcal{A}) \to [0, +\infty]$ which coincides with μ_0 on $\mathcal{A}$? Is it unique? The answer to the first question is affirmative if μ_0 is σ-additive on $\mathcal{A}$. In this case μ is constructed by means of the Caratheodory's procedure. For the details we refer to [57, 66] and we limit ourselves to stating the main result and give some clues about its derivation.

Theorem 2.5. *Let Ω be a set and μ a finitely additive measure on an algebra $\mathcal{A}$ of subsets of Ω. If μ is σ-additive on $\mathcal{A}$ then there exists at least one σ-additive measure on $\sigma(\mathcal{A})$ which coincides with μ on $\mathcal{A}$.*

The procedure leading to the construction of the extension relies on the following steps

(1) First a set function $\mu_0^* : \mathcal{P}(\Omega) \to [0, +\infty]$ is defined as

$$\mu_0^*(E) := \inf \sum_n \mu_0(A_n),$$

where the infimum is evaluated over all sequences $\{A_n\} \subset \mathcal{A}$ such that $E \subset \cup_n A_n$. In fact, μ_0^* is an *outer measure*, i.e. a set function enjoying the following properties:

- $\mu_0^*(E) \in [0, +\infty]$ for all $E \in \mathcal{P}(\Omega)$;
- $\mu_0^*(\emptyset) = 0$;
- $\mu_0^*(E_1) \le \mu_0^*(E_2)$ for all $E_1, E_2 \in \mathcal{P}(\Omega)$ such that $E_1 \subset E_2$;
- $\mu_0^*(\cup_n E_n) \le \sum_n \mu_0^*(E_n)$ for all countable sequences $\{E_n\} \in \mathcal{P}(\Omega)$.

(2) The class $\mathcal{M} \subset \mathcal{P}(\Omega)$ of μ_0^*-measurable sets is considered, where a set $A \subset \Omega$ is defined μ_0^*-measurable if for any $E \subset \Omega$:

$$\mu_0^*(E) = \mathbb{P}_0^*(E \cap A) + \mu_0^*(E \cap A^c)$$

(3) The main step is the proof that $\mathcal{M}$ is a σ-algebra including $\mathcal{A}$ and that μ_0^* is σ-additive on $\mathcal{M}$. Moreover $\mu_0(E) = \mu_0^*(E)$ for all $E \in \mathcal{A}$.
(4) The previous step yields $\sigma(\mathcal{A}) \subset \mathcal{M}$. Eventually, the measure $\mu :$ $\sigma(\mathcal{A}) \to [0, +\infty]$ is defined as the restriction of μ_0^* to $\sigma(\mathcal{A})$.

Concerning the problem of the uniqueness of the extension, this can be addressed by means of Theorem 2.1. In particular, a sufficient condition for the uniqueness of the extension of a finite-additive measure μ on an algebra $\mathcal{A} \subset \mathcal{P}(\Omega)$ is the existence of a countable family $\{E_n\} \subset \mathcal{A}$ such that $\Omega = \cup_n E_n$ and $\mu(E_n) < +\infty$ for all $n \in \mathbb{N}$. If this condition is fulfilled, then μ is called σ-*finite*.

Theorem 2.6. *Let μ be a σ-finite and sigma additive measure on an algebra $\mathcal{A}$. There exists a unique σ-additive measure on $\sigma(\mathcal{A})$ which coincides with μ on $\mathcal{A}$.*

Since any algebra is also a π-system, the proof is an straightforward application of Theorem 2.1.

2.4 Measurable functions

Definition 2.13. Let $(\Omega, \mathcal{F})$ and $(\Omega', \mathcal{F}')$ be two measurable spaces. A function $T : \Omega \to \Omega'$ is said *measurable* if for any $E \in \mathcal{F}'$ the set $T^{-1}(E) = \{\omega \in \Omega : T(\omega) \in E\}$ belongs to $\mathcal{F}$.

By applying Definition 2.13 it is rather simple to see that any constant function $T : \Omega \to \Omega'$ between two measurable spaces $(\Omega, \mathcal{F})$ and $(\Omega', \mathcal{F}')$ is measurable. Further, given three measurable spaces $(\Omega, \mathcal{F})$, $(\Omega', \mathcal{F}')$ and $(\Omega'', \mathcal{F}'')$ and a pair of measurable functions $T_1 : \Omega \to \Omega'$ and $T_2; \Omega' \to \Omega''$, the composition $T_2 \circ T_1 : \Omega \to \Omega''$ is measurable.

The following theorem provides a useful measurability criterion.

Theorem 2.7. *Let $T : \Omega \to \Omega'$ be a function between two measurable spaces $(\Omega, \mathcal{F})$ and $(\Omega', \mathcal{F}')$. Let $\mathcal{C} \subset \mathcal{P}(\Omega')$ be a family of subsets of Ω' such that $\sigma(\mathcal{C}) = \mathcal{F}'$. If for any $A \in \mathcal{C}$ the set $T^{-1}(A)$ belongs to $\mathcal{F}$, then T is measurable.*

Proof. Let us consider the collection of sets $\mathcal{G} \subset \mathcal{P}(\Omega')$ defined by

$$\mathcal{G} = \{E \subset \Omega' : T^{-1}(E) \in \mathcal{F}\}.$$

By assumption $\mathcal{C} \subset \mathcal{G}$. Moreover $\mathcal{G}$ is a σ-algebra, indeed:

- $\Omega' \in \mathcal{G}$ since $T^{-1}(\Omega') = \Omega \in \mathcal{F}$.
- if $E \in \mathcal{G}$ then $E^c \in \mathcal{G}$. Indeed $T^{-1}(E^c) = \left(T^{-1}(E)\right)^c$.
- given a countable family $\{E_n\}_n \subset \mathcal{G}$ then $\cup_n E_n \in \mathcal{G}$, since $T^{-1}(\cup_n E_n) = \cup_n T^{-1}(E_n)$.

By definition $\sigma(\mathcal{C})$ is the intersection of all σ-algebras containing $\mathcal{C}$, hence $\sigma(\mathcal{C}) \subset \mathcal{G}$. □

The following results are straightforward applications of Theorem 2.7

Theorem 2.8. *Let Ω, Ω' be two topological spaces and let $\mathcal{F}, \mathcal{F}'$ be the corresponding Borel σ-algebras. Every continuous function $T : \Omega \to \Omega'$ is measurable.*

Proof. It is sufficient to consider the collection $\mathcal{C} \subset \mathcal{P}(\Omega')$ of open sets in Ω' and apply Theorem 2.7. □

Moreover, if $(\Omega, \mathcal{F})$, $(X, \mathcal{F}_X)$ and $(Y, \mathcal{F}_Y)$ are measurable spaces and $T_1 : \Omega \to X$ and $T_2 : \Omega \to Y$ are measurable maps, then $T_3 : \Omega \to X \times Y$ defined by

$$T_3(\omega) := (T_1(\omega), T_2(\omega))$$

is a measurable mapping from $(\Omega, \mathcal{F})$ to $(X \times Y, \mathcal{F}_X \times \mathcal{F}_Y)$. In this case it is sufficient to recall that, by definition, $\mathcal{F}_X \times \mathcal{F}_Y$ is generated by the collection of sets of the form $I \times J$, with $I \in \mathcal{F}_X$ and $J \in \mathcal{F}_Y$.

Similarly, given two measurable functions f, g from a measurable space $(\Omega, \mathcal{F})$ to $(\mathbb{R}, \mathcal{B}(\mathbb{R}))$ then

(1) for every $\alpha, \beta \in \mathbb{R}$, the map $\alpha f + \beta g$ is measurable from $(\Omega, \mathcal{F})$ to $(\mathbb{R}, \mathcal{B}(\mathbb{R}))$,

(2) fg is measurable from $(\Omega, \mathcal{F})$ to $(\mathbb{R}, \mathcal{B}(\mathbb{R}))$,

(3) $f + ig$ is a measurable function from $(\Omega, \mathcal{F})$ to $(\mathbb{C}, \mathcal{B}(\mathbb{C}))$.

Furthermore, if $\{f_n\}_n$ is a sequence of measurable functions from $(\Omega, \mathcal{F})$ to the extended real line[2] then the functions $\sup f_n$, $\inf f_n$, $\limsup f_n$ and $\liminf f_n$ defined pointwise as

$$(\sup f_n)(\omega) := \sup\{f_n(\omega)\}, \quad (\inf f_n)(\omega) := \inf\{f_n(\omega)\},$$

---

[2]For *extended real line* is meant the topological space $[-\infty, +\infty]$ with a topology generated by sets of the form (a, b), $[-\infty, a)$ and $(b, +\infty]$, with $a, b \in \mathbb{R}$.

$$(\limsup f_n)(\omega) := \limsup\{f_n(\omega)\}, \quad (\liminf f_n)(\omega) := \liminf\{f_n(\omega)\},$$

are measurable. Moreover the limit of every pointwise convergent sequence of complex measurable functions is measurable. For details we refer, e.g., [293].

2.5   Image measure

Theorem 2.9. *Let* $T : \Omega \to \Omega'$ *be a measurable mapping between two measurable spaces* $(\Omega, \mathcal{F})$ *and* $(\Omega', \mathcal{F}')$. *The collection of sets* $\mathcal{G} := \{T^{-1}(E) : E \in \mathcal{F}'\} \subset \mathcal{F}$ *is a* σ-*algebra.*

The σ-algebra $\mathcal{G}$ defined in Theorem 2.9 is called σ-*algebra generated by* T and denoted with the symbol $\sigma(T)$. In fact it is the smallest σ-algebra making T a measurable function, in the sense that T is measurable from $(\Omega, \mathcal{F})$ to $(\Omega', \mathcal{F}')$ if and only if $\sigma(T) \subset \mathcal{F}$.

Let us consider now a measure space $(\Omega, \mathcal{F}, \mu)$, a measurable space $(\Omega', \mathcal{F}')$ and a measurable map $T : \Omega \to \Omega'$. Let $\mu_T : \mathcal{F}' \to [0, +\infty]$ be the set function defined by

$$\mu_T(E) := \mu(T^{-1}(E)), \qquad E \in \mathcal{F}' \tag{2.11}$$

It is rather simple to prove that μ_T is a σ-additive measure on $\mathcal{F}'$. It is usually called *image measure* or *pushforward measure* of μ under the mapping T. In addition, μ_T it is a probability measure if μ is. It is rather simple to check that the construction of image measure satisfies the following composition law

$$\mu_{T_2 \circ T_1} = (\mu_{T_1})_{T_2} \tag{2.12}$$

which holds for any pair of measurable mappings $T_1 : \Omega \to \Omega'$ and $T_2; \Omega' \to \Omega''$ defined on three measurable spaces $(\Omega, \mathcal{F})$, $(\Omega', \mathcal{F}')$ and $(\Omega'', \mathcal{F}'')$.

Let us consider the d-dimensional euclidean space $\mathbb{R}^d$ endowed with the corresponding Borel σ-algebra $\mathcal{B}(\mathbb{R}^d)$. The σ-algebra $\mathcal{B}(\mathbb{R}^d)$ can be proved to be generated by the class $\mathcal{C}$ of *plurirectangles* $R_{a_1,b_1,\ldots,a_d,b_d}$ of the form

$$R_{a_1,b_1,\ldots,a_d,b_d} = \{x \in \mathbb{R}^d : a_1 < x_i \leq b_i, \, i = 1, \ldots, d\}$$

for some $2d$-tuples of real numbers $a_1, b_1, \ldots, a_d, b_d$ of real numbers such that $a_i \leq b_i$, $i = 1, \ldots, d$. The *Lebesgue measure* $\mathcal{L}^d$ on $\mathcal{B}(\mathbb{R}^d)$ is defined on the sets in $\mathcal{C}$ as

$$\mathcal{L}^d(R_{a_1,b_1,\ldots,a_d,b_d}) = \Pi_{i=1}^d (b_i - a_i). \tag{2.13}$$

Since the class $\mathcal{C}$ is closed under finite intersections, by Theorem 2.1 the measure $\mathcal{L}^d$ is uniquely defined by (2.13).

For any vector $y \in \mathbb{R}^d$, let $\tau_y : \mathbb{R}^d \to \mathbb{R}^d$ be the translation mapping defined as

$$\tau_y(x) := x + y, \qquad x \in \mathbb{R}^d.$$

τ_y is continuous hence measurable and for any Borel measure $\mu : \mathcal{B}(\mathbb{R}^d) \to [0, +\infty]$ on R^d we can define its image measure $(\mu)_{\tau_y}$ under the action of the mapping τ_y.

Definition 2.14. A Borel measure $\mu : \mathcal{B}(\mathbb{R}^d) \to [0, +\infty]$ on $\mathbb{R}^d$ is said to be *translation invariant* if $\mu = (\mu)_{\tau_y}$ for all $y \in \mathbb{R}^d$.

The following result gives an interesting characterisation of the Lebesgue measure $\mathcal{L}^d$

Theorem 2.10. *Let* $C \subset \mathbb{R}^d$ *be the unit cube, i.e.* $C = \{x \in \mathbb{R}^d : 0 < x_i \leq 1, i = 1, \ldots, d\}$. *Every translation invariant Borel measure* $\mu : \mathcal{B}(\mathbb{R}^d) \to [0, +\infty]$ *on* R^d *such that* $\mu(C) < +\infty$ *has the form* $\mu = c\mathcal{L}^d$ *for some real constant* $c \geq 0$.

Proof. Let us set $c := \mu(C)$. If $c = 0$ then, let us consider the countable family of disjoint cubes $C_{k_1, \ldots, k_d} := \{x \in \mathbb{R}^d : k_i < x_i \leq k_i + 1, i = 1, \ldots, d\}$ with $k_i \in \mathbb{Z}$, $i = 1, \ldots, d$. Since $\mathbb{R}^d$ is covered by the cubes $C_{k_1, \ldots, k_d}$ and all of them can be obtained by the translation of C, by the translation invariance of μ we get $\mu = 0$.

In the case $c > 0$ let us consider the set $C_n := \{x \in \mathbb{R}^d : 0 < x_i \leq 1/n, i = 1, \ldots, d\}$. It is easy to check that the unit cube C is the disjoint union of n^d smaller cubes obtained as the translation of C_n, hence $\mu(C_n) = c/n^d = c\mathcal{L}^d(C_n)$. By the same argument we obtain the following equality

$$\mu(R_{a_1, b_1, \ldots, a_d, b_d}) = c\mathcal{L}^d(R_{a_1, b_1, \ldots, a_d, b_d})$$

valid for all plurirectangles $R_{a_1, b_1, \ldots, a_d, b_d}$ with rational coordinates $a_1, b_1, \ldots, a_d, b_d$. Since these collection of sets is closed under finite intersection and generates the Borel σ-algebra $\mathcal{B}(\mathbb{R}^d)$, by Theorem 2.1 we eventually obtain $\mu(A) = c\mathcal{L}^d(A)$ for all $A \in \mathcal{B}(\mathbb{R}^d)$. $\square$

A map $U : \mathbb{R}^d \to \mathbb{R}^d$ is said to be *orthogonal* or *unitary* if it preserves the inner product $\langle \, , \, \rangle$ in $\mathbb{R}^d$, i.e. if for all $x, y \in \mathbb{R}^d$ the following holds:

$$\langle x, y \rangle = \langle U(x), U(y) \rangle$$

Definition 2.15. A Borel measure $\mu : \mathcal{B}(\mathbb{R}^d) \to [0, +\infty]$ on R^d is said to be *invariant under unitary transformation* if $\mu = (\mu)_U$ for all orthogonal transformations $U : \mathbb{R}^d \to \mathbb{R}^d$.

Theorem 2.11. *The Lebesgue measure* $\mathcal{L}^d : \mathcal{B}(\mathbb{R}^d) \to [0, +\infty]$ *is invariant under unitary transformations.*

Proof. Let $U : \mathbb{R}^d \to \mathbb{R}^d$ be an arbitrary unitary transformation and let us set $\mu := (\mathcal{L})_U$. Let $y \in \mathbb{R}^d$ be an arbitrary vector and consider the translation map and the corresponding image measure $(\mu)_{\tau_y}$ of μ. By the explicit form of the linear transformations U and τ_y we can easily obtain the following composition property:

$$\tau_y \circ U(x) = U(x) + y = U(x + U^{-1}(y)) = U \circ \tau_{U^{-1}(y)}(x).$$

Hence, by Eq. (2.12) and the translation invariance of the Lebesgue measure $\mathcal{L}^d$ we have:

$$(\mu)_{\tau_y} = ((\mathcal{L}^d)_{\tau_{U^{-1}(y)}})_U = (\mathcal{L}^d)_U = \mu$$

which shows that μ is translation invariant. Further, if $C = \{x \in \mathbb{R}^d : 0 < x_i \le 1, i = 1, \ldots, d\}$ is the unit cube, then $\mu(C) = \mathcal{L}^d(U^{-1}(C)) < +\infty$, hence by Theorem 2.10, we have $\mu = c\mathcal{L}^d$ for a suitable constant c. In order to determine the value of c, it is sufficient to consider the unit ball $B = \{x \in \mathbb{R}^d : \|x\| \le 1\}$. Since $B = U^{-1}B$ for any unitary map U, we obtain:

$$c\mathcal{L}^d(B) = \mu(B) = \mathcal{L}(U^{-1}(B)) = \mathcal{L}(B)$$

hence $c = 1$ and $\mathcal{L} = (\mathcal{L})_U$ for all unitary mappings U. $\qquad\square$

These result can be generalized to the case the d-dimensional Euclidean space is replaced by a locally compact group G. In this case the role of Lebesgue measure is played by the *Haar measure* [96]. Given a locally compact group G, a Borel measure $\mu : \mathcal{B}(G) \to [0, +\infty]$ is said to be *invariant under left translations* or simply *left invariant* if for all $x \in G$ and for all $E \in \mathcal{B}(G)$ the following holds

$$\mu(E) = \mu(xE),$$

where the set xE is defined as the set of elements $y \in G$ of the form $y = xx'$ for some $x' \in E$. A non-zero regular left invariant measure is called *left Haar measure* or simply *Haar measure*. Its existence and uniqueness up to multiplication constants are assured by the following theorems.

Theorem 2.12. *If G is a locally compact group, then there exists a left invariant Haar measure.*

Theorem 2.13. *If G is a locally compact group and let μ and ν left invariant Haar measures on G. Then there is a positive constant $c \in \mathbb{R}$ such that $\mu = c\nu$.*

2.6 Integration of measurable functions

Let $(\Omega, \mathcal{F})$ a measurable space. A *simple function* $s : \Omega \in \mathbb{R}$ is a finite linear combination of indicator functions of measurable sets, i.e. a mapping of the form

$$s = \sum_{i=1}^{N} c_i \chi_{E_i}, \tag{2.14}$$

for some $N \in \mathbb{N}$, $c_1, \dots, c_N \in \mathbb{R}$, $E_1, \dots, E_N \in \mathcal{F}$ and χ_E is defined as

$$\chi_E(\omega) := \begin{cases} 1 & \omega \in E \\ 0 & \omega \notin E. \end{cases}$$

The following results shows that every positive measurable function can be pointwise approximated by an increasing sequence of simple functions.

Theorem 2.14. *Let $f : \Omega \to [0, +\infty]$ be a measurable function. There exists a sequence $\{s_n\}_n$ of simple functions such that:*

- $0 \leq s_n \leq f$ *and* $s_n \leq s_{n+1}$ *for all* $n \in \mathbb{N}$,
- $\lim_{n \to \infty} s_n(\omega) = f(\omega)$ *for all* $\omega \in \Omega$.

Proof. For any positive real number $t \in \mathbb{R}$ and any strictly positive integer $n \in \mathbb{N}$, let us consider the positive integer $k(n, t) \in \mathbb{N}$ defined as $k(n, t) := \lfloor 2^n t \rfloor$, where the symbol $\lfloor x \rfloor$ denotes the integer part of x. In particular, the following holds:

$$\frac{k(n, t)}{2^n} \leq t < \frac{k(n, t) + 1}{2^n}.$$

Let us define the sequence of functions $g_n : [0, +\infty] \to \mathbb{R}$ as

$$g_n(t) := \begin{cases} \frac{k(n,t)}{2^n} & t \in [0, n) \\ n & t \in [n, +\infty). \end{cases}$$

Clearly, for every $n \geq 1$ the mapping g_n is Borel measurable and is a simple function. Moreover, by construction we have

- $0 \leq g_n(t) \leq t$ *for all* $t \in [0, +\infty]$ and $g_n \leq g_{n+1}$ for all $n \in \mathbb{N}$,
- $\lim_{n \to \infty} g_n(t) = t$ for all $t \in [0, +\infty]$.

By setting $s_n := g_n \circ f$ we obtain the thesis. $\qquad \square$

Given a positive measure $\mu : \mathcal{F} \to [0, +\infty]$, a simple function of the form (2.14) and a measurable set $E \in \mathcal{F}$, the *integral of s on E with respect to the measure μ* is defined as:

$$\int_E s \, d\mu := \sum_{i=1}^{N} c_i \mu(E_i \cap E),$$

with the convention $0 \cdot \infty = 0$ in order to handle the cases where $c_i = 0$ and $\mu(E_i \cap E) = +\infty$.

Given a non-negative measurable function $f : \Omega \to [0, +\infty]$, the *Lebesgue integral of f on the measurable set $E \in \mathcal{F}$ with respect to the measure μ* is defined as:

$$\int_E f \, d\mu := \sup \int_E s \, d\mu$$

where the supremum is taken over all simple measurable functions s such that $0 \le s \le f$.

A real valued function $f : \Omega \to \mathbb{R}$ is said to be *summable* or *absolutely integrable* if $\int_\Omega |f| \, d\mu < +\infty$. In this case its Lebesgue integral $\int_E f \, d\mu$ is defined as:

$$\int_E f \, d\mu := \int_E f^+ \, d\mu - \int_E f^- \, d\mu \, ,$$

where the non-negative measurable mappings f^+ and f^- are the *positive part* and the *negative part* of f respectively, defined pointwise for any $\omega \in \Omega$ as:

$$f^+(\omega) := \sup\{f(\omega), 0\}, \qquad f^-(\omega) := \sup\{-f(\omega), 0\} \, . \tag{2.15}$$

Analogously, the Lebesgue integral of a complex-valued measurable function $f : \Omega \to \mathbb{C}$ such that $\int_\Omega |f| \, d\mu < +\infty$ is defined as

$$\int_E f \, d\mu := \int_E f_R \, d\mu + i \int_E f_I \, d\mu \, ,$$

where f_R and f_I are the real and the imaginary part of f respectively, defined for every $\omega \in \Omega$ by

$$f_R(\omega) := \frac{f(\omega) + \bar{f}(\omega)}{2}, \qquad f_I(\omega) := \frac{f(\omega) - \bar{f}(\omega)}{2} \, .$$

The integral is in fact a linear functional on the set of complex valued summable functions. Indeed if c_1, c_2 are complex numbers and f_1, f_2 complex valued summable functions, then the map $g : \Omega \to \mathbb{C}$ defined as

$$g(\omega); = c_1 f_1(\omega) + c_2 f_2(\omega), \qquad \omega \in \Omega \, ,$$

is summable and

$$\int_\Omega g \, d\mu = c_1 \int_\Omega f_1 \, d\mu + c_2 \int_\Omega f_2 \, d\mu. \tag{2.16}$$

Lebesgue integral enjoys some important continuity properties stated in the following theorems.

Theorem 2.15 (Monotone convergence theorem). *Let $\{f_n\}_n$ be a sequence of complex-valued measurable functions on $(\Omega, \mathcal{F})$ such that*

- $\lim_{n\to\infty} f_n(\omega) = f(\omega)$ *for every* $\omega \in \Omega$,
- $f_n \geq 0$ *and* $f_n \leq f_{n+1}$ *for all* $n \in \mathbb{N}$.

then f is measurable and

$$\lim_{n\to\infty} \int_\Omega f_n \, d\mu = \int_\Omega f \, d\mu.$$

Theorem 2.16 (Dominated convergence theorem). *Let $\{f_n\}_n$ be a sequence of complex-valued measurable functions on $(\Omega, \mathcal{F})$ such that*

- $\lim_{n\to\infty} f_n(\omega) = f(\omega)$ *for every* $\omega \in \Omega$,
- *there exists a summable function g such that $|f_n| \leq g$ for all $n \in \mathbb{N}$.*

Then the limit function is summable and

$$\lim_{n\to\infty} \int_\Omega |f_n - f| \, d\mu = 0,$$

$$\lim_{n\to\infty} \int_\Omega f_n \, d\mu = \int_\Omega f \, d\mu.$$

Let $T : \Omega \to \Omega'$ a measurable mapping between two measurable spaces $(\Omega, \mathcal{F})$ and $(\Omega', \mathcal{F}')$, let μ be a positive measure on $(\Omega, \mathcal{F})$ and μ_T its image measure under T on $(\Omega', \mathcal{F}')$. The following transformation of variables formula connects integrals with respect to μ_T with integrals with respect to μ.

Theorem 2.17. *A real-valued measurable function $f : \Omega' \to \mathbb{R}$ is summable with respect to μ_T iff the function $f \circ T : \Omega \to \mathbb{R}$ is summable with respect to μ and in this case the integrals coincide:*

$$\int_{\Omega'} f(y) d\mu_T(y) = \int_\Omega f(T(x)) d\mu(x). \tag{2.17}$$

Proof. If $f = \chi_E$ is the indicator function of a measurable set $E \in \mathcal{F}'$ then the left-hand side of (2.17) is

$$\int_{\Omega'} f(y) d\mu_T(y) = \mu_T(E),$$

while the right-hand side is

$$\int_{\Omega} \chi_E(T(x)) d\mu(x) = \int_{\Omega} \chi_{T^{-1}(E)}(x) d\mu(x) = \mu(T^{-1}(E))$$

and (2.17) follows from the definition (2.11) of image measure. By linearity the same result holds when f is a simple function. If f is a non-negative function, then (2.17) can be proved by approximating f with an increasing sequence of simple functions (see Theorem 2.14) and applying monotone convergence Theorem 2.15. Eventually, for a general measurable $f : \Omega' \to \mathbb{R}$, the conclusion follows by writing $f = f^+ - f^-$ as the linear combination of its positive and negative part (see (2.15)). $\qquad\square$

2.7 Complex measures

Let $(\Omega, \mathcal{F})$ be a measurable space. A *complex measure* μ on $(\Omega, \mathcal{F})$ is a mapping $\mu : \mathcal{F} \to \mathbb{C}$ such that for all sets $E \in \mathcal{F}$ and for all partitions[3] $\{E_n\} \subset \mathcal{F}$ of E the following holds

$$\mu(E) = \sum_n \mu(E_n). \tag{2.18}$$

The absolute convergence of the series on the right-hand side of (2.18) is part of the definition since it is necessary that any rearrangement of the terms in the series $\sum_n \mu(E_n)$ doesn't change the value of the sum. Contrary to the case of positive measures, where the value $\mu(E) = +\infty$ is admissible, in the case of complex measure it is required that $\mu(E)$ is a complex number for any $E \in \mathcal{F}$.

Given a complex measure μ on $(\Omega, \mathcal{F})$, let us define the positive set function $|\mu| : \mathcal{F} \to [0, +\infty]$ as

$$|\mu|(E) := \sup \sum_n |\mu|(E_n) \tag{2.19}$$

where the supremum is taken over all partitions $\{E_n\}$ of E.

By construction the set function $|\mu|$ dominates μ, indeed

$$|\mu|(E) \geq |\mu(E)| \qquad \forall E \in \mathcal{F}. \tag{2.20}$$

---

[3] A partition of a set E is a countable collection $\{E_n\}$ of mutually disjoint sets such that $E = \cup_n E_n$.

In fact $|\mu|$ is a positive finite measure, as stated in the following theorem:

Theorem 2.18. *For μ is a complex measure on $(\Omega, \mathcal{F})$, then $|\mu| : \mathcal{F} \to [0, +\infty]$ is a positive measure and $|\mu|(\Omega) < +\infty$.*

The mapping $|\mu| : \mathcal{F} \to [0, +\infty]$ is called *total variation measure* of the complex measure μ. The number $|\mu|(\Omega)$ is called *total variation of μ* and denoted with the symbol $\|\mu\|$.

If Ω is a topological space and $\mathcal{F}$ is the Borel σ-algebra, a complex measure μ is said to be *regular* if $|\mu|$ is regular in the sense of Definition 2.9.

We point out that the set $\mathcal{M}(\Omega)$ of complex measures on a measurable space $(\Omega, \mathcal{F})$ is a complex vector space. Indeed, for any pair of complex measures μ_1, μ_2 and for any constant $c \in \mathbb{C}$, the set functions $\mu_1 + \mu_2$ and $c\mu_1$ defined as

$$(\mu_1 + \mu_2)(E) := \mu_1(E) + \mu_2(E), \quad (c\mu_1)(E) := c\mu_1(E), \qquad E \in \mathcal{F}$$

are complex measures as well. Moreover their total variation enjoys the following properties:

$$|\mu_1 + \mu_2| \le |\mu_1| + |\mu_2|, \qquad |c\,\mu_1| = |c|\,|\mu_1|.$$

More specifically, the map $\| \ \| : \mathcal{M}(\Omega) \to [0, +\infty)$ defined on any complex measure $\mu \in \mathcal{M}(\Omega)$ as the total variation of μ:

$$\|\mu\| := |\mu|(\Omega) \tag{2.21}$$

is a norm and $\mathcal{M}(\Omega)$ endowed with $\| \ \|$ becomes a normed linear space.

We address now the problem of the relation between a complex measure μ and the corresponding total variation measure $|\mu|$.

Definition 2.16. Given a positive measure μ and a complex measure ν on a measurable space $(\Omega, \mathcal{F})$, the measure ν is said to be *absolutely continuous* with respect to μ if $\nu(E) = 0$ for all $E \in \mathcal{F}$ such that $\mu(E) = 0$.

A fundamental result of measure theory is the following theorem.

Theorem 2.19 (Radon-Nikodym). *Let μ be a $\sigma-$ finite positive measure and ν a complex measure on $(\Omega, \mathcal{F})$ such that ν is absolutely continuous with respect to μ. Then there exist a summable complex-valued measurable function $h : \Omega \to \mathbb{C}$ such that*

$$\nu(E) = \int_E h \, d\mu, \qquad \forall E \in \mathcal{F}. \tag{2.22}$$

Further h is unique up to μ-null sets, i.e. if $h' : \Omega \to \mathbb{C}$ is another measurable function fulfilling (2.22) then $\mu(\{\omega \in \Omega : h(\omega) \ne h'(\omega)\}) = 0$.

The map h whose existence is given by Theorem 2.19 is called the *Radon-Nikodym derivative of μ with respect to μ* and denoted with the symbol $\frac{d\nu}{d\mu}$.

The inequality (2.20) shows that any complex measure μ is absolutely continuous with respect to its total variation measure $|\mu|$, hence it is possible to construct the corresponding Radon-Nicodym derivative $h = \frac{d\mu}{d|\mu|}$. Moreover in this case h is a complex valued function such that $|h| = 1$ up to μ-null sets. Indeed, if α is a positive real number, $A_\alpha := \{\omega \in \Omega : |h(\omega)| < \alpha\}$ the set where $|h|$ attains values less than α and $\{E_j\}$ is a generic partition of A_α, we have:

$$\sum_j |\mu(E_j)| = \sum_j \left| \int_{E_j} h \, d|\mu| \right| \leq \alpha \sum_j |\mu|(E_j) = \alpha |\mu|(A_\alpha).$$

Hence, by taking the supremum over all partitions of A_α, we obtain the inequality $|\mu|(A_\alpha) \leq \alpha |\mu|(A_\alpha)$. If $\alpha < 1$ then necessary $|\mu|(A_\alpha) = 0$. This shows that $|h| \geq 1$ almost everywhere. On the other hand, by inequality (2.20) for any measurable set $E \in \mathcal{F}$ such that $\mu(E) > 0$ the following holds

$$\frac{| \int_E h \, d|\mu| \, |}{|\mu|(E)} \leq 1,$$

showing that the ratio $\frac{\int_E h \, d|\mu|}{|\mu|(E)}$ is a complex number belonging to the circle $C = \{z \in \mathbb{C} : |z| \leq 1\}$. This necessary gives $|h| \leq 1$ almost everywhere. Indeed, given a point α in the open set C^c and taken an open ball $B(\alpha, \epsilon) = \{z \in \mathbb{C} : |z - \alpha| < \epsilon\}$ contained in C^c, let us show that the set $E_{\alpha,\epsilon} := h^{-1}(B(\alpha, \epsilon))$ must be a $|\mu|$-null set. If it were not true, then

$$\epsilon < \left| \frac{\int_{E_{\alpha,\epsilon}} h \, d|\mu|}{|\mu|(E_{\alpha,\epsilon})} - \alpha \right| = \frac{\left| \int_{E_{\alpha,\epsilon}} (h - \alpha) \, d|\mu| \right|}{|\mu|(E_{\alpha,\epsilon})} \leq \frac{\int_{E_{\alpha,\epsilon}} |h - \alpha| \, d|\mu|}{|\mu|(E_{\alpha,\epsilon})} < \epsilon$$

obtaining a contradiction.

Remark 2.4. Let us consider a positive measure μ on the measurable space $(\Omega, \mathcal{F})$ and a complex-valued measurable function $f : \Omega \to \mathbb{C}$ such that $\int_\Omega |f| \, d\mu < +\infty$. Then the complex-valued set function $\nu : \mathcal{F} \to \mathbb{C}$ defined by

$$\nu(E) := \int_E f \, d\mu$$

defines a complex measure on $(\Omega, \mathcal{F})$. Its total variation measure $|\nu|$ is given by

$$|\nu|(E) := \int_E |f| \, d\mu.$$

Indeed, by the Radon-Nikodym theorem $f d\mu = d\nu = h d|\nu|$. Hence $d|\nu| = \bar{h} f d\mu$. Since μ is a positive measure, then $\bar{h} f$ must be a positive real number. In addition $|h| = 1$, hence we obtain $\bar{h} f = |f|$.

We end this section with the theory of integration with respect to complex measures.

Let μ be a complex measure on a measurable space $(\Omega, \mathcal{F})$ and let $|\mu|$ be its total variation measure. By the Radon-Nikodym Theorem 2.19 there exists a complex measurable function h such that $|h| = 1$ and $d\mu = h d|\mu|$. The integral of a complex-valued measurable function $f : \Omega \to \mathbb{C}$ with respect to μ is defined as:

$$\int_E f \, d\mu := \int_E f h \, d|\mu| \, .$$

Equality (2.16) naturally generalizes to the case where μ is a complex measure, showing that the application $f \mapsto \int f \, d\mu$ is a linear functional to the linear space of complex valued measurable functions $f : \Omega \to \mathbb{C}$. Clearly the study of any further continuity property of the functional requires the introduction of a suitable topology on the set of complex valued functions $f : \Omega \to \mathbb{C}$. To this end, in the following we shall restrict ourselves to the case where Ω is a locally compact Hausdorff space, $\mathcal{F}$ is the Borel σ-algebra and $\mu : \mathcal{F} \to \mathbb{C}$ is a complex regular Borel measure.

Let $C_0(\Omega)$ be the space of continuous functions $f : \Omega \to \mathbb{C}$ that *vanish at* ∞, i.e. the set of continuous mappings f such that for every $\epsilon > 0$ there exists a compact set $K_\epsilon \subset \Omega$ such that $|f| < \epsilon$ outside K_ϵ. The set $C_0(\Omega)$ endowed with the $\| \ \|_\infty$ norm:

$$\|f\|_\infty := \sup_{\omega \in \Omega} |f(\omega)|, \qquad f \in C_0(\Omega),$$

is a complete separable Banach space [293, 292]. Further, the linear functional $L_\mu : C_0(\Omega) \to \mathbb{C}$

$$L_\mu(f) := \int f \, d\mu, \tag{2.23}$$

is continuous and its norm coincides with the total variation of μ:

$$\|L_\mu\| = \|\mu\| \, .$$

The following theorem shows that every continuous linear functional on $C_0(\Omega)$ has exactly the form (2.23). For the detailed proof we refer to [293].

Theorem 2.20 (Riesz-Markov representation theorem). *If Ω be locally compact Hausdorff space and $L : C_0(\Omega) \to \mathbb{C}$ a linear continuous functional on the Banach space $C_0(\Omega)$, then there exists a unique regular complex Borel measure μ on Ω such that*

$$L(f) = \int f \, d\mu \, .$$

In addition $\|L\| = \|\mu\|$.

2.8　Fourier transform of complex measures on $\mathbb{R}^n$

Let us consider now the measurable space $(\mathbb{R}^n, \mathcal{B}(\mathbb{R}^n))$ and the set $\mathcal{M}(\mathbb{R}^n)$ of complex measures μ on $(\mathbb{R}^n, \mathcal{B}(\mathbb{R}^n))$, the $\mathcal{M}(\mathbb{R}^n)$-norm being defined as the total variation (see Eq. (2.21)).

For any $\mu \in \mathcal{M}(\mathbb{R}^n)$ let us define the complex-valued function $\hat{\mu} : \mathbb{R}^n \to \mathbb{C}$ as

$$\hat{\mu}(x) := \int_{\mathbb{R}^n} e^{i\,x\cdot y} d\mu(y) \qquad x \in \mathbb{R}^n \, , \tag{2.24}$$

where i is the imaginary unit and $x \cdot y$ denotes the inner product between $x, y \in \mathbb{R}^n$. The mapping $\hat{\mu}$ is well defined for any $x \in \mathbb{R}^n$ as the integral of a bounded complex-valued function with respect to a measure (with finite total variation), in particular

$$|\hat{\mu}(x)| \le \|\mu\|$$

In addition, a straightforward application of Theorem 2.16 shows that $\hat{\mu}$ is a continuous function.

On the basis of formula (2.24), $\hat{\mu}$ is called *Fourier transform* of the measure μ.

In the following we shall denote with the symbol $\mathcal{F}(\mathbb{R}^n)$ the space of Fourier transforms of measures $\mu \in \mathcal{M}(\mathbb{R}^n)$. It is important to stress that there is a one-to one correspondence between $\mathcal{M}(\mathbb{R}^n)$ and $\mathcal{F}(\mathbb{R}^n)$ as stated in the following theorem.

Theorem 2.21. *For any pair $\mu, \nu \in \mathcal{M}(\mathbb{R}^n)$ of complex Borel measures on $\mathbb{R}^n$ the following assertion are equivalent.*

　i. $\mu = \nu$.
　ii. $\hat{\mu}(x) = \hat{\nu}(x) \; \forall x \in \mathbb{R}^n$.

Proof. The implication $i. \Rightarrow ii.$ is trivial.

Concerning the implication $ii. \Rightarrow i.$, by Riesz-Markov representation theorem it is sufficient to show that for any $f \in C_0(\mathbb{R}^n)$ the following holds

$$\int_{\mathbb{R}^n} f(x)d\lambda(x) = 0 \tag{2.25}$$

where $\lambda \in \mathcal{M}(\mathbb{R}^n)$ is the complex function defined as $\lambda := \mu - \nu$.

If f is a trigonometric polynomial, i.e. f is of the form

$$f(x) = \sum_{j=1}^{m} c_j e^{i\, x \cdot y_j}, \qquad x \in \mathbb{R}^n$$

for some $m \geq 1$, $c_1, \ldots, c_m \in \mathbb{C}$, $y_1, \ldots, y_m \in \mathbb{R}^n$, then equality (2.25) follows from condition $i.$ and the definition (2.24) of Fourier transform of a complex measure. If f is continuous and compactly supported, equality (2.25) follows from Stone-Weirstrass approximation theorem, which allows to uniformly approximate f with trigonometric polynomials. Eventually, for a general $f \in C_0(\mathbb{R}^n)$, given an $\epsilon > 0$ there exists a compact set $K_\epsilon \subset \mathbb{R}^N$ such that

$$|f(x)| < \sqrt{\epsilon} \qquad \forall x \notin K_\epsilon.$$

Moreover, since $|\lambda|$ is a finite positive measure, by its continuity from below there exists a compact set $\tilde{K}_\epsilon$ such that

$$|\lambda|(\tilde{K}_\epsilon^c) < \sqrt{\epsilon}.$$

If we consider now the compact set $H_\epsilon := K_\epsilon \cup \tilde{K}_\epsilon$, we can estimate $\int f \, d\lambda$ as:

$$\left| \int f(x)\, d\lambda(x) \right| \leq \left| \int f(x)\chi_{H_\epsilon}(x)d\lambda(x) \right| + \left| \int f(x)\chi_{H_\epsilon^C}(x)d\lambda(x) \right|.$$

The first term on the right-hand side vanishes since the mapping $f\chi_{H_\epsilon}$ is compactly supported. Concerning the second term we have

$$\left| \int f(x)\chi_{H_\epsilon^C}(x)d\lambda(x) \right| \leq \int_{H_\epsilon^C} |f(x)|\, d|\lambda|(x) \leq \epsilon.$$

and by the arbitrariness on the choice of ϵ we obtain (2.25). $\qquad\square$

If we restrict ourselves to probability measures μ on $(\mathbb{R}^n, \mathcal{B}(\mathbb{R}^n))$, then there exists an interesting and useful characterization of their Fourier transform $\hat{\mu} : \mathbb{R}^n \to \mathbb{C}$. Indeed, the following theorem is a fundamental result of probability theory. For the proof an its generalization to locally compact abelian groups see [291].

Theorem 2.22 (Bochner). *A function $\phi : \mathbb{R}^n \to \mathbb{C}$ is the Fourier transform of a probability measure on $(\mathbb{R}^n, \mathcal{B}(\mathbb{R}^n))$ if and only if it enjoys the following properties:*

(1) $\phi(0) = 1$.
(2) ϕ is continuous in 0.
(3) ϕ is positive semi-definite, i.e. $\forall N \geq 1$, $\forall t_1, ..., t_N \in \mathbb{R}^n$ and $\forall z_1, ..., z_N \in \mathbb{C}$ the following holds:

$$\sum_{j,k=1}^{N} \phi(t_j - t_k) z_j \bar{z}_k \geq 0 \tag{2.26}$$

As remarked in Section 2.7, the set $\mathcal{M}(\mathbb{R}^n)$ endowed with the total variation norm $\| \ \|$ is a normed linear space. In addition, if we introduce in $\mathcal{M}(\mathbb{R}^n)$ the multiplication $* : \mathcal{M}(\mathbb{R}^n) \times \mathcal{M}(\mathbb{R}^n) \to \mathcal{M}(\mathbb{R}^n)$ defined as the convolution

$$\mu * \nu(E) = \int \int \chi_E(x + y) \, d\mu(x) d\nu(y) \,,$$

then $\mathcal{M}(\mathbb{R}^n)$ becomes a Banach algebra with unit 1, the latter being the delta point measure δ_0 concentrated at $0 \in \mathbb{R}^n$ (see (2.6)). Indeed for any bounded continuous function $f : \mathbb{R}^n \to \mathbb{R}$, we have

$$\int f(x) d(\mu * \nu)(x) = \int \int f(x + y) \, d\mu(x) d\nu(y) \,,$$

which yields the inequality $\|\mu * \nu\| \leq \|\mu\| \|\nu\|$.

By Theorem 2.21, the map $\mu \mapsto \hat{\mu}$ is a one-to-one correspondence between $\mathcal{M}(\mathbb{R}^n)$ and $\mathcal{F}(\mathbb{R}^n)$. It is rather easy to verify that it enjoys the following properties:

- $\widehat{\alpha\mu + \beta\nu} = \alpha\hat{\mu} + \beta\hat{\nu}$ for all $\mu, \nu \in \mathcal{M}(\mathbb{R}^n)$, $\alpha, \beta \in \mathbb{C}$.
- $\widehat{\mu * \nu} = \hat{\mu}\,\hat{\nu}$, for all $\mu, \nu \in \mathcal{M}(\mathbb{R}^n)$.

By endowing $\mathcal{F}(\mathbb{R}^n)$ with the norm $\| \ \|$

$$\hat{\mu} := \|\mu\|$$

we obtain that $(\mathcal{F}(\mathbb{R}^n), \| \ \|)$ is a Banach function algebra of continuous bounded function. In addition, it is easy to verify that

$$\|\hat{\mu}\|_\infty = \sup_{x \in \mathbb{R}^n} |\hat{\mu}(x)| \leq \|\mu\| = \|\hat{\mu}\| \,.$$

Chapter 3

Measures on infinite dimensional spaces

In the present chapter we describe the main topics of infinite dimensional measure and integration theory that are relevant in the mathematical definition of Feynman path integrals.

The first section is devoted to the proof of a number of no-go results, showing why in an infinite dimensional setting it is actually impossible to define a reasonable and useful analogue of Lebesgue measure on $\mathbb{R}^n$. This result rules out the naive attempt to define the heuristic "Feynman measure" (1.31) as an absolutely continuous measure with respect to a Lebesgue type measure $d\gamma$ on the path space with a density of the form

$$\gamma \mapsto \frac{e^{\frac{i}{\hbar} S_t(\gamma)}}{\int e^{\frac{i}{\hbar} S_t(\gamma)} d\gamma}.$$

Actually the lack of a "reference measure", such as, e.g., the Lebesgue measure, is an important issue which has to be tackled not only in the definition of Feynman integrals, but more generally in the theory of stochastic processes, where the problem of defining a probability space $(\Omega, \mathcal{F}, \mu)$ where an infinite number of random variables are simultaneously defined arises. This issue was brilliantly tackled by Kolmogorov, who proved a fundamental result which allows to construct a probability measure on an infinite dimensional space starting from its finite dimensional approximations. Kolmogorov theorem is one of the cornerstones of probability theory and will be detailed proved in the second section. Section 3.3 presents the construction of Wiener measure in detail, along with Feynman-Kac formula. In section 3.4 we present a generalization of Kolmogorov theorem on projective limit spaces due to S. Bochner. At this point the reader might be wondering whether similar techniques, in particular those leading to the construction of Wiener measure, can be applied to the construction of a complex measure on the path space $\mathbb{R}^{[0,+\infty)}$ which can play for the Schrödinger equation the same role that Wiener measure plays for the heat equation. That's why in Section 3.5 we present a generalization of Kolmogorov theorem to the case of projective systems of complex measures (instead of probability measures). In particular we shall prove a rather restrictive condition for the existence of projective limit measures that, as we shall see, cannot be fulfilled in the most interesting cases. This result shows that in an infinite dimensional setting the construction of non-trivial complex measures is a rather difficult task, and it is impossible in the cases we are interested in. In particular, we shall obtain as a byproduct Cameron's result [77] on the non-existence of Wiener measure with complex variance.

The final sections contain some technical results which play an important role in the construction of infinite dimensional Fresnel and oscillatory integrals. In particular, Section 3.6 deals with the Banach algebra of complex valued functions that are Fourier transforms of complex Borel measures on real separable Banach or Hilbert spaces. Eventually, Section 3.7 gives a concise account of the theory of Gaussian measures on Hilbert spaces as well as of abstract Wiener spaces.

3.1   Non-existence of Lebesgue measure

This section presents some classical no-go result on the construction of Lebesgue measure on infinite dimensional linear spaces. We start by considering the particular case of an infinite dimensional real separable Hilbert space $(\mathcal{H}, \langle \, , \, \rangle)$, since it is a natural analogue of $\mathbb{R}^n$ in an infinite dimensional setting, but the whole discussion can be generalized to infinite dimensional topological vector spaces or infinite dimensional groups (see [335]).

Let $\mathcal{B}(\mathcal{H})$ be the Borel σ-algebra on $\mathcal{H}$, i.e. the σ-algebra generated by the open balls $B_\rho(\bar{x})$, with $\rho \in \mathbb{R}^+$ and $\bar{x} \in \mathcal{H}$. Given a vector $\bar{x} \in \mathcal{H}$, let $\tau_{\bar{x}} : \mathcal{H} \to \mathcal{H}$ denote the translation of $\bar{x}$, i.e. the map defined by $\tau_{\bar{x}}(x) = x + \bar{x}$. A measure μ on $\mathcal{B}(\mathcal{H})$ is said *translation invariant* if $\forall \bar{x} \in \mathcal{H}$ the image measure $\mu_{\tau_{\bar{x}}}$ defined by

$$\mu_{\tau_{\bar{x}}}(A) := \mu(\tau_{\bar{x}}^{-1}(A)) = \mu(\{x \in \mathcal{H} : x + \bar{x} \in A\}),$$

is equal to μ, i.e.

$$\forall \bar{x} \in \mathcal{H}, A \in \mathcal{B}(\mathcal{H}) \qquad \mu(A) = \mu(\{x \in \mathcal{H} : x + \bar{x} \in A\}),$$

A linear bounded operator $U : \mathcal{H} \to \mathcal{H}$ is called unitary if $\langle Ux, Uy \rangle = \langle x, y \rangle$ for all $x, y \in \mathcal{H}$. A measure μ on $\mathcal{B}(\mathcal{H})$ is said *invariant under isometries* if for any unitary operator $U : \mathcal{H} \to \mathcal{H}$ the image measure μ_U defined by

$$\mu_U(A) := \mu(U^{-1}(A)) = \mu(\{x \in \mathcal{H} : U(x) \in A\}),$$

is equal to μ.

In Section 2.5 we proved that if $\mathcal{H}$ is finite dimensional, than a Borel measure μ assigning a finite value to bounded sets is translation invariant if and only if $\mu = c\mathcal{L}$, where $\mathcal{L}$ is the Lebesgue measure on $\mathbb{R}^n$ and $c \in \mathbb{R}^+$ is a positive constant. On the other hand, the following theorem shows that if $\mathcal{H}$ is infinite dimensional a reasonable Lebesgue-type measure cannot exists.

Theorem 3.1. *Let $(\mathcal{H}, \langle \, , \, \rangle)$ be a real separable infinite dimensional Hilbert space. Then it cannot exists a Borel measure μ such that*

(1) it assigns a strictly positive finite measure to open balls;
(2) it is translation invariant.

Proof. By contradiction, let us assume that such a measure exists. Let $\{e_i\}$ be a complete orthonormal basis and let us consider the family of open balls $B_{1/2}(e_i)$ with center e_i and radius $1/2$. It is simple to see that they are disjoint, i.e. $B_{1/2}(e_i) \cap B_{1/2}(e_j) = \emptyset$ if $i \neq j$. Moreover, considering the open ball $B_2(0)$ centered at 0 with radius 2, we have that $B_{1/2}(e_i) \subset B_2(0)$ for all i. We then have $\cup_i B_{1/2}(e_i) \subset B_2(0)$, hence

$$\mu(\cup_i B_{1/2}(e_i)) = \sum_i \mu(B_{1/2}(e_i)) \leq \mu(B_2(0)) \tag{3.1}$$

Furthermore, for any choice of i, j, we can map the ball $B_{1/2}(e_i)$ into the ball $B_{1/2}(e_j)$ by means of the translation $\tau_{e_j - e_i}$. Hence, by condition 2, $B_{1/2}(e_i) = c$ for all i and, by condition 1, $0 < c < \infty$. Since $B_{1/2}(e_i) = c$ for all i by assumption 1, we have $\mu(\cup_i B_{1/2}(e_i)) = \sum_i \mu(B_{1/2}(e_i)) = +\infty$. On the other hand, by assumption 1, $\mu(B_2(0))$ has a finite value, but, by (3.1), it must be infinite since $\mu(B_2(0)) \geq \sum_j c = +\infty$. $\square$

The same argument can be used also in the proof of the following result.

Theorem 3.2. *Let $(\mathcal{H}, \langle \, , \, \rangle)$ be a real separable infinite dimensional Hilbert space. It cannot exists a Borel measure μ such that*

(1) it assigns a strictly positive finite measure to open balls;
(2) it is invariant under isometries.

In particular Theorem 3.2 forbids not only the existence of Lebesgue measure on an infinite dimensional Hilbert space, but also the existence of the centered Gaussian measure with covariance operator equal to the identity (which would be invariant under isometries).

The simple argument presented above allows the reader to grasp the role that infinite dimensionality plays in the proof of non-existence of a reasonable version of Lebesgue measure on $\mathcal{B}(\mathcal{H})$. In fact a stronger result can be proved. Given a topological vector space X endowed with the Borel σ-algebra, we say that a measure $\mu : \mathcal{B}(X) \to [0, +\infty]$ is *quasi-invariant* if μ is equivalent to μ_{τ_x} for any $x \in X$. More specifically, if for any $\bar{x} \in X$ and for any $I \in \mathcal{B}(X)$

$$\mu(I) = 0 \quad \Leftrightarrow \quad \mu(I + \bar{x}) = 0 \,,$$

where the set $I + \bar{x}$ is defined as the set of all elements of X of the form $x + \bar{x}$ for some $x \in I$.

Clearly the trivial measure which assigns to any Borel set the value zero is translation invariant. When X is infinite dimensional linear space, under rather general assumption the only sigma-finite quasi-invariant Borel measure is the trivial one. Before stating the next theorem, it is convenient to give the following definition.

Definition 3.1. Let Y be a subset of a linear space X. The *absolutely convex hull* of Y is the set of all finite linear combinations $c_1 y_1 + \cdots + c_n y_n$, with $n \in \mathbb{N}$, $n \geq 1$, $y_1, \ldots, y_n \in Y$ and $c_1, \ldots, c_n \in \mathbb{R}$ such that $\sum_{k=1}^{n} |c_k| \leq 1$.

Theorem 3.3. *Let X be a complete metric linear space containing a countable dense set and such that the absolutely convex hull of any compact set K in X is nowhere dense in X[1]. Then the only σ-finite quasi invariant Borel measure μ on X is the identically zero measure.*

Proof.   Without loss of generality it is sufficient to prove the thesis under the assumption that μ is a finite measure. Indeed, if $\mu(X) = +\infty$ then by the assumption of σ-finiteness there exists a countable partition $\{X_n\}_n$ of X such that $0 < \mu(X_n) < +\infty$ for every n. Let $f : X \to \mathbb{R}$ be defined as

$$f(x) := \frac{1}{2^n \mu(X_n)}, \quad x \in X_n$$

and consider the measure ν on $\mathcal{B}(X)$ defined by

$$\nu(E) := \int_E f \, d\mu \,. \tag{3.2}$$

---

[1]A set $Y \subset X$ of a topological space X is said to be *nowhere dense* if its closure has empty interior.

It is easy to check that ν is a finite measure since by construction $\nu(X) = 1$. Furthermore, by Eq. (3.2) it turns out that ν is quasi invariant if μ is.

By the discussion above we can restrict ourselves to the case where μ is a probability measure. Since by assumption X is a Polish space, by Theorem 2.2 the Borel measure μ is regular. In particular for every $n \in \mathbb{N}$ there exists a compact set K_n such that

$$\mu(K_n) \geq 1 - 1/n. \tag{3.3}$$

Furthermore, by the continuity from below of the measure μ and from (3.3) we obtain that the countable union $Y = \cup_n K_n$ has measure $\mu(Y) = 1$ and, consequently, its (non-closed) linear span $\hat{Y}$ has measure $\mu(\hat{Y}) = 1$.

Let us show now that $\hat{Y}$ cannot coincide with X. Indeed, let us consider for any finite n the compact set $Y_n := \cup_{k=1}^{n} K_k$ and let $\tilde{Y}_n$ be the absolutely convex hull of Y_n. By assumption $\tilde{Y}_n$ is nowhere dense, hence the sets $mY_n := \{my : y \in Y_n\}$ are nowhere dense for any $m, n \in \mathbb{N}$. Further, the set $\hat{Y}$ can be represented as $\hat{Y} = \cup_{m,n} mY_n$, hence we can deduce that $X \neq \hat{Y}$ since by the Baire category theorem [292], a complete metric space cannot be a countable union of nowhere dense sets.

If we consider now a point $\bar{x} \in X$ such that $\bar{x} \notin \hat{Y}$, then the sets $\bar{x} + \hat{Y}$ and $\hat{Y}$ are disjoint since $\hat{Y}$ is a linear space. Similarly, since by construction $Y \subset \hat{Y}$ and correspondingly $\bar{x} + Y \subset \bar{x} + \hat{Y}$, we have that $\bar{x} + Y$ is included in the complement of Y. Hence $\mu(\bar{x} + Y) = 0$ while $\mu(Y) = 1$ and we conclude that μ is not quasi-invariant. $\qquad\square$

The assumptions of Theorem 3.3 seem to be quite involved, but in fact it is quite simple to find interesting example where they are satisfied, as the following result shows.

Corollary 3.1. *Let X be a complete normed linear space containing a countable dense set. Then the only σ-finite quasi invariant Borel measure μ on X is the identically zero measure.*

Proof. It is sufficient to show that the absolutely convex hull of any compact set K in X is nowhere dense in X.

Let $K \subset X$ be a compact set and let $\tilde{K}$ be its absolutely convex hull. We have to prove that $\tilde{K}$ has no inner points, i.e. that for any $x_0 \in X$ and any positive radius r, the open ball $B(x_0, r)$ cannot lie in $\tilde{K}$. To this end, let us consider the open covering of K made of the open balls $B(x, r/2)$ with center in the points $x \in K$ and radius $r/2$. By compactness we can extract a finite covering $B(x_i, r/2)$, $i = 1, \ldots, n$. Clearly the absolutely convex hull $\tilde{K}$ of K is contained in the absolutely convex hull $\tilde{B}$ of the set $\cup_{i=1}^{n} B(x_i, 1/2)$. In particular, any element x of $\tilde{B}$ can be written in the form $x = v + u$, where v belongs to the linear span $\mathrm{span}(x_1, \ldots, x_n)$ of $x_1, \ldots, x_n$, while $\|u\| < r/2$. Indeed, by the definition of $\tilde{B}$, any $x \in \tilde{B}$ can be represented as

$$x = c_1(x_1 + u_1) + \cdots + c_n(x_n + u_n)$$
$$= (c_1 x_1 + \cdots + c_n x_n) + (c_1 u_1 + \cdots + c_n u_n)$$

with $|c_1| + \cdots + |c_n| \leq 1$ and $|u_i| \leq 1/2$, for all $i = 1, \ldots, n$.

It is now simple to show [259] that the closure of $\tilde{B}$ cannot contain the ball $B(x_0, r)$. Indeed, every element $x \in B(x_0, r)$ is of the form $x = x_0 + y$, with $\|y\| \leq r$, and we can always find a vector $y \in B(0, r)$ that cannot be of the form $y = v - x_0 + u$ with $v \in \mathrm{span}(x_1, \ldots, x_n)$ and $\|u\| \leq r/2$. Indeed, let us consider the closed vector space $V = \mathrm{span}(x_0, x_1, \ldots, x_n)$ and a vector $x' \notin V$. Let $\alpha \in \mathbb{R}^+$ be the distance $d(x', V)$ of x' to V where

$$d(x', V) := \inf_{x \in V} \|x' - x\|.$$

Clearly α is strictly positive since $x' \notin V$ and V is closed. Let us choose a vector $w \in V$ such that $\alpha < \|x' - w\| < 2\alpha$. By construction the vector $y := r(x' - w)/\|x' - w\|$ has norm $\|y\| = r$. Moreover $d(y, V) > r/2$ since $d(x', V) = d(x' - w, V)$. $\qquad\square$

It is interesting to point out that in the no-go results stated above, the requirement that the measure μ is σ-finite plays a fundamental role. If this assumption is dropped, it is possible to provide examples of translation invariants measures on the Borel σ-algebra of $\mathbb{R}^{\mathbb{N}}$, the latter being endowed with the coarser topology making all projections $P_n : \mathbb{R}^N \to \mathbb{R}^n$

$$P_n(x_1, x_2, \ldots) := (x_1, \ldots, x_n) \qquad (x_i)_i \in \mathbb{R}^{\mathbb{N}},$$

continuous. For a detailed description of this topic see [335, 163].

3.2   Kolmogorov theorem

Kolmogorov theorem is one of the basic tools for the construction of probability measures on infinite dimensional spaces and it is one of the cornerstones of the theory of stochastic processes. It was originally established by Kolmogorov in the case where the set Ω where the probability space $(\Omega, \mathcal{F}, \mu)$ is constructed coincides with the set $\Omega = \mathbb{R}^{[0,+\infty)}$ of all real valued paths $\omega : [0, +\infty) \to \mathbb{R}$. The theorem was later generalized to projective limit spaces by Bochner (see [68] and Section 3.4 below). For a detailed discussion of this topic see, e.g., [335, 68]. In the following we shall present the detailed proof of a version which is sufficiently powerful for our purposes.

Let S be a topological space that in the following we shall assume to be a complete separable metric space. Simple examples of such spaces are: $\mathbb{R}^d$ endowed with the Euclidian topology, any Riemannian manifold M with metric tensor g, and any separable Banach space. Let $T \subset [0, +\infty)$ be a subset of the half-real line and us consider the set $\Omega = S^T$ of real-valued maps $\omega : T \to S$ from the "time set" T into S. Typical examples of interesting time sets T are $T = \mathbb{N}$ and $T = [0, +\infty)$. In the following the points $\omega \in \Omega$ will be called *paths*.

Let $\mathcal{I}$ be the set of all finite subsets of the set T, endowed with the partial order relation $\leq$ defined by

$$J \leq K \quad \text{if} \quad J \subseteq K.$$

In fact $\mathcal{I}$ has the structure of a *directed set*, in the sense that for any $J, K \in \mathcal{I}$ there is an $H \in \mathcal{I}$ such that $J \leq H$ and $K \leq H$.

Given a $J \in \mathcal{I}$, with $J = \{t_1, t_2, ..., t_n\}$, $0 \leq t_1 < t_2 \cdots < t_n$, let S^J denote the set of all maps from J to S. Any element of S^J can be identified with an n-ple of real numbers $(\omega(t_1), \omega(t_2), ..., \omega(t_n))$, hence in the following we shall identify S^J with the n-Cartesian product of n copies of S, n being the cardinality of J. Correspondingly, we shall consider on S^J the product σ-algebra of n-copies of the Borel σ-algebra $\mathcal{B}(S)$, and shall denote it with the symbol $\mathcal{B}(S^J)$ since it coincides with the Borel σ-algebra associated to the product topology on S^J.

Given an element $J \in \mathcal{I}$, let $\Pi_J : \Omega \to S^J$ be the mapping which assigns to each path $\omega \in \Omega$ its values at the points of J:

$$\Pi_J(\omega) := (\omega(t_1), \omega(t_2), ..., \omega(t_n)), \qquad \omega \in S^T, J = \{t_1, t_2, ..., t_n\}$$

Similarly, given a pair $J, K \in \mathcal{I}$ such that $J \leq K$, let us consider the map $\Pi_J^K : S^K \to S^J$ that assigns to a general $\eta \in S^K$ the function $\Pi_J^K(\eta) \in S^J$ defined by

$$\Pi_J^K(\eta)(t) := \eta(t), \qquad t \in J. \tag{3.4}$$

By construction, the mappings $(\Pi_J^K)_{J \leq K}$ are continuous, hence Borel measurable, and if $J = K$ then the map Π_K^K is the identity on S^K. Furthermore, the mappings $\{\Pi_J^K\}$ are surjective and satisfy the following composition law whenever $J \leq K \leq H$

$$\Pi_J^K \circ \Pi_K^H = \Pi_J^H. \tag{3.5}$$

Moreover, if $J \leq K$ then

$$\Pi_J^K \circ \Pi_K = \Pi_J. \tag{3.6}$$

A key ingredient in formulating Kolmogorov theorem is the collection of *cylinder sets*, i.e. the subsets of Ω of the form $\Pi_J^{-1}(B_J)$ for some $J \in \mathcal{I}$ and some Borel set $B_J \in \mathcal{B}(S^J)$. In fact, if $J = \{t_1, t_2, ..., t_n\}$, $0 \leq t_1 < t_2 \cdots < t_n$, the cylinder set $\Pi_J^{-1}(B_J)$ is the set of all paths $\omega \in S^T$ such that $(\omega(t_1), \omega(t_2), ..., \omega(t_n)) \in B_J$. In the following we shall denote $\mathcal{C}$ the set of all cylinder sets, and $\mathcal{F} = \sigma(\mathcal{C})$ the σ-algebra generated by $\mathcal{C}$. In fact $\sigma(\mathcal{C})$ can be characterised as the smallest σ-algebra making all the mappings $\{\Pi_J\}_{J \in \mathcal{I}}$ measurable.

Given a measure μ on $(\Omega, \mathcal{F})$, for every $J \in \mathcal{I}$ it is possible to construct a measure μ_J on $(S^J, \mathcal{B}(S^J))$ as the image measure $\mu_J := (\mu)_{\Pi_J}$ of μ under the action of the mapping Π_J:

$$\mu_J(B_J) := \mu(\Pi_J^{-1}(B_J)), \qquad B_J \in \mathcal{B}(S^J).$$

Equivalently, the μ_J-measure of a Borel set $B_J \in \mathcal{B}(S^J)$ is defined as the μ-measure of the cylinder set $\Pi_J^{-1}(B_J)$. Note that the definition of μ_J is well posed since Π_J is a measurable map from $(\Omega, \mathcal{F})$ to $(S^J, \mathcal{B}(S^J))$.

Given two elements $J, K \in \mathcal{I}$, with $J \leq K$, the measures μ_J on $(S^J, \mathcal{B}(S^J))$ and μ_K on $(S^K, \mathcal{B}(S^K))$ are related by the identity

$$\mu_J = (\mu_K)_{\Pi_J^K} \qquad \forall J, K \in \mathcal{I}, \ J \leq K \tag{3.7}$$

which can be equivalently written in the following form

$$\mu_J(B_J) := \mu_k((\Pi_J^K)^{-1}(B_J)), \qquad B_J \in \mathcal{B}(S^J), \tag{3.8}$$

as one can verify by means of the identity (3.6).

Definition 3.2. A family of measures $\{\mu_J\}_{J \in \mathcal{I}}$ satisfying the compatibility condition (3.7) is called a *projective* or *consistent* family of measures.

Now the converse is our problem. Namely, given a family of measures $\{\mu_J\}_{J \in \mathcal{I}}$ satisfying the compatibility condition (3.7), does it exist a measure μ on $(S^T, \mathcal{F})$ such that for any $J \in \mathcal{I}$ one has that $\mu_J = \Pi_J(\mu)$? In other words, is it possible to construct a measure on the (infinite dimensional) space of paths $\Omega = S^T$ by means of its "finite dimensional approximations"? In the case of probability measures the answer is affirmative.

Theorem 3.4 (Kolmogorov). *Let S be a complete separable metric space and $T \subset [0, +\infty)$. For any projective family $\{\mu_J\}_{J \in \mathcal{I}}$ of probability measures on $(S^J, \mathcal{B}(S^J))$ there exists a unique probability measure μ on the σ-algebra generated by the cylinder sets such that*

$$\mu_J = \Pi_J(\mu). \tag{3.9}$$

Proof. The proof relies on Theorem 2.5 and it is divided in three steps.

(1) *The family $\mathcal{C}$ of cylinder sets is an algebra.*
Let us first remark that a cylinder set admits several equivalent representations. Indeed, given $J \in \mathcal{I}$ and $B_J \in \mathcal{B}(S^J)$, then by identity (3.6) we have that for any $K \in \mathcal{I}$ such that $J \leq K$ the following holds

$$\Pi_J^{-1}(B_J) = \Pi_K^{-1}(B_K), \qquad \text{where } B_K = (\Pi_J^K)^{-1}(B_J). \tag{3.10}$$

Now the proof that $\mathcal{C}$ enjoys all the properties of an algebra of sets is straightforward. First, $\Omega \in \mathcal{C}$. Indeed, it is sufficient to choose a generic $J \in \mathcal{I}$ and set $B_J = S^J$. Further, the relation $(\Pi_J^{-1}(B_J))^c = \Pi_J^{-1}(B_J^c)$ shows that $\mathcal{C}$ is closed under complementation. Eventually, given two cylinder sets $\Pi_J^{-1}(B_J)$ and $\Pi_K^{-1}(B_K)$, let us choose a set $H \in \mathcal{I}$ such that $J \leq H$ and $K \leq H$ and consider the two Borel sets $\tilde{B}_J, \tilde{B}_K \in \mathcal{B}(S^H)$ defined as $\tilde{B}_J := (\Pi_J^H)^{-1}(B_K)$ and $\tilde{B}_K := (\Pi_K^H)^{-1}(B_K)$. By using the equivalent representations of the two cylinder sets $\Pi_J^{-1}(B_J)$ and $\Pi_K^{-1}(B_K)$:

$$\Pi_J^{-1}(B_J) = \Pi_H^{-1}(\tilde{B}_J), \qquad \Pi_K^{-1}(B_K) = \Pi_H^{-1}(\tilde{B}_K) \tag{3.11}$$

it is simple to see that $\mathcal{A}$ is closed under finite union, since

$$\Pi_J^{-1}(B_J) \cup \Pi_K^{-1}(B_K) = \Pi_H^{-1}(\tilde{B}_J) \cup \Pi_H^{-1}(\tilde{B}_K) = \Pi_H^{-1}(\tilde{B}_J \cup \tilde{B}_K).$$

(2) *Definition of a finitely additive probability measure μ on $\mathcal{C}$.*
Let $\mu : \mathcal{C} \to [0, 1]$ be the set function defined on a generic cylinder set $\Pi_J^{-1}(B_J)$ as

$$\mu(\Pi_J^{-1}(B_J)) := \mu_J(B_J). \tag{3.12}$$

In fact, μ is unambiguously defined by (3.12) thanks to the consistency condition (3.7). Indeed, considered an equivalent representation of the cylinder set $\Pi_J^{-1}(B_J)$ of the form (3.10), identity (3.7) gives

$$\mu(\Pi_J^{-1}(B_J)) = \mu_J(B_J) = \mu_K(B_K) = \mu(\Pi_K^{-1}(B_K)), \quad B_K = (\Pi_J^K)^{-1}(B_J).$$

Further, μ is a finitely additive measure. To see this we have to show that for any pair of mutually disjoint cylinder sets $\Pi_J^{-1}(B_J)$ and $\Pi_K^{-1}(B_K)$ we have

$$\mu(\Pi_J^{-1}(B_J)) + \mu(\Pi_K^{-1}(B_K)) = \mu(\Pi_J^{-1}(B_J) \cup \Pi_K^{-1}(B_K)).$$

Taking a set $H \in \mathcal{I}$ such that hat $J \leq H$ and $K \leq H$, let us consider the equivalent representations $\Pi_J^{-1}(B_J) = \Pi_H^{-1}(\tilde{B}_J)$ and $\Pi_K^{-1}(B_K) = \Pi_H^{-1}(\tilde{B}_K)$, where the two Borel sets $\tilde{B}_J = (\Pi_J^H)^{-1}B_J$ and $\tilde{B}_K = (\Pi_K^H)^{-1}B_K$ turn out to be disjoint (since $\Pi_H^{-1}(\tilde{B}_J)$ and $\Pi_H^{-1}(\tilde{B}_K)$

are disjoint by assumption). Hence, by using the finite additivity of the measure μ_H we obtain:

$$\mu(\Pi_J^{-1}(B_J) \cup \Pi_K^{-1}(B_K)) = \mu(\Pi_H^{-1}(\tilde{B}_J) \cup \Pi_H^{-1}(\tilde{B}_K))$$
$$= \mu_H(\tilde{B}_J \cup \tilde{B}_K)$$
$$= \mu_H(\tilde{B}_J) + \mu_H(\tilde{B}_K)$$
$$= \mu(\Pi_H^{-1}(\tilde{B}_J)) + \mu(\Pi_H^{-1}(\tilde{B}_K))$$
$$= \mu(\Pi_J^{-1}(B_J)) + \mu(\Pi_K^{-1}(B_K))$$

(3) *Proof of the σ-additivity of μ on $\mathcal{C}$.*

By Theorem 2.3 it is sufficient to show that μ is continuous at $\emptyset$, i.e. that for any sequence $\{E_n\} \subset \mathcal{C}$ of cylinder sets such that $E_{n+1} \subset E_n$ and $\cap_n E_n = \emptyset$, the following holds

$$\lim_{n \to \infty} \mu(E_n) = 0$$

We prove it "ad absurdum". Assuming that $\mu(E_n) \not\to 0$, then there is a strictly positive constant $\varepsilon \in (0, 1]$, such that

$$\mu(E_n) \geq \varepsilon > 0, \quad \forall n \in \mathbb{N}. \tag{3.13}$$

Since E_n is a cylinder set, it can be represented as $E_n = (\Pi_{J_n})^{-1}(B_n)$ for some $J_n \in \mathcal{I}$ and $B_n \in \mathcal{B}(S^{J_n})$. Without loss of generality, we can assume $J_n \subset J_{n+1}$ and $J_n = \{t_1, \ldots, t_n\}$, with $t_i \in [0, T]$ (otherwise we would replace J_n by $J'_n = \{J_1 \cup \cdots \cup J_n\}$ and take $B'_n \in \mathcal{B}(\mathbb{R}^{J'_n})$ equal to $B'_n = (\Pi_{J_n}^{J'_n})^{-1}(B_n)$). From the definition of μ and by assumption (3.13), we have

$$\mu_{J_n}(B_n) = \mu(E_n) \geq \varepsilon > 0 \quad \forall n \in \mathbb{N}.$$

On the other hand, by Theorem 2.2 for every $n \in \mathbb{N}$ the Borel measure μ_{J_n} is regular (as any probability measure on the Borel σ-algebra of a complete separable metric space, see Theorem 2.2), hence there exist a compact set K_n such that $K_n \subset B_n$ and

$$\mu_{J_n}(B_n \setminus K_n) \leq \frac{\varepsilon}{2^n}. \tag{3.14}$$

Let us consider the sequences of cylinder sets $G_n = (\Pi_{J_n})^{-1}(K_n)$ and $L_n \equiv \cap_{m \leq n} G_m$. Clearly, by construction we have the following chain of inclusions

$$L_n \subset G_n \subset E_n \subset S^T$$

and $L_n \downarrow \varnothing$, since $E_n \downarrow \varnothing$ by assumption. The definitions of μ, E_n, and G_n imply, together with (3.14),

$$\mu(E_n \setminus G_n) = \mu_{J_n}(B_n \setminus K_n) \leq \frac{\varepsilon}{2^n} \quad \forall n \in \mathbb{N},$$

and therefore from the definition of L_n, together with $E_n \subset E_m$ for $m \leq n$ and the subadditivity of the measure of a union of a finite number of sets, it follows

$$\mu(E_n \setminus L_n) = \mu \left(\bigcup_{m \leq n} (E_n \setminus G_m) \right)$$

$$\leq \mu \left(\bigcup_{m \leq n} (E_m \setminus G_m) \right) \leq \sum_{m=1}^{n} \mu(E_m \setminus G_m) < \varepsilon,$$

whereas by (3.13) we have

$$\mu(E_n \setminus L_n) = \mu(E_n) - \mu(L_n) \geq \varepsilon - \mu(L_n).$$

Hence $\mu(L_n) > 0$ for all $n \in \mathbb{N}$, thus L_n is not empty. We can conclude that for all $n \in \mathbb{N}$ there exists (at least) a path $\omega^{(n)}$ belonging to L_n. Therefore, since $L_n \subset G_m$ for all $m \leq n$ by construction, we have

$$(\omega^{(n)}(t_1), ..., \omega^{(n)}(t_m)) \in K_m, \qquad \forall m \leq n.$$

Consider now the sequence of points $\{\omega^{(n)}(t_1)\}_n \subset K_1$. By the compactness of K_1 it is possible to extract a convergent subsequence $\{\omega^{(n_{1,m})}(t_1)\}_m$ converging to a point $l(t_1) \in K_1$. Let us consider the sequence $\{\omega^{(n_{1,m})}(t_2)\}_m$. Since the vectors $(\omega^{(n_{1,m})}(t_1), \omega^{(n_{1,m})}(t_2))$ belong to K_2, then it is possible to extract from $\{\omega^{(n_{1,m})}(t_2)\}_m$ a convergent subsequence $\{\omega^{(n_{2,m})}(t_2)\}_m$, converging to a limit $l(t_2)$. Further the subsequence $(\omega^{(n_{2,m})}(t_1), \omega^{(n_{2,m})}(t_2))$ converges to $(l(t_1), l(t_2)) \in K_2$. By iterating this procedure, we can construct for any $k \in \mathbb{N}$, $k \geq 1$, a subsequence $(\omega^{(n_{k,m})}(t_1), ..., \omega^{(n_{k,m})}(t_k))$ converging to a point $(l(t_1), ..., l(t_k)) \in K_k$. Note that the increasing sequences $\{n_{1,m}\}_m, \{n_{2,m}\}_m, ..., \{n_{k,m}\}_m$ can be ordered in an infinite matrix of the following form

$$n_{1,1} \; n_{1,2} \; n_{1,3} \cdots$$
$$n_{2,1} \; n_{2,2} \; n_{2,3} \cdots$$
$$\vdots \qquad \vdots \qquad \vdots \qquad \vdots$$
$$n_{r,1} \; n_{r,2} \; n_{r,3} \cdots$$

where each row is an increasing sequence of positive integers and the r-th row is a subsequence of the $(r-1)$-th row. If we consider now the sequence $\{n_m\}_m$ defined as $n_m := n_{m,m}$ then, except for a finite number of initial terms, the sequence $\{\omega^{(n_m)}(t_k)\}_m$ is a subsequence of $\{\omega^{(n_{k,m})}(t_k)\}_m$ and it converges to $l(t_k)$. We have thus obtained for any k that

$$(\omega^{(n_m)}(t_1), ..., \omega^{(n_m)}(t_k)) \to (l(t_1), ..., l(t_k)) \in K_k$$

If we define a path $\bar{\omega} : T \to S$ such that $\bar{\omega}(t_k) = l(t_k)$ for any $k \in \mathbb{N}$, then we have that $\bar{\omega} \in B_k \subset A_k$ for all k, hence $\cap_k E_k \neq \emptyset$ and we obtain a contradiction.

By Theorem 2.5 we can then conclude that μ can be extended to a σ-additive probability measure on $\mathcal{F} = \sigma(\mathcal{A})$.

The uniqueness of the extension follows by Theorem 2.6 which holds since μ is a finite measure. □

Remark 3.1. If the index set T is countable then the result holds without any topological assumption on S. This is a theorem of C. Ionescu Tulcea, see, e.g., [267].

The measure μ described by Kolmogorov's theorem is said the *projective limit* of the projective family $\{\mu_J\}$. By construction, for any $J \in \mathcal{I}$ the measure μ_J turns out to be the image measure of μ under the action of the measurable map Π_J.

Remark 3.2. In fact, the σ-algebra $\mathcal{F}$ can be regarded as the σ-algebra $\sigma(X_t, t \in T)$ generated by the measurable functions $X_t : \Omega \to S$ defined by

$$X_t(\omega) := \omega(t), \qquad \omega \in \Omega. \tag{3.15}$$

Equivalently, $\mathcal{F}$ is the σ-algebra generated by the collection $\mathcal{C}'$ of sets of the form $X_t^{-1}(I)$, for some $t \in T$ and $I \in \mathcal{B}(S)$. Indeed, it is easy to verify that $\mathcal{C}' \subset \mathcal{C}$, since any set in $\mathcal{C}'$ is in fact a particular cylinder set, hence $\sigma(\mathcal{C}') \subset \sigma(\mathcal{C})$. On the other hand, $\mathcal{C} \subset \sigma(\mathcal{C}')$. Indeed, any cylinder set $\Pi_J^{-1}(H)$ with $J = \{t_1, ..., t_n\}$ and $H \in \mathcal{B}(S^J)$, belongs to the σ-algebra generated by the mappings $X_{t_1}, ..., X_{t_n}$ [2]. Hence $\sigma(\mathcal{C}) \subset \sigma(\mathcal{C}')$.

---

[2] It is sufficient to realize that, fixed the set $J = \{t_1, ..., t_n\} \in \mathcal{I}$, there is a one-to-one correspondence between the class of cylinder sets $\{\Pi_J^{-1}(H), H \in \mathcal{B}(S^J)\}$ and the Borel σ-algebra $\mathcal{B}(S^J)$. In addition, the correspondence $H \mapsto \Pi_J^{-1}(H)$ preserves the set

An interesting class of functions $f : \Omega \to \mathbb{R}$ that can be easily integrated with respect to μ are the so-called *cylinder functions*, i.e. the maps of the form $f = g \circ \Pi_J$ for some $J = \{t_1, \ldots, t_n\} \in \mathcal{I}$ and $g : \mathbb{R}^n \to \mathbb{R}$ Borel measurable. In other words, a cylinder function is a map of the form

$$f(\omega) = g(\omega(t_1), \ldots, \omega(t_n)), \qquad \omega \in \Omega.$$

A straightforward application of formula (2.17) gives

$$\int_\Omega f(\omega) d\mu(\omega) = \int_\Omega g(\Pi_J(\omega)) d\mu(\omega) = \int_{\mathbb{R}^n} g(x_1, \ldots, x_n) d\mu_J(x_1, \ldots, x_n).$$

Example 3.1. Kolmogorov theorem assures the existence measures obtained as the product of an infinite number of probability measures. Indeed, if for any $t \in T$ it is given a probability measure μ_t on $(S, \mathcal{F})$, setting for all $J \in \mathcal{F}(T)$ the measure $\mu_J : \mathcal{F}_J \to [0, 1]$ as $\mu_J := \times_{t \in J} \mu_t$ we obtain a consistent family of probability measures. Its projective limit μ can be denoted with the symbol $\mu = \times_{t \in T} \mu_t$.

From a probabilistic point of view, this construction provides the probability space where an infinite (countable or uncountable) family of independent random variables are defined.

Example 3.2. An instructive example of set $E \in \mathcal{F}$ can be easily constructed. Let $T = \mathbb{N}$ and for any $J \in \mathcal{F}(\mathbb{N})$, $J = \{m_1, \ldots, m_n\}$, let $\mu_J \equiv \mu_{m_1, \ldots, m_n}$ be the product measure $\mu_{m_1, \ldots, m_n} := \times_{n \in J} \mu_G$ of n copies of the standard Gaussian measure μ_G is on $\mathcal{B}(\mathbb{R})$. It is easy to verify that the family $\{\mu_J\}_{J \in \mathcal{F}(\mathbb{N})}$ is consistent, hence Kolmogorov Theorem 3.4 allows the construction of a probability measure μ on the measurable space $(\Omega, \mathcal{F})$, where $\Omega = \mathbb{R}^\mathbb{N}$ is the set of all real-valued sequences $\{a_n\}_{n \in \mathbb{N}}$ while the cylinder sets generating the σ-algebra $\mathcal{F}$ are of the form

$$C_{m_1, \ldots, m_n; H} = \{(a_n) \in \mathbb{R}^\mathbb{N} : (a_{m_1}, \ldots, a_{m_n}) \in H\}$$

for some $J = \{m_1, \ldots, m_n\} \in \mathcal{F}(\mathbb{N})$ and $H \in \mathcal{B}(\mathbb{R}^n)$.

Let us now consider the set $l_2 \subset \Omega$ of square integrable sequences:

$$l_2 = \left\{(a_n) \in \mathbb{R}^\mathbb{N} : \sum_n a_n^2 < \infty\right\}$$

---

theoretic operations:

$$\Pi_J^{-1}(H \cap H') = \Pi_J^{-1}(H) \cap \Pi_J^{-1}(H')$$
$$\Pi_J^{-1}(H \cup H') = \Pi_J^{-1}(H) \cup \Pi_J^{-1}(H')$$
$$\Pi_J^{-1}(H^c) = (\Pi_J^{-1}(H))^c$$

This allows to see, e.g., that any cylinder set $\Pi_J^{-1}(H)$, with $H \in \mathcal{B}(S^J)$, belongs to the σ-algebra generated by the cylinder sets of the form $\Pi_J^{-1}(I_1 \times \cdots \times I_n)$ with $I_1, \ldots, I_n \in \mathcal{B}(S)$

First of all we can prove that $l_2 \in \mathcal{F}$, since it can be obtained as

$$l_2 = \bigcup_{N \in \mathbb{N}} \bigcap_{M \in \mathbb{N}} E_{M,N}, \qquad E_{M,N} := \left\{ (a_n) \in \mathbb{R}^{\mathbb{N}} : \sum_{n=1}^{M} a_n^2 \leq N \right\},$$

and it is easy to see that for any choice of $M, N \in N$ the set $E_{M,N}$ belongs to $\mathcal{F}$. This follows from the representation $E_{M,N} = Y^{-1}((-\infty, N])$, where $Y : \Omega \to \mathbb{R}$ is the measurable function defined by

$$Y = \sum_{n=1}^{M} X_n^2.$$

while the measurable mappings $X_n : \mathbb{R}^{\mathbb{N}} \to \mathbb{R}$ are just defined as the projections onto the n-th component:

$$X_n(\omega) = \omega_n \quad \omega \in \Omega.$$

Moreover we can prove that $\mathbb{P}(l_2) = 0$. To this end, it is sufficient to prove that $\mu(\bigcap_{M \in \mathbb{N}} E_{M,N}) = 0$ for any $N \in \mathbb{N}$. Indeed, for any $N \in \mathbb{N}$ the sequence of events $\{E_{M,N}\}_M$ is decreasing, i.e. $E_{M+1,N} \subset E_{M,N} \; \forall M \in \mathbb{N}$. Hence, by the continuity from above of μ (see Eq (2.8)) we have

$$\mu\left(\bigcap_M E_{M,N}\right) = \lim_{M \to \infty} \mu(E_{M,N}).$$

Hence, we can write

$$\mu(E_{M,N}) = \int_{\Omega} 1_{E_{M,N}}(\omega) \, d\mu(\omega)$$

$$= \int_{\Omega} 1_{(-\infty, N]}(X_1^2(\omega) + \cdots + X_M^2(\omega)) d\mu(\omega)$$

$$= \int_{\mathbb{R}^M} 1_{(-\infty, N]}(x_1^2 + \cdots + x_M^2) d\mu_{1,\ldots,M}(x_1, \ldots, x_M)$$

$$= \int_{B(0, \sqrt{N})} \frac{e^{-\frac{1}{2}\sum_{n=1}^{M} x_n^2}}{(2\pi)^{M/2}} dx_1 \ldots dx_M,$$

where $B(0, \sqrt{N}) = \{(x_1, \ldots, x_M) \in \mathbb{R}^M : \sum_{n=1}^{M} x_n^2 \leq N\}$ is the close ball in $\mathbb{R}^M$ with radius $\sqrt{N}$ centered at 0. In other words the value of $\mu(E_{M,N})$ is the measure of $B(0, N)$ with respect to the standard Gaussian measure on $\mathbb{R}^M$. Since $B(0, \sqrt{N}) \subset [-\sqrt{N}, \sqrt{N}]^M$ we have:

$$\int_{B(0, \sqrt{N})} \frac{e^{-\frac{1}{2}\sum_{n=1}^{M} x_n^2}}{(2\pi)^{M/2}} dx_1 \ldots dx_M \leq \int_{[-\sqrt{N}, \sqrt{N}]^M} \frac{e^{-\frac{1}{2}\sum_{n=1}^{M} x_n^2}}{(2\pi)^{M/2}} dx_1 \ldots dx_M,$$

$$= \left(\int_{-\sqrt{N}}^{\sqrt{N}} \frac{e^{-\frac{1}{2}x^2}}{(2\pi)^{1/2}} dx\right)^M = c(N)^M$$

Since $c(N) = \int_{-\sqrt{N}}^{\sqrt{N}} \frac{e^{-\frac{1}{2}x^2}}{(2\pi)^{1/2}} dx < 1$ we have:

$$0 \leq \lim_{M\to\infty} \mathbb{P}(A_M) \leq \lim_{M\to\infty} c(N)^M = 0.$$

3.3 Wiener measure

In this section we shall restrict ourselves to the case where $T = \mathbb{R}^+$ and $S = \mathbb{R}$ and apply Kolmogorov Theorem 3.4 to the construction of probability measures on the space of real value paths $\Omega = \mathbb{R}^{[0,+\infty)}$ endowed with the σ algebra $\mathcal{F}$ generated by the cylinder sets.

Definition 3.3. A *Markov transition function* on $(\mathbb{R}, \mathcal{B}(\mathbb{R}))$ is a map $p : D \times \mathbb{R} \times \mathcal{B}(\mathbb{R}) \to \mathbb{R}$

$$(s, t, x, I) \mapsto p(s, t, x, I), \qquad (s, t) \in D, x \in \mathbb{R}, I \in \mathcal{B}(\mathbb{R}),$$

where $D \subset \mathbb{R}^+ \times \mathbb{R}^+$ is defined as $D := \{(s, t) \in \mathbb{R}^2 : 0 \leq s \leq t\}$, such that

i. for fixed $(s, t, I) \in D \times \mathcal{B}(\mathbb{R})$ the map $x \mapsto p(s, t, x, I)$ is Borel measurable;

ii. for fixed $(s, t, x) \in D \times \mathbb{R}$ the map $I \mapsto p(s, t, x, I)$ is a probability measure on $\mathcal{B}(\mathbb{R})$;

iii. the function p satisfies the *Chapman-Kolmogorov equation* $\forall s \leq u \leq t$, $\forall x \in \mathbb{R}$ and $\forall I \in \mathcal{B}(\mathbb{R})$:

$$p(s, t, x, I) = \int_{\mathbb{R}} p(u, t, y, I) p(s, u, x, dy) \tag{3.16}$$

Given a Markov transition function p, it is possible to construct a projective family of probability measures. Indeed, let $\mathcal{I} = \mathcal{F}(\mathbb{R}_+)$ be the directed set of finite subsets of $\mathbb{R}_+$, let ν a probability measure on $(\mathbb{R}, \mathcal{B}(\mathbb{R}))$ and for any $J \in \mathcal{I}$, with $J = \{t_1, t_2, ..., t_n\}$ and $0 < t_1 < t_2 < ... < t_n < +\infty$, let μ_J be the probability measure on $(\mathbb{R}^J, \mathcal{B}(\mathbb{R}^J))$ given on a product set $B_1 \times B_2 \times ... \times B_n \in \mathcal{B}(\mathbb{R}^n)$, with $B_1, B_2, \ldots, B_n \in \mathcal{B}(\mathbb{R})$ by

$$\mu_J(B_1 \times B_2 \times ... \times B_n) = \int_{\mathbb{R}} \int_{B_1} ... \int_{B_n} p(t_{n-1}, t_n, x_{n-1}, dx_n)$$
$$... p(t_1, t_2, x_1, dx_2) p(0, t_1, x_0, dx_1) d\nu(x_0) \tag{3.17}$$

By the Chapman-Kolmogorov Eq. (4.29), one can easily verify that $(\mu_J)_{J\in\mathcal{I}}$ forms a projective system of probability measures, i.e., for any if $J, K \in \mathcal{I}$, with $J \leq K$, one has $\mu_J = \pi_J^K \circ \mu_K$. In fact it is sufficient to verify this conditions for $J, K \in \mathcal{I}$ such that $K \setminus J$ consists of exactly one element, since the case where $K \setminus J$ consists of $m > 1$ points can be studied by constructing a chain $J \subset K_1 \subset K_2 \subset \dots \subset K_m = K$ where each different $K_i \setminus K_{i-1}$ is a singleton, and using the identity

$$\pi_J^K = \pi_{K_{m-1}}^K \circ \pi_{K_{m-2}}^{K_{m-1}} \circ \dots \circ \pi_J^{K_1}.$$

Let $J = \{t_1, t_2, \dots, t_n\}$ and $K \setminus J = t'$, with $t_i < t' < t_{i+1}$. We have to show that for any $B \in \mathcal{B}(\mathbb{R}^J)$, $\mu_J(B) = \mu_K((\pi_J^K)^{-1}(B))$. By Theorem 2.1 it is sufficient to show this equality for $B \in \mathcal{B}(\mathbb{R}^J)$ of the form $B = B_1 \times B_2 \times \dots \times B_n$, with $B_i \in \mathcal{B}(\mathbb{R})$ for all $i = 1, \dots, n$, that is

$$\int_{\mathbb{R}} \int_{B_1} .. \int_{B_i} \int_{B_{i+1}} \dots \int_{B_n} p(t_{n-1}, t_n, x_{n-1}, dx_n) \dots p(t_{i-1}, t_i, x_{i-1}, dx_i)$$

$$p(t_{i-2}, t_{i-1}, x_{i-2}, dx_{i-1}) \dots p(0, t_1, x_0, dx_1) d\nu(x_0)$$

$$= \int_{\mathbb{R}} \int_{B_1} .. \int_{B_i} \int_{\mathbb{R}} \int_{B_{i+1}} \dots \int_{B_n} p(t_{n-1}, t_n, x_{n-1}, dx_n) \dots p(t', t_i, x', dx_i)$$

$$p(t_{i-1}, t', x_{i-1}, dx') p(t_{i-2}, t_{i-1}, x_{i-2}, dx_{i-1}) \dots p(0, t_1, x_0, dx_1) d\nu(x_0),$$
$$(3.18)$$

which holds by Eq. (4.29).

A particular example of Markov transition function is provided by the fundamental solution of the heat equation. Indeed, for $s, t \in \mathbb{R}^+$, $s \le t$, $x \in \mathbb{R}$ and $B \in \mathcal{B}(\mathbb{R})$ let $p(s, t, x, B)$ be defined by

$$p(s, t, x, B) = \int_B G_{t-s}(x, y) dy. \qquad (3.19)$$

where G is the fundamental solution of the parabolic equation

$$\frac{\partial}{\partial t} u(t, x) = \frac{1}{2} \frac{\partial^2}{\partial x^2} u(t, x) \quad t \in [0, +\infty), \, x \in \mathbb{R}, \qquad (3.20)$$

and it is equal to

$$G_t(x, y) = \begin{cases} \dfrac{e^{-\frac{(y-x)^2}{2t}}}{\sqrt{2\pi t}} & t > 0 \\ \delta(x - y) & t = 0 \end{cases}$$

In particular, for any Borel and bounded function $u_0 : \mathbb{R} \to \mathbb{R}$ the solution of the initial value problem

$$\begin{cases} \frac{\partial}{\partial t} u(t, x) = \frac{1}{2} \frac{\partial^2}{\partial x^2} u(t, x) \\ u(0, x) = u_0(x) \end{cases} \qquad (3.21)$$

is given by

$$u(t, x) = \int G_t(x, y) u_0(y) \, dy. \tag{3.22}$$

Conditions i and ii of Definition 3.3 are trivially satisfied. Concerning Chapman-Kolmogorov Eq. (4.29), this follows by the identity

$$G_{t+s}(x, y) = \int_{\mathbb{R}} G_t(x, z) G_s(z, y) \, dz, \tag{3.23}$$

which can be easily proved by taking the Fourier transform of both sides of (3.23).

Given the measure $\nu = \delta_0$ on $(\mathbb{R}, \mathcal{B}(\mathbb{R}))$ concentrated at the point $0 \in \mathbb{R}$ and considering the consistent family of probability measure $\{\mu_J\}_{J \in \mathcal{F}(\mathbb{R}^+)}$ defined by Eq. (4.30), Kolmogorov Theorem assures the existence of a probability measure μ on $(\mathbb{R}^{[0,+\infty)}, \mathcal{F})$ such that for any choice of $n \geq 1$, $0 \leq t_1 \leq \cdots \leq t_n < +\infty$ and $I_1, \ldots, I_n \in \mathcal{B}(\mathbb{R})$ the following holds

$$\mu\left(\omega \in \Omega : \omega(t_1) \in I_1, \ldots, \omega(t_n) \in I_n\right)$$

$$= \int_{I_1 \times \cdots \times I_n} \frac{e^{-\frac{1}{2} \sum_{i=1}^{n} \frac{(x_i - x_{i-1})^2}{t_i - t_{i-1}}}}{(2\pi)^{n/2} \sqrt{t_1 (t_2 - t_1) \cdots (t_n - t_{n-1})}} dx_1 \ldots dx_n \tag{3.24}$$

where $x_0 := 0$ and $t_0 := 0$. Analogously, given a cylinder function $F : \mathbb{R}^{[0,+\infty)} \to \mathbb{R}$ of the form

$$F(\omega) := f(\omega(t_1), \ldots, \omega(t_n)) \qquad \omega \in \mathbb{R}^{[0,+\infty)}$$

for some $n \geq 1, 0 \leq t_1 \leq \cdots \leq t_n < +\infty$ and a Borel and bounded function $f : \mathbb{R}^n \to \mathbb{R}$, the integral of F with respect to μ can be computed in terms of the finite dimensional Gaussian integral on the right-hand side of the following identity

$$\int_{\Omega} F(\omega) \, d\mu(\omega) = \int_{\mathbb{R}^n} \frac{f(x_1, \ldots, x_n) e^{-\frac{1}{2} \sum_{i=1}^{n} \frac{(x_i - x_{i-1})^2}{t_i - t_{i-1}}}}{(2\pi)^{n/2} \sqrt{t_1 (t_2 - t_1) \cdots (t_n - t_{n-1})}} dx_1 \ldots dx_n.$$

We point out that μ is not what is commonly called *Wiener measure* W, the latter being defined on the measurable space $(\mathcal{C}, \mathcal{B}(\mathcal{C}))$, where $\mathcal{C} = \mathcal{C}(\mathbb{R}^+, \mathbb{R})$ is the set of real valued continuous paths on $\mathbb{R}^+$:

$$\mathcal{C} = \{\gamma : \mathbb{R}^+ \to \mathbb{R} \text{ continuous}\} \tag{3.25}$$

endowed with the topology induced by the uniform convergence on the compact[3] sets $K \subset \mathbb{R}^+$, and $\mathcal{B}(\mathcal{C})$ is the corresponding Borel σ-algebra.

---

[3] A sequence $\{\gamma_n\} \subset \mathcal{C}$ converges to $\gamma \in \mathcal{C}$ iff for all compact sets $K \subset \mathbb{R}^+$ we have $\sup_K |\gamma_n - \gamma| \to 0$ as $n \to \infty$.

It is important to stress that W cannot be simply constructed as the restriction of μ on $\mathcal{C} \subset \Omega$ since $\mathcal{C}$ is not measurable! In order to prove that $\mathcal{C}$ does not belong to the σ-algebra $\mathcal{F}$ generated by the cylinder sets we present the following characterisation of the elements $A \in \mathcal{F}$.

Theorem 3.5. *For any set $A \in \mathcal{F}$ there exists a countable subset $T \subset [0, +\infty)$ such that if $\omega \in A$ and $\omega' \in \Omega$ with $\omega'(t) = \omega(t)$ for all $t \in T$, then $\omega' \in A$.*

The proof of the theorem relies the characterisation of $\mathcal{F}$ as the σ-algebra generated by the measurable mappings $X_t : \Omega \to \mathbb{R}$, $t \in [0, +\infty)$, defined as

$$X_t(\omega) := \omega(t) \qquad \omega \in \Omega, \tag{3.26}$$

(see Remark 3.2) and on the following two lemmas.

Lemma 3.1. *Let $T \subset [0, +\infty)$ and let $A \in \sigma(X_t : t \in T)$. If $\omega \in A$ and $\omega' \in \Omega$ with $\omega'(t) = \omega(t)$ for all $t \in T$, then $\omega' \in A$.*

Proof. Let us define the set $\Omega' = \mathbb{R}^T$ as the set of all maps $\gamma : T \to \mathbb{R}$. Let $\mathcal{F}'$ be the σ algebra in Ω' generated by the cylinder subsets of Ω'. Let $\xi : \Omega \to \Omega'$ be the map defined as $\xi(\omega)(t) = \omega(t)$, $t \in T$. In fact $\xi(\omega)$ is the path $\gamma : T \to \mathbb{R}$ which is the restriction to T of the path $\omega : \mathbb{R}^+ \to \mathbb{R}$. It is easy to prove that $\xi : (\Omega, \mathcal{F}) \to (\Omega', \mathcal{F}')$ is measurable. Hence, it is possible to define the sub-σ-algebra $\tilde{\mathcal{F}} \subset \mathcal{F}$ of all sets $A \in \mathcal{F}$ of the form $\xi^{-1}(M) = \{\omega \in \Omega : \xi(\omega) \in M\}$ for some set $M \in \mathcal{F}'$. It is easy to see that $\tilde{\mathcal{F}}$ contains the sets of the form $\{\omega(t) \in I\} = \{X_t(\omega) \in I\}$ for all possible choices of $t \in T$ and $I \in \mathcal{B}(\mathbb{R})$. This means that $\tilde{\mathcal{F}}$ contains the σ-algebra $\sigma(X_t, t \in T)$ generated by the random variables X_t, $t \in T$. Hence, given a set $A \in \sigma(X_t : t \in T)$, then A will be of the form $A = \xi^{-1}(M)$ for some $M \in \mathcal{F}'$. In particular, a path $\omega \in A$ iff $\xi(\omega) \in M$. This means that if $\omega \in A$ and $\omega' \in \Omega$ coincides with ω on T, i.e. $\omega'(t) = \omega(t)$ for all $t \in T$, then $\xi(\omega') = \xi(\omega) \in M$. Hence $\omega' \in M$. $\qquad\square$

Lemma 3.2. *If $A \in \sigma(X_t : t \in \mathbb{R}^+)$ then $A \in \sigma(X_t : t \in T)$ for some countable subset $T \subset \mathbb{R}^+$.*

Proof. For $T \subset \mathbb{R}^+$, let us define the σ-algebra $\mathcal{F}_T := \sigma(X_t : t \in T)$. Let $\mathcal{G} = \cup_T \mathcal{F}_T$ where the union extends to all *countable* subsets $T \subset \mathbb{R}^+$. If we show that $\mathcal{G} = \mathcal{F}$ the lemma is proved.

The family $\mathcal{G} \subseteq \mathcal{F}$ is actually a σ-algebra, indeed:

- $\Omega \in \mathcal{G}$ since $\Omega \in \mathcal{F}_T$ for all $T \subset \mathbb{R}^+$ countable.
- If $A \in \mathcal{G}$ then $A^c \in \mathcal{G}$. Indeed if $A \in \mathcal{G}$ then $A \in \mathcal{F}_T$ for some $T \subset \mathbb{R}^+$ countable. Hence $A^c \in \mathcal{F}_T \subset \mathcal{G}$.
- Given a countable family $\{A_n\} \subset \mathcal{G}$, then $\cup_n A_n \in \mathcal{G}$. Indeed if $A_n \in \mathcal{G}$ then there exists a countable subset $T_n \subset \mathbb{R}^+$ such that $A_n \in \mathcal{F}_{T_n}$. Let us consider the set $\tilde{T} := \cup_n T_n$, which is a countable subset of $\mathbb{R}^+$. We have that $\mathcal{F}_{T_n} \subset \mathcal{F}_{\tilde{T}}$ for all n, hence $A_n \in \mathcal{F}_{\tilde{T}}$ for all n and, since $\mathcal{F}_{\tilde{T}}$ is a σ algebra, $\cup_n A_n \in \mathcal{F}_{\tilde{T}} \subset \mathcal{G}$.

Moreover the σ-algebra $\mathcal{G}$ includes $\mathcal{F} = \sigma(X_t : t \geq 0)$ since it includes all the events of the form $\{X_t \in I\}$ for some $t \in \mathbb{R}^+$ and $I \in \mathcal{B}(\mathbb{R})$ which generate $\mathcal{F}$. We can eventually conclude that $\mathcal{G} = \mathcal{F}$ and the lemma is proved. □

Theorem 3.5 shows that many interesting sets lie outside $\mathcal{F}$, in particular the set $\mathcal{C}$ of continuous paths cannot belong to $\mathcal{F}$. Indeed, if *ab absurdum* $\mathcal{C}$ belonged to $\mathcal{F}$ then there would exist a countable set $T \subset \mathbb{R}^+$ such that if $\omega \in C$ is a continuous paths, any path $\omega' \in \Omega$ coinciding with ω on T will be continuous. However, whatever the countable set T is, it is always possible to construct a discontinuous path ω' which coincides with a continuous path ω on T.

In order to get around to this no-go result and construct a Borel probability measure on the space $\mathcal{C}$ of continuous paths we need to introduce a theorem which plays a fundamental role in the theory of stochastic processes. In the jargon of probability theory, the family $(X_t)_{t \geq 0}$ of measurable functions defined on the probability space $(\Omega, \mathcal{F}, \mu)$ by (3.26) is a *stochastic process*. For every $\omega \in \Omega$ we can define *trajectory* of the process $(X_t)_{t \geq 0}$ as the map $\gamma_\omega : [0, +\infty) \to \mathbb{R}$ given by $\gamma_\omega(t) := X_t(\omega)$. In fact in the case of the measurable maps (3.26) the trajectories γ_ω coincide with the corresponding paths $\omega \in \Omega$. Indeed, it is easy to verify that, by construction, $\gamma_\omega(t) = \omega(t)$ for all $\omega \in \Omega$ and $t \geq 0$. As a result of the above discussion, the set of points $\omega \in \Omega$ corresponding to continuous trajectories γ_ω is not even measurable. However, we are going to show that it is possible to construct an alternative family $(X'_t)_{t \geq 0}$ of measurable functions defined on the probability space $(\Omega, \mathcal{F}, \mu)$ such that

$$\mu\left(\{\omega \in \Omega \colon X_t(\omega) = X'_t(\omega)\}\right) = 0 \qquad \forall t \in [0, +\infty). \qquad (3.27)$$

A collection of measurable maps $(X'_t)_{t \geq 0}$ fulfilling (3.27) is called a *modification* or *version* of the process $(X_t)_{t \geq 0}$. It is important to point out that condition (3.27) yields for any choice of $n \geq 1$, $t_1, \ldots, t_n \in [0, +\infty)$ the equality between the measures $\mu_{t_1,\ldots,t_n}$ and $\mu'_{t_1,\ldots,t_n}$ obtained respectively as the image of μ under the mappings $T_{t_1,\ldots,t_n} : \Omega \to \mathbb{R}^n$ and $T_{t_1,\ldots,t_n} : \Omega \to \mathbb{R}^n$ given by

$$T_{t_1,\ldots,t_n}(\omega) := (X_{t_1}(\omega),\ldots,X_{t_n}(\omega))$$
$$T'_{t_1,\ldots,t_n}(\omega) := \left(X'_{t_1}(\omega),\ldots,X'_{t_n}(\omega)\right)$$

Indeed, the following holds[4] for any choice of Borel sets $I_1,\ldots,I_n \in \mathcal{B}(\mathbb{R})$:

$$\mu\left(\{\omega \in \Omega : (X_{t_1}(\omega),\ldots,X_{t_n}(\omega)) \in I_1 \times \cdots \times I_n\}\right)$$
$$= \mu\left(\{\omega \in \Omega : (X'_{t_1}(\omega),\ldots,X'_{t_n}(\omega)) \in I_1 \times \cdots \times I_n\}\right). \quad (3.28)$$

In this context it is possible to provide a sufficient condition for a stochastic process to admit a modification with continuous trajectories. In fact, the regularity of the trajectories can be described in more detail.

Definition 3.4. A function $f : I \to \mathbb{R}$ defined on an interval $I \subset \mathbb{R}$ of the real line is said to be *locally Hölder continuous of order* $\gamma > 0$ if any $t \in I$ has a neighborhood U such that

$$\frac{|f(t) - f(s)|}{|t - s|^\gamma}$$

is bounded for $s, t \in U \cap I, s \neq t$.

Theorem 3.6 (Kolmogorov-Chentsov continuity theorem).   *Let* $(X_t)_{t \geq 0}$ *be a real valued stochastic process on* $(\Omega, \mathcal{F}, \mu)$. *If there exist positive constants* α, β, C *such that for all* $s, t \in \mathbb{R}^+$ *the following inequality holds:*

$$\int_\Omega |X_t(\omega) - X_s(\omega)|^\alpha \, d\mu(\omega) \leq C|t - s|^{1+\beta} \quad (3.29)$$

then there exists a modification $(X'_t)_{t \geq 0}$ *of* $(X_t)_{t \geq 0}$ *with locally Hölder continuous trajectories of order* $\gamma < \frac{\beta}{\alpha}$.

---

[4]Denoting $E_{t_1,\ldots,t_n} := \{(X_{t_1},\ldots,X_{t_n}) \in I_1 \times \ldots \times I_n\}$, $E'_{t_1,\ldots,t_n} := \{(X'_{t_1},\ldots,X'_{t_n}) \in I_1 \times \cdots \times I_n\}$, we have to show that $\mu(E_{t_1,\ldots,t_n}) = \mu(E'_{t_1,\ldots,t_n})$. Denoting $N_{t_i} = \{X_{t_j} \neq X'_{t_j}\}$, we have that for any $\omega \in \cap_{j=1}^n N_{t_j}^c$ the values of $X_{t_j} = X'_{t_j}$ coincide for any $j = 1,\ldots,n$, i.e. $\cap_{j=1}^n N_{t_j}^c = \{X_{t_j} = X'_{t_j}, j = 1,\ldots,n\}$. By decomposing $E_{t_1,\ldots,t_n}$ and $E'_{t_1,\ldots,t_n}$ into a finite disjoint union of measurable sets we obtain:

$$E_{t_1,\ldots,t_n} = \left(E_{t_1,\ldots,t_n} \cap (\cap_{j=1}^n N_{t_j}^c)\right) \cup \left(E_{t_1,\ldots,t_n} \cap (\cup_{j=1}^n N_{t_j})\right),$$

$$E'_{t_1,\ldots,t_n} = \left(E'_{t_1,\ldots,t_n} \cap (\cap_{j=1}^n N_{t_j}^c)\right) \cup \left(E'_{t_1,\ldots,t_n} \cap (\cup_{j=1}^n N_{t_j})\right)$$

Since $\mu(\cup_{j=1}^n N_{t_j}) = 0$ and $\mu\left(E_{t_1,\ldots,t_n} \cap (\cap_{j=1}^n N_{t_j}^c)\right) = \mathbb{P}\left(E'_{t_1,\ldots,t_n} \cap (\cap_{j=1}^n N_{t_j}^c)\right)$ we obtain $\mu(E_{t_1,\ldots,t_n}) = \mu(E'_{t_1,\ldots,t_n})$.

For a detailed proof we refer to, e.g., [56]. Here we limit ourselves to show that the measurable mappings $(X_t)_{t \geq o}$ defined in (3.26) fulfil all the conditions for the application of Kolmogorov-Chentsov theorem. Indeed for any $\alpha \geq 0$ we have:

$$\int_\Omega |X_t(\omega) - X_s(\omega)|^\alpha \, d\mu(\omega) = \int_\mathbb{R} |x|^\alpha \frac{e^{-\frac{x^2}{2(t-s)}}}{\sqrt{2\pi(t-s)}} dx$$

$$= |t-s|^{\alpha/2} \int_\mathbb{R} |x|^\alpha \frac{e^{-\frac{x^2}{2}}}{\sqrt{2\pi}} dx$$

hence, it is sufficient to take $C = \int_\mathbb{R} |x|^\alpha \frac{e^{-\frac{x^2}{2}}}{\sqrt{2\pi}} dx$ and $\beta = \frac{\alpha}{2} - 1$ for obtaining the existence of a modification with locally Hölder continuous trajectories of order $\gamma < \frac{1}{2} - \frac{1}{\alpha}$. By the arbitrariness in the choice of α, the trajectories of the modification $(X'_t)_{t \geq 0}$ can be taken locally Hölder continuous of any $\gamma < 1/2$

Let us consider now the space $\mathcal{C}$ of real valued continuous paths (3.25) endowed with the topology induced by the uniform convergence on the compact sets. A system of open neighborhoods in this topology is provided by the sets of the form

$$U_{\eta,T,\epsilon} = \{\gamma \in \mathcal{C}: \sup_{t \in [0,T]} |\gamma(t) - \eta(t)| < \epsilon\} \tag{3.30}$$

for some $\eta \in \mathcal{C}, T > 0, \epsilon > 0$. Let us denote with $\mathcal{B}(\mathcal{C})$ the Borel σ-algebra. In fact $\mathcal{B}(\mathcal{C})$ is generated by sets of the form

$$\bar{U}_{\eta,T,\epsilon} = \{\gamma \in \mathcal{C}: \sup_{t \in [0,T]} |\gamma(t) - \eta(t)| \leq \epsilon\} \tag{3.31}$$

Indeed any open neighborhood of the form (3.30) can be obtained as the countable union of sets of the form (3.31):

$$U_{\eta,T,\epsilon} = \bigcup_n \bar{U}_{\eta,T,\epsilon-\frac{1}{n}}$$

Given the modification with continuous trajectories $(X'_t)_{t \geq 0}$ on the probability space $(\Omega, \mathcal{F}, \mu)$, let us define the mapping $\xi : \Omega \to \mathcal{C}$ which assign to any element $\omega \in \Omega$ of the sample space Ω its trajectory:

$$\xi(\omega)(t) := X'_t(\omega), \qquad t \in \mathbb{R}^+$$

In fact the map $\xi : (\Omega, \mathcal{F}) \to (\mathcal{C}, \mathcal{B}(\mathcal{C}))$ is measurable. In order to prove this property, by Theorem 2.7 it is sufficient to check that the set $\xi^{-1}(\bar{U}_{\eta,T,\epsilon})$ belongs to $\mathcal{F}$ for any choice of $\eta \in \mathcal{C}, T > 0, \epsilon > 0$, since, as discussed above, the sets of the form (3.31) generate the σ-algebra $\mathcal{B}(\mathcal{C})$. This is given by the following chain of equalities:

$$\xi^{-1}(\bar{U}_{\eta,T,\epsilon}) = \{\omega \in \Omega: \sup_{t \in [0,T]} |X_t(\omega) - \eta(t)| \leq \epsilon\}$$

$$= \{\omega \in \Omega: \sup_{t \in [0,T] \cap \mathbb{Q}} |X_t(\omega) - \eta(t)| \leq \epsilon\}$$

$$= \bigcap_{t \in [0,T] \cap \mathbb{Q}} \{\omega \in \Omega: |X_t(\omega) - \eta(t)| \leq \epsilon\}$$

$$= \bigcap_{t \in [0,T] \cap \mathbb{Q}} \{\omega \in \Omega: X_t(\omega) \in [\eta(t) - \epsilon, \eta(t) + \epsilon]\}$$

where the second equality follows from the continuity of the trajectories $t \mapsto X_t(\omega)$. The last line above shows that the set $\xi^{-1}(\bar{U}_{\eta,T,\epsilon})$ belongs to $\mathcal{F}$ since it is the countable intersection of sets belonging to $\mathcal{F}$.

Since $\xi: (\Omega, \mathcal{F}, \mathbb{P}) \to (\mathcal{C}, \mathcal{B}(\mathcal{C}))$ is measurable, it is possible to define on $(\mathcal{C}, \mathcal{B}(\mathcal{C}))$ the probability measure μ_ξ as the image measure (or push-forward measure) of μ under the action of ξ. In particular μ_ξ is defined on Borel sets $E \in \mathcal{B}(\mathcal{C})$ as:

$$\mu_\xi(E) := \mu(\xi^{-1}(E)) = \mathbb{P}(\{\omega \in \Omega: \xi(\omega) \in E\})$$

The probability measure μ_ξ on $(\mathcal{C}, \mathcal{B}(\mathcal{C}))$ is called *Wiener measure* and denoted with the symbol W. By construction and by Eq. (3.28) we can easily check the following formula for the Wiener measure of a general cylinder set in $\mathcal{C}$:

$$W(\{\gamma \in \mathcal{C}: \gamma(t_1) \in I_1, \ldots, \gamma(t_n) \in I_n\})$$

$$= \int_{I_1 \times \cdots \times I_n} \frac{e^{-\frac{1}{2} \sum_{i=1}^n \frac{(x_i - x_{i-1})^2}{t_i - t_{i-1}}}}{(2\pi)^{n/2} \sqrt{t_1(t_2 - t_1) \cdots (t_n - t_{n-1})}} dx_1 \ldots dx_n \quad (3.32)$$

valid for any choice of $n \in \mathbb{N}$, $0 < t_1 < \cdots < t_n < +\infty$ and $I_1, \ldots, I_n \in \mathcal{B}(\mathbb{R})$. Moreover, given a cylinder function $f: \mathcal{C} \to \mathbb{R}$ of the form

$$f(\gamma) = F(\gamma(t_1), \ldots, \gamma(t_n)) \qquad \gamma \in \mathcal{C},$$

for some $n \geq 1$, $0 < t_1 < \cdots < t_n < +\infty$ and $F: \mathbb{R}^n \to \mathbb{R}$ bounded and Borel-measurable, the integral of f with respect to the Wiener measure W can be computed as

$$\int_\mathcal{C} F \, dW = \int_{\mathbb{R}^n} f(x_1, \ldots, x_n) \frac{e^{-\frac{1}{2} \sum_{i=1}^n \frac{(x_i - x_{i-1})^2}{t_i - t_{i-1}}}}{(2\pi)^{n/2} \sqrt{t_1(t_2 - t_1) \cdots (t_n - t_{n-1})}} dx_1 \ldots dx_n.$$

$$(3.33)$$

Let us consider again the heat Eq. (3.20). By formula (3.33), the solution (3.22) of the initial value problem (3.21) can be represented as the Wiener integral

$$u(t,x) = \int_C u_0(x + \gamma(t))dW(\gamma) \qquad t > 0, x \in \mathbb{R}.$$

As proved for the first time by M. Kac in 1949 [211], this result can be generalized to the case where a perturbation potential is included in Eq. (3.20):

$$\begin{cases} \frac{\partial}{\partial t} u(t,x) = \frac{1}{2} \frac{\partial^2}{\partial x^2} u(t,x) - V(x)u(t,x) \\ u(0,x) = u_0(x) \end{cases} \tag{3.34}$$

Indeed, it $u_0 : \mathbb{R} \to \mathbb{R}$ and $V : \mathbb{R} \to \mathbb{R}$ are continuous, V is bounded from below and $|u_0|$ has at most polynomial growth for $|x| \to \infty$, then the solution of (3.34) can be represented by the following integral

$$u(t,x) = \int_C u_0(x + \gamma(t))e^{-\int_0^t V(\gamma(s)+x)\,ds}dW(\gamma) \qquad t > 0, x \in \mathbb{R}. \tag{3.35}$$

For the proof and generalizations to a larger class of potentials V we refer to, e.g., [52, 302].

The representation formula (3.35) is the celebrated *Feynman-Kac formula*. It is actually the first and most famous example of theory [52, 194, 131] that nowadays is extensively developed and connects parabolic equations associated to second order elliptic operators

$$\begin{cases} \frac{\partial}{\partial t} u(t,x) = \frac{1}{2}\text{Tr}[\sigma(x)\sigma^*(x)D_x^2 u(t,x)] + \langle b(x), D_x u(t,x) \rangle + V(x)u(t,x) \\ u(0,x) = u_0(x) \end{cases}$$
$$\tag{3.36}$$

to stochastic Markov processes $X^x = (X_t^x)_{t \geq 0}$, solutions of the stochastic differential equations of the following form

$$\begin{cases} dX_t^x = b(X_t^x)dt + \sigma(X_t^x)dW_t, \\ X(0) = x, \qquad x \in \mathbb{R}^d, \end{cases} \tag{3.37}$$

Indeed, if $\sigma : \mathbb{R}^d \to L(\mathbb{R}^d, \mathbb{R}^d)$ and $b : \mathbb{R}^d \to \mathbb{R}^d$ are Lipschitz maps the solution of (3.36) can be represented by the following formula (see [122, 143])

$$u(t,x) = \mathbb{E}\left[u(0, X_t^x)e^{-\int_0^t V(X_s^x)ds}\right], \qquad t \geq 0, x \in \mathbb{R}^d. \tag{3.38}$$

3.4 Projective limit spaces

Kolmogorov theorem has been generalized [68] to the case where Ω is a projective limit space. In this section we present this rather abstract formulation which allows to construct non-trivial probability measures not only on spaces of paths of the form S^T, but also, e.g., on topological vector spaces.

Definition 3.5. A semi-ordered set $\mathcal{I}$ with partial ordering $\leq$ is called a *directed set* if for any $J, K \in \mathcal{I}$ there exists $H \in \mathcal{I}$ such that $J \leq H$ and $K \leq I$.

Definition 3.6. Let $\{S_J\}_{J \in \mathcal{I}}$ be a collection of (non-empty) sets S_J labelled by the elements of a non-empty directed set $\mathcal{I}$, called index set. Let us assume that for any $J, K \in \mathcal{I}$, $J \leq K$, there exists a subjective map $\pi_J^K : S_K \to S_J$ such that π_K^K is the identity on S_K and for all $J \leq K \leq H$, $J, K, H \in \mathcal{I}$, the following *consistency property* holds:

$$\pi_J^H = \pi_J^K \circ \pi_K^H .$$

Such a family $\{S_J, \pi_J^K\}_{J,K \in \mathcal{I}}$ is called a *projective (or inverse) family of sets*.

The projective family $\{S_J, \pi_J^K\}_{J,K \in \mathcal{I}}$ is called *topological* if each S_J, $J \in \mathcal{I}$, is a topological space and the maps $\pi_J^K : S_K \to S_J$, $J \leq K$, are continuous.

Example 3.3. Let $(S_t^0)_{t \in T}$ be a collection of non-empty sets labelled by the elements of a non-empty set T and let $\mathcal{I} = \mathcal{F}(T)$ be the totality of non-empty subsets of T with a finite number of elements. The partial ordering $\leq$ in $\mathcal{F}(T)$ is then defined in terms of inclusions of sets, i.e., if $J, J' \in \mathcal{F}(T)$ and $J \subseteq J'$ then $J \leq J'$. Define for each $J \in \mathcal{F}(T)$

$$S_J = \times_{t \in J} S_t^0,$$

i.e., S_J is the "Cartesian product of the sets S_t^0 along J". We also say that S_J is a product space (constructed from the S_t^0). In particular, we have $S_{\{t\}} = S_t^0$.

If $J \leq K$, $J, K \in \mathcal{F}(T)$, then we can take π_J^K to be the canonical projection mapping from S_K to S_J, given by

$$\pi_J^K \omega_K = \omega_J, \tag{3.39}$$

with $\omega_K := (\omega_t)_{t \in K}$ understood as the element in S_K with component ω_t in S_t^0.

If the spaces $(S_t^0)_{t \in I}$ are topological, then for any $J \in \mathcal{F}(T)$ we endow the product space S_J with the product topology, i.e., the coarsest topology for which all the projections $\pi_{t'} : \times_{t \in J} S_t^0 \to S_{t'}^0$, $t' \in J$, are continuous. For $J, K \in \mathcal{F}(T)$, $J \leq K$, the projection π_J^K from S_K to S_J results to be continuous. By constructions $(E_J, \pi_J^K)_{J,K \in \mathcal{F}(I)}$ is a topological projective family.

Example 3.4. Let $(\mathcal{H}, \langle\ ,\ \rangle)$ be a real Hilbert space. Let us consider the set $\mathcal{I}$ of orthogonal projection operators P onto finite-dimensional subspaces $P(\mathcal{H})$ of $\mathcal{H}$. $\mathcal{I}$ is a directed set, with partial order given by $P \leq Q$, $P, Q \in \mathcal{I}$, iff $P(\mathcal{H}) \subseteq Q(\mathcal{H})$. Set $S_P := P(\mathcal{H})$. For $P \leq Q$, let $\pi_P^Q : Q(\mathcal{H}) \to P(\mathcal{H})$ be the projection $\pi_P^Q := P|_{Q(\mathcal{H})}$, where $P|_{Q(\mathcal{H})}$ stands for the restriction of P to $Q(\mathcal{H})$. This procedure produces a projective family of sets $(S_P, \pi_P^Q)_{P,Q \in A}$. Moreover, if the finite-dimensional vector spaces $P(\mathcal{H})$ are endowed with the natural (Euclidean) topology, the projections π_P^Q are continuous and $(S_P, \pi_P^Q)_{P,Q \in \mathcal{I}}$ forms a topological projective family of sets.

Definition 3.7. The *projective (or inverse) limit* $S_\mathcal{I} := \varprojlim S_J$ of the projective family of sets $\{S_J, \pi_J^K\}_{J,K \in \mathcal{I}}$ is defined as the following subset of the Cartesian product of the family $\{S_J\}_{J \in \mathcal{I}}$:

$$S_\mathcal{I} := \{(x_J) \in \times_{J \in \mathcal{I}} S_J, \ : \ x_J = \pi_J^K(x_K) \text{ for all } J \leq K, \ J, K \in \mathcal{I}\}$$

In the case of Example 3.3 the projective limit $\varprojlim S_J$ is isomorphic to the product space $\times_{t \in I} S_t^0$.

In the case of Example 3.4, with $\mathcal{H}$ separable Hilbert space, we have that $\mathcal{H}$ is strictly included in the projective limit of its finite-dimensional subspaces. To see this, let us consider the particular case where $\mathcal{H} = l_2$. The projective limit of the family of finite-dimensional subspaces of l_2 is the space $\mathbb{R}^\mathbb{N}$ of all sequences, that strictly includes l_2.

Given a projective family of sets $\{S_J, \pi_J^K\}_{J,K \in \mathcal{I}}$, let $\tilde{S}$ be the Cartesian product $\tilde{S} := \otimes_{J \in \mathcal{I}} S_J$ and for any $J \in \mathcal{I}$ let $\tilde{\pi}_J : \tilde{S} \to S_J$ be the coordinate projection[5] of $\tilde{S}$ into S_J and its restriction $\pi_J := \tilde{\pi}_J|_{S_\mathcal{I}}$ to $S_\mathcal{I}$. It is easy to verify that for any pair $J, K \in \mathcal{I}$, with $J \leq K$, the following composition property holds:

$$\pi_J = \pi_J^K \circ \pi_K. \tag{3.40}$$

If $(S_J, \pi_J^K)_{J,K \in \mathcal{I}}$ is a topological projective family, then $S_\mathcal{I} = \varprojlim S_J$ will be endowed with the coarsest topology making all the projection maps $\pi_J : S_\mathcal{I} \to S_J$ continuous. This is also called *initial* or *inductive* topology.

---

[5] If $\tilde{\omega} = (\omega_J)_{J \in A} \in \tilde{S}$ then $\tilde{\pi}_J(\tilde{\omega}) = \omega_J$.

Remark 3.3. Given a general projective family $\{S_J, \pi_J^K\}_{J,K\in\mathcal{I}}$, two problems may occur:

(1) Even if $S_J \neq \emptyset$ for all $J \in \mathcal{I}$, it might happen that $S_{\mathcal{I}} = \emptyset$. See, e.g., [159].
(2) Even if all the projections $(\pi_J^K)_{J,K\in\mathcal{I}}$ are surjective, the maps $\pi_J : S_{\mathcal{I}} \to S_J$ may fail to be subjective. See, e.g., [241].

Definition 3.8. A projective family $\{S_J, \pi_J^K\}_{J,K\in\mathcal{I}}$ is called *perfect inverse system* if for all $J \in \mathcal{I}$, $\omega_J \in S_J$, there exist an $\omega \in S_{\mathcal{I}}$ such that $\omega_J = \pi_J\omega$. In this case all the projections are subjective.

One can easily verify that the projective families presented in examples 3.3 and 3.4 are perfect inverse systems. In the terminology of [68] a perfect inverse system $\{S_J, \pi_J^K\}_{J,K\in\mathcal{I}}$ is called *simply maximal*. Moreover, it is called *sequentially maximal* if for any increasing sequence $\{J_n\} \subseteq \in \mathcal{I}$ and any choice of points $s_n \in S_{J_n}$ such that $s_n = \pi_{J_n}^{J_m} s_m$, $n \leq m$, there exists a point $\tilde{s} \in S_{\mathcal{I}}$ such that $s_n = \pi_{J_n}\tilde{s}$ for all n.

Let us consider now the particular case where each element $(S_J)_{J\in\mathcal{I}}$ of a projective family $\{S_J, \pi_J^K\}$ is endowed with a σ-algebra $\mathcal{F}_J$ of subsets of S_J and that all the maps $\pi_J^K : S_K \to S_J$, $J \leq K$, are measurable. In this case the collection $\{S_J, \mathcal{F}_J, \pi_J^K)\}$ of measurable spaces $(S_J, \mathcal{F}_J)$ endowed with the family of measurable maps is called a *projective family of measurable spaces* [335]. If in addition for each measurable space $(S_J, \mathcal{F}_J)$ it is defined a probability measure $\mu_J : \mathcal{F}_J \to [0,1]$ and if the collection of measures $\{\mu_J\}$ satisfies the following consistency property:

$$\mu_J = (\mu_K)_{\pi_J^k}, \qquad J \leq K \qquad (3.41)$$

then it is called a *consistent* or *projective family of probability measures*.

Given the collection of σ-algebras $\{\mathcal{F}_J\}_{J\in\mathcal{I}}$, let $\mathcal{F}_{\mathcal{I}}$ be the σ-algebra on the projective limit space $S_{\mathcal{I}}$ generated by the projection maps $\pi_J : S_{\mathcal{I}} \to S_J$. In fact $\mathcal{F}_{\mathcal{I}}$ coincides with the σ-algebra generated by the collection of cylinder sets of the form $\pi_J^{-1}(H)$ for some $J \in \mathcal{I}$ and some $H \in \mathcal{F}_J$.

Let us restrict ourselves to the case $\{S_J, \pi_J^K\}$ is a topological projective family and for every $J \in \mathcal{I}$ the collection $\mathcal{F}_J$ is the Borel σ-algebra of $\mathcal{F}_J$. In this case the collection $\{S_J, \mathcal{F}_J, \pi_J^K\}$ will be called a *topological projective family of measurable spaces*. In this context it is possible to formulate the following version of Kolmogorov theorem [68].

Theorem 3.7. *Let $\{S_J, \mathcal{F}_J, \pi_J^K\}$ be a topological and sequentially maximal projective family of measurable spaces, where for every $J \in \mathcal{I}$ the space S_J is*

Hausdorff. Let $\{\mu_J\}$ be a projective family of inner regular Borel probability measures on $\{(S_J, \mathcal{F}_J)\}$. Then there exists a unique probability measure μ on the measurable space $(S_\mathcal{I}, \mathcal{F}_\mathcal{I})$ such that $\mu_J = \mu_{\pi_J}$ for all $J \in \mathcal{I}$

The measure μ whose existence is stated in Theorem 3.7 is called *the projective limit* of the projective family of probability measures $\{\mu_J\}$.

For the details of the proof we refer to [68]. We only limit ourselves to point out that the argument is completely similar to that used in the proof of Theorem 3.4.

3.5 Non-existence of Feynman's measure

If $\{\mu_J\}_{J \in A}$ is a projective family of probability measures defined on a topological projective family of measurable spaces, then Kolmogorov's existence Theorem 3.7 assures the existence and uniqueness of the projective limit μ. In this section we are going to see that in the case the measures $\{\mu_J\}_{J \in A}$ of the projective family are not real and positive, i.e., if we consider a projective family of *signed* or *complex* measures, then in general their projective limit μ cannot exists.

Given a projective family of measurable spaces $\{S_J, \mathcal{F}_J, \pi_J^K)\}$, a collection $\{\mu_J\}_{J \in \mathcal{I}}$ of complex measures on $\{(S_J, \mathcal{F}_J)\}_{J \in \mathcal{I}}$ satisfying the consistency property (3.41) is called a *consistent* or *projective family of complex measures*. A complex measure μ on the measurable space $(S_\mathcal{I}, \mathcal{F}_\mathcal{I})$, where $S_\mathcal{I} = \varprojlim S_J$ and $\mathcal{F}_\mathcal{I}$ is the σ-algebra generated by the cylinder sets, is called the *projective limit* of the projective family of complex measures $\{\mu_J\}_{J \in \mathcal{I}}$ if

$$\mu_J = \mu_{\pi_J}, \qquad \forall J \in \mathcal{I}. \tag{3.42}$$

The following theorem shows that a necessary condition for the existence of μ is a uniform bound on the total variation of the complex measures $\{\mu_J\}_{J \in \mathcal{I}}$.

Theorem 3.8. *Let $(S_J, \mathcal{F}_J, \pi_J^K)_{J, K \in \mathcal{I}}$ be a projective family of measure spaces and let $\{\mu_J\}_{J \in A}$ be a projective family of complex measures. If there exists a complex measure μ on $(S_\mathcal{I}, \mathcal{F}_\mathcal{I})$ satisfying the relation (3.42) then*

$$\sup_{J \in \mathcal{I}} |\mu_J| < +\infty \tag{3.43}$$

where $|\mu_J|$ denotes the total variation of the measure μ_J.

Proof. Let μ be a complex measure on $(S_\mathcal{I}, \mathcal{F}_\mathcal{I})$ satisfying (3.42). If $\{A_i\} \subset \mathcal{F}_J$ is a measurable partition of S_J then $\{\pi_J^{-1}(A_i)\} \subset \mathcal{F}_\mathcal{I}$ is a partition of $S_\mathcal{I}$ and by (3.42) we have

$$\sum_i |\mu(\pi_J^{-1}(A_i))| = \sum_i |\mu_J(A_i)|.$$

Hence for any $J \in \mathcal{I}$ the total variation $|\mu_J|$ of the measure μ_J is bounded from above by the total variation $|\mu|$ of the measure μ:

$$|\mu_J| \le |\mu| \qquad \forall J \in \mathcal{I},$$

Hence:

$$\sup_{J \in \mathcal{I}} |\mu_J| \le |\mu| < +\infty.$$

$\square$

In the case of a projective family of probability measures the condition (3.43) is trivially satisfied. In the general case of complex measures satisfying (3.43), sufficient conditions for the existence of the measure μ on $(S_\mathcal{I}, \mathcal{F}_\mathcal{I})$ have been given in [319, 70], where Kolmogorov existence theorem for projective limits of probability measures has been generalized to the case of complex measures.

Theorem 3.9. *Let $(S_J, \pi_J^K)_{J,K \in \mathcal{I}}$ be a projective system of Hausdorff topological spaces and let $(\mu_J)_{J \in \mathcal{I}}$ be a consistent family of complex regular measures. Further assume that the spaces S_J and the space $S_\mathcal{I} = \varprojlim S_J$ are completely regular. Then there exists a complex regular measure μ on $S_\mathcal{I}$ satisfying (3.42) if and only if the following two conditions are fulfilled:*

i. $\sup_J |\mu_J| < +\infty$,

ii. *for every $\epsilon > 0$ there exists a compact $K \subset S_\mathcal{I}$ such that if $K_J := \pi_J(K)$, the following holds*

$$|\mu_J|(S_J \setminus K_J) < \epsilon, \qquad \forall J \in \mathcal{I}$$

Under these assumptions μ is uniquely determined and $|\mu| = \sup_J |\mu_J|$.

For a complete proof see [319].

On the other hand, if μ_J are signed or complex bounded measures, in many interesting cases condition (3.43) cannot be satisfied. A rather easy example of this situation is the case where $\mathcal{I} \equiv \mathcal{F}(\mathbb{N})$ is the direct set of finite subsets of $\mathbb{N}$ and let $\{(S_n, \mathcal{F}_n, \mu_n)\}_{n \in \mathbb{N}}$ is a sequence of measure spaces, where for any $n \in \mathbb{N}$, μ_n is a complex measure such that $\int_{E_n} d\mu_n = 1$. For any $J \in \mathcal{I}$, let S_J be the product space $S_J \equiv \times_{n \in J} S_n$, endowed with the product σ-algebra $\mathcal{F}_J \equiv \times_{n\in J}\mathcal{F}_n$ and the product measure $\mu_J = \times_{n\in J}\mu_n$. By construction $(S_J, \mathcal{F}_J, \mu_J)_{J\in\mathcal{I}}$ turns out to be a projective system of measure spaces, as one has easily verify, and the projective limit measure space $(S_{\mathcal{I}}, \mathcal{F}_{\mathcal{I}})$ is naturally isomorphic to the product space $(\times_{n\in\mathbb{N}}S_n, \otimes_{n\in\mathbb{N}}\mathcal{F}_n)$. If a complex bounded variation measure μ satisfying (3.42) existed, then for any $N \in \mathbb{N}$ its total variation $|\mu|$ would be greater than the finite product $\Pi_{n=1}^N \|\mu_n\|$. Thus a necessary condition for the existence of the projective limit measure is

$$\Pi_n|\mu_n| < +\infty. \tag{3.44}$$

Equivalently one has to require the convergence of the series $\sum_n \log(\|\mu_n\|)$. This condition cannot be satisfied, for instance, in the case where for any $n \in \mathbb{N}$, one has that $(S_n, \mathcal{F}_n, \mu_n) = (E, \mathcal{F}, \nu)$ and ν is a complex bounded measure such that $\int_E d\mu = 1$ and $\|\mu\| = c > 1$.

Analogously, in the case of a continuous product space, i.e., if $\mathcal{I} \equiv \mathcal{F}(\mathbb{R}^+)$, then the necessary condition for the existence of a signed or complex bounded product measure is

$$\sup_{J\in A}\{\sum_{t\in J} \log \|\mu_t\|\} < +\infty. \tag{3.45}$$

One of the first attempts to define Feynman path integrals was based on the construction of a Wiener measure with complex variance σ^2, i.e. by a measure on the space of continuous paths $\mathcal{C}$ such that its value on an arbitrary cylinder set is given by

$$W_\sigma\left(\{\gamma \in \mathcal{C}\colon \gamma(t_1) \in I_1, \ldots, \gamma(t_n) \in I_n\}\right)$$

$$= \int_{I_1\times\cdots\times I_n} \frac{e^{-\frac{1}{2\sigma^2}\sum_{i=1}^n \frac{(x_i-x_{i-1})^2}{t_i-t_{i-1}}}}{(2\pi\sigma^2)^{n/2}\sqrt{t_1(t_2-t_1)\cdots(t_n-t_{n-1})}}dx_1\ldots dx_n \tag{3.46}$$

valid for any choice of $n \in \mathbb{N}$, $0 < t_1 < \cdots < t_n < +\infty$ and $I_1, \ldots, I_n \in \mathcal{B}(\mathbb{R})$. Clearly, for $\sigma^2 \in \mathbb{R}^+$, W_σ turns out to be as a Borel probability measure. Formally, if the parameter σ^2 is set to be purely imaginary, in particular if $\sigma^2 = i\hbar$, then formula (4.1) should give the value of a complex measure on the cylinder sets of $\mathcal{C}$. It this measure existed, then it could be recognized as a putative "Feynman measure" and the fundamental building block for the mathematical definition of Feynman path integrals. Unfortunately, if σ^2 is a complex number lying outside the half real line $\mathbb{R}^+$, then W_σ cannot exist. Indeed, let $\sigma^2 = \alpha + i\beta$, with $\alpha, \beta \in \mathbb{R}$, $\alpha > 0$ and $\beta \neq 0$. By retracing the first steps leading to the definition of the Wiener measure W, let us consider the direct set $\mathcal{I}$ of finite subsets of $[0, +\infty)$, the space

$\Omega = \mathbb{R}^{[0,+\infty)}$ of real-valued paths endowed with the $\sigma-$algebra generated by the cylinder sets and the projective family of complex measures $(\mu_J)_{J\in\mathcal{I}}$, given by

$$\mu_J(I_1 \times \cdots \times I_n)$$

$$= \int_{I_1 \times \cdots \times I_n} \frac{e^{-\frac{1}{2(\alpha+i\beta)}\sum_{i=1}^{n}\frac{(x_i-x_{i-1})^2}{t_i-t_{i-1}}}}{(2\pi(\alpha+i\beta))^{n/2}\sqrt{t_1(t_2-t_1)\cdots(t_n-t_{n-1})}} dx_1 \ldots dx_n$$

where $J = \{t_1,\ldots,t_n\}$, $I_1,\ldots,I_n \in \mathcal{B}(\mathbb{R})$. By Remark 2.4 for any $J \in \mathcal{I}$ the total variation of the complex measure μ_J is equal to

$$|\mu_J| = \int_{\mathbb{R}^n} \left| \frac{e^{-\frac{1}{2(\alpha+i\beta)}\sum_{j=1}^{n}\frac{(x_j-x_{j-1})^2}{t_j-t_{j-1}}}}{(2\pi(\alpha+i\beta))^{n/2}\sqrt{t_1(t_2-t_1)\cdots(t_n-t_{n-1})}} \right| dx_1 \ldots dx_n$$

$$= \int_{\mathbb{R}^n} \left| \frac{e^{-\frac{1}{2(\alpha+i\beta)}\sum_{j=1}^{n}x_j^2}}{(2\pi(\alpha+i\beta))^{n/2}} \right| dx_1 \ldots dx_n$$

$$= \left(\int_{\mathbb{R}} \left| \frac{e^{-\frac{x^2}{2(\alpha+i\beta)}}}{\sqrt{2\pi(\alpha+i\beta)}} \right| dx \right)^n .$$

It is now simple to see that $\sup_{J\in\mathcal{I}} |\mu_J| = +\infty$ if $\beta \neq 0$. Indeed, in this case the integral $\int_{\mathbb{R}} \left| \frac{e^{-\frac{x^2}{2(\alpha+i\beta)}}}{\sqrt{2\pi(\alpha+i\beta)}} \right| dx$ is strictly greater than 1 and by Theorem 3.8 there cannot exists a complex measure μ projective limit of the consistent family $(\mu_J)_{J\in\mathcal{I}}$.

3.6   Complex Borel measures on Banach spaces and their Fourier transforms

Let us consider a real separable Banach space $(X, \| \ \|_X)$ and let $\mathcal{B}(X)$ be the Borel σ-algebra generated by the open sets in X.

The following lemma shows that $\mathcal{B}(X)$ is in fact generated by the collection of open (or closed) balls.

Lemma 3.3.   *Let X be a separable Banach space and $\mathcal{C} \subset \mathcal{B}(X)$ the collection of open (closed) balls of X. Then $\sigma(\mathcal{C}) = \mathcal{B}(X)$.*

Proof.   Let $\mathcal{D} \subset X$ be a countable dense set in X. Given an open set $U \subset X$, let us consider for any $x \in U$ an open (resp. closed) ball $B(x,r) = \{y \in X : \|y - x\|_X < r\}$ such that $r \in \mathbb{Q}$ and $B(x,r) \subset U$ and take a

$y_x \in \mathcal{D} \cap B(x, r/3)$. In particular we have:

$$x \in B(y_x, r/2) \subset B(x, r)$$

If we now introduce the notation $r_x \equiv r/2$ we have

$$U = \cup_{x \in U} \{x\} \subset \cup_{x \in U} B(y_x, r_x)$$

where the union on the right-hand side contains in fact a countable number of open (resp. closed) balls. This shows that $U \in \sigma(\mathcal{C})$ and, consequently, that $\mathcal{B}(X) \in \sigma(\mathcal{C})$. $\qquad\square$

Corollary 3.2. *Let $\mathcal{C}' \subset \mathcal{B}(X)$ be a collection of Borel sets such that $\sigma(\mathcal{C}')$ contains all open (closed) balls, then $\sigma(\mathcal{C}') = \mathcal{B}(X)$.*

Proof. This follows from Lemma 3.3 and Eq. (2.5). $\qquad\square$

Another relevant class of sets which plays an important role in the construction of non-trivial measures on X are the cylinder sets [70, 71, 335]. Let X^* be the topological dual space of X. A *cylinder set* in X is a set of the form

$$C_{\theta_1,\dots,\theta_n;H} := \{x \in X : (\theta_1(x), \dots, \theta_n(x)) \in H\} \qquad (3.47)$$

for some $n \geq 1$, $\theta_1, \dots, \theta_n \in X^*$ and $H \in \mathcal{B}(\mathbb{R}^n)$ Borel set in $\mathbb{R}^n$. Equivalently, by considering the linear mapping $L_{(\theta_1,\dots,\theta_n)} : X \to \mathbb{R}^n$ defined as

$$L_{(\theta_1,\dots,\theta_n)}(x) := (\theta_1(x), \dots, \theta_n(x)) \qquad x \in X, \qquad (3.48)$$

the cylinder set (3.47) can be represented as

$$C_{\theta_1,\dots,\theta_n;H} = L^{-1}_{(\theta_1,\dots,\theta_n)}(H).$$

A third geometrical representation can be given by considering the closed subspace $W_{(\theta_1,\dots,\theta_n)} \subset X$ of X defined as $W_{(\theta_1,\dots,\theta_n)} := \mathrm{Ker}\, L_{(\theta_1,\dots,\theta_n)}$ and the finite dimensional linear space $V_{(\theta_1,\dots,\theta_n)} \subset X$ which gives X as the direct sum with $W_{(\theta_1,\dots,\theta_n)}$:

$$X = V_{(\theta_1,\dots,\theta_n)} \oplus W_{(\theta_1,\dots,\theta_n)}.$$

The cylinder set (3.47) can be written in the form

$$C_{\theta_1,\dots,\theta_n;H} = L^{-1}_{(\theta_1,\dots,\theta_n)}(H) = A \oplus W_{(\theta_1,\dots,\theta_n)}$$

where A is a Borel set in the finite dimensional linear space $V_{(\theta_1,\dots,\theta_n)}$. If we denote with the symbol $\tilde{L}_{(\theta_1,\dots,\theta_n)} : V_{(\theta_1,\dots,\theta_n)} \to \mathbb{R}^n$ the restriction of $L_{(\theta_1,\dots,\theta_n)}$ to $V_{(\theta_1,\dots,\theta_n)}$, then the Borel set A can be obtained as $A = \tilde{L}^{-1}_{(\theta_1,\dots,\theta_n)}(H)$.

We shall denote $\mathcal{F}_X$ the σ-algebra generated by the cylinder sets in X. It coincides with the Borel σ-algebra, as stated in the following theorem.

Theorem 3.10. *If X is a real separable Banach space, then $\mathcal{F}_X = \mathcal{B}(X)$.*

This result can be actually generalized to a larger class of topological vector spaces. We refer to [70, 71] for a detailed presentation of these topics and limit ourselves to prove Theorem 3.10 in the case X is a separable Banach space admitting a Schauder basis.

Definition 3.9. A sequence $\{e_n\} \subset X$ is called a Schauder basis for X if $\|e_n\|_X = 1$ for all n, and for all vector $x \in X$ there exists a unique sequence $\{x_n\} \subset \mathbb{R}$ such that

$$\lim_{n \to \infty} \left\| x - \sum_{k=1}^{n} x_k e_k \right\|_X = 0 \,. \tag{3.49}$$

Proof (of Theorem 3.10). By Corollary 3.2 it is sufficient to show that any closed ball $B(x, r) = \{y \in X \colon \|y - x\|_X \leq r\}$ is contained in $\mathcal{F}_X$. Let $\{e_k\}$ be a Schauder basis for X and let us consider for every $n \geq 1$ the finite dimensional linear space $X_n := \mathrm{span}(x_1, \ldots, x_n)$ and the projection $P_n : X \to X_n$ sending any $x \in X$, $x = \sum_{k=1}^{\infty} x_k e_k$ to the vector $\sum_{k=1}^{n} x_k e_k$. Given a general ball $B \equiv B(x, r)$, let us consider the sets $B_n \subset X_n$ defined as $B_n := P_n B$ and the cylinder sets $C_n := B_n + Y_n$, where $Y_n = \ker P_n$. By using property (3.49) it is easy to see that $B = \cap_n C_n$. Hence $B \in \mathcal{F}_X$. $\square$

Let $\mathcal{M}(X)$ the Banach algebra of complex Borel measures on X, where the multiplication of two measures μ and ν is defined as their convolution [71]

$$\mu * \nu(E) = \int \chi_E(x + y) \, d\mu(x) d\nu(y), \qquad E \in \mathcal{B}(X),$$

the unit 1 is the δ point measure at $0 \in X$

$$1 = \delta_0$$

and the norm of μ is defined as its total variation

$$\|\mu\| := |\mu| \,.$$

For any $\mu \in \mathcal{M}(X)$ let us define the map $\hat{\mu} : X^* \to \mathbb{C}$ as

$$\hat{\mu}(\theta) := \int_X e^{i\langle \theta, x \rangle} d\mu(x), \qquad \theta \in X^*, \tag{3.50}$$

where $\langle \theta, x \rangle$ stands for the pairing between $\theta \in X^*$ and $x \in X$. Similarly to the case where X is a finite dimensional space the map $\hat{\mu}$ is continuous and bounded. Indeed

$$|\hat{\mu}(\theta)| \leq \|\mu\|, \qquad \forall \theta \in X^*$$

and by applying Lebesgue dominated convergence we get:

$$|\hat{\mu}(\theta_1) - \hat{\mu}(\theta_2)| = \left| \int_{\mathbb{R}^n} (e^{i\langle\theta_1,x\rangle} - e^{i\langle\theta_2,x\rangle}) d\mu(x) \right|$$

$$\leq \int_{\mathbb{R}^n} |e^{i\langle\theta_1,x\rangle} - e^{i\langle\theta_2,x\rangle}| d|\mu|(x)$$

$$= \int_{\mathbb{R}^n} |1 - e^{i\langle\theta_2-\theta_1,x\rangle}| d|\mu|(x) \xrightarrow{\theta_2 \to \theta_1} 0 .$$

Analogously to the finite dimensional case, there is a one-to-one correspondence between a measure and its Fourier transform.

Theorem 3.11. *If $\mu, \nu \in \mathcal{M}(X)$ and $\hat{\mu} = \hat{\nu}$ then $\mu = \nu$.*

Proof. The identity $\hat{\mu}(\theta) = \hat{\nu}(\theta)$ for all $\theta \in X^*$ gives the equality of μ and ν when restricted to the algebra of cylinder sets. Indeed, chosen $n \geq 1$, $\theta_1, \ldots, \theta_n \in X^*$, let us consider the linear mapping $L_{(\theta_1,\ldots,\theta_n)} : X \to \mathbb{R}^n$ defined by (3.48) and the image measures $\mu_{(\theta_1,\ldots,\theta_n)}$ and $\nu_{(\theta_1,\ldots,\theta_n)}$ under the action of $L_{(\theta_1,\ldots,\theta_n)}$. Since for any choice of $\alpha_1, \ldots, \alpha_n \in \mathbb{R}$ the following holds

$$\hat{\mu}(\alpha_1\theta_1 + \cdots + \alpha_n\theta_n) = \hat{\nu}(\alpha_1\theta_1 + \cdots + \alpha_n\theta_n) ,$$

it is easy to see that $\hat{\mu}_{(\theta_1,\ldots,\theta_n)} = \hat{\nu}_{(\theta_1,\ldots,\theta_n)}$ and by Theorem 2.21 we obtain $\mu_{(\theta_1,\ldots,\theta_n)} = \nu_{(\theta_1,\ldots,\theta_n)}$. In particular, this means that for any cylinder set $C_{\theta_1,\ldots,\theta_n;H} = L_{(\theta_1,\ldots,\theta_n)}^{-1}(H)$, with $H \in \mathcal{B}(\mathbb{R})$, we have:

$$\mu\left(C_{\theta_1,\ldots,\theta_n;H}\right) = \nu\left(C_{\theta_1,\ldots,\theta_n;H}\right) .$$

By theorem (2.1) and Theorem 3.10 we eventually obtain that $\mu(E) = \nu(E)$ for all $E \in \mathcal{B}(X)$. $\square$

The set of mappings $\hat{\mu} : X^* \to \mathbb{C}$ of the form (3.50) for some $\mu \in \mathcal{M}(X)$ will be denoted with the symbol $\mathcal{F}(X)$. By Theorem 3.11 there is a one-to-one correspondence between $\mathcal{M}(X)$ and $\mathcal{F}(X)$. In particular, if $\mathcal{F}(X)$ is endowed with the norm $\| \ \|$ defined as

$$\|\hat{\mu}\| := \|\mu\|$$

then $\mathcal{F}(X)$ becomes a Banach algebra of functions (the multiplication is the pointwise one) and the mapping $\mu \mapsto \hat{\mu}$ a Banach-algebra isomorphism.

In the case where X is a real separable Hilbert space $(\mathcal{H}, \langle \ , \ \rangle)$, then the natural isomorphism between $\mathcal{H}$ and $\mathcal{H}^*$ allows to represent the Fourier transform $\hat{\mu}$ of a measure $\mu \in \mathcal{M}(\mathcal{H})$ as a mapping $\hat{\mu} : \mathcal{H} \to \mathbb{C}$ defined as

$$\hat{\mu}(x) = \int_{\mathcal{H}} e^{i\langle x,y \rangle} d\mu(y) \qquad x \in \mathcal{H}, \tag{3.51}$$

where $\langle x, y \rangle$ denotes the inner product between $x, y \in \mathcal{H}$.

If we restrict ourselves to the study of probability measures μ, Bochner Theorem 2.22 can be generalized to an infinite dimensional setting. To this end, it is important to introduce the definition of nuclear operator.

A trace class operator A on a (complex or real) separable Hilbert space $\mathcal{H}$ is by definition a bounded linear operator on $\mathcal{H}$ such that $\mathrm{Tr}\,|A| := \sum_{n=1}^{\infty} \langle |A|\, e_n, e_n \rangle < \infty$, for any orthonormal basis $\{e_n\}_{n \in \mathbb{N}}$ in $\mathcal{H}$. The symbol $|A|$ denotes the square root of A^*A (which is well defined since A^*A is a linear positive, bounded, self-adjoint operator). It is also easy to see that $\mathrm{Tr}\,|A|$ is independent of the choice of the basis used for its definition. Indeed, for any other orthonormal basis $(f_m)_{m \in \mathbb{N}}$ the following holds

$$\sum_m \langle |A|\, f_m, f_m \rangle = \sum_m \| |A|^{\frac{1}{2}} f_m \|^2 = \sum_m \sum_n \left| \langle |A|^{\frac{1}{2}} f_m, e_n \rangle \right|^2,$$

where we used Bessel-Parseval's equality. Using the symmetry of $|A|^{\frac{1}{2}}$ and interchanging the order of the summations we get that this is equal to $\sum_n \langle e_n\, |A|\, e_n \rangle = \mathrm{Tr}\,|A|$. $\mathrm{Tr}\,|A|$ is $\mathbb{R}_+$-linear and monotonic, as is also easily seen (see, e.g., [280, 281, 259]).

Definition 3.10. An operator N on a separable Hilbert space $\mathcal{H}$ is called *nuclear operator* if

(i) N is a trace class operator;
(ii) N is positive, i.e. $N \geq 0$;
(iii) N is self-adjoint, i.e. $N^* = N$.

Theorem 3.12 (Bochner-Milnos-Sazonov). *A function $\phi : \mathcal{H} \to \mathbb{C}$ is the Fourier transform of a probability measure on $(\mathcal{H}, \mathcal{B}(\mathcal{H}))$ if and only if it enjoys the following properties:*

(1) $\phi(0) = 1$.
(2) ϕ *is positive semi-definite, i.e.* $\forall n \geq 1$, $\forall x_1, ..., x_n \in \mathcal{H}$ *and* $\forall z_1, ..., z_n \in \mathbb{C}$ *the following holds:*

$$\sum_{j,k=1}^{n} \phi(x_j - x_k) z_j \bar{z}_k \geq 0 \tag{3.52}$$

(3) for any $\varepsilon > 0$ there exists a nuclear operator $N_\varepsilon \in \mathcal{N}(\mathcal{H})$ such that

$$1 - \mathrm{Re}\,\varphi(x) < \langle N_\varepsilon x, x \rangle + \varepsilon, \quad \forall x \in \mathcal{H}.$$

For the proof and a discussion of this topic see, e.g., [38].

3.7 Gaussian measures on Hilbert spaces and abstract Wiener spaces

Definition 3.11. A probability measure μ on $(\mathbb{R}, \mathcal{B}(\mathbb{R}))$ is said to be a *Gaussian measure* if its Fourier transform $\hat{\mu}$ has the following form

$$\hat{\mu}(x) = e^{imx} e^{-\frac{1}{2}\sigma^2 x^2} \qquad x \in \mathbb{R}, \tag{3.53}$$

for some constants $m, \sigma^2 \in \mathbb{R}$, $\sigma^2 \geq 0$. The constants m and σ^2 are called the *mean* and the *variance* of μ respectively.

In σ^2 is a strictly positive constant, then μ is absolutely continuous with respect to the Lebesgue measure on $(\mathbb{R}, \mathcal{B}(\mathbb{R}))$ with a Radon-Nicodym derivative $h : \mathbb{R} \to \mathbb{R}$ equal to

$$h(x) = \frac{e^{-\frac{1}{2\sigma^2}(x-m)^2}}{\sqrt{2\pi\sigma^2}} \qquad x \in \mathbb{R},$$

while in the case $\sigma^2 = 0$, the measure μ is the δ_m point measure concentrated at $m \in \mathbb{R}$ (see Eq. (2.6)). In particular, the following identities hold

$$m = \int x \, d\mu(x) \qquad \sigma^2 = \int (x - m)^2 \, d\mu(x). \tag{3.54}$$

Definition 3.12. A probability measure μ on the Borel σ-algebra of a real separable Banach space X is called *Gaussian measure* if for any $\theta \in X^*$ the image measure μ_θ of μ under the action of θ is a Gaussian measure on the real line.

By the change of variable formula (2.17), it turns out that the Fourier transform $\hat{\mu}$ of a Gaussian measure on X has the following form

$$\hat{\mu}(\theta) = e^{il(\theta)} e^{-\frac{1}{2}q(\theta)^2} \qquad \theta \in X^*, \tag{3.55}$$

where $l : X^* \to \mathbb{R}$ is the linear functional defined by

$$l(\theta) = \int_X \theta(x) \, d\mu(x),$$

while $q : X^* \to \mathbb{R}$ is the quadratic form associated to the positive bilinear map $Q : X^* \times X^* \to \mathbb{R}$ defined as

$$Q(\theta, \eta) := \int_X (x - l(\theta))(x - l(\eta)) d\mu(x) \qquad \theta, \eta \in X^*.$$

Analogously to the finite-dimensional case, the bilinear map Q is called the *covariance* of the measure μ.

Conversely, if μ is a probability measure on $\mathcal{B}(X)$ such that its Fourier transform has the form (3.55) then μ is Gaussian since for any $\theta \in X^*$ the Fourier transform of the image measure μ_θ will have the form (3.53). It is important to point out that in an infinite dimensional setting not all positive bilinear map $Q : X^* \times X^* \to \mathbb{R}$ are the covariance of a Gaussian measure. In the particular case where X is a real separable Hilbert space $\mathcal{H}$, Theorem 3.12 allows to get a characterization of the bilinear maps $Q : X^* \times X^* \to \mathbb{R}$ which are the covariance of a Gaussian measure. In this setting it is convenient to exploit the natural isomorphism between $\mathcal{H}$ and $\mathcal{H}^*$ and identify them. In particular, by Riesz theorem it is possible to define a vector $m_\mu \in \mathcal{H}$ such that $l(x) = \langle m_\mu, x \rangle$, which is called the *mean vector* of the measure μ. Similarly we can uniquely associate to the positive bilinear map $Q : \mathcal{H} \times \mathcal{H} \to \mathbb{R}$ a bounded symmetric positive operator $S_\mu : \mathcal{H} \to \mathcal{H}$ such that

$$Q(x,y) = \langle x, S_\mu y \rangle \qquad x, y \in \mathcal{H},$$

which is called the *covariance operator* of μ.

Theorem 3.13 (Prokhorov). *Let $\mathcal{H}$ be a real separable Hilbert space. A bounded symmetric positive operator $S : \mathcal{H} \to \mathcal{H}$ is the covariance operator of a Gaussian measure μ on $\mathcal{B}(\mathcal{H})$ if and only if S is nuclear.*

For additional details we refer to [69, 70, 38].

In particular, Prokhorov theorem shows that on an infinite dimensional Hilbert space the standard Gaussian measure (with mean $m = 0$ and covariance operator S equal to the identity $I : \mathcal{H} \to \mathcal{H}$) cannot exists since the identity I is not trace-class. This result can be also obtained as a consequence of Theorem 3.2. Indeed, if a Gaussian measure on $\mathcal{B}(\mathcal{H})$ with $m = 0$ and $S = I$ existed, then it would be invariant under isometries.

It turns out that a useful substitute for the standard Gaussian measure on $\mathcal{H}$ is a cylinder measure. By definition a *cylinder measure* on $\mathcal{H}$ is a positive and finitely-additive set function ν defined on the algebra $\mathcal{A}_\mathcal{H}$ of cylinder sets in $\mathcal{H}$. As remarked above, in a Hilbert space every cylinder set can be represented in the form

$$P^{-1}(C) = \{x \in \mathcal{H} : Px \in C\},$$

where $P : \mathcal{H} \to \mathcal{H}$ is an orthogonal projection operator onto a finite dimensional subspace $P\mathcal{H} \subset \mathcal{H}$ and $C \subset P\mathcal{H}$ is a Borel set in $P\mathcal{H}$.

Let us consider the cylinder measure $\nu : \mathcal{A}_\mathcal{H} \to \mathbb{R}$ defined on a general cylinder set $P^{-1}(C)$ as

$$\nu\left(P^{-1}(C)\right) = (2\pi)^{-n/2} \int_C e^{-\frac{1}{2}\|Px\|^2} \, dx,$$

where n is the dimension of $P\mathcal{H}$ and dx denotes the Lebesgue measure on $P\mathcal{H}$. ν is called *standard Gaussian measure associated with* $\mathcal{H}$.

If $\mathcal{H}$ is infinite dimensional, by Prokhorov's theorem the standard Gaussian measure associated with $\mathcal{H}$ cannot be σ-additive on $\mathcal{A}_{\mathcal{H}}$, since in this case it could be extended to a σ-additive probability measure μ on $\mathcal{B}(\mathcal{H})$ invariant under isometries. Indeed, if such a μ existed, its Fourier transform would be given by

$$\hat{\mu}(x) = e^{-\frac{1}{2}\|x\|^2} \qquad x \in \mathcal{H}.$$

as it is easy to check since $\hat{\mu}$ is uniquely determined by the value of μ on the cylinder sets.

However, if $\mathcal{H}$ is embedded in a suitable Banach space $(X, |\ |)$, then ν can determine a unique Gaussian measure on $\mathcal{B}(X)$ [237, 169, 168] according to the procedure described below.

Definition 3.13. A norm $|\ |$ (or a semi-norm) on $\mathcal{H}$ is called *measurable* if for every $\epsilon > 0$ there exist a finite-dimensional projection $P_\epsilon : \mathcal{H} \to \mathcal{H}$, such that for all $P \perp P_\epsilon$ one has:

$$\nu(\{x \in \mathcal{H}|\ |P(x)| > \epsilon\}) < \epsilon,$$

where P and P_ϵ are called orthogonal ($P \perp P_\epsilon$) if their ranges are orthogonal in $(\mathcal{H}, \langle\ , \rangle)$.

It turns out that $|\ |$ is weaker than $\|\ \|$. Indeed there exists a $c \in \mathbb{R}^+$ such that [237]

$$|x| \leq c\|x\|, \qquad \forall x \in \mathcal{H}. \tag{3.56}$$

In the following we shall denote X the Banach space constructed as the completion of $\mathcal{H}$ in the $|\ |$-norm and i the inclusion of $\mathcal{H}$ in X, which is continuous by (3.56). Analogously, the dual map $i^* : X^* \to \mathcal{H}^*$, which is given by restriction, i.e. $i^*(x) = x_{|\mathcal{H}}$, is continuous. Identifying $\mathcal{H} \equiv \mathcal{H}^*$ we have the following chain of densely embedded subspaces

$$X^* \subset \mathcal{H} \subset X.$$

The triple $(i, \mathcal{H}, X)$ is called an *abstract Wiener space*.

Given a cylinder set $C_{\theta_1,\ldots,\theta_n;H}$ in X, (with $\theta_1,\ldots,\theta_n \in X^*$ and $H \in \mathcal{B}(\mathbb{R}^n)$) (see Eq. (3.47)), the intersection $C_{\theta_1,\ldots,\theta_n;H} \cap \mathcal{H}$ is a cylinder set in $\mathcal{H}$. We can define the cylinder set measure μ_0 on the algebra $\mathcal{A}_X$ of cylinder sets in X as:

$$\mu_0\left(C_{\theta_1,\ldots,\theta_n;H}\right) := \nu\left(C_{\theta_1,\ldots,\theta_n;H} \cap \mathcal{H}\right),$$

A fundamental result of L. Gross is the following statement, which is actually a consequence of the Definition 3.13 of measurable norm [169].

Theorem 3.14. *μ_0 is σ-additive on $\mathcal{A}_X$. Hence it defines a unique probability measure μ on the Borel σ-algebra of X, in such a way that for any $\theta \in X^* \subset \mathcal{H}$:*

$$\int_X e^{i\theta(x)}\, d\mu(x) = e^{-\frac{1}{2}\|\theta\|^2}. \tag{3.57}$$

Example 3.5 (The classical Wiener space). *Let us consider the Sobolev space $\mathcal{H}^{1,2}([0,t], \mathbb{R}^d)$, i.e. the space of absolutely continuous functions $\gamma : [0,t] \to \mathbb{R}^d$ such that $\gamma(0) = 0$ and $\dot{\gamma} \in L^2([0,t])$, where $\dot{\gamma}$ denotes the distributional derivative of the function γ, endowed with the inner product*

$$\langle \gamma_1, \gamma_2 \rangle = \int_0^t \dot{\gamma}_1(s)\dot{\gamma}_2(s)ds.$$

Let $(C_0([0,t], \mathbb{R}^d), |\cdot|)$ be the Banach space of continuous functions $\omega : [0,t] \to \mathbb{R}^d$ such that $\omega(0) = 0$, endowed with the supremum norm:

$$|\omega| := \sup_{s \in [0,t]} |\omega(s)|.$$

It is possible to prove (see [237] for a additional details) that $\mathcal{H}^{1,2}([0,t], \mathbb{R}^d)$ is dense in $C_0([0,t], \mathbb{R}^d)$ with respect to the supremum norm $|\cdot|$.

Moreover, considered the standard Gaussian measure ν associated with $\mathcal{H}^{1,2}([0,t], \mathbb{R}^d)$, it is possible to prove that the supremum norm $|\cdot|$ is a ν-measurable norm on $\mathcal{H}^{1,2}([0,t], \mathbb{R}^d)$. This makes $(i, \mathcal{H}^{1,2}([0,t], \mathbb{R}^d), C_0([0,t], \mathbb{R}^d))$ an abstract Wiener space, which is called classical Wiener space *while μ is the* Wiener measure

In particular, it is possible to prove that the norm $\|i\|$ of the continuous embedding

$$i : \mathcal{H}^{1,2}([0,t], \mathbb{R}^d) \to C_0([0,t], \mathbb{R}^d)$$

is equal to $\|i\| = \sqrt{t}$.

By Theorem 3.14, each element $\theta \in \mathcal{B}^*$ can be regarded as a measurable function $n(\theta)$ on $(X, \mathcal{B}(X), \mu)$. By adopting the standard terminology of probability theory, $n(\theta)$ is a random variable on the probability space $(X, \mathcal{B}(X), \mu)$. Its distribution, i.e. the probability measure on $(\mathbb{R}, \mathcal{B}(\mathbb{R}))$ obtained as the image measure of μ under the action of $n(\theta)$, is a centered

Gaussian measure with covariance $\|\theta\|^2$. More generally, given $\theta_1, \theta_2 \in X^*$, one has

$$\int_X n(\theta_1)(x)n(\theta_2)(x)\,d\mu(x) = \langle \theta_1, \theta_2 \rangle. \tag{3.58}$$

Equation (3.58) and the density of X^* in $\mathcal{H}$ allow the extension of the map $n : X^* \to L^2(X, \mu)$ to a map $n : \mathcal{H} \to L^2(X, \mu)$ (with an abuse of notation we denote the map n on X^* and its extension to $\mathcal{H}$ with the same symbol)[169]. Moreover, given a complete orthonormal system $\{e_i\}$ in $\mathcal{H}$, for any $h \in \mathcal{H}$ the sequence of random variables

$$\sum_{i=1}^{n} h_i n(e_i), \qquad h_i = \langle e_i, h \rangle,$$

converges in $L^2(X, \mu)$ and by subsequences μ-almost everywhere to the random variable $n(h)$.

The family of measurable mappings $\{n(h)\}_{h \in \mathcal{H}}$ plays an important role in the study of transformation properties of the measure μ under translation. In particular, the following result shows that μ is quasi invariant under translations in the direction of vectors $h \in \mathcal{H}$.

Theorem 3.15. [169, 237] *Let $(i, \mathcal{H}, \mathcal{B})$ be an abstract Wiener space. For every vector $y \in \mathcal{H}$ and every function $f \in L^1(\mathcal{B}, \mu)$, the following holds*

$$\int_{\mathcal{B}} f(x)\,d\mu(x) = \int_{\mathcal{B}} f(x + y)\rho(y, x)\,d\mu(x),$$

where ρ is given by

$$\rho(y, x) = e^{-\frac{1}{2}\|y\|^2 - n(y)(x)}.$$

Given an orthogonal projection operator $P : \mathcal{H} \to \mathcal{H}$ with

$$P(h) = \sum_{i=1}^{n} \langle e_i, h \rangle e_i \qquad h \in \mathcal{H},$$

for some orthonormal $e_1, \ldots, e_n \in \mathcal{H}$, let us define the *stochastic extension* $\tilde{P} : X \to \mathcal{H}$ of P as the random variable

$$\tilde{P}(x) = \sum_{i=1}^{n} n(e_i)(x) e_i \qquad x \in X.$$

More generally, this procedure allows the extension of a larger class of mappings f from $\mathcal{H}$ to X. We present here the approach proposed in [278]. In the following $f : \mathcal{H} \to X_1$ will be a function with values in a real separable Banach space $(X_1, \| \; \|_{X_1})$.

Definition 3.14. A function $f : \mathcal{H} \to X_1$ is said to determine a random variable $\tilde{f} : X \to X_1$ if for any sequence $\{P_n\}$ of finite dimensional projection operators converging strongly to the identity on $\mathcal{H}$ the sequence of random variables $\{f \circ \tilde{P}_n\}$ converges in probability[6] on X. In this case the random variable $\tilde{f}$ will be called the *stochastic extension* of f.

The following important result holds [169]:

Theorem 3.16. *If $g : X \to X_1$ is continuous and $f := g|_{\mathcal{H}}$, then the stochastic extension of f is well defined and it is equal to g μ-almost everywhere.*

Other interesting classes of functions on $\mathcal{H}$ which admit a stochastic extension have been studied. For the application to the mathematical theory of Feynman path integrals a relevant role is played by the functions $f \in \mathcal{F}(\mathcal{H})$, i.e. by the mappings $f : \mathcal{H} \to \mathbb{C}$ of the form

$$f(h) = \int_{\mathcal{H}} e^{i\langle h, h' \rangle} d\mu(h') \qquad h \in \mathcal{H}, \tag{3.59}$$

for some complex Borel measure μ on $\mathcal{H}$. The following result shows that every function $f \in \mathcal{F}(\mathcal{H})$ admits a stochastic extension. We refer to [215] for the proof.

Theorem 3.17. *Let $f : \mathcal{H} \to \mathbb{C}$ be a function of the form* (3.59). *Then f admits a stochastic extensions $\tilde{f} : X \to \mathbb{C}$ equal to*

$$\tilde{f}(x) = \int_{\mathcal{H}} e^{in(h')(x)} d\mu(h') \qquad x \in X, .$$

Moreover $\tilde{f} \in L^1(X, \mu)$.

A similar result can be proved if we consider a quadratic form $q : \mathcal{H} \to \mathbb{R}$ on $\mathcal{H}$ associated to an Hilbert-Schmidt operator $K : \mathcal{H} \to \mathcal{H}$

$$q(h) := \langle h, Kh \rangle \qquad h \in \mathcal{H}. \tag{3.60}$$

Actually the mapping q can be extended to a random variable $\tilde{q}$ on X provided that suitable normalization counterterms are introduced.

---

[6]We recall that a sequence of random variables $\{g_n\}$ on a probability space $(\Omega, \mathcal{F}, \mu)$ is said to converge in probability to a random variable g on $(\Omega, \mathcal{F}, \mu)$ if the following holds

$$\forall \epsilon > 0 \quad \lim_{n \to \infty} \mu\left(\{\omega \in \Omega : |g_n(\omega) - g(\omega)| > \epsilon\}\right) = 0$$

Theorem 3.18. *Let $K : \mathcal{H} \to \mathcal{H}$ be an Hilbert Schmidt operator, $\{e_n\}$ an orthonormal basis in $\mathcal{H}$ and let $\{f_n\}$ be the sequence of random variables on X defined by*

$$f_n(x) := \sum_{j=1}^{n} \left(\widetilde{(e_i K)}(x) n(e_i)(x) - \langle e_j, K e_j \rangle \right)$$

Then the sequence $\{f_n\}$ converges in $L^2(X, \mu)$ to a random variable $\tilde{f}$ which is independent of the choice of the basis $\{e_n\}_n$.

For the proof we refer to the original paper [278]. The random variable $\tilde{f}$ whose existence is given by Theorem 3.18 is denoted with the symbol $\tilde{f} = \langle x, Kx \rangle - \mathrm{Tr}K$. Clearly, if the operator K is not trace class, the sequence $\mathrm{Tr}P_n K = \sum_{j=1}^{n} \langle e_j, K e_j \rangle$ doesn't converge.

In the case the operator $K : \mathcal{H} \to \mathcal{H}$ is trace class and self-adjoint, then it is possible to construct a stochastic extension $\tilde{q}$ of the quadratic form (3.60). Indeed, if $\{\lambda_j\}_j$ are the eigenvalues and $\{e_j\}_j$ the corresponding eigenvectors of K, then the mapping $\tilde{q} : X \to \mathbb{R}$ can be obtained as:

$$\tilde{q}(x) := \begin{cases} \lim_{n \to \infty} \sum_{j=1}^{n} \lambda_j \left(n(e_j)(x) \right)^2 & \text{if the limit exists} \\ 0 & \text{otherwise} \end{cases}$$

the mapping $\tilde{q}$ is also denoted with an abuse of notation in the following way

$$\tilde{q}(x) = \langle x, Kx \rangle \qquad x \in X.$$

Moreover the following holds:

Theorem 3.19. *Let $K : \mathcal{H} \to \mathcal{H}$ be a self-adjoint trace class operator and $\mu \in \mathcal{M}(\mathcal{H})$ a complex Borel measure on $\mathcal{H}$. The function $f : \mathcal{H} \to \mathbb{C}$*

$$f(h) = e^{\frac{i}{2} \langle h, Kh \rangle} \int_{\mathcal{H}} e^{i \langle h, h' \rangle} \, d\mu(h')$$

admits a stochastic extension $\tilde{f} : X \to \mathbb{C}$ equal to

$$\tilde{f}(x) = e^{\frac{i}{2} \langle x, Kx \rangle} \int_{\mathcal{H}} e^{in(h')(x)} \, d\mu(h')$$

We end this section with a useful integrability result.

Theorem 3.20. *Let $K : \mathcal{H} \to \mathcal{H}$ be a self-adjoint trace class operator such that $(I - B)$ is strictly positive and let $y \in \mathcal{H}$. Then the random variable on $(\mathcal{B}, \mu)$ defined by*

$$g(x) = e^{n(y)(x)} e^{\frac{1}{2} \langle x, Kx \rangle}$$

is μ-summable and

$$\int_B g\,d\mu = (\det(I - K))^{-1/2} e^{\frac{1}{2}\langle y,(I-B)^{-1}y\rangle},$$

where $\det(I - K)$ *denotes the Fredholm determinant of the operator* $I - K$.

For a proof of these results we refer to [237, 215].

Chapter 4

Projective systems of functionals and the Fourier transform approach

As explained in Section 3.5 it is impossible to construct a reasonable Feynman measure. In particular, the integrals appearing in the piecewise linear approximations of Feynman formula

$$\int_{\mathbb{R}^{dn}} \frac{e^{\frac{i}{\hbar} \sum_{j=1}^{n} \left(\frac{(x_j - x_{j-1})^2}{2(t/n)^2} - V(x_j) \right) \frac{t}{n}}}{(2\pi i \hbar t / n)^{nd/2}} \psi_0(x_0) dx_0 \ldots dx_{n-1}$$

cannot be interpreted as integrals of a cylinder function with respect to a complex measure μ_F on the space of paths $\Omega : \{\gamma : [0, +\infty) \to \mathbb{R}\}$ whose values on cylinder set is given by

$$\mu_F \left(\{\gamma \in \Omega : \gamma(t_1) \in I_1, \ldots, \gamma(t_n) \in I_n\} \right)$$

$$= \int_{I_1 \times \cdots \times I_n} \frac{e^{\frac{i}{2\hbar} \sum_{i=1}^{n} \frac{(\tau_i - \tau_{i-1})^2}{t_i - t_{i-1}}}}{(2\pi i \hbar)^{n/2} \sqrt{t_1 (t_2 - t_1) \cdots (t_n - t_{n-1})}} dx_1 \ldots dx_n \quad (4.1)$$

(where $n \in \mathbb{N}$, $0 < t_1 < \cdots < t_n < +\infty$ and $I_1, \ldots, I_n \in \mathcal{B}(\mathbb{R})$).

At this point it is clear that the construction of Feynman integration theory requires to change point of view.

Let us recall that by Riesz-Markov Theorem 2.20, on locally compact Hausdorff topological spaces X there is a one-to-one correspondence between complex regular Borel measures μ and linear continuous functionals $L_\mu : C_0(X) \to \mathbb{C}$, given by the identity

$$L_\mu(f) = \int_X f(x) \, d\mu(x) \qquad f \in C_0(X). \quad (4.2)$$

On the other hand this result relies on the assumption of local compactness of the space X which is actually a rather strong condition which cannot be easily fulfilled in the case of infinite dimensional topological vector spaces. It cannot hold, for instance, in the case of Hilbert and Banach spaces.

In fact, in general the concept of linear functional on a suitable domain is wider than the concept of Lebesgue integral and can be used to give a mathematical meaning to Feynman formulas, in the spirit of Daniell's approach to integration (see [274]). This chapter presents a systematic implementation of infinite dimensional integration theory based on the concept of *projective system of functionals* instead of projective systems of measures [39]. Similarly to the construction proposed in Kolmogorov Theorem 3.7, the functional will be constructed out of its finite dimensional approximations. This point of view allows to provide a unified view of infinite dimensional integration which covers both the case of probabilistic and oscillatory integration. In particular, according to this approach the Feynman path integral will be constructed as a linear functional $L_F : D(L_F) \to \mathbb{C}$ defined on a suitable domain of "integrable functions" $f : \Gamma \to \mathbb{C}$ defined over a space of paths Γ. The property characterizing the functional as "Feynman integral" will be the following action over the cylinder functions $f \in D(L_F)$

$$L_F(f) = \int_{\mathbb{R}^n} \frac{g(x_1, \ldots, x_n) e^{\frac{i}{2\hbar} \sum_{i=1}^{n} \frac{(x_i - x_{i-1})^2}{t_i - t_{i-1}}}}{(2\pi i \hbar)^{n/2} \sqrt{t_1(t_2 - t_1) \cdots (t_n - t_{n-1})}} dx_1 \ldots dx_n$$

with $f(\gamma) = g(\gamma(t_1), \ldots, \gamma(t_n))$, for some $n \geq 1$, $0 < t_1 < \cdots < t_n < +\infty$ and $g : \mathbb{R}^n \to \mathbb{C}$.

Sections 4.1 and 4.2 present the general theory of projective systems of functionals, while Section 4.3 focuses on those associated to projective systems of complex measures that do not admit a projective limit because of Theorem 3.8, also proposing some examples. Section 4.4 describes a particular construction of a functional associated to a projective systems of complex measures based on a Fourier transform approach. These results are the basis for the definition of *infinite dimensional Fresnel integrals*, which will be extensively described along with their applications to Feynman path integrals in Section 4.5. Eventually, Section 4.6 presents a generalization of infinite dimensional Fresnel integrals which allows to construct a class of Feynman-Kac type formulae for the representation of the solution of high-order heat-type equations.

4.1   Projective systems of functionals

Let $(\mathcal{I}, \leq)$ be a directed set and $\{S_J, \pi_J^K\}_{J,K \in \mathcal{I}}$ a projective family of sets. In the following we shall assume that $\{S_J, \pi_J^K\}_{J,K \in \mathcal{I}}$ is a perfect inverse system (in the sense of Definition 3.8). Let us denote with the symbol $\hat{S}_J$ the set of complex-valued functions $f_J : S_J \to \mathbb{C}$. For any pair $J, K \in \mathcal{I}$ such that $J \leq K$ let $\mathcal{E}_J^K : \hat{S}_J \to \hat{S}_K$ be the function which maps any element $f_J \in \hat{S}_J$ to its extension $\mathcal{E}_J^K(f_J)$ to $\hat{S}_K$, defined as

$$\mathcal{E}_J^K(f_J)(\omega_K) := f_J\big(\pi_J^K(\omega_K)\big), \qquad \omega_K \in S_K. \tag{4.3}$$

In particular, if $\{S_J, \pi_J^K\}_{J,K \in A}$ is a topological projective family and $f_J \in \hat{E}_J$ is a continuous function, then for any $K \in \mathcal{I}$ such that $J \leq K$, the extension $\mathcal{E}_J^K(f_J)$ is a continuous function on S_K.

We shall denote with the symbol $Map(\hat{S}_J)$ the family of linear maps $L_J : \hat{S}_J^0 \to \mathbb{C}$ from subspaces $\hat{S}_J^0 \subseteq \hat{S}_J$ to $\mathbb{C}$, called *functionals*. The set $\hat{S}_J^0$ is called the *domain of* L_J and it is required to be a linear space.

For $J, K \in \mathcal{I}$, such that $J \leq K$, let us define the map $\hat{\pi}_J^K : Map(\hat{S}_K) \to Map(\hat{S}_J)$ as the transport of functionals $L_K \in Map(\hat{S}_K)$ induced by the map $\mathcal{E}_J^K$ from $\hat{S}_J$ to $\hat{S}_K$, which is actually given by:

$$\hat{\pi}_J^K(L_K)(f_J) := L_K(\mathcal{E}_J^K(f_J)), \qquad L_K \in Map(\hat{S}_K), \tag{4.4}$$

where the domain of $\hat{\pi}_J^K(L_K)$ is given by

$$Dom(\hat{\pi}_J^K(L_K)) = \{f_J \in \hat{E}_J, |\mathcal{E}_J^K(f_J) \in \hat{S}_K^0\}.$$

Definition 4.1. A family of functionals $\{L_J, \hat{S}_J^0\}_{J \in \mathcal{I}}$ labelled by the elements of the directed set $\mathcal{I}$ is called a *projective system of functionals* if for all $J, K \in \mathcal{I}$ with $J \leq K$ the projective (or coherence or compatibility) conditions hold

$$\mathcal{E}_J^K(f_J) \in \hat{S}_K^0, \qquad \forall f_J \in \hat{S}_J^0,$$

$$\hat{\pi}_J^K(L_K)(f_J) = L_J(f_J), \qquad \forall f_J \in \hat{S}_J^0. \tag{4.5}$$

Let $S_\mathcal{I} := \varprojlim S_J$ be the projective limit of the projective family of sets $\{S_J, \pi_J^K\}_{J,K \in \mathcal{I}}$ and let $\hat{S}_\mathcal{I}$ denote the set of complex-valued functions on $S_\mathcal{I}$. A function $f_J \in \hat{S}_J$, $J \in \mathcal{I}$, can be extended to a function $\mathcal{E}_J^\mathcal{I} f_J \in \hat{S}_\mathcal{I}$ defined by

$$\mathcal{E}_J^\mathcal{I} f_J(\omega) := f_J(\pi_J \omega), \qquad \omega \in S_\mathcal{I}.$$

By Eq. (3.40), the extension maps $\mathcal{E}_J^\mathcal{I} : \hat{S}_J \to \hat{S}_\mathcal{I}$ satisfy the following condition for any $J, K \in \mathcal{I}$, with $J \leq K$:

$$\mathcal{E}_J^\mathcal{I} = \mathcal{E}_K^\mathcal{I} \circ \mathcal{E}_J^K. \tag{4.6}$$

In particular, if $(S_J, \pi_J^K)_{J,K \in \mathcal{I}}$ is a topological projective family, then all the extensions $\mathcal{E}_J^K : \hat{S}^J \to \hat{S}_\mathcal{I}$ and $\mathcal{E}_J^\mathcal{I} : \hat{S}^J \to \hat{S}_\mathcal{I}$ map continuous functions into continuous functions.

In this context we shall call *cylindrical (or cylinder) functions* the mappings $f : S_\mathcal{I} \to \mathbb{C}$ of the form $\mathcal{E}_J^\mathcal{I} f_J$ for some $J \in \mathcal{I}$ and some $f_J \in \hat{S}_J$. The family of cylinder functions will be denoted with the symbol $\mathcal{C} = \cup_{J \in \mathcal{I}} \mathcal{E}_J^\mathcal{I}(\hat{S}_J)$.

Remark 4.1. If $\{S_J, \pi_J^K\}_{J,K \in A}$ is a perfect inverse system, then it is simple to prove the injectivity of the extension map $\mathcal{E}_J^\mathcal{I} : \hat{S}_J \to \hat{S}_\mathcal{I}$. Indeed for any $J \in \mathcal{I}$, $f, g \in \hat{S}_J$:

$$f = g \quad \Leftrightarrow \quad \mathcal{E}_J^\mathcal{I} f = \mathcal{E}_J^\mathcal{I} g.$$

Given a projective system of functionals $\{L_J, \hat{S}_J^0\}_{J \in \mathcal{I}}$, we shall denote by $\mathcal{C}_0 \subset \mathcal{C}$ the set of those cylindrical functions obtained as extensions to $S_\mathcal{I}$ of functions belonging to the domain of L_J for some $J \in \mathcal{I}$, i.e.:

$$\mathcal{C}_0 := \cup_{J \in \mathcal{I}} \mathcal{E}_J^\mathcal{I}(\hat{S}_J^0) = \{f \in \mathcal{C} : f = \mathcal{E}_J^\mathcal{I} f_J, \text{ for some } J \in \mathcal{I}, f_J \in \hat{S}_J^0\}. \quad (4.7)$$

Definition 4.2. A *projective extension* $(L, D(L))$ of a projective system of functionals $\{L_J, \hat{S}_J^0\}_{J \in \mathcal{I}}$ is a functional $L : D(L) \to \mathbb{C}$, with $D(L) \subset \hat{E}_\mathcal{I}$ such that

- $\mathcal{C}_0 \subseteq D(L)$,
- $L(\mathcal{E}_J^\mathcal{I} f_J) = L_J(f_J)$, for all $f_J \in \hat{S}_J^0$.

Theorem 4.1. *Let $\{S_J, \pi_J^K\}_{J,K \in \mathcal{I}}$ be a perfect inverse system and let $\{L_J, \hat{S}_J^0\}_{J \in \mathcal{I}}$ be a projective system of functionals. Then the functional $(L, D(L))$ defined by:*

$$D(L) := \mathcal{C}_0 \quad (4.8)$$

$$L(f) := L_J(f_J), \qquad f = \mathcal{E}_J^\mathcal{I} f_J, \, f_J \in \hat{E}_J^0. \quad (4.9)$$

is a projective extension of $\{L_J, \hat{S}_J^0\}_{J \in \mathcal{I}}$.

Proof. The functional L is unambiguously defined. Indeed let us consider function $f \in \mathcal{C}_0$ that can be obtained both as the extension of $f_J \in \hat{S}_J^0$ and of $f_K \in \hat{S}_K^0$, i.e., $f = \mathcal{E}_J^\mathcal{I} f_J = \mathcal{E}_K^\mathcal{I} f_K$, $J, K \in \mathcal{I}$. By the very definition of directed set, there exists an $H \in \mathcal{I}$ such that $J \leq H$ and $K \leq H$. If we define the two functions $f_H := \mathcal{E}_J^H f_J$ and $g_H := \mathcal{E}_K^H f_K$, it is easy to see that $\mathcal{E}_H^\mathcal{I} f_R = \mathcal{E}_H^\mathcal{I} g_R = f$. Thus, by Remark 4.1, $f_R = g_R$ and $L_R(f_R) = L_R(g_R)$. On the other hand, by the projectivity condition (4.5) it follows that:

$$L_J(f_J) = L_H(\mathcal{E}_J^H f_J) = L_H(f_H) = L_H(g_H) = L_H(\mathcal{E}_K^H f_K) = L_K(f_K).$$

Thus L is unambiguously defined by $L(f) = L_J(f_J) = L_k(f_K)$. $\qquad \square$

The functional $(L, D(L))$ defined by (4.8) and (4.9) is called the *minimal extension* of the projective system of $\{L_J, \hat{S}_J^0\}_{J \in \mathcal{I}}$ since any other extension $(L', D(L'))$ of $\{L_J, \hat{S}_J^0\}_{J \in \mathcal{I}}$ enjoys the following properties

$$D(L) \subseteq D(L') \tag{4.10}$$

$$L'(f) = L(f) \quad \forall f \in D(L). \tag{4.11}$$

In the following the minimal extension will be denoted with the symbol $(L_{min}, D(L_{min}))$. We shall also write $L_{min} = \varprojlim L_J$.

An interesting characterisation of the domain of the minimal extension of a projective system of functionals $\{L_J, \hat{E}_J^0\}_{J \in A}$ relies on the concept of direct or inductive limit of a direct system. We recall here their definitions (see also [290]).

Definition 4.3. Let $(\mathcal{I}, \leq)$ be a directed set. A *direct system* on $\mathcal{I}$ is a collection $(S_J, F_{JK})_{J,K \in \mathcal{I}}$, where $\{S_J\}_{J \in \mathcal{I}}$ is a family of (non-empty) sets indexed by the elements of $\mathcal{I}$ and $\{F_{JK}\}_{J \leq K}$ a collection of mappings $F_{JK} : E_J \to E_K$ defined for all pairs $J, K \in \mathcal{I}$, $J \leq K$, and enjoying the following properties:

- for every $J \in \mathcal{I}$, F_{JJ} is the identity of E_J,
- $F_{KH} \circ F_{JK} = F_{JH}$ for all $J \leq K \leq H$.

Let us consider on the disjoint union $\cup_J E_J$ the equivalence relation $\sim$ as $\omega_J \sim \omega_K$, with $\omega_J \in E_J$ and $\omega_K \in E_K$, if there is some $H \in \mathcal{I}$ such that $J \leq H$, $K \leq H$, and $F_{JH}(\omega_J) = F_{KH}(\omega_K)$.

Definition 4.4. The *direct (or inductive) limit* $\varinjlim S_J$ of the direct system $(S_J, F_{JK})_{J,K \in \mathcal{I}}$ is the quotient set $\varinjlim E_J = \cup_J E_J / \sim$ of the equivalence relation $\sim$.

If the sets $(S_J)_{J \in \mathcal{I}}$ are topological spaces and the maps $(F_{JK})_{J,K \in \mathcal{I}}$ are continuous, the family $(S_J, F_{JK})_{J,K \in \mathcal{I}}$ is called a *topological direct system*. Its direct limit is the space $\varinjlim E_J$ endowed with the finest topology making all the maps $F_J : E_J \to \varinjlim E_J$ continuous.

Given a direct system $(S_J, F_{JK})_{J,K \in \mathcal{I}}$ and its direct limit $\varinjlim S_J$, let $F_J : S_J \to \varinjlim S_J$ be the function which maps each element of S_J into its equivalence class. It is easy to verify that the maps $\{F_J)_{J \in \mathcal{I}}$ and $\{F_{JK}\}_{J \leq K}$ enjoy the following composition property

$$F_J = F_K \circ F_{JK} \qquad J, K \in \mathcal{I}, J \leq K.$$

An application of the concept of Definitions 4.3 and 4.4 is found in considering a projective system of functionals $(L_J, \hat{S}_J^0)_{J \in \mathcal{I}}$. Indeed, the family of sets $\{\hat{S}_J^0\}_{J \in \mathcal{I}}$ and the collection $\{\mathcal{E}_J^K\}_{J \leq K}$ of extension maps

$$\mathcal{E}_J^K : \hat{S}_J^0 \to \hat{S}_K^0, \qquad J \leq K, J, K \in A$$

form a direct system of sets. Its direct limit $\varinjlim \hat{S}_J^0$ is isomorphic to the set $\mathcal{C}_0$ of cylindrical functions defined in (4.7), as stated in the following theorem.

Theorem 4.2. *Let* $\{S_J, \pi_J^K\}_{J,K \in \mathcal{I}}$ *be a perfect inverse system, let* $\{L_J, \hat{S}_J^0\}_{J \in \mathcal{I}}$ *be a projective system of functionals and* $(L_{min}, D(L_{min}))$ *its minimal extension. Then* $D(L_{min})$ *is isomorphic to the direct limit* $\varinjlim \hat{S}_0^J$ *of the direct system* $\{\hat{S}_J^0, \mathcal{E}_J^K\}_{J,K \in \mathcal{I}}$.

Proof. We have to prove a one-to-one correspondence between $D(L_{min}) = \mathcal{C}_0$ and $\varinjlim \hat{S}_0^J$. Let us consider an arbitrary element $[f] \in \varinjlim \hat{E}_0^J$, i.e. an equivalence class of functions such that for any two elements f_J and f_K belonging to $[f]$ there is an $H \in \mathcal{I}$, with $J \leq H$ and $K \leq H$, such that $\mathcal{E}_J^H(f_J) = \mathcal{E}_K^H(f_K)$. In fact, all the elements belonging to this equivalence class $[f]$ define a unique cylindrical function. Indeed, given two points $f_J, f_K \in [f]$ and considering the associated cylindrical functions $\mathcal{E}_J^\mathcal{I} f_J$ and $\mathcal{E}_K^\mathcal{I} f_K$, it is simple to check that for any $\omega \in S_\mathcal{I}$ we have $\mathcal{E}_J^A f_J(\omega) = \mathcal{E}_K^A f_R(\omega)$. In order to show this equality it is sufficient to take an element $H \in \mathcal{I}$, with $J \leq H$ and $K \leq H$, such that $\mathcal{E}_J^H(f_J) = \mathcal{E}_K^H(f_K)$ (such an H exists as f_J and f_K belong to the same equivalence class g), and apply Eq. (4.6), which gives

$$\mathcal{E}_J^\mathcal{I} f_J(\omega) = \mathcal{E}_H^\mathcal{I} \circ \mathcal{E}_J^H f_J(\omega) = \mathcal{E}_H^\mathcal{I} \circ \mathcal{E}_K^H f_K(\omega) = \mathcal{E}_K^\mathcal{I} f_K(\omega) \,.$$

Conversely, under the assumption that $\{S_J, \pi_J^K\}_{J,K \in \mathcal{I}}$ is a perfect inverse system, to any cylindrical function belonging to $\mathcal{C}_0$ it is possible to associate a unique element of $\varinjlim \hat{E}_0^J$.

Indeed, let $f \in \mathcal{C}_0$ of the form $f = \mathcal{E}_J^\mathcal{I} f_J$, $J \in A$, $f_J \in \hat{S}_J^0$, and let $[f_J] \in \varinjlim \hat{E}_0^J$ be the equivalence class of f_J. If the cylindrical function f can be also be obtained as $f = \mathcal{E}_K^\mathcal{I} f_K$, for some $K \in \mathcal{I}$, $f_K \in \hat{S}_K^0$, then necessarily the following equality holds $[f_K] = [f_J]$, i.e., there exists an $H \in \mathcal{I}$, with $J \leq H$ and $K \leq H$ such that $\mathcal{E}_J^H f_J = \mathcal{E}_K^H f_K$. In order to prove this, it is sufficient to consider a set $H \in \mathcal{I}$ with $J \leq H$ and $K \leq H$ and the functions $f_H, f_H' \in \hat{S}_H^0$ of the form $f_H := \mathcal{E}_J^H f_J$ and $f_H' := \mathcal{E}_K^H f_K$. By Eq. (4.6) it follows that $\mathcal{E}_H^\mathcal{I} f_H = \mathcal{E}_H^A f_H'$, indeed $\mathcal{E}_H^\mathcal{I} f_H = \mathcal{E}_H^\mathcal{I} \mathcal{E}_J^H f_J = \mathcal{E}_J^\mathcal{I} f_J = f$ and analogously $\mathcal{E}_H^\mathcal{I} f_R' = \mathcal{E}_H^\mathcal{I} \mathcal{E}_K^H f_K = \mathcal{E}_K^\mathcal{I} f_K = f$, and by Remark 4.1 it follows that $f_R = f_R'$ (since we assumed that (E_J, π_J^K) is a perfect inverse system). $\square$

Let $\mathcal{L} = \{(L, D(L))\}$ be the set of all possible projective extension of a projective system of functionals $\{L_J, \hat{S}_J^0\}_{J \in \mathcal{I}}$. Under the assumption that $(E_J, \pi_J^K)_{J,K \in A}$ is a perfect inverse system, the set $\mathcal{L}$ is not empty since in this case the minimal extension exists. In addition, it is possible to define a partial order relation $\leq$ on $\mathcal{L}$ as

$$L \leq L' \ \text{ if } \ D(L) \subset D(L') \text{ and } L(f) = L'(f) \ \forall f \in D(L). \tag{4.12}$$

Besides the study of the minimal extension, it is reasonable to investigate the existence of a *maximal extension* of a projective system of functionals $\{L_J, \hat{E}_J^0\}_{J \in A}$, i.e., a functional $(L_{max}, D(L_{max}))$ which cannot be further extended. More precisely, an extension $(L_{max}, D(L_{max})) \in \mathcal{L}$ is called maximal if the following holds

$$L_{max} \leq L \ \Rightarrow L = L_{max}.$$

Similarly, we can define a *maximum extension* $(\bar{L}, D(\bar{L}))$, as the functional such that $L \leq \bar{L}$ for all $L \in \tilde{L}$. The latter problem is closely related to the uniqueness property of the extensions of a projective system. Indeed, if there existed two extensions $(L, D(L))$ and $(L', D(L'))$ such that $L(f) \neq L'(f)$ for some $f \in D(L) \cap D(L')$, then it would be impossible to construct $\bar{L}$, as it would be ambiguously defined on the element f. In fact, this property is also related to the existence of $(L_{max}, D(L_{max}))$, as is stated in the following theorem.

Theorem 4.3. *Let* $\{L_J, \hat{E}_J^0\}_{J \in A}$ *be a projective system of functionals,* $\mathcal{L} = \{(L, D(L)\} \subset \mathcal{L}$ *be the family of its projective extensions and let us consider the following statements:*

(1) Existence of a maximum extension: *there exists a functional* $(\bar{L}, D(\bar{L})) \in \mathcal{L}$ *such that* $L \leq \bar{L}$ *for all* $L \in \mathcal{L}$.
(2) Uniqueness property: *if* $(L, D(L)), (L', D(L')) \in \tilde{\mathcal{L}}$ *and* $f \in D(L) \cap D(L')$, *then* $L(f) = L'(f)$.
(3) *Existence of a maximal extension:* *there exists a functional* $(L_{max}, D(L_{max})) \in \mathcal{L}$ *such that if* $L \in \mathcal{L}$ *and* $L_{max} \leq L$ *then* $L = L_{max}$.

Then $1 \Rightarrow 2 \Rightarrow 3$.

Proof. The implication $1 \Rightarrow 2$ is trivial, as remarked above. Concerning the implication $2 \Rightarrow 1$, it is sufficient to realise that $\mathcal{L}$ is a nonvoid partially ordered set, with partial order $\leq$ given by (4.12). Moreover any chain $\{L_\alpha, D(L_\alpha)\}_\alpha \subset \mathcal{L}$ has an upper bound $(\bar{L}; D(\bar{L}))$ given by

$$D(\bar{L})) = \cup_\alpha D(L_\alpha) \subset \hat{S}_{\mathcal{I}}, \qquad \bar{L}(f) = L_\alpha(f), \ f \in D(L_\alpha),$$

(the set inclusion being clear since all $D(L_\alpha)$ are in $\hat{S}_\mathcal{I}$). By Zorn's lemma we can deduce that $\mathcal{L}$ has a maximal element. $\square$

Given two extensions $(L, D(L))$ and $(L', D(L'))$ of a projective system of functionals $\{L_J, \hat{E}^0_J\}_{J \in A}$, in general it is not always true that they coincide on the intersection of their domains, i.e., that for any $f : E_A \to \mathbb{C}$ such that $f \in D(L) \cap D(L')$, one has that $L(f) = L'(f)$. Indeed let us consider the following counterexample.

Example 4.1. Let us consider Example 3.3 in the particular case where $I = \mathbb{N}$, hence $\mathcal{I} = \mathcal{F}(\mathbb{N})$ is the totality of non-empty finite subsets of $\mathbb{N}$. Given an element $J \in \mathcal{I}$, let S_J be given by $S_J = \mathbb{R}^J$. We also identify $S_\mathcal{I}$ with the set $\mathbb{R}^\mathbb{N}$ of all real-valued sequences.
Let μ denote a probability measure on the Borel σ-algebra $\mathcal{B}(\mathbb{R})$ and let $(L_J, \hat{S}^0_J)$ be the family of functionals defined by

$$\hat{S}^0_J = B_b(S_J)$$
$$L_J(f) = \int_{S_J} f \, d\mu_J, \qquad f \in \hat{S}^0_J$$

where $B_b(S_J)$ is the set of Borel bounded functions on S_J and $\mu_J = \times_{n \in J} \mu$ is the product measure of $|J| < \infty$ copies of μ ($|J|$ denoting the cardinality of $J \in \mathcal{I}$). By Kolmogorov's existence Theorem 3.4 there exists a probability measure $\mu_\mathcal{I}$ on the σ-algebra generated by the cylindrical sets in $S_\mathcal{I} = \mathbb{R}^\mathbb{N}$ obtained as the product $\times_{n \in \mathbb{N}} \mu$. In particular, if μ is the standard centered Gaussian measure on $\mathcal{B}(\mathbb{R})$, the subset $l_2 = \{\omega \in \mathbb{R}^\mathbb{N}, \, | \sum_n |\omega_n|^2 < \infty\}$ has zero $\mu_\mathcal{I}$-measure (see Example 3.2).
Let us consider the following two different extensions $(L, D(L))$ and $(L', D(L'))$ of the projective system of functionals $\{L_J, \hat{E}^J_0\}_{J \in A}$. The functional L is defined by:

$$D(L) = B_b(\mathbb{R}^\mathbb{N})$$
$$L(f) = \int_{\mathbb{R}^\mathbb{N}} f(\omega) \, d\mu_\mathcal{I}(\omega), \quad f \in D(L) \qquad (4.13)$$

where $B_b(\mathbb{R}^\mathbb{N})$ denotes the Borel bounded functions over $\mathbb{R}^\mathbb{N}$.
Denote by $\pi_n : \mathbb{R}^\mathbb{N} \to \mathbb{R}^\mathbb{N}$ the projection of a sequence $\omega \in \mathbb{R}^\mathbb{N}$ to the sequence $\pi_n \omega$ given by

$$(\pi_n \omega)_m = \begin{cases} \omega_m & \text{if } m \leq n \\ 0 & \text{otherwise.} \end{cases}$$

Let us consider the functional L' given by

$$D(L') = \{f : \mathbb{R}^\mathbb{N} \to \mathbb{C} \mid \exists \lim_{n \to \infty} \int f(\pi_n \omega) \, d\mu_\mathcal{I}(\omega)\},$$

$$L'_J(f) = \lim_{n \to \infty} \int f(\pi_n \omega) d\mu_\mathcal{I}(\omega). \tag{4.14}$$

Let us now consider the function $f : \mathbb{R}^\mathbb{N} \to \mathbb{C}$ defined as

$$f(\omega) = \begin{cases} 0 & \text{if } \omega \text{ has an infinite number of non-vanishing components} \\ 1 & \text{otherwise.} \end{cases}$$

In fact, f is the characteristic function of the set

$$B = \{\omega \in \mathbb{R}^\mathbb{N} \mid \omega \text{ has a finite number of non-vanishing components } \}. \tag{4.15}$$

One can easily verify that B is a measurable subset of $\mathbb{R}^\mathbb{N}$, indeed it belongs to the σ-algebra generated by the cylindrical sets, as:

$$B = \cup_{N \in \mathbb{N}} \cap_{n > N} B_n$$

where

$$B_n := \{\omega \in \mathbb{R}^\mathbb{N} \mid \omega_n = 0\}.$$

Then one has, for $\mu_\mathcal{I}$ the above Gaussian measure

$$L(f) = \mu_\mathcal{I}(B)$$
$$< \mu_\mathcal{T}(l_2) = 0$$

while

$$L'(f) = \lim_{n \to \infty} \int_{\mathbb{R}^n} f(\pi_n \omega) \, d\mu_\mathcal{I}(\omega) = \lim_{n \to \infty} 1 = 1.$$

4.2 Continuous extensions of projective systems of functionals

In the following we shall always assume, unless otherwise stated, that the projective family $(S_J, \pi_J^K)_{J,K \in \mathcal{I}}$ is a perfect inverse system, in such a way that, by Theorem 4.1, the minimal extension $(L_{min}, D(L_{min}))$ of a projective system of functionals $\{L_J, \hat{S}_J^0\}_{J \in \mathcal{I}}$ is well-defined. We shall study its possible continuous extensions.

Let us consider the set of cylindrical functions $\mathcal{C}_0$ associated to a projective system of functionals $\{L_J, \hat{S}_J^0\}_{J \in \mathcal{I}}$ and let us endow the space $\hat{S}_\mathcal{I}$ of complex-valued functions on the projective limit $S_\mathcal{I}$ with a topology τ

making $(\hat{S}_\mathcal{I}, \tau)$ a topological vector space.[1] We say that the minimal extension $(L_{min}, D(L_{min}))$ is *closable* in τ if the closure of the graph $\mathcal{G}(L_{min})$ of L_{min}

$$\mathcal{G}(L_{min}) := \{(f, L_{min}(f)) \in \hat{S}_\mathcal{I} \times \mathbb{C} : f \in D(L_{min})\}$$

in $\hat{S}_\mathcal{I} \times \mathbb{C}$ with respect to the product topology $\tau \times \tau_\mathbb{C}$, $\tau_\mathbb{C}$ denoting the standard topology in $\mathbb{C}$, is the graph of a well-defined functional. In this case we define the closure of L_{min} in τ as the functional $\bar{L}_\tau$ whose graph satisfies $\mathcal{G}(\bar{L}_\tau) = \overline{\mathcal{G}(L_{min})}$.

If $(\hat{S}_\mathcal{I}, \tau)$ satisfies the first axiom of countability, i.e., if any $f \in \hat{S}_\mathcal{I}$ has a countable neighbourhood basis, then the closability condition is equivalent to the requirement that for any sequence $f_n \in \mathcal{C}_0$ converging to 0 and such that $L_{min}(f_n) \to z$, $z \in \mathbb{C}$, it follows that $z = 0$.

In this case the closure $(\bar{L}_\tau, D(\bar{L}_\tau))$ of $(L_{min}, \mathcal{C}_0)$ in the τ-topology is given by:

$$D(\bar{L}_\tau) := \{f \in \hat{S}_\mathcal{I} \mid \text{ there exist } f_n \in \mathcal{C}_0, f_n \to f, L_{min} f_n \to z\}$$
$$\bar{L}_\tau(f) := \lim_n L_{min}(f_n) = z, \quad f \in D(\bar{L}_\tau).$$

$\bar{L}_\tau$ is well-defined by the closability condition, since the value of $\bar{L}_\tau(f)$ for $f \in D(\bar{L}_\tau)$ does not depend on the sequence $\{f_n\} \subset \mathcal{C}_0$ converging to f.

Let us now assume that $L_{min} : \mathcal{C}_0 \to \mathbb{C}$ is continuous[2] in the τ-topology on $\mathcal{C}_0$, i.e., if for any $f \in \mathcal{C}_0$ and for any $\epsilon > 0$ there exists a neighbourhood U of f (depending of f and ϵ) such that for any $g \in U \setminus \{f\}$, with $g \in \mathcal{C}_0$, one has $|L_{min}(f) - L_{min}(g)| < \epsilon$. Then it is easy to verify that L_{min} is closable, as stated in the following theorem.

Theorem 4.4. *Let $L_{min} : \mathcal{C}_0 \to \mathbb{C}$ be continuous in the τ-topology on $\mathcal{C}_0$. Then L_{min} is closable.*

Proof. Let us consider two points $(g, \alpha), (g, \beta) \in \overline{\mathcal{G}(L_{min})}$, and assume *ad absurdum* that $\alpha \neq \beta$ and set $\epsilon \equiv \frac{|\alpha - \beta|}{2}$. In this case for any neighborhood $U(0)$ of 0, it would be possible to find an element $\tilde{f} \in U(0) \cap \mathcal{C}_0$ such that $L(\tilde{f}) > \epsilon$, thus contradicting the assumed continuity of the functional L_{min}. By the properties of the topological vector spaces, given the neighborhood $U(0)$ of 0, then it is possible to find two neighborhoods $V(0), W(0)$ of 0 such that if $f_1 \in V(0)$ and $f_2 \in W(0)$, then $f_1 - f_2 \in U(0)$. Let

---

[1] We require a topological vector space in order to assure later on continuity under limits of sums and products by scalars.

[2] We recall that in a topological vector space if a functional is continuous in one point, then it is continuous everywhere

$V(g) := V(0) + g$ and $W(g) := W(0) + g$ be the neighborhoods of g obtained by translating $V(0)$ and $W(0)$ by g. Since $(g, \alpha), (g, \beta) \in \overline{\mathcal{G}(L_{min})}$, there is an $f \in V(g) \cap \mathcal{C}_0$ and an $f' \in W(g) \cap \mathcal{C}_0$ such that $|L(f) - \alpha| < \epsilon$ and $|L(f') - \beta| < \epsilon$. Then we have that $(f - g) \in V(0)$, $f' - g \in W(0)$ and $(f - g) - (f' - g) = f - f' \in U(0)$. Moreover $L(f - f') > \epsilon$, and L cannot be continuous in 0, against the assumption. $\square$

Let us consider now two different topologies τ_1 and τ_2 on the set $\mathcal{C}_0$, with τ_1 coarser than τ_2. If $(L_{min}, D(L_{min}))$ is continuous in the τ_1-topology, then it is continuous also in the τ_2-topology and the closability in τ_1 implies the closability in τ_2. More generally, if τ_1 is coarser than τ_2, then $\overline{\mathcal{G}(L_{min})^{\tau_2}} \subset \overline{\mathcal{G}(L_{min})^{\tau_1}}$ and if $\overline{\mathcal{G}(L_{min})^{\tau_1}}$ is the graph of a well-defined functional, then $\bar{L}_{\tau_1}$ and $\bar{L}_{\tau_2}$ are both well-defined and $\bar{L}_{\tau_1}$ is a extension of $\bar{L}_{\tau_2}$.

Conversely, if τ_1 and τ_2 are not comparable, then there might exists a $f \in D(L_1) \cap D(L_2)$ such that $L_1(f) \neq L_2(f)$. This is the case of Example 4.1. Indeed, in this case the functional L defined by (4.13) is continuous on $L^1(\mathbb{R}^{\mathbb{N}}, \mu)$, while the second functional (4.14) is continuous on the space $D(L')$ endowed with the norm

$$\|f\|_{L'} := \lim_{n \to \infty} \int |f(\pi_n \omega)| d\mu(\omega).$$

One can easily see that these two norms are not comparable, as if we consider the function $f = \chi_B$, with B defined by (4.15) in Example 4.1, then $\|f\|_{L^1(\mathbb{R}^{\mathbb{N}}, \mu_A)} = 0$ and $\|f\|_{L'} = 1$, while if we consider the function $g - \chi_{\mathbb{R}^{\mathbb{N}}} - \chi_B$, then $\|g\|_{L^1(\mathbb{R}^{\mathbb{N}}, \mu_A)} = 1$ and $\|g\|_{L'} = 0$.

4.3   Projective systems of complex measure spaces and associated functionals

Let us consider a projective family of measurable spaces $\{S_J, \mathcal{F}_J, \pi_J^K)\}$ and assume that for any $J \in \mathcal{I}$ a complex measure $\mu_J : \mathcal{F}_J \to \mathbb{C}$ is defined. We shall adopt the usual notation and denote $L^1(S_J, \mathcal{F}_J, \mu_J)$ the subset of $\hat{S}_J$ consisting of (real resp. complex) functions on S_J which are μ_J-integrable.

Let us consider the family of functionals $\{L_J, \hat{S}_J^0\}_{J \in \mathcal{I}}$, given by

$$\hat{S}_J^0 := L^1(S_J, \mathcal{F}_J, \mu_J), \qquad L_J(f) := \int_{S_J} f \, d\mu_J, \, f \in \hat{S}_J^0. \tag{4.16}$$

In this case L_J is a linear functional (real- resp. complex-valued).

Remark 4.2. We point out that for any choice of $J, K \in \mathcal{I}$, with $J \leq K$, the function $\mathcal{E}_J^K : \hat{S}_J \to \hat{S}_K$ maps a $\mathcal{F}_J$-measurable function $f_J \in \hat{S}_J$ into a $\mathcal{F}_K$-measurable function $\mathcal{E}_J^K f_J \in \hat{S}_K$. However in general it is not true that

$$\mathcal{E}_J^K(\hat{S}_J^0) \subseteq \hat{S}_K^0 \qquad \forall J, K \in \mathcal{I} \; J \leq K . \tag{4.17}$$

The condition (4.17) is automatically fulfilled in the case where for every $J \in \mathcal{I}$ the domain $\hat{S}_J^0$ is defined as the set of all bounded measurable functions on $(S_J, \mathcal{F}_J)$. On the other hand, if $\hat{S}_J^0 = L^1(S_J, \mathcal{F}_J, \mu_J)$, a sufficient condition for (4.17) to be fulfilled is the consistency property (3.41) of the family of measures $\{\mu_J\}_{J \in \mathcal{I}}$.

According to Definition 4.1, the collection $\{L_J, \hat{S}_J^0\}_{J \in \mathcal{I}}$ is a projective system of functionals if for all $J \leq K$, $K, J \in \mathcal{I}$, $f_J \in \hat{S}_J^0$ the following compatibility conditions hold:

$$\mathcal{E}_J^K(f_J) \in \hat{S}_K^0, \tag{4.18}$$

$$\hat{\pi}_J^K(L_K)(f_J) = L_J(f_J) . \tag{4.19}$$

Due to the relation between L_J and μ_J and (4.3), one can easily see that (4.19) implies

$$\int_{S_J} f_J \, d\mu_J = \int_{S_K} f_J \circ \pi_J^K \, d\mu_K, \qquad f_J \in L^1(E_J, \Sigma_J, \mu_J). \tag{4.20}$$

For f_J taken to be the characteristic function of a set $A_J \in \mathcal{F}_J$ this gives

$$(\mu_K)_{\pi_J^K} = \mu_J, \tag{4.21}$$

which is the usual projectivity property (3.41) for measures on projective spaces. Hence $\{\mu_J\}_{J \in A}$ turns out to be a projective family of measures. Conversely, if (4.21) holds, by Remark 4.2 both (4.18) and (4.19) are fulfilled.

By Theorem 3.7, if $\{\mu_J\}_{J \in \mathcal{I}}$ is a consistent family of inner regular Borel probability measures on a topological and sequentially maximal projective family of Hausdorff measurable spaces $\{S_J, \mathcal{F}_J, \pi_J^K\}$, then there exists a probability measure μ on $(S_\mathcal{I}, \mathcal{F}_\mathcal{I})$ such that $\mu_J = \mu_{\pi_J}$ for all $J \in \mathcal{I}$. In this case the projective system of functionals $\{L_J, \hat{S}_J^0\}_{J \in \mathcal{I}}$ given by (4.16) admits a projective extension L given by

$$D(L) := L^1(S_\mathcal{I}, \mathcal{F}_\mathcal{I}, \mu)$$

$$L(f) := \int_{S_\mathcal{I}} f \, d\mu, \qquad f \in D(L) . \tag{4.22}$$

The same result holds if $\{\mu_J\}_{J \in \mathcal{I}}$ is a consistent family of complex measures fulfilling the assumptions of Theorem 3.9.

On the other hand, if the consistent family $\{\mu_J\}_{J \in \mathcal{I}}$ does not enjoy the following property

$$\sup_{J \in \mathcal{I}} |\mu_J| < +\infty$$

by Theorem 3.8 there cannot exists a measure μ on $(S_\mathcal{I}, \mathcal{F}_\mathcal{I})$ such that $\mu_J = \mu_{\pi_J}$ for all $J \in \mathcal{I}$. In this case the projective system of functionals $\{L_J, \hat{S}_J^0\}_{J \in \mathcal{I}}$ defined by (4.16) cannot admit a projective extension $(L, D(L))$ of the form (4.22).

This problem has been discussed in Section 3.5, where the impossibility of constructing a "Feynman measure" is proved. In the following we shall discuss in detail two examples, both deeply related to the mathematical definition of Feynman path integrals

4.3.1   *Fresnel integrals*

Let us consider the function $h : \mathbb{R}^n \to \mathbb{C}$ defined as

$$h(x) := \frac{e^{\frac{i}{2} \|x\|^2}}{(2\pi i)^{n/2}} \qquad x \in \mathbb{R}^n \,,$$

Clearly h is not integrable with respect the Lebesgue measure dx on $\mathbb{R}^n$ and, strictly speaking, it cannot exists a complex measure μ_n absolutely continuous with respect to dx with density h. On the other hand, $\chi_B h \in L^1(\mathbb{R}^n, dx)$ for every bounded Borel set $B \in \mathcal{B}(\mathbb{R}^n)$, χ_B being the indicator function of B, hence for every $R > 0$ the function $\chi_{[-R,R]^n} h$ is the density of a complex measure $\mu_{n,R}$ absolutely continuous with respect to the Lebesgue measure:

$$\mu_{n,R} := \chi_{[-R,R]^n} h \, dx \,. \tag{4.23}$$

In particular, the total variation $|\mu_{n,R}|$ is given by

$$|\mu_{n,R}| = \int |\chi_{[-R,R]^n}(x) h(x)| \, dx = (2R)^n \,.$$

Given a complex bounded Borel function $f : \mathbb{R}^n \to \mathbb{C}$, we shall define the integral $\int f h \, dx$ as the limit $\lim_{R \to \infty} \int f \, d\mu_{n,R}$, if this limit exists. In this case it will be denoted

$$\widetilde{\int}_{\mathbb{R}^n} f(x) \frac{e^{\frac{i}{2} \|x\|^2}}{(2\pi i)^{n/2}} \, dx$$

Clearly, whenever $f \in L^1(\mathbb{R}^n, dx)$ Lebesgue dominated convergence theorem gives

$$\widetilde{\int}_{\mathbb{R}^n} f(x) \frac{e^{\frac{i}{2} \|x\|^2}}{(2\pi i)^{n/2}} \, dx = \int f(x) h(x) \, dx \,.$$

Moreover, taking $f(x) = 1$, we obtain

$$\int_{\mathbb{R}^n} \frac{e^{\frac{i}{2}\|x\|^2}}{(2\pi i)^{n/2}}\, dx = 1\,.$$

For any $n \geq 1$ the complex measure $\mu_{n,R}$ can be represented as the product of n copies of the complex measure μ_R on $\mathbb{R}$, where $d\mu_R(x) := \chi_{[-R,R]}(x) \frac{e^{\frac{i}{2}\|x\|^2}}{\sqrt{2\pi i}} dx$.

Let $\mathcal{I} = \mathcal{F}(\mathbb{N})$ be the directed set of finite subsets of $\mathbb{N}$ and let us consider for any $J \in \mathcal{I}$ the set $S_J := \mathbb{R}^J$ endowed with the Borel σ-algebra and for fixed $R > 0$ the complex measure $\times_{n \in J} d\mu_R$. Let us define for any $J \in \mathcal{I}$ the functional $L_J : \hat{S}_J^0 \to \mathbb{C}$, given by:

$$\hat{S}_J^0 := \left\{ f \in \mathcal{B}_b(S_J) \ : \ \exists \lim_{R \to +\infty} \int f \times_{n \in J} d\mu_R \right\}$$

$$L_J(f) := \lim_{R \to +\infty} \int f \times_{n \in J} d\mu_R\,, \quad f \in \hat{S}_J^0.$$

One can easily verify that $(L_J, \hat{S}_J^0)_{J \in \mathcal{I}}$ is a projective system of functionals, however it is impossible to construct a projective extension on $S_{\mathcal{I}} = \varprojlim S_J \equiv \mathbb{R}^{\mathbb{N}}$ in terms of a (Lebesgue type) improper integral. Indeed, contrary to the case of finite dimension, if the infinite product measure $\times_{n \in \mathbb{N}} d\mu_R$ existed, then its total variation would be infinite for any $R > 1$.

More generally, let $(\mathcal{H}, \langle\,,\,\rangle)$ be a real separable Hilbert space and let $\mathcal{I}$ be the directed set of its finite-dimensional subspaces, where $V \leq W$ if V is a subspace of W, and $\pi_V^W : W \to V$ for $V \leq W$ is the orthogonal projection from W onto V. If for any $V \in \mathcal{I}$ we set $\mathcal{F}_V$ to be the Borel σ-algebra on V, then the collection $(V, \mathcal{F}_V, \pi_V^W)_{V,W \in A}$ provides a projective system of measure spaces.

For any $V \in A$ let $L_V : D(L_V) \to \mathbb{C}$ be the linear functional defined by

$$D(L_V) := \{ f \in \mathcal{B}_b(V) \ : \ \exists \lim_{R \to +\infty} \int_{[-R,R]^{|V|}} f(x) e^{\frac{i}{2}\|x\|^2} dx \}$$

$$L_V(f) := \lim_{R \to +\infty} \int_{[-R,R]^{|V|}} f(x) \frac{e^{\frac{i}{2}\|x\|^2}}{(2\pi i)^{|V|/2}} dx\,, \quad f \in \hat{D}(L_V).$$

where dx denotes the Lebesgue measure, $\|\,\|$ the norm and $|V|$ the dimension of V.

It is easy to verify that the family $(L_V, D(L_V))_{V \in A}$ is a projective system of linear functionals, but as remarked above it is not possible to define on the projective limit space $\varprojlim V$ a complex measure with finite variation on bounded sets obtained as the projective limit of the complex measures

$d\mu_V(x) := \chi_{[-R,R]^{|V|}}(x)\frac{e^{\frac{i}{2}\|x\|^2}}{(2\pi i)^{|V|/2}}dx$, $x \in V$. Consequently there cannot be an extension of the projective system of functionals $(L_V, D(L_V))_{V \in \mathcal{I}}$ of the form $L(f) = \int_{E_A} f(x)\mu(dx)$, even if f is supported in a product of bounded sets.

4.3.2 Feynman functionals

Let us consider the fundamental solution $G_t \in \mathcal{S}'(\mathbb{R}^d)$ of the Schrödinger equation for a non-relativistic quantum particle (with mass $m = 1$) moving freely in the d–dimensional Euclidean space:

$$i\hbar\frac{\partial}{\partial t}\psi(t,x) = -\frac{\hbar^2}{2}\Delta\psi(t,x) \qquad t \in \mathbb{R}, \, x \in \mathbb{R}^d \qquad (4.24)$$

($\hbar$ being the reduced Planck constant). For an initial datum $\psi(0,x) \equiv \psi_0(x)$ belonging to the space $\mathcal{S}(\mathbb{R}^d)$ of Schwartz test function, the solution of (4.24) is given by

$$\psi(t,x) = \int_{\mathbb{R}^d} G_t(x-y)\psi_0(y)dy,$$

where $G_0(x-y) = \delta(x-y)$, while for $t \neq 0$, G_t is the $C^\infty(\mathbb{R}^d)$ function

$$G_t(x-y) = \frac{e^{\frac{i}{2t\hbar}|x-y|^2}}{(2\pi it\hbar)^{d/2}}. \qquad (4.25)$$

In particular, for any $s, t \in \mathbb{R}$ the following equality holds:

$$\int_{\mathbb{R}^d} G_t(x-z)G_s(z-y)dz = G_{t+s}(x-y), \qquad (4.26)$$

where the integral on the left-hand side is meant as an improper Riemann integral.

Analogously to the construction of Wiener measure and Wiener integrals, let us consider the directed set $\mathcal{I} = \mathcal{F}(\mathbb{R}_+)$ of finite subsets of $\mathbb{R}_+$ and the projective family of sets (S_J, π_J^K), where $S_J := (\mathbb{R}^d)^J$ and for any $J \leq K$ the mapping π_J^K is the canonical projection from $\mathbb{R}^K$ to $\mathbb{R}^J$, as defined in Eq. (3.39).

For any $J \in \mathcal{I}$, $J = \{t_1, \ldots, t_n\}$ with $0 \leq t_1 \leq \cdots \leq t_n < \infty$, let us define the linear functional L_J with domain $\hat{S}_J^0 = L^1(S_J, dx)$ equal to the set of integrable functions with respect to the Lebesue measure dx on $(\mathbb{R}^d)^J$ and defined by :

$$L_J(f) := \int_{S_J} f(x_1, \ldots, x_n)G_{t_1}(x_1-x_0)\cdots G_{t_n-t_{n-1}}(x_n-x_{n-1})dx_1 \ldots dx_n,$$
$$(4.27)$$

where $x_0 \equiv 0$.

By using Eq. (4.26), it is easy to check that $(L_J, \hat{S}_J^0)$ is a projective family of functionals, however by Theorem 3.8 there cannot exist a projective extension $(L, D(L))$ of $(L_J, \hat{E}_J^0)$ of the form

$$D(L) = L^1((\mathbb{R}^d)^{[0,+\infty)}, d\mu),$$

$$L(f) = \int_{(\mathbb{R}^d)^{[0,+\infty)}} f \, d\mu \quad f \in D(L)$$

for a complex measure μ on $(\mathbb{R}^d)^{[0,+\infty)}$.

4.3.3 *Complex Markov kernels, pseudoprocesses and high-order heat-type equations*

The problems arising in the mathematical definition of Feynman path integrals and, more precisely, in the representation of the solution of Schödinger equation, pop up also in the generalization of Feynman-Kac formula to evolution equations that, contrary to the heat equation, do not admit a real and positive fundamental solution G_t. In order to tackle this issue in full generality, let us introduce the concept of complex Markov kernels, which generalizes the one of *Markov (resp. sub-Markov) kernel* (see, e.g., [56]).

Definition 4.5. Let $(S, \mathcal{F}_S)$ be a measurable space. A *complex kernel* G on $(S, \mathcal{F}_S)$ is a map $G : S \times \mathcal{F}_S \to \mathbb{C}$ with the following properties

   i. the map $x \in S \mapsto G(x, B)$ is measurable for each $B \in \mathcal{F}_S$,
  ii. the map $B \in \mathcal{F}_S \mapsto G(x, B)$ is a complex measure on $\mathcal{F}_S$ for each $x \in S$.

Given two complex kernels G_1, G_2 on (E, Σ), their composition $G_1 \circ G_2$ is defined as the complex kernel given by:

$$G_1 \circ G_2(x, B) := \int_S G_1(x, dx') G_2(x', B), \qquad x \in E, B \in \mathcal{F}_S.$$

Definition 4.6. A family of complex kernels $\{G_t\}_{t \geq 0}$ on a measurable space $(S, \mathcal{F}_S)$ with index set $\mathbb{R}_+$ is called a *semigroup of complex kernels* if

$$G_{t+s} = G_t \circ G_s \quad \forall t, s \in \mathbb{R}_+. \tag{4.28}$$

The semigroup law can be explicitly written in the following form

$$G_{t+s}(x, B) = \int_S G_t(x, dx') G_s(x', B), \qquad x \in S, B \in \mathcal{F}_S. \tag{4.29}$$

Given a semigroup of complex kernels and a complex measure ν on $(S, \mathcal{F}_S)$ such that $\nu(S) = 1$, it is possible to construct a projective family of complex measures. Let $\mathcal{I} = \mathcal{F}(\mathbb{R}_+)$ be the directed set of finite subsets of $\mathbb{R}_+$ and for any $J \in \mathcal{I}$ let us consider the Cartesian product $S_J := \times_{t \in J} S$ of $|J|$ copies of the set S, endowed with the product σ-algebra $\mathcal{F}_J := \times_{t \in J} \mathcal{F}_S$. Given a pair $J, K \in \mathcal{I}$ such that $J \leq K$, the projection map $\pi_J^K : S_K \to S_J$ will be defined by (3.39).

For $J = \{t_1, t_2, \ldots, t_n\}$, with $0 < t_1 < t_2 < \cdots < t_n < +\infty$, let μ_J be the complex measure on $(S_J, \mathcal{F}_J)$ given by

$$\mu_J(B_1 \times B_2 \times \cdots \times B_n) = \int_S \int_{B_1} \cdots \int_{B_n} G_{t_n - t_{n-1}}(x_{n-1}, dx_n)$$
$$\cdots G_{t_2 - t_1}(x_1, dx_2) G_{t_1}(x_0, dx_1) d\nu(x_0). \quad (4.30)$$

The semigroup property (4.29) allows to prove that or any if $J, K \in \mathcal{I}$, with $J \leq K$, one has $\mu_J = (\mu_K)_{\pi_J^K}$, i.e. $\{\mu_J\}$ is a projective family of complex measures. To this end it is sufficient to prove this conditions for $J, K \in \mathcal{I}$ such that $K \setminus J$ consists of exactly one element, since the case where $K \setminus J$ consists of $m > 1$ points can be obtained by constructing a chain $J \subset K_1 \subset K_2 \subset \ldots \subset K_m = K$ where each difference $K_i \setminus K_{i-1}$ is a singleton, and using the identity

$$\pi_J^K = \pi_{K_{m-1}}^K \circ \pi_{K_{m-2}}^{K_{m-1}} \circ \cdots \circ \pi_J^{K_1}.$$

Hence, let us consider a finite index set $J = \{t_1, t_2, \ldots, t_n\}$ and $K = J \cup \{t'\}$, with $t_i < t' < t_{i+1}$. We have to show that for any $B \in \mathcal{F}_J$ of the form $B = B_1 \times B_2 \times \cdots \times B_n$, with $B_i \in \Sigma$ for all $i = 1, \ldots, n$, the equality $\mu_J(B) = \pi_J^K \circ \mu_K(B)$ holds. Indeed, by the semigroup property (4.29) we have:

$$\int_S \int_{B_1} \cdots \int_{B_i} \int_{B_{i+1}} \cdots \int_{B_n} G_{t_n - t_{n-1}}(x_{n-1}, dx_n) \ldots G_{t_i - t_{i-1}}(x_{i-1}, dx_i)$$
$$G_{t_{i-1} - t_{i-2}}(x_{i-2}, dx_{i-1}) \ldots G_{t_1}(x_0, dx_1) d\nu(x_0)$$
$$= \int_S \int_{B_1} \cdots \int_{B_i} \int_S \int_{B_{i+1}} \cdots \int_{B_n} G_{t_n - t_{n-1}}(x_{n-1}, dx_n) \ldots G_{t_i - t'}(x', dx_i)$$
$$G_{t' - t_{i-1}}(x_{i-1}, dx') G_{t_{i-1} - t_{i-2}}(x_{i-2}, dx_{i-1}) \ldots G_{t_1}(x_0, dx_1) d\nu(x_0). \quad (4.31)$$

According to Theorem 3.8, a necessary condition for the existence of a complex measure μ projective limit of the projective family $\{\mu_J\}$ is a uniform bound on the total variation of the measures μ_J of the form

$$\sup_{J \in \mathcal{I}} |\mu_J| < \infty. \quad (4.32)$$

If K_t is a Markov kernel, i.e. if for any $t \in \mathbb{R}^+$ the map $B \in \mathcal{F}_S \mapsto G_t(x, B)$ is a probability measure on $\mathcal{F}_S$ for each $x \in S$, then condition (4.32) is trivially satisfied since in this case $|\mu_J| = 1$ for every $J \in \mathcal{I}$. In particular, if $(S, \mathcal{F}_S)$ is a Polish space then Kolmogorov Theorem 3.4 yields the existence of the projective limit measure μ on $(S^{\mathbb{R}^+}, \mathcal{F})$ associated to the projective family $(\mu_J)_{J \in A}$ of compatible probability measures. In fact in this case μ plays the role of the distribution of a Markov process $(X_t)_{t \geq 0}$ [52, 56]. In particular, the random variables $X_t : S^{\mathbb{R}^+} \to \mathbb{R}$ are defined as

$$X_t(\omega) := \omega(t) \qquad \omega \in S^{\mathbb{R}^+}, t \in \mathbb{R}^+, \tag{4.33}$$

and the measures $\{\mu_J\}$ are the finite dimensional distributions of the process $(X_t)_{t \geq 0}$ in the sense that for any $J \in \mathcal{I}$, with $J = \{t_1, \ldots, t_n\}$, the following holds

$$\mu(\pi_J^{-1}(B_1 \times \cdots \times B_n)) = \mu(X_{t_1} \in B_1, \ldots, X_{t_n} \in B_n) \tag{4.34}$$

$$= \mu_J(B_1 \times \cdots \times B_n), \qquad B_1, \ldots, B_n \in \mathcal{F}_S \tag{4.35}$$

In the case of complex (or signed) kernels G, under the assumptions of Theorem 3.9 the projective limit measure μ exists and one can also introduce the concept of *pseudoprocesses* $(X_t)_{t \geq 0}$, with the family of "finite-dimensional distributions" given by the μ_J in the sense that (4.33) and (4.34) hold again (without of course a probabilistic interpretation).

On the other hand, if $G_t(x, \cdot)$, with $x \in E$ and $t \in \mathbb{R}^+$, are general complex or signed measures, in most interesting cases condition (4.32) is not satisfied.

This kind of problems arise for instance when one tries to construct a Feynman-Kac type formula for the solution of evolution equations of the following form:

$$\frac{\partial}{\partial t} u(t, x) = (-1)^{N+1} (\Delta_x)^N u(t, x), \qquad x \in \mathbb{R}, t \in \mathbb{R}^+, \tag{4.36}$$

where the Laplacian appearing in the heat equation is replaced by one of its powers Δ^N, $N \geq 2$. In this case the fundamental solution $G_t(x, y)$ of the equation, i.e. the distribution such that

$$u(t, x) = \int_{\mathbb{R}} G_t(x, y) u(0, y) dy,$$

can be explicitly computed in terms of the following Fourier integral:

$$G_t(x, y) = g_t(x - y) \tag{4.37}$$

$$g(x) = \frac{1}{2\pi} \int_{\mathbb{R}} e^{ik(x-y)} e^{-tk^{2N}} dk, \tag{4.38}$$

where $N \in \mathbb{N}$, $N \geq 2$. By the Fourier integral representation (4.38), it follows that for $t > 0$ the distribution $G_t \in \mathcal{S}'(\mathbb{R})$ is a $C^{\infty}(\mathbb{R})$ real-valued function, whereas G_0 is Dirac's measure concentrated at 0. Moreover an analysis of the asymptotic behaviour of $g_t(x)$ as $|x| \to +\infty$ shows that it belongs to $L^1(\mathbb{R})$ whenever $t > 0$. In particular, if $N \geq 2$ then g is not positive but it rather has an oscillatory behaviour, changing sign an infinite number of times [185, 253].

Formula (4.38) yields the following convolution identity

$$g_{t+s}(x) = \int g_t(x - y) g_s(y) \, dy, \qquad x \in \mathbb{R}, s, t > 0$$

which show that the family of mappings $G_t : \mathbb{R} \times \mathcal{B}(\mathbb{R}) \to \mathbb{C}$ defined for any $t \in \mathbb{R}^+$ by

$$G_t(x, B) := \int_B g_t(x - y) dy,$$

is a semigroup of complex kernels. The associated projective family of complex measures $\{\mu_J\}$ defined by Eq. (4.30) is explicitly given in the case where $\nu = \delta_0$ by

$$\mu_J(E) = \int_{B_1} \cdots \int_{B_n} g_{t_1}(x_1) g_{t_2-t_1}(x_2 - x_1) \cdots g_{t_n-t_{n-1}}(x_n - x_{n-1}) \, dx_1 \cdots dx_n,$$

$$(4.39)$$

where $J = \{t_1, t_2, \ldots, t_n\}$, with $0 < t_1 < t_2 < \cdots < t_n < +\infty$, and $E = B_1 \times \cdots \times B_n$, $B_1, \ldots, B_n \in \mathcal{B}(\mathbb{R})$.

By Remark 2.4 and Eq. (4.39), it is easy to verify that the total variation of the measure μ_J is equal to

$$|\mu_J| = \Pi_{j=1}^n \int_{\mathbb{R}} |g_{t_j-t_{j-1}}(x)| dx.$$

On the other hand, by Eq. (4.38), for any $t > 0$ we have

$$\int_{\mathbb{R}} |g_t(x)| dx = \int_{\mathbb{R}} t^{-1/2N} |g_1(t^{-1/2N} x)| dx = \int_{\mathbb{R}} |g_1(y)| dy,$$

hence

$$|\mu_J| = \left(\int_{\mathbb{R}} |g_1(y)| dy \right)^n.$$

Since for $N > 1$ the smooth function g has an oscillatory behaviour and $\int_{\mathbb{R}} g_1(y) dy = 1$, one concludes that for $\int_{\mathbb{R}} |G_1(y)| dy = c > 1$ and $|\mu_J| = c^n \to +\infty$ as $n \to +\infty$ (for $N = 1$ there is no such result since $g_t \geq 0$ and c would be equal to 1). We can conclude that there cannot exists a signed measure μ on $\mathbb{R}^{[0,+\infty)}$ whose values on cylinder sets $I_k \subset \Omega = \{x : [0,\infty) \to \mathbb{R}\}$ of the form

$$I_k := \{\omega \in \Omega : \omega(t_j) \in [a_j, b_j], j = 1, \ldots k\}, \quad 0 < t_1 < t_2 < \cdots t_k,$$

is equal to

$$\mu(I_k) = \int_{a_1}^{b_1} \cdots \int_{a_k}^{b_k} \prod_{j=0}^{k-1} g_{t_{j+1}-t_j}(x_{j+1} - x_j) dx_1 \ldots dx_k. \qquad (4.40)$$

Hence, the projective system of functionals $(L_J, \hat{E}_J^0)$, defined by

$$\hat{E}_J^0 = L^1(S_J, |\mu_J|), \qquad S_J = \mathbb{R}^J,$$

$$L_J(f) = \int_{S_J} f d\mu_J$$

cannot have a projective extension $(L, D(L))$ of the form

$$D(L) = L^1(S_{\mathcal{I}}, |\mu|), \qquad S_{\mathcal{I}} = \mathbb{R}^{[0,+\infty)},$$

$$L(f) = \int_{S_{\mathcal{I}}} f d\mu.$$

For a detailed analysis of this problem see, e.g., [234, 185, 252, 72, 253, 306].

4.4   The Fourier transform approach

In this section we present two examples of possible extensions of projective system of functionals of the form (4.16), associated to a projective system of complex measures.

Let $\mathcal{H}$ be a real separable infinite-dimensional Hilbert space and let $\mathcal{I}$ be the directed set of its finite-dimensional subspaces, ordered by inclusion. For $V, W \in \mathcal{I}$, with $V \leq W$, let $\pi_V^W : W \to V$ be the orthogonal projection from W onto V and $i_V^W : V \to W$ be the inclusion map. One has that $(V, \pi_V^W)_{V,W \in \mathcal{I}}$ is a projective family of sets, while $(V, i_V^W)_{V,W \in \mathcal{I}}$ forms a direct system on $\mathcal{I}$. Let us consider the projective limit space $E_{\mathcal{I}} := \varprojlim_{V \in A} V$, the direct limit $\tilde{E}_{\mathcal{I}} := \varinjlim_{V \in \mathcal{I}} V$, the projection $\pi_V : E_{\mathcal{I}} \to V$ and the inclusion maps $i_V : V \to \tilde{E}_{\mathcal{I}}$. Considered on each $V \in \mathcal{I}$ the topology induced by the finite-dimensional Hilbert space structure of V, the space $\varprojlim_{V \in \mathcal{I}} V$ is endowed with the weakest topology making all the projections $\pi_V : E_{\mathcal{I}} \to V$ continuous, while the space $\varinjlim_{V \in \mathcal{I}} V$ is endowed with the *final topology*, i.e., the strongest topology making all the inclusion maps $i_V : V \to \tilde{E}_{\mathcal{I}}$ continuous.

The inverse system $(V, \pi_V^W)_{V,W \in \mathcal{I}}$ and the direct system $(V, i_V^W)_{V,W \in \mathcal{I}}$ are linked by dualization. Indeed if we identify the dual of a finite-dimensional vector space V with V itself, we have that the inclusion $i_V^W : V \to W$, $V \le W$, can be identified with the transpose map $(\pi_V^W)^* : V^* \to W^*$ of the projection $\pi_V^W : W \to V$. Further the direct limit space $\tilde{E}_{\mathcal{I}}$ can be naturally identified with a subspace of $(E_{\mathcal{I}})^*$, since any $\eta \in \tilde{E}_{\mathcal{I}}$ can be associated with the element of $(E_{\mathcal{I}})^*$ whose action on any $\omega \in E_{\mathcal{I}}$ is given by

$$\eta(\omega) := \langle v, \pi_V \omega \rangle, \qquad (4.41)$$

$v \in V$ being any representative of the equivalence class of vectors associated to η. The definition (4.41) is well posed, indeed if we choose a different representative of the equivalence class η, i.e., a vector $v' \in V'$ such that there exists a $W \in \mathcal{I}$, with $V \le W$, $V' \le W$ and $i_V^W v = i_{V'}^W v'$, we get:

$$\langle v, \pi_V \omega \rangle = \langle v, \pi_V^W \circ \pi_W \omega \rangle = \langle i_V^W v, \pi_W \omega \rangle = \langle i_{V'}^W v', \pi_W \omega \rangle = \langle v', \pi_{V'} \omega \rangle.$$

Furthermore, the explicit form (4.41) of the functional η shows its continuity on $E_{\mathcal{I}}$.

Analogously the transpose map $\pi_V^* : V^* \to E_{\mathcal{I}}^*$ can be identified with the map $i_V : V \to \tilde{E}_{\mathcal{I}}$, giving:

$$\langle v, \pi_V \omega \rangle = {}_{E_{\mathcal{I}}^*}\langle i_V v, \omega \rangle_{E_{\mathcal{I}}},$$

where the symbol $\langle \, , \, \rangle$ on the left-hand side denotes the inner product in V, while the symbol ${}_{E_{\mathcal{I}}^*}\langle \, , \, \rangle_{E_{\mathcal{I}}}$ denotes the dual pairing between $E_{\mathcal{I}}$ and $E_{\mathcal{I}}^*$.

Let us consider on any $V \in \mathcal{I}$ the Borel σ-algebra $\mathcal{B}(V)$ and a complex measure $\mu_V : \mathcal{B}(V) \to \mathbb{C}$ in such a way that the family $(\mu_V)_{V \in A}$ satisfies the compatibility condition (3.41). Let us also consider, for any $V \in \mathcal{I}$, the Fourier transform $\hat{\mu}_V : V \to \mathbb{C}$ of the measure μ_V, i.e.,

$$\hat{\mu}_V(v) = \int_V e^{i\langle v', v \rangle} \mu_V(dv'), \qquad v \in V.$$

By the projectivity condition (3.41) of the family of measures $(\mu_V)_{V \in \mathcal{I}}$, one deduces the following compatibility relation among the Fourier transforms $\{\hat{\mu}_V\}_{V \in \mathcal{I}}$:

$$\hat{\mu}_V(v) = \hat{\mu}_W(i_V^W v), \qquad V \le W. \qquad (4.42)$$

Let us now define the map $F : \tilde{E}_{\mathcal{I}} \to \mathbb{C}$ by:

$$F(\eta) := \hat{\mu}_V(v), \qquad \eta \in \tilde{E}_{\mathcal{I}},$$

where $v \in V$ is any representative of the equivalence class $\eta \in \tilde{E}_{\mathcal{I}}$. F is unambiguously defined, indeed given a $v' \sim v$, with $v' \in V'$, there exists a $W \in \mathcal{I}$, with $V \leq W$ and $V' \leq W$, such that $i_V^W v = i_{V'}^W v'$. By the compatibility condition (4.42)

$$\hat{\mu}_V(v) = \hat{\mu}_W(i_V^W v) = \hat{\mu}_W(i_{V'}^W v') = \hat{\mu}_{V'}(v').$$

Further, the map F is continuous on $\tilde{E}_{\mathcal{I}}$ in the final topology.

If there existed a measure μ on $E_{\mathcal{I}}$ such that $\mu_V = \pi_V \circ \mu$ for all $V \in \mathcal{I}$, then its Fourier transform $\hat{\mu}$ would coincide with F on $\tilde{E}_{\mathcal{I}}$ and

$$\|\hat{\mu}\|_\infty = \sup_{\eta \in (E_{\mathcal{I}})*} |\hat{\mu}| \leq |\mu|,$$

where $|\mu|$ is the total variation of μ.

Let us consider the projective system of functionals $(L_V, D(L_V))_{V \in A}$, where $D(L_V) \equiv \mathcal{F}(V)$ is the space of functions $f : V \to \mathbb{C}$ of the form

$$f(v) = \int_V e^{i\langle v', v \rangle} \, d\nu_f(v')$$

for some complex bounded measure ν_f on V. $D(L_V)$ is a Banach algebra of functions, where the multiplication is the pointwise one and the norm of a function $f \in \mathcal{F}(V)$ is the total variation of the associated measure ν_f. Let $L_V : D(L_V) \to \mathbb{C}$ be the linear functional defined by

$$D(L_V) := \mathcal{F}(V)$$
$$L_V(f) := \int_V \hat{\mu}_V(v) \, d\nu_f(v), \tag{4.43}$$

One can easily verify that L_V is continuous in the $\mathcal{F}(V)-$norm. Indeed for any $V \in \mathcal{I}$ one has that the map $\hat{\mu}_V$ is uniformly continuous and bounded, since $\|\hat{\mu}_V\|_\infty \leq |\mu_V|$, $|\mu_V|$ denoting the total variation of the measure μ_V. Hence

$$L_V(f) \leq \|\hat{\mu}_V(v)\|_\infty |\nu_f| \leq |\mu_V| \|f\|_{\mathcal{F}(V)}.$$

On the other hand by Fubini's theorem one has:

$$\int_V f(v) d\mu_V(v) = \int_V \hat{\mu}_V(v) \, d\nu_f(v), \tag{4.44}$$

both integrals being absolutely convergent. The family $(L_V, D(L_V))_{V \in A}$ forms a projective system of functionals.

If $\sup_{V \in \mathcal{I}} |\mu_V| = +\infty$, according to Theorem 3.8, there cannot exist a complex bounded measure μ on $E_{\mathcal{I}}$ which is the projective limit of the measures $(\mu_V)_{V \in \mathcal{I}}$. Hence there cannot exists a projective extension $(L, D(L))$ of the projective system $(L_V, D(L_V))_{V \in \mathcal{I}}$ of the form

$$D(L) := L^1(E_{\mathcal{I}}, |\mu|)$$
$$L(f) := \int_{E_{\mathcal{I}}} f(\omega)\mu(d\omega). \tag{4.45}$$

However, even if μ does not exists, the map $F : \tilde{E}_{\mathcal{I}} \to \mathbb{C}$ is still well-defined, and can be used in the construction of a projective extension of $(L_V, D(L_V))_{V \in \mathcal{I}}$ alternative to (4.45).

Consider on $\tilde{E}_{\mathcal{I}}$ the Borel $\sigma-$algebra $\mathcal{B}(\tilde{E}_{\mathcal{I}})$, then one has that the continuous map $F : \tilde{E}_{\mathcal{I}} \to \mathbb{C}$ is measurable. If the condition:

$$\sup_{V \in A} \|\hat{\mu}_V\|_\infty < +\infty \tag{4.46}$$

is satisfied, then the functional $L : D(L) \to \mathbb{C}$ given by

$$D(L) := \mathcal{F}(\tilde{E}_{\mathcal{I}})$$
$$L(f) = \int_{\tilde{E}_{\mathcal{I}}} F(\eta) \, d\nu_f(\eta)$$

is well-defined on the Banach algebra $\mathcal{F}(\tilde{E}_{\mathcal{I}})$ of functions $f : E_{\mathcal{I}} \to \mathbb{C}$ of the form $f(\omega) = \int_{\tilde{E}_{\mathcal{I}}} e^{i\langle \eta, \omega \rangle} d\nu_f(\eta)$ for some complex bounded measure ν_f on $\tilde{E}_{\mathcal{I}}$. L is a projective extension of the projective system of functionals (4.43). Indeed considered a cylindrical function $f : E_{\mathcal{I}} \to \mathbb{C}$ of the form $f(\omega) := g(\pi_V \omega)$, for some $V \in \mathcal{I}$ and $g \in \mathcal{F}(V)$, $g \equiv \hat{\nu}_g$, then $f \in \mathcal{F}(\tilde{E}_{\mathcal{I}})$ with $\nu_f = (\nu_g)_{i_V}$ is the image measure of ν_g under the action of the mapping $i_V : V \to \tilde{E}_{\mathcal{I}}$. Indeed we have:

$$f(\omega) = g(\pi_V \omega)$$
$$= \int_V e^{i\langle v', \pi_V \omega \rangle} d\nu_g(v')$$
$$= \int_V e^{i\langle i_V v', \omega \rangle} d\nu_g(v')$$
$$= \int_{\tilde{E}_{\mathcal{I}}} e^{i\langle \eta, \omega \rangle} d\nu_f(\eta),$$

hence

$$L(f) = \int_{\tilde{E}_A} f(\eta)\nu_f(d\eta)$$

$$= \int_V f(i_v v)\nu_g(dv)$$

$$= \int_V \hat{\mu}_V(v)\nu_g(dv)$$

$$= L_V(g)$$

However, for a general $f \in \mathcal{F}(\tilde{E}_\mathcal{I})$, contrary to the finite-dimensional case, the equality (4.44) does not make sense and the action of the functional on f cannot be described in terms of an integral with respect to a measure.

Depending on the regularity properties of the function $F : \tilde{E}_\mathcal{I} \to \mathbb{C}$ one can define the functional L on different domains. Let $\mathcal{B}$ be a Banach space where $\tilde{E}_\mathcal{I}$ is densely embedded and let F be continuous with respect to the $\mathcal{B}$-norm. Then F can be extended to a function $\tilde{F} : \mathcal{B} \to \mathbb{C}$, with $\tilde{F}(\eta) = F(\eta)$ for all $\eta \in \tilde{E}_\mathcal{I}$. Let $\mathcal{F}(\mathcal{B})$ be the Banach algebra of functions $f : \mathcal{B}^* \to \mathbb{C}$ of the form

$$f(x) = \int_\mathcal{B} e^{i\langle y,x \rangle} d\nu_f(y), \quad x \in \mathcal{B}^*,$$

for some complex bounded variation measure ν_f on $\mathcal{B}$. Then the functional $L' : \mathcal{F}(\mathcal{B}) \to \mathbb{C}$ defined by

$$D(L') := \mathcal{F}(\mathcal{B})$$

$$L'(f) = \int_\mathcal{B} \tilde{F}(x) d\nu_f(x)$$

is an alternative projective extension of the system of functionals (4.43).

4.5　Infinite dimensional Fresnel integrals

In the 60s a couple of papers by K. Itô [195, 196] proposed a particular approach for the mathematical definition of Feynman integrals based on techniques of Fourier analysis. Ito's ideas where further developed by S. Albeverio and R. Høegh-Krohn [17, 18] and led to the development of the theory of *infinite dimensional Fresnel integrals*. The main idea is a generalization of the classical Parseval equality

$$\int_{\mathbb{R}^n} \phi(x) f(x) dx = \int_{\mathbb{R}^n} \hat{\phi}(x) \hat{f}(x) dx, \tag{4.47}$$

where $\hat{\phi}$ and $\hat{f}$ denote the Fourier transform of the Schwartz test functions $\phi, f \in S(\mathbb{R}^n)$. Distribution theory allows to extend equality (4.47) to the case where ϕ is the bounded smooth function

$$\phi(x) = (2\pi i\epsilon)^{-n/2} e^{\frac{i}{2\epsilon}\|x\|^2} \qquad x \in \mathbb{R}^n$$

where $\epsilon \in \mathbb{R}^+$ is a positive constant. In fact ϕ admits a Fourier transform $\hat{\phi}$ in the sense of tempered distributions:

$$\hat{\phi}(x) = (2\pi i\epsilon)^{-n/2} \int e^{i\langle x, y\rangle} e^{\frac{i}{2\epsilon}\|y\|^2} \, dy = e^{-\frac{i\epsilon}{2}\|x\|^2} \qquad x \in \mathbb{R}^n .$$

Let us consider now the case where the mapping $f : \mathbb{R}^n \to \mathbb{C}$ is the Fourier transform of a complex Borel measure μ_f on $\mathbb{R}^n$, i.e. f can be represented as

$$f(x) = \int_{\mathbb{R}^n} e^{ikx} d\mu_f(k), \qquad x \in \mathbb{R}^n , \tag{4.48}$$

and define the *Fresnel integral* of f, heuristically denoted as

$$\widetilde{\int}_{\mathbb{R}^n} e^{\frac{i}{2\epsilon}\|x\|^2} f(x) \, dx \tag{4.49}$$

as the absolutely convergent Lebesgue integral

$$\int_{\mathbb{R}^n} e^{-\frac{i\epsilon}{2}\|x\|^2} d\mu_f(x) .$$

As remarked in Section 2.8, the set $\mathcal{F}(\mathbb{R}^n)$ of Fourier transforms of complex Borel measures on $\mathbb{R}^n$ is a Banach algebra of functions, where the norm of an element $f \in \mathcal{F}(\mathbb{R}^n)$ is defined as the total variation $|\mu_f|$ of the measure μ_f associated to f through relation (4.48). In this setting the Fresnel integral can be defined as the linear functional on $\mathcal{F}(\mathbb{R}^n)$ given by the right-hand side of the following equality

$$\widetilde{\int}_{\mathbb{R}^n} e^{\frac{i}{2\epsilon}\|x\|^2} f(x) := \int_{\mathbb{R}^n} e^{-\frac{i\epsilon}{2}\|x\|^2} d\mu_f(x) . \tag{4.50}$$

By construction, it is continuous in the $\mathcal{F}(\mathbb{R}^n)$ norm, since

$$\left| \int_{\mathbb{R}^n} e^{-\frac{i\epsilon}{2}\|x\|^2} d\mu_f(x) \right| \leq |\mu_f| = \|f\| .$$

We point out that the functional depends explicitly on the parameter ϵ. It is interesting to tackle this issue in particular in the study of the asymptotic behaviour as $\epsilon \downarrow 0$. We postpone this discussion to Chapter 6.5 and in the remainder of this section we shall consider the case $\epsilon = 1$.

The relation (4.50) defining the Fresnel integral on $\mathbb{R}^n$ is the starting point for the generalisation to an infinite dimensional setting. We can now replace $\mathbb{R}^n$ with a real separable infinite dimensional Hilbert space $(\mathcal{H}, \langle \, , \, \rangle)$. Indeed, while the left-hand side is no longer meaningful for infinite dimensional Hilbert spaces due to the lack of a reasonable Lebesgue measure dx, the right-hand side still makes sense within Lebesgue traditional integration theory since it is the integral of a bounded continuous function with respect to a complex Borel measure (with finite total variation).

Given a real separable Hilbert space $(\mathcal{H}, \langle \, , \, \rangle)$, let us consider the Banach algebra $\mathcal{F}(\mathcal{H})$ of functions $f : \mathcal{H} \to \mathbb{C}$ of the form

$$f(x) = \int_{\mathcal{H}} e^{i\langle x,y \rangle} d\mu_f(y), \qquad x \in \mathcal{H, \tag{4.51}$$

for some complex Borel measure μ_f on $\mathcal{H}$ (see Section 3.6). In the following, in order to simplify the notation we shall restrict ourselves to the case where $\epsilon = 1$, but the whole argument can be extended to arbitrary values of $\epsilon > 0$

Definition 4.7. The *infinite dimensional Fresnel integral* on $\mathcal{H}$ is defined as the linear functional $I_F : \mathcal{F}(\mathcal{H}) \to \mathbb{C}$ whose value on functions $f \in \mathcal{F}(\mathcal{H})$ of the form (4.51) is

$$I_F(f) := \int_{\mathcal{H}} e^{-\frac{i}{2}\langle x,x \rangle} d\mu_f(x). \tag{4.52}$$

It is also denoted with the symbol $\widetilde{\int_{\mathcal{H}}} e^{\frac{i}{2}\|x\|^2} f(x) dx$.

By construction, the functional $I_F : \mathcal{F}(\mathcal{H}) \to \mathbb{C}$ enjoys some remarkable properties. First of all, analogously to the finite-dimensional case, it is continuous in the $\mathcal{F}(\mathcal{H})$-norm since

$$\left| \int_{\mathcal{H}} e^{-\frac{i}{2}\|x\|^2} d\mu_f(x) \right| \leq |\mu_f| = \|f\|_{\mathcal{F}}.$$

A function $f \in \mathcal{F}(\mathcal{H})$ is said to be a *finitely based function* if there exists a finite dimensional projection operator $P : \mathcal{H} \to \mathcal{H}$ such that $f(x) = f(Px)$ for all $x \in \mathcal{H}$. In this case its infinite dimensional Fresnel integral coincides with the (finite dimensional) Fresnel integral on $P\mathcal{H}$ defined in (4.50). More precisely, the following holds:

$$I_F(f) = \widetilde{\int_{P\mathcal{H}}} e^{\frac{i}{2}\|x\|^2} g(x) dx, \tag{4.53}$$

where $g := f|_{P\mathcal{H}}$. Indeed, if f is finitely based with base $P\mathcal{H}$, then one can easily see that the measure μ_f coincides with its image $(\mu_f)_P$ under the action of P since

$$f(x) = \int_{\mathcal{H}} e^{i\langle x,y \rangle} d\mu_f(y)$$
$$= f(Px)$$
$$= \int_{\mathcal{H}} e^{i\langle Px,y \rangle} d\mu_f(y)$$
$$= \int_{\mathcal{H}} e^{i\langle x,Py \rangle} d\mu_f(y)$$
$$= \int_{\mathcal{H}} e^{i\langle x,y \rangle} d(\mu_f)_P(y).$$

In particular, the equality $\mu_f = (\mu_f)_P$ implies that the support of f is contained in $P\mathcal{H}$, hence

$$I_F(f) = \int_{\mathcal{H}} e^{-\frac{i}{2}\langle x,x \rangle} d\mu_f(x) = \int_{P\mathcal{H}} e^{-\frac{i}{2}\langle x,x \rangle} d(\mu_f)_P(x).$$

Let us consider now the group of Euclidean transformations $E(\mathcal{H})$, i.e. the group of mappings $T : \mathcal{H} \to \mathcal{H}$ of the form

$$T(x) := Ox + a \qquad x \in \mathcal{H},$$

where $a \in \mathcal{H}$ is a vector and O is an orthogonal transformation of $\mathcal{H}$ onto $\mathcal{H}$. In fact the space of Fresnel integrable functions $\mathcal{F}(\mathcal{H})$ is invariant under $E(\mathcal{H})$, and $E(\mathcal{H})$ is a group of isometries of $\mathcal{F}(\mathcal{H})$. If $O : \mathcal{H} \to \mathcal{H}$ and $f \in \mathcal{F}(\mathcal{H})$ then the mapping $f_O : \mathcal{H} \to \mathbb{C}$ defined as

$$f_O(x) := f(Ox) \qquad x \in \mathcal{H}$$

belongs to $\mathcal{F}(\mathcal{H})$. Indeed:

$$f(Ox) = \int_{\mathcal{H}} e^{i\langle Ox,y \rangle} d\mu_f(y)$$
$$= \int_{\mathcal{H}} e^{i\langle x,O^*y \rangle} d\mu_f(y)$$
$$= \int_{\mathcal{H}} e^{i\langle x,y \rangle} d(\mu_f)_{O^*}(y)$$

with $(\mu_f)_{O^*}$ denoting the image measure of μ_f under the action of the adjoint operator $O^* : \mathcal{H} \to \mathcal{H}$. We can now easily verify that $I_F(f) =$

$I_F(f_O)$ since

$$
\begin{aligned}
I_F(f_O) &= \int_{\mathcal{H}} e^{-\frac{i}{2}\langle x,x\rangle} d(\mu_f)_{O^*}(x) \\
&= \int_{\mathcal{H}} e^{-\frac{i}{2}\langle O^*x, O^*x\rangle} d\mu_f(x) \\
&= \int_{\mathcal{H}} e^{-\frac{i}{2}\langle x, OO^*x\rangle} d\mu_f(x) \\
&= \int_{\mathcal{H}} e^{-\frac{i}{2}\langle x,x\rangle} d\mu_f(x) \,,
\end{aligned}
$$

which can be heuristically written as

$$
\widetilde{\int_{\mathcal{H}}} e^{\frac{i}{2}\|x\|^2} f(Ox) dx = \widetilde{\int_{\mathcal{H}}} e^{\frac{i}{2}\|x\|^2} f(x) dx \,. \tag{4.54}
$$

Similarly, if we consider a vector $a \in \mathcal{H}$, a map $f \in \mathcal{F}(\mathcal{H})$ and construct the translated function $f_a : \mathcal{H} \to \mathbb{C}$ defined by

$$
f_a(x) := f(x+a) \qquad x \in \mathcal{H} \,,
$$

then it is easy to verify that $f \in \mathcal{F}(\mathcal{H})$ since:

$$
f(x+a) = \int_{\mathcal{H}} e^{i\langle x,y\rangle} e^{i\langle a,y\rangle} d\mu_f(y) \,.
$$

Moreover, the infinite dimensional Fresnel integral of f_a is given by

$$
\begin{aligned}
I_F(f_a) &= \int_{\mathcal{H}} e^{-\frac{i}{2}\langle x,x\rangle} e^{i\langle a,y\rangle} d\mu_f(y) \\
&= e^{\frac{i}{2}\|a\|^2} \int_{\mathcal{H}} e^{-\frac{i}{2}\langle x-a, x-a\rangle} d\mu_f(y) \\
&= e^{\frac{i}{2}\|a\|^2} \int_{\mathcal{H}} e^{-\frac{i}{2}\langle x,x\rangle} d(\mu_f)_{\tau_a}(y),
\end{aligned}
$$

where $(\mu_f)_{\tau_a}$ is the image measure of μ_F under the action of the translation map $\tau_a : \mathcal{H} \to \mathcal{H}$ defined as $\tau_a(x) := x - a$, and its Fourier transform is the mapping

$$
\begin{aligned}
\widetilde{(\mu_f)_{\tau_a}}(x) &= \int e^{i\langle x,y\rangle} d(\mu_f)_{\tau_a}(y) \\
&= \int e^{i\langle x, y-a\rangle} d\mu_f(y) \\
&= e^{i\langle x,a\rangle} f(x) \,.
\end{aligned}
$$

This gives the following equality

$$\widetilde{\int_{\mathcal{H}}} e^{\frac{i}{2}\|x\|^2} f(x+a)dx = e^{\frac{i}{2}\|a\|^2} \widetilde{\int_{\mathcal{H}}} e^{\frac{i}{2}\|x\|^2} e^{i\langle x,a\rangle} f(x)dx$$

$$= \widetilde{\int_{\mathcal{H}}} e^{\frac{i}{2}\|x-a\|^2} f(x)dx$$

which, together with Eq. (4.54), is consistent with the heuristic idea that the symbol dx behaves a measure on $\mathcal{H}$ invariant under Euclidean transformations.

In this setting it is also possible to prove a Fubini-type theorem. Let us assume that $\mathcal{H} = \mathcal{H}_1 \oplus \mathcal{H}_2$ is the orthogonal sum of two subspaces $\mathcal{H}_1$ and $\mathcal{H}_2$ and for $f(x) \in \mathcal{F}(\mathcal{H})$ let us set $f(x_1, x_2) = f(x_1 \oplus x_2)$ with $x_1 \in \mathcal{H}_1$ and $x_2 \in \mathcal{H}_2$. Then for fixed x_2, the mapping $f_{x_2} : \mathcal{H}_1 \to \mathbb{C}$ defined by

$$f_{x_2}(x_1) := f(x_1, x_2) \qquad x_1 \in \mathcal{H}_1$$

belongs to $\mathcal{F}(\mathcal{H}_1)$. Since the spaces $\mathcal{H}_1 \oplus \mathcal{H}_2$ and $\mathcal{H}_1 \times \mathcal{H}_2$ are equivalent metrics space, the measure μ_f associated with f through (4.51) can be handled as a measure on the product space $\mathcal{H}_1 \times \mathcal{H}_2$. Hence we can write

$$f_{x_2}(x_1) = \int e^{i\langle x_1, y_1\rangle} e^{i\langle x_2, y_2\rangle} d\mu_f(y_1, y_2).$$

If we consider now the measure $\mu_{x_2} \in \mathcal{M}(\mathcal{H}_1)$ defined by

$$\mu_{x_2}(E) = \int \chi_E(x_1) e^{i\langle x_2, y_2\rangle} d\mu_f(y_1, y_2) \qquad E \in \mathcal{B}(\mathcal{H}_1).$$

one can easily verify that $f_{x_2} = \widehat{\mu_{x_2}}$ and the Fresnel integral

$$g(x_2) = \widetilde{\int_{\mathcal{H}_1}} e^{\frac{i}{2}\|x_1\|^2} f_{x_2}(x_1)dx_1 \qquad (4.55)$$

is well defined. Moreover the mapping $g : \mathcal{H}_2 \to \mathbb{C}$ defined by Eq. (4.55) belongs to $\mathcal{F}(\mathcal{H}_2)$. Indeed, by considering the measure $\nu \in \mathcal{M}(\mathcal{H}_2)$ defined by

$$\nu(E) := \int_{\mathcal{H}} \chi_E(y_2) e^{-\frac{i}{2}\|y_1\|^2} d\mu(y_1, y_2) \qquad E \in \mathcal{B}(\mathcal{H}_2),$$

one can easily verify that $g = \hat{\nu}$. We eventually have

$$\widetilde{\int_{\mathcal{H}_2}} e^{\frac{i}{2}\|x_2\|^2} g(x_2)\, dx_2 = \int e^{-\frac{i}{2}\|x_2\|^2} e^{-\frac{i}{2}\|y_1\|^2} d\mu(y_1, y_2)$$

and by the traditional Fubini theorem

$$\widetilde{\int_{\mathcal{H}_2}} e^{\frac{i}{2}\|x_2\|^2} g(x_2)\, dx_2 = \widetilde{\int_{\mathcal{H}}} e^{\frac{i}{2}\|x\|^2} g(x)\, dx.$$

This formula can be equivalently written in the following more intuitive form:

$$\widetilde{\int_{\mathcal{H}_2}} e^{\frac{i}{2}\|x_2\|^2} g(x_2) dx_2 = \widetilde{\int_{\mathcal{H}_2}} e^{\frac{i}{2}\|x_2\|^2} \left(\widetilde{\int_{\mathcal{H}_1}} e^{\frac{i}{2}\|x_1\|^2} f(x_1, x_2) dx_1 \right) dx_2$$

$$= \widetilde{\int_{\mathcal{H}}} e^{\frac{i}{2}\|x\|^2} f(x) dx .$$

A systematic implementation of the theory of infinite dimensional Fresnel integrals as long as their applications to the mathematical definition of Feynman path integrals has been presented in the first edition of the pioneering book [18] and in the fundamental paper [17] by Sergio Albeverio and Raphael Høegh-Krohn. The main application of this functional is the representation of the solution of Schrödinger Eq. (1.4).

Let us consider the Cameron-Martin Hilbert space $\mathcal{H}_t$, i.e. the space of absolutely continuous paths $\gamma : [0, t] \to \mathbb{R}^d$ with fixed end point $\gamma(t) = 0$ and weak derivative $\dot{\gamma} \in L^2(\mathbb{R}^d)$, endowed with the inner product

$$\langle \gamma, \eta \rangle = \frac{m}{\hbar} \int_0^t \dot{\gamma}(s) \dot{\eta}(s) ds, \qquad \gamma, \eta \in \mathcal{H}_t.$$

According to Theorem 4.5 below, under suitable conditions on the potential V and the initial datum ψ_0 the solution of the Schrödinger Eq. (1.4) admits a representation in terms of an infinite dimensional Fresnel integral on the Cameron-Martin space $\mathcal{H}_t$. This particular Hilbert space will be extensively studied in Chapter 5.5. Here we limit ourselves to show how the theory of infinite dimensional Fresnel integrals on $\mathcal{H}_t$ provides a realisation of the Feynman functional as well as mathematical definition for the Feynman path integral representation for the solution of the Schrödinger Eq. (1.4).

Let us start by pointing out that when applied to particular cylinder functions, the infinite dimensional Fresnel integral assumes the form (4.27). Indeed, let $f : \mathcal{H}_t \to \mathbb{C}$ be a complex-valued function on $\mathcal{H}_t$ of the following form

$$f(\gamma) = F(\gamma(t_1), \dots, \gamma(t_n))$$

for some $n \geq 1$, $0 \leq t_1 < t_2 < \cdots < t_n < t$ and $F \in S(\mathbb{R}^{nd})$ a Schwartz test function. Denoted with $\hat{F}$ the Fourier transform of F:

$$F(x_1, \dots, x_n) = \int_{\mathbb{R}^d \times \cdots \times \mathbb{R}^d} e^{i(x_1 \cdot y_1 + \cdots + x_n \cdot y_n)} \frac{\hat{F}(y_1, \dots, y_n)}{(2\pi)^{nd}} \, dy_1 \dots dy_n$$

it is rather easy to check that $f \in \mathcal{F}(\mathcal{H}_t)$. Indeed, f is the Fourier transform of the measure ν defined by

$$\nu(E) := \int_{\mathbb{R}^d \times \cdots \times \mathbb{R}^d} \chi_E(y_1 \cdot v_{t_1} + \cdots + y_n \cdot v_{t_n}) \frac{\hat{F}(y_1, \ldots, y_n)}{(2\pi)^{nd}} \, dy_1 \ldots dy_n \, .$$

For $y \in \mathbb{R}^d$ and $\tau \in [0, t]$, the symbol $y \cdot v_\tau$ stands for the vector in $\mathcal{H}_t$ defined by $y \cdot v_\tau = \sum_{j=1}^d y_j v_{\tau,j}$, where $v_{\tau,j} \in \mathcal{H}_t$ is given to

$$v_{\tau,j}(s) \cdot \hat{e}_k = \frac{\hbar}{m} \delta_j^k(t - \tau \vee s) \qquad s \in [0, t], k = 1, \ldots, d$$

(with $\{\hat{e}_k\}_{k=1,\ldots d}$ being the canonical basis of $\mathbb{R}^d$). In particular, for any $\gamma \in \mathcal{H}_t$, with $\gamma(s) = \sum_{j=1}^d \gamma_j(s)\hat{e}_j$, the following holds

$$\langle \gamma, v_{s,j} \rangle = \gamma_j(s) \qquad s \in [0, t], j = 1, \ldots, d \, .$$

By Eq. (4.52) the infinite dimensional Fresnel integral of f is equal to

$$\widetilde{\int_{\mathcal{H}}} e^{\frac{i}{2}\|x\|^2} f(x) dx = \int_{\mathbb{R}^d \times \cdots \times \mathbb{R}^d} e^{-\frac{i}{2}\|y_1 \cdot v_{t_1} + \cdots + y_n \cdot v_{t_n}\|^2} \frac{\hat{F}(y_1, \ldots, y_n)}{(2\pi)^{nd}} \, dy_1 \ldots dy_n$$

$$= \int_{\mathbb{R}^d \times \cdots \times \mathbb{R}^d} e^{-\frac{i\hbar}{2m} \sum_{j,k=1}^n y_j \cdot y_k(t - t_j \vee t_k)} \frac{\hat{F}(y_1, \ldots, y_n)}{(2\pi)^{nd}} \, dy_1 \ldots dy_n$$

On the other hand the Fourier transform of the Schwartz distribution defined by the smooth map $g : \mathbb{R}^{nd} \to \mathbb{C}$ defined by

$$g(y_1, \ldots, y_n) := e^{-\frac{i\hbar}{2m} \sum_{j,k=1}^n y_j \cdot y_k(t - t_j \vee t_k)}, \qquad (y_1, \ldots, y_n) \in \mathbb{R}^d \times \cdots \times \mathbb{R}^d$$

it equal to

$$\hat{g}(x_1, \ldots, x_n) = \prod_{j=1}^n \frac{e^{\frac{im}{2\hbar} \frac{|x_{j+1} - x_j|}{t_{j+1} - t_j}}}{(2\pi i\hbar/m)^{d/2}} \qquad (x_1, \ldots, x_n) \in \mathbb{R}^d \times \cdots \times \mathbb{R}^d$$

and we eventually get

$$\widetilde{\int_{\mathcal{H}}} e^{\frac{i}{2}\|x\|^2} f(x) dx = \int_{\mathbb{R}^d \times \cdots \times \mathbb{R}^d} F(x_1, \ldots x_n) \prod_{j=1}^n G_{t_{j+1} - t_j}(x_{j+1} - x_j) dx_1 \cdots dx_n \, .$$

This result is actually one of the main building block for the construction of the Feynman path integral representation of the solution of Schrödinger equation.

Theorem 4.5. *Let $V \in \mathcal{F}(\mathbb{R}^d)$ and $\psi_0 \in \mathcal{F}(\mathbb{R}^d) \cap L^2(\mathbb{R}^d)$ be Fourier transforms of complex Borel measures in $\mathbb{R}^d$. Then the complex-valued function $f : \mathcal{H}_t \to \mathbb{C}$ on the Cameron-Martin space $\mathcal{H}_t$ defined by*

$$f(\gamma) = e^{-\frac{i}{\hbar} \int_0^t V(\gamma(s) + x) ds} \psi_0(\gamma(0) + x) \qquad \gamma \in \mathcal{H}_t \qquad (4.56)$$

belongs to $\mathcal{F}(\mathcal{H}_t)$ *and the solution of the Schrödinger Eq. (1.4) is given by the infinite dimensional Fresnel integral*

$$\psi(t, x) = \widetilde{\int_{\mathcal{H}_t}} e^{\frac{i}{2\hbar}\|\gamma\|^2} f(\gamma)d\gamma. \tag{4.57}$$

We do not provide the explicit proof of this result here since a more general case will be presented in the next section (see Theorem 4.6 below).

We end this section by mentioning that in [18] the definition (4.52) of infinite dimensional Fresnel integral is generalized to the case where the quadratic exponent $\frac{\|x\|^2}{2}$ is replaced by a quadratic form on $\mathcal{H}$

$$x \mapsto \langle x, Bx \rangle, \qquad x \in D(B) \tag{4.58}$$

associated to a densely defined symmetric linear operator $B : D(B) \subset \mathcal{H} \to \mathcal{H}$. In this case one has to assume that there exists a dense subspace $D \subset \mathcal{H}$ containing the range of B and a symmetric bilinear form $\Delta : D \times D \to \mathbb{C}$ such that $Im(\Delta(x, x)) \leq 0$ for all $x \in D$ and satisfying

$$\Delta(x, By) = \langle x, y \rangle, \qquad \forall x \in D, y \in D(B). \tag{4.59}$$

Moreover, D is assumed to be a Banach space and endowed with a norm $\| \ \|_D$ which is s stronger than the norm $\| \ \|$ of the Hilbert space $\mathcal{H}$. This yields the chain of continuous embeddings

$$D \subset \mathcal{H} \subset D^*,$$

and for any fixed $x \in D$ the map $y \mapsto \Delta(x, y)$ is an element of D^*. This actually gives by (4.59) a mapping from D into D^* which can be considered a left inverse of B.

Eventually, the space $\mathcal{F}(D^*)$ is defined as the set of mappings $f : D^* \to \mathbb{C}$ of the form

$$f(x) = \int_D e^{i\langle x, y \rangle} d\mu_f(y), \qquad x \in D^*, \tag{4.60}$$

where $\langle x, y \rangle$ stands for the dual pairing between $x \in D^*$ and $y \in D$ while μ_f is a complex Borel measure on D.

Definition 4.8. The *Fresnel integral with respect to* Δ, denoted with the symbol

$$\int^{\Delta} e^{\frac{i}{2}\langle x, Bx \rangle} f(x)dx,$$

is the linear functional on $\mathcal{F}(D^*)$ defined on functions $f \in \mathcal{F}(D^*)$ of the form (4.60) by

$$\int^{\Delta} e^{\frac{i}{2}\langle x, Bx \rangle} f(x)dx := \int_D e^{-\frac{i}{2}\Delta(x, x)} d\mu_f(x).$$

Besides the Schrödinger equation with harmonic oscillator potential, the main applications of this generalisation of definition (4.52) can be found in field theory. Already in the 1976 edition of the book [18] a construction of the Feynman path integral for the relativistic quantum boson field is presented, both in the case of free field and in the presence of a regularised interaction term, constructed out of a map $V \in \mathcal{F}(\mathbb{R})$. These pioneering results haven't been developed later as they deserved, since they provided the first construction of the functional integral for the bosonic field in real time, regardless any Euclidean approach or analytic continuation procedure. Another interesting application of the Fresnel integral can be found in Chern-Simons theory, for further details see [41, 42, 16, 40].

4.6   Infinite dimensional Fresnel integrals with polynomial phase

As described in the previous section, the infinite dimensional Fresnel integrals provide the realisation of a functional which extends the projective system of linear functionals associated to Fresnel integrals described in Section 4.3.1 on the one hand, and on the other hand it is also able to give an explicit construction of the Feynman functional introduced in Section 4.3.2. In the present section we tackle the issue introduced in Section 4.3.3 and propose a generalisation of Definition 4.7 to the case where the exponent $\|x\|^2$ appearing in the integrand on the right-hand side of (4.52) is replaced by homogeneous functions of any integer order $p \geq 1$. In addition, we shall see how these *generalized Fresnel integrals* can be applied to the functional integral representation of the solution of higher order heat-type equations of the form (4.36).

Let us start by considering a real separable Banach space $(\mathcal{B}, \| \, \|)$ and the linear space $\mathcal{M}(\mathcal{B})$ of complex Borel measures on $\mathcal{B}$, endowed with the total variation norm. As discussed in Section 3.6, $\mathcal{M}(\mathcal{B})$ is a Banach algebra under convolution. Moreover the linear space $\mathcal{F}(\mathcal{B})$ of complex-valued functions $f : \mathcal{B}^* \to \mathbb{C}$ of the form

$$f(x) = \int_{\mathcal{B}} e^{i\langle x,y \rangle} d\mu_f(y), \qquad x \in \mathcal{B}^*, \tag{4.61}$$

(where $\langle \, , \, \rangle$ denotes the dual pairing between $\mathcal{B}$ and the topological dual space $\mathcal{B}^*$, while $\mu_f \in \mathcal{M}(\mathcal{B})$) endowed with the norm

$$\|f\|_{\mathcal{F}} := |\mu_f|$$

is a Banach algebra of complex-valued functions with unit.

The following definition generalizes the construction of infinite dimensional Fresnel integrals in terms of a class of linear continuous functionals on $\mathcal{F}(\mathcal{B})$.

Definition 4.9. Let $p \in \mathbb{N}$ and let $\Phi_p : \mathcal{B} \to \mathbb{C}$ be a continuous homogeneous map of order p, i.e. such that:

(1) $\Phi_p(\lambda x) = \lambda^p \Phi_p(x)$, for all $\lambda \in \mathbb{R}$, $x \in \mathcal{B}$,
(2) $\mathrm{Re}(\Phi_p(x)) \leq 0$ for all $x \in \mathcal{B}$.

The functional $I_{\Phi_p} : \mathcal{F}(\mathcal{B}) \to \mathbb{C}$, given on functions $f \in \mathcal{F}(\mathcal{B})$ of the form (4.61) by

$$I_{\Phi_p}(f) := \int_{\mathcal{B}} e^{\Phi_p(x)} d\mu_f(x), \tag{4.62}$$

is called *infinite dimensional Fresnel integral on $\mathcal{B}^*$ with phase function Φ_p.*

Since by assumption $Re(\Phi_p(x)) \leq 0$, the integrand in the definition of $I_{\Phi_p}(f)$ is uniformly bounded, i.e.

$$|e^{\Phi_p(x)}| \leq 1 \quad \forall x \in \mathcal{B},$$

hence the following inequality

$$|I_{\Phi_p}(f)| \leq \int_{\mathcal{B}} |e^{\Phi_p}| d|\mu|(x) \leq \|\mu\| = \|f\|_{\mathcal{F}}$$

yields the continuity of I_{Φ_p} in the $\| \ \|_{\mathcal{F}}$-norm. Moreover the functional is normalized, i.e. $I_{\Phi_p}(1) = 1$.

We can now give an interesting example of infinite dimensional Fresnel integral with polynomial phase function.

Fixed a $p \in \mathbb{N}$, with $p \geq 2$, let us consider the Banach space $\mathcal{B}_p$ of absolutely continuous maps $\gamma : [0, t] \to \mathbb{R}$, with $\gamma(t) = 0$ and weak derivative $\dot{\gamma}$ belonging to $L^p([0, t])$. The space $\mathcal{B}_p$ will be endowed with the norm:

$$\|\gamma\|_{\mathcal{B}_p} = \left(\int_0^t |\dot{\gamma}(s)|^p ds \right)^{1/p}.$$

Clearly the transformation $T : \mathcal{B}_p \to L^p([0, t])$ mapping an element $\gamma \in \mathcal{B}_p$ to its weak derivative $\dot{\gamma}$ is an isomorphism and its inverse $T^{-1} : L^p([0, t]) \to \mathcal{B}_p$ is given by

$$T^{-1}(v)(s) = -\int_s^t v(u)du \qquad v \in L^p([0, t]). \tag{4.63}$$

Analogously, the dual space $\mathcal{B}_p^*$ is isomorphic to $L_q([0,t]) = (L_p([0,t]))^*$, with $\frac{1}{p} + \frac{1}{q} = 1$, and the pairing between an element $\eta \in \mathcal{B}_p^*$ and $\gamma \in \mathcal{B}_p$ is given by

$$\langle \eta, \gamma \rangle = \int_0^t \dot{\eta}(s)\dot{\gamma}(s)ds \qquad \dot{\eta} \in L_q([0,t]), \gamma \in \mathcal{B}_p.$$

In fact, the map (4.63) allows to verify that $\mathcal{B}_p^*$ is isomorphic to $\mathcal{B}_q$.

Let $\mathcal{F}(\mathcal{B}_p)$ be the space of complex mappings $f : \mathcal{B}_q \to \mathbb{C}$ of the form

$$f(\eta) = \int_{\mathcal{B}_p} e^{i \int_0^t \dot{\eta}(s)\dot{\gamma}(s)ds} d\mu_f(\gamma), \qquad \eta \in \mathcal{B}_q, \mu_f \in \mathcal{M}(\mathcal{B}_p),$$

and let $\Phi_p : \mathcal{B}_p \to \mathbb{C}$ be the "phase function"

$$\Phi_p(\gamma) := (-1)^p \alpha \int_0^t \dot{\gamma}(s)^p ds \qquad \gamma \in \mathcal{B}$$

where $\alpha \in \mathbb{C}$ is assumed to be a complex constant such that

- $\mathrm{Re}(\alpha) \leq 0$ if p is even,
- $\mathrm{Re}(\alpha) = 0$ if p is odd.

In particular, in the case where p is even $\Phi_p(\gamma)$ turns out to be proportional to the $\mathcal{B}_p$-norm of the vector γ. Under this set of assumptions, we can define and compute the infinite dimensional Fresnel integral with phase function Φ_p of any function $f \in \mathcal{F}(\mathcal{B}_p)$, $f = \hat{\mu}_f$, and it is given by

$$I_{\Phi_p}(f) = \int_{\mathcal{B}_p} e^{(-1)^p \alpha \int_0^t \dot{\gamma}(s)^p ds} d\mu_f(\gamma). \qquad (4.64)$$

Let us now consider the initial value problem associated to the following p−order partial differential equation

$$\begin{cases} \dfrac{\partial}{\partial t} u(t,x) = (-i)^p \alpha \dfrac{\partial^p}{\partial x^p} u(t,x) \\ u(0,x) = u_0(x), \qquad x \in \mathbb{R}, t \in [0,+\infty) \end{cases} \qquad (4.65)$$

where $p \in \mathbb{N}$, $p \geq 2$, and $\alpha \in \mathbb{C}$ is a complex constant fulfilling the assumptions above. In particular, this implies the following inequality

$$|e^{\alpha t x^p}| \leq 1 \qquad \forall x \in \mathbb{R}, \forall t \in [0,+\infty).$$

In the case where $p = 2$ and $\alpha \in \mathbb{R}$, $\alpha < 0$, we obtain the heat equation, while for $p = 2$ and $\alpha = i$ Eq. (4.65) is the Schrödinger equation. In the following we shall focus ourselves on the case where $p \geq 3$.

Let us consider the fundamental solution $G_t^p(x,y)$ of Eq. (4.65), which for $t > 0$ is given by $G_t^p(x-y) = g_t^p(x-y)$,, where $g_t^p : \mathbb{R} \to \mathbb{C}$ is the mapping defined by the following Fourier integral:

$$g_t^p(x) := \frac{1}{2\pi} \int e^{ikx} e^{\alpha t k^p} dk, \qquad x \in \mathbb{R}. \tag{4.66}$$

The following lemma states some regularity properties of the distribution g_t^p that will be used later.

Lemma 4.1. *For $p \geq 3$ and $t > 0$ the tempered distribution $g_t^p \in S'(\mathbb{R})$ defined by (4.66) belongs to $C^\infty(\mathbb{R}) \cap L^1(\mathbb{R})$.*

For the proof of this technical result we refer to [253].

The following result creates an interesting connection between the functional I_{Φ_p} and the high order pde (4.65).

Lemma 4.2. *Let $f : \mathcal{B}_q \to \mathbb{C}$ be a cylindric function of the following form:*

$$f(\eta) = F(\eta(t_1), \eta(t_2), \ldots, \eta(t_n)), \qquad \eta \in \mathcal{B}_q,$$

with $0 \leq t_1 < t_2 < \ldots < t_n < t$ and $F : \mathbb{R}^n \to \mathbb{C}$ a function belonging to the space $F \in \mathcal{F}(\mathbb{R}^n)$ of Fourier transforms of complex Borel measures on $\mathbb{R}^n$, with:

$$F(x_1, x_2, \ldots, x_n) = \int_{\mathbb{R}^n} e^{i \sum_{k=1}^n y_k x_k} d\nu_F(y_1, \ldots, y_n), \qquad \nu_F \in \mathcal{M}(\mathbb{R}^n).$$

Then $f \in \mathcal{F}(\mathcal{B}_p)$ and its infinite dimensional Fresnel integral with phase function Φ_p is given by

$$I_{\Phi_p}(f) = \int_{\mathbb{R}^n} F(x_1, x_2, \ldots, x_n) \Pi_{k=1}^n G_{t_{k+1}-t_k}^p(x_{k+1}, x_k) dx_k, \tag{4.67}$$

where $x_{n+1} \equiv 0$, $t_{n+1} \equiv t$ and G_s^p is the fundamental solution (4.38) of the high order heat-type Eq. (4.65).

Remark 4.3. One can easily see that Lemma 4.1, stating the L^1-summability of the function g_t^p defined by (4.66), assures that the integral in (4.67) is absolutely convergent. Indeed, we have:

$$\int_{\mathbb{R}^n} |F(x_1, x_2, \ldots, x_n)| \, \Pi_{k=1}^n |G_{t_{k+1}-t_k}^p(x_{k+1}, x_k)| dx_k$$

$$\leq \|F\|_{\mathcal{F}(\mathbb{R}^n)} \int_{\mathbb{R}^n} \Pi_{k=1}^n |g_{t_{k+1}-t_k}^p(y_k)| dy_k < \infty \tag{4.68}$$

where $y_k = x_{k+1} - x_k$ and we have used the inequality $\|F\|_\infty \leq \|F\|_{\mathcal{F}(\mathbb{R}^n)}$.

Proof. By the representation

$$f(\eta) = F(\eta(t_1), \eta(t_2), \ldots, \eta(t_n)) = \int_{\mathbb{R}^n} e^{i\sum_{k=1}^n y_k \eta(t_k)} d\nu_F(y_1, \ldots, y_n)), \quad \eta \in \mathcal{B}_q,$$

one can easily verify that $f \in \mathcal{F}(\mathcal{B}_p)$, indeed:

$$e^{iy\eta(s)} = \int_{\mathcal{B}_p} e^{i\langle \eta, \gamma \rangle} \delta_{yv_s}(\gamma),$$

where $v_s \in \mathcal{B}_p$ is the element of $\mathcal{B}_p$ defined by

$$\langle \eta, v_s \rangle = \eta(s), \qquad \forall \eta \in \mathcal{B}_q,$$

and it is equal to

$$v_s(\tau) = \chi_{[0,s]}(t-s) + \chi_{(s,t]}(t-\tau)s \qquad \tau \in [0,t].$$

We then have:

$$\begin{aligned}
I_{\Phi_p}(f) &= \int_{\mathbb{R}^n} e^{(-1)^p \alpha \int_0^t \left(\sum_{k=1}^n y_k \dot{v}_{t_k}(\tau) \right)^p d\tau} d\nu_F(y_1, \ldots, y_n) \\
&= \int_{\mathbb{R}^n} e^{\alpha \int_0^t \left(\sum_{k=1}^n y_k \chi_{(t_k, t]}(\tau) \right)^p d\tau} d\nu_F(y_1, \ldots, y_n) \\
&= \int_{\mathbb{R}^n} e^{\alpha \int_0^t \left(\sum_{k=1}^n \chi_{(t_k, t_{k+1}]}(\tau) \sum_{j=1}^k y_j \right)^p d\tau} d\nu_F(y_1, \ldots, y_n) \\
&= \int_{\mathbb{R}^n} e^{\alpha \sum_{k=1}^n (\sum_{j=1}^k y_j)^p (t_{k+1} - t_k)} d\nu_F(y_1, \ldots, y_n). \qquad (4.69)
\end{aligned}$$

On the other hand the last line of Eq. (4.69) coincides with

$$\int_{\mathbb{R}^n} F(x_1, x_2, \ldots, x_n) \Pi_{k=1}^n G_{t_{k+1}-t_k}^p(x_{k+1}, x_k) dx_k,$$

indeed, by a change of variables and Fubini's theorem, the latter is equal to

$$\begin{aligned}
&\int_{\mathbb{R}^n} \left(\int_{\mathbb{R}^n} e^{-i\sum_{k=1}^n \xi_k \sum_{l=1}^k y_l} \Pi_{k=1}^n g_{t_{k+1}-t_k}^p(\xi_k) d\xi_k \right) d\nu_F(y_1, \ldots, y_n) \\
&= \int_{\mathbb{R}^n} e^{\alpha \sum_{k=1}^n (\sum_{l=1}^k y_l)^p (t_{k+1}-t_k)} d\nu_F(y_1, \ldots, y_n). \qquad \square
\end{aligned}$$

The following result is a direct consequence of Lemma 4.2.

Corollary 4.1. *Let $u_0 \in \mathcal{F}(\mathbb{R})$. Then the cylindric function $f_0 : \mathcal{B}_q \to \mathbb{C}$ defined by*

$$f_0(\eta) := u_0(x + \eta(0)), \qquad x \in \mathbb{R}, \eta \in \mathcal{B}_q,$$

belongs to $\mathcal{F}(\mathcal{B}_p)$ and its infinite dimensional Fresnel integral with phase function Φ_p provides a representation for the solution of the Cauchy problem (4.65), in the sense that the function $u(t,x) := I_{\Phi_p}(f_0)$ has the form $u(t,x) = \int_{\mathbb{R}} G_t(x,y) u_0(y) dy$.

Let us consider now the Cauchy problem of the form

$$\begin{cases} \dfrac{\partial}{\partial t}u(t,x) = (-i)^p \alpha \dfrac{\partial^p}{\partial x^p}u(t,x) + V(x)u(t,x) \\ u(0,x) = u_0(x), \qquad x \in \mathbb{R}, t \in [0,+\infty), \end{cases} \tag{4.70}$$

where $p \in \mathbb{N}$, $p \geq 2$, and $\alpha \in \mathbb{C}$ as above (in such a way that $|e^{\alpha t x^p}| \leq 1$ for all $x \in \mathbb{R}, t \in [0,+\infty)$), while $V : \mathbb{R} \to \mathbb{C}$ is a bounded continuous function. Under these assumption the Cauchy problem (4.70) is well posed on $L^2(\mathbb{R})$. Indeed the operator $\mathcal{D}_p : D(\mathcal{D}_p) \subset L^2(\mathbb{R}) \to L^2(\mathbb{R})$ defined by

$$D(\mathcal{D}_p) := H^p = \{u \in L^2(\mathbb{R}), k \mapsto k^p \hat{u}(k) \in L^2(\mathbb{R})\},$$
$$\widehat{\mathcal{D}_p u}(k) := k^p \hat{u}(k),\ u \in D(\mathcal{D}_p),$$

($\hat{u}$ denoting the Fourier transform of u) is self-adjoint. Since by assumption the equality $|e^{\alpha t x^p}| \leq 1$ holds for all $x \in \mathbb{R}$ and $t \in [0,+\infty)$, one has that the operator $\Lambda := \alpha \mathcal{D}_p$ generates a strongly continuous semigroup $(e^{tA})_{t \geq 0}$ on $L^2(R)$. By denoting with $B : L^2(\mathbb{R}) \to L^2(\mathbb{R})$ the bounded multiplication operator defined by

$$Bu(x) = V(x)u(x), \qquad u \in L^2(\mathbb{R}),$$

one has that the operator sum $A + B : D(A) \subset L^2(\mathbb{R}) \to L^2(\mathbb{R})$ generates a strongly continuous semigroup $(T(t))_{t \geq 0}$ on $L^2(\mathbb{R})$. Moreover, given a $u \in L^2(\mathbb{R})$, the vector $T(t)u$ can be computed by means of the convergent (in the $L^2(\mathbb{R})$-norm) Dyson series (see [184, Th. 13.4.1]):

$$T(t)u = \sum_{n=0}^{\infty} S_n(t)u, \tag{4.71}$$

where $S_0(t)u = e^{tA}u$ and $S_n(t)u = \int_0^t e^{(t-s)A}V S_{n-1}(s)u\,ds$. By passing to a subsequence, the series (4.71) converges also a.e. in $x \in \mathbb{R}$ giving

$$T(t)u(x)$$
$$= \sum_{n=0}^{\infty} \int \cdots \int_{0 \leq s_1 \leq \cdots \leq s_n \leq t} \int_{\mathbb{R}^{n+1}} V(x_1)\dots V(x_n)G_{t-s_n}(x,x_n)G_{s_n-s_{n-1}}(x_n,x_{n-1})$$
$$\dots G_{s_1}(x_1,x_0)u_0(x_0)dx_0\dots dx_n\,ds_1\dots ds_n, \qquad a.e.\ x \in \mathbb{R}. \tag{4.72}$$

Under suitable assumptions on the initial datum u_0 and the potential V, we are going to construct a representation of the solution of Eq. (4.70) in $L^2(\mathbb{R})$ in terms of an infinite dimensional oscillatory integral with polynomial phase.

Theorem 4.6. *Let $u_0 \in \mathcal{F}(\mathbb{R}) \cap L^2(\mathbb{R})$ and $V \in \mathcal{F}(\mathbb{R})$, with $u_0(x) = \int_{\mathbb{R}} e^{ixy} d\mu_0(y)$ and $V(x) = \int_{\mathbb{R}} e^{ixy} d\nu(y)$, $\mu_0, \nu \in \mathcal{M}(\mathbb{R})$. Then the function $f_{t,x} : \mathcal{B}_q \to \mathbb{C}$ defined by*

$$f_{t,x}(\eta) := u_0(x + \eta(0)) e^{\int_0^t V(x+\eta(s))ds}, \qquad x \in \mathbb{R}, \eta \in \mathcal{B}_q, \qquad (4.73)$$

belongs to $\mathcal{F}(\mathcal{B}_p)$ and its infinite dimensional Fresnel integral with phase function Φ_p provides a representation for the solution of the Cauchy problem (4.70).

Remark 4.4. By Plancherel's theorem the assumption that $u_0 \in \mathcal{F}(\mathbb{R}) \cap L^2(\mathbb{R})$ is equivalent to the fact that u_0 is the Fourier transform of a function belonging to $L^1(\mathbb{R}) \cap L^2(\mathbb{R})$.

Proof.

Let $\mu_V \in \mathcal{M}(\mathcal{B}_p)$ be the measure defined by

$$\int_{\mathcal{B}_p} f(\gamma) d\mu_V(\gamma) = \int_0^t \int_{\mathbb{R}} e^{ixy} f(y\, v_s) d\nu(y) ds, \qquad f \in C_b(\mathcal{B}_p),$$

where $v_s \in \mathcal{B}_p$ is the function $v_s(\tau) = \chi_{[0,s]}(\tau)(t-s) + \chi_{(s,t]}(t-\tau)s$. One can easily verify that $\|\mu_V\|_{\mathcal{M}(\mathcal{B}_p)} \leq t\|\nu\|_{\mathcal{M}(\mathbb{R})}$ and the map

$$\eta \in \mathcal{B}_q \mapsto \int_0^t V(x+\eta(s))ds$$

is the Fourier transform of μ_V. Analogously the map

$$\eta \in \mathcal{B}_q \mapsto \exp\left(\int_0^t V(x+\eta(s))ds\right)$$

is the Fourier transform of the measure $\nu_V \in \mathcal{M}(\mathcal{B}_p)$ given by $\nu_V = \sum_{n=0}^{\infty} \frac{1}{n!}\mu_V^{*n}$, where $*$ stands for convolution and μ_V^{*n} denotes the n-fold convolution of μ_V with itself, i.e. for any $f \in C_b(\mathcal{B}_p)$:

$$\int_{\mathcal{B}_p} f(\gamma) d\nu_V(\gamma) = \sum_{n=0}^{\infty} \frac{1}{n!} \int_{\mathcal{B}_p \times \mathcal{B}_p \times ... \times \mathcal{B}_p} f(\gamma_1+...+\gamma_n) d\mu_V(\gamma_1) \cdots d\mu_V(\gamma_n).$$

The series is convergent in the $\mathcal{M}(\mathcal{B}_p)$-norm and one has $\|\nu_V\|_{\mathcal{M}(\mathcal{B}_p)} \leq e^{t\|\nu\|_{\mathcal{M}(\mathbb{R})}}$. Further, by Lemma 4.2 the cylindric function $\eta \mapsto u_0(x + \eta(0))$, $\eta \in \mathcal{B}_q$, is an element of $\mathcal{F}(\mathcal{B}_p)$. More precisely, it is the Fourier transform of the measure ν_{u_0} determined by

$$\int_{\mathcal{B}_p} f(\gamma) d\nu_{u_0}(\gamma) = \int_{\mathbb{R}} e^{ixy} f(y\, v_0) d\mu_0(y), \qquad f \in C_b(\mathcal{B}_p).$$

We can then conclude that the map $f_{t,x} : \mathcal{B}_q \to \mathbb{C}$ defined by (4.73) belongs to $\mathcal{F}(\mathcal{B}_p)$ and its infinite dimensional Fresnel integral $I_{\Phi_p}(f_{t,x})$ with phase function Φ_p is given by

$$\sum_{n=0}^{\infty} \frac{1}{n!} \int_{\mathcal{B}_p} e^{(-1)^p \alpha \int_0^y \dot{\gamma}(s)^p ds} d\nu_{u_0} * \mu_V * \cdots * \mu_V$$

$$= \sum_{n=0}^{\infty} \frac{1}{n!} \int_0^t \cdots \int_0^t I_{\Phi_p} \big(u_0(x+\eta(0))V(x+\eta(s_1)) \ldots V(x+\eta(s_n)) \big) ds_1 \cdots ds_n .$$

By the symmetry of the integrand the latter is equal to

$$\sum_{n=0}^{\infty} \int \cdots \int_{0 \leq s_1 \leq \cdots \leq s_n \leq t} I_{\Phi_p} \big(u_0(x+\eta(0))V(x+\eta(s_1)) \cdots V(x+\eta(s_n)) \big) ds_1 \cdots ds_n .$$

By Lemma 4.2 we finally obtain

$$\sum_{n=0}^{\infty} \int \cdots \int_{0 \leq s_1 \leq \cdots \leq s_n \leq t} \int_{\mathbb{R}^{n+1}} u_0(x+x_0)V(x+x_1) \cdots V(x+x_n)G_{s_1}(x_1,x_0)$$

$$G_{s_2-s_1}(x_2,x_1) \cdots G_{t-s_n}(0,x_n)dx_0 dx_1 \cdots dx_n ds_1 \cdots ds_n,$$

which coincides with the Dyson series (4.72) for the solution of the high-order PDE (4.70), as one can easily verify by means of a change of variables argument. $\square$

Chapter 5

Infinite dimensional oscillatory integrals

In the theory of infinite dimensional Fresnel integrals on real separable Hilbert spaces $\mathcal{H}$ described in Section 4.5, one of the main issues is the restriction on the class of integrable functions, in other words the domain of the functional. Indeed, the infinite dimensional Fresnel integral of a mapping $f : \mathcal{H} \to \mathbb{C}$

$$\widetilde{\int_{\mathcal{H}}} e^{\frac{i}{2}\|x\|^2} f(x) dx$$

can be defined only for f belonging to the Banach algebra $\mathcal{F}(\mathcal{H})$ of Fourier transform of complex Borel measures on $\mathcal{H}$. In the application of this theory to the mathematical definition of Feynman path integrals, such restrictions limit the possibility of constructing a representation of the solution of the Schrödinger equation to the case in which the potential V in Eq. (1.4) belongs to the set $\mathcal{F}(\mathbb{R}^d)$ of Fourier transform of Borel measures. This class is actually rather narrow and does not include most of the physically relevant potentials. In addition, aiming to the functional integral construction of non-trivial interacting quantum field theories, the discussion of polynomially growing potentials would be of fundamental importance. It is thus important to provide an alternative definition of Fresnel integrals in infinite dimensions which allows to enlarge the domain of the functionals defined in Section 4.5. The first step to accomplish this task was done in the 80s by D. Elworthy and A. Truman [129]. Their approach was extensively developed in the 90s by S. Albeverio and Z. Brzezniak [7], leading to the definition of *infinite dimensional oscillatory integrals*. The leading idea of this approach is to extend the definition and the main properties of classical oscillatory integrals on $\mathbb{R}^n$

$$\int_{\mathbb{R}^n} e^{\frac{i}{\hbar}\Phi(x)} f(x) dx$$

to the case the integration domain is a real separable (infinite dimensional) Hilbert space $\mathcal{H}$

$$\int_{\mathcal{H}} e^{\frac{i}{\hbar}\Phi(x)} f(x)dx$$

This definition has several advantages, in particular it allows to expand the class of integrable functions including those with polynomial growth, as proved by S. Albeverio and S. Mazzucchi in 2004 [26, 28], and to construct the Feynman path integral representation of the solution of the Schrödinger equation with polynomially growing potentials. It is important to point out that a byproduct of these result is the proof of a deep relation between infinite dimensional oscillatory integrals and Gaussian integrals, paving the way for the application of techniques of stochastic analysis to the theory of Feynman integration.

5.1   Finite dimensional oscillatory integrals

The study of oscillatory integrals of the form

$$\int_{\mathbb{R}^n} e^{\frac{i}{\hbar}\Phi(x)} f(x)dx, \tag{5.1}$$

(where $\Phi : \mathbb{R}^n \to \mathbb{R}$ is a smooth phase function, $f : \mathbb{R}^n \to \mathbb{C}$ is the function which is integrated and $\hbar \in \mathbb{R} \setminus \{0\}$) is a classical topic of investigation, largely developed in connection with several applications in mathematics, such as the theory of Fourier integral operators [186, 187], and in physics (such as in optics).

Well known examples of integrals of the above form are the Fresnel integrals

$$\int e^{ix^2} dx, \tag{5.2}$$

applied in the theory of wave diffraction [308] and the Airy integrals

$$\int e^{ix^3} dx, \tag{5.3}$$

applied in the theory of rainbow.

In the mathematical literature, a particular interest has been devoted to the study of the asymptotic behavior of integrals (5.1) when $\hbar$ is regarded as a small parameter converging to 0. Originally introduced by Stokes and Kelvin and successively developed by several mathematicians, in particular van der Corput, the "stationary phase method" provides a powerful tool to handle the asymptotics of (5.1) as $\hbar \downarrow 0$. According to it, the main contribution to the asymptotic behavior of the integral should come from those points $x_c \in \mathbb{R}^n$ which belong to the critical manifold:

$$\mathcal{C}(\Phi) := \{x \in \mathbb{R}^n, \mid \Phi'(x) = 0\}, \tag{5.4}$$

that is the points which make stationary the phase function Φ.

Beautiful mathematical work on oscillatory integrals and the method of stationary phase is connected with the mathematical classification of singularities of algebraic and geometric structures (Coxeter indices, catastrophe theory), see, e.g. [44, 47, 125]. We shall give more details on this topic in Chapter 7.

If the function f in Eq. (5.1) is summable, the integral (5.1) is well defined as a convergent Lebesgue integral, but in several interesting cases, as the examples (5.2) and (5.3) show, this condition is not satisfied. Indeed for several applications it is convenient to introduce a definition which allows to handle a general class of functions Φ and f. As in the case of the convergence of some improper Riemann integrals, the sign of the function which is integrated plays an important role. In particular, in the convergence of oscillatory integrals the cancellations due to the oscillatory behaviour of the integrand $e^{\frac{i}{\hbar}\Phi(x)}$ are fundamental and have to be taken into account in proposing a definition of Eq. (5.1) which allows to handle not only integrals of the form (5.2) and (5.3), but also expression as

$$\int e^{ix^2} x^m dx, \qquad m \in \mathbb{N}.$$

It is worthwhile to recall that this particular feature makes oscillatory integrals the suitable mathematical tool describing the physical concept of coherent superposition and interference.

We give here a definition which is a generalization of the one proposed in [129], due to Hörmander [186, 187], and propose a related more general definition of *oscillatory integral in the Σ-sense*.

Definition 5.1. Let Φ be a continuous real-valued function on $\mathbb{R}^n$. The oscillatory integral on $\mathbb{R}^n$, with $\hbar \in \mathbb{R} \setminus \{0\}$,

$$\int_{\mathbb{R}^n} e^{\frac{i}{\hbar}\Phi(x)} f(x) dx,$$

is well defined if for each test function $\phi \in \mathcal{S}(\mathbb{R}^n)$, such that $\phi(0) = 1$, the limit of the sequence of absolutely convergent integrals

$$\lim_{\epsilon \downarrow 0} \int_{\mathbb{R}^n} e^{\frac{i}{\hbar}\Phi(x)} \phi(\epsilon x) f(x) dx,$$

exists and is independent of ϕ. In this case the limit is denoted by

$$\int_{\mathbb{R}^n}^{o} e^{\frac{i}{\hbar}\Phi(x)} f(x) dx.$$

If the same holds only for ϕ such that $\phi(0) = 1$ and $\phi \in \Sigma$, for some subset Σ of $\mathcal{S}(\mathbb{R}^N)$, we say that the oscillatory integral exists in the Σ-sense and we shall denote it by the same symbol.

Hörmander in his work on Fourier integral operators [186, 187] gives a detailed treatment of oscillatory integrals and describes a large class of integrable functions, called *symbols*.

Definition 5.2. A C^∞ map $f : \mathbb{R}^n \to \mathbb{C}$ belongs to the space of symbols $S_\lambda^N(\mathbb{R}^n)$, where N, λ are two real numbers and $0 < \lambda \le 1$, if for each $\alpha = (\alpha_1, \ldots, \alpha_n) \in \mathbb{N}_0^n$ there exists a constant $C_\alpha \in \mathbb{R}$ such that

$$\left| \frac{d^{\alpha_1}}{dx_1^{\alpha_1}} \cdots \frac{d^{\alpha_n}}{dx_n^{\alpha_n}} f \right| \le C_\alpha (1 + |x|)^{N - \lambda|\alpha|}, \qquad |x| \to \infty, \qquad (5.5)$$

where $|\alpha| = \alpha_1 + \alpha_2 + \cdots + \alpha_n$.

One can prove that S_λ^N is a Fréchet space under the topology defined by taking as seminorms $|f|_\alpha$ the best constants C_α in (5.5) (see [186]). The space increases as N increases and λ decreases. If $f \in S_\lambda^N$ and $g \in S_\lambda^M$, then $fg \in S_\lambda^{N+M}$. We denote

$$\bigcup_N S_\lambda^N \equiv S_\lambda^\infty.$$

It is quite simple to verify that S_λ^N, with $\lambda = 1$, includes for instance the homogeneous polynomials of degree N. The following theorem shows that, under rather general assumptions on the phase functions Φ, if f belong to S_λ^∞ for some $\lambda \in (0, 1]$, the oscillatory integral $\int^o e^{\frac{i}{\hbar}\Phi(x)} f(x) dx$ is well defined.

Theorem 5.1. *Let Φ be a real-valued C^2 function on $\mathbb{R}^n$ with the critical set $\mathcal{C}(\Phi)$ (defined by (5.4)) being finite. Let us assume that for each $N \in \mathbb{N}$ there exists a $k \in \mathbb{N}$ such that $\frac{|x|^{N+1}}{|\nabla\Phi(x)|^k}$ is bounded for $|x| \to \infty$. Let $f \in S_\lambda^M$, with $M, \lambda \in \mathbb{R}$, $0 < \lambda \le 1$. Then the oscillatory integral $\int^o e^{\frac{i}{\hbar}\Phi(x)} f(x) dx$ exists for each $\hbar \in \mathbb{R} \setminus \{0\}$.*

Proof. We follow the method of Hörmander [186], see also [129, 7].

Let us assume that the phase function $\Phi(x)$ has l stationary points $c_1, \ldots, c_l$, that is

$$\nabla\Phi(c_i) = 0, \qquad i = 1, \ldots, l.$$

Let us choose a suitable partition of unity $1 = \sum_{i=0}^{l} \chi_i$, where χ_i, $i = 1, \ldots, l$, are $C_0^\infty(\mathbb{R}^n)$ functions constant equal to 1 in a open ball centered in the stationary point c_i respectively and $\chi_0 = 1 - \sum_{i=1}^{l} \chi_i$. Each of the integrals

$$I_i(f) \equiv \int_{\mathbb{R}^n} e^{\frac{i}{\hbar}\Phi(x)} \chi_i(x) f(x) dx, \qquad i = 1, \ldots, l,$$

is well defined in Lebesgue sense since $f\chi_i \in C_0(\mathbb{R}^n)$.

Let

$$I_0 \equiv \int_{\mathbb{R}^n} e^{\frac{i}{\hbar}\Phi(x)} \chi_0(x) f(x) dx.$$

To see that I_0 is a well defined oscillatory integral let us introduce the operator L^+ given by

$$L^+ g(x) = -i\hbar \frac{\chi_0(x)}{|\nabla\Phi(x)|^2} \nabla\Phi(x) \nabla g(x),$$

while its adjoint in $L^2(\mathbb{R}^n)$ is given by

$$Lf(x) = i\hbar \frac{\chi_0(x)}{|\nabla\Phi(x)|^2} \nabla\Phi(x) \nabla f(x) + i\hbar \, \mathrm{div}\, \left(\frac{\chi_0(x)}{|\nabla\Phi(x)|^2} \nabla\Phi(x) \right) f(x).$$

Let us choose $\phi \in \mathcal{S}(\mathbb{R}^n)$, such that $\phi(0) = 1$. It is easy to see that if $f \in S_\lambda^M$ then f_ϵ, defined as $f_\epsilon(x) := \phi(\epsilon x) f(x)$, belongs to $S_\lambda^{M+1} \cap \mathcal{S}(\mathbb{R}^n)$ for any $\epsilon > 0$. By iterated application of the Stokes formula, we have:

$$\int_{\mathbb{R}^n} e^{\frac{i}{\hbar}\Phi(x)} \phi(\epsilon x) f(x) \chi_0(x) dx = \int_{\mathbb{R}^n} L^+(e^{\frac{i}{\hbar}\Phi(x)}) \phi(\epsilon x) f(x) dx$$

$$= \int_{\mathbb{R}^n} e^{\frac{i}{\hbar}\Phi(x)} L f_\epsilon(x) dx. \quad (5.6)$$

By iterating the procedure k times, for k sufficiently large, one obtains an absolutely convergent integral and it is possible to pass to the limit for $\epsilon \to 0$ by the Lebesgue dominated convergence theorem.

Considering $\sum_{i=0}^{l} I_i(f)$ we have, by the existence result proved for I_0 and the additivity property of oscillatory integrals, that $\int_{\mathbb{R}^n}^o e^{\frac{i}{\hbar}\Phi(x)} f(x) dx$ is well defined and equal to $\sum_{i=0}^{l} I_i(f)$. $\qquad\square$

Remark 5.1. If $\mathcal{C}(\Phi)$ has countably many non-accumulating points $\{x_c^i\}_{i\in\mathbb{N}}$, the same method yields $\int_{\mathbb{R}^n}^o e^{\frac{i}{\hbar}\Phi(x)} f(x) dx = \sum_{i=0}^{\infty} I_i(f)$ provided this sum converges.

There are partial extensions of the above construction in the case of critical points which form a submanifold in $\mathbb{R}^n$ [125], or are degenerate [44], see also [111].

In particular we have proved the existence for $f \in S_\lambda^\infty$, $0 < \lambda \leq 1$, of the oscillatory integrals $\int e^{ix^M} f(x)dx$, with M arbitrary. For $M = 2$ one has the Fresnel integral of [17, 18], for $M = 3$ one has integrals called, for $n = 1$, Airy integrals [187].

5.2   The Parseval type equality

This section is devoted to the study of the *Fresnel integrals*, a particular class of oscillatory integrals with quadratic phase function:

$$\int_{\mathbb{R}^n}^{o} e^{\frac{i}{\hbar}\|x\|^2} f(x)dx. \tag{5.7}$$

We shall see that in this case it is possible to identify a subclass of integrable functions f, such that the corresponding Fresnel integral can be explicitly computed in terms of a Parseval type equality. This property is particular important because, as we shall see below, it allows the generalization of the definition to an infinite dimensional setting.

In the following we shall denote by $\mathcal{H}$ an (finite or infinite dimensional) real separable Hilbert space. The norm will be denoted by $\|\ \|$ and the inner product with $\langle\ ,\ \rangle$. If the dimension of $\mathcal{H}$ is finite, $\dim(\mathcal{H}) = n$, we shall identify it with $\mathbb{R}^n$.

Let us consider the Banach algebra $\mathcal{M}(\mathcal{H})$ of complex Borel measures on $\mathcal{H}$ endowed with the total variation norm, and the corresponding Banach algebra $\mathcal{F}(\mathcal{H})$ of functions $f : \mathcal{H} \to \mathbb{C}$ which are the Fourier transforms of measures $\mu_f \in \mathcal{M}(\mathcal{H})$:

$$f(x) = \hat{\mu}_f(x) := \int_{\mathcal{H}} e^{i\langle y,x\rangle} d\mu_f(y), \qquad \mu_f \in \mathcal{M}(\mathcal{H}), x \in \mathcal{H}.$$

(see Sections 2.8 and 3.6).

A complete characterizations of the functions belonging to $\mathcal{F}(\mathcal{H})$ is not easy. The elements in $\mathcal{F}(\mathcal{H})$ are bounded uniformly continuous functions on $\mathcal{H}$ and the following inequality holds:

$$|f|_\infty \leq \|f\|, \qquad f \in \mathcal{F}(\mathcal{H}),$$

where $|f|_\infty$ denotes the sup-norm. Moreover, since each $\mu \in \mathcal{M}(\mathcal{H})$ is a finite linear complex combination of positive probability measures, every $f \in \mathcal{F}(\mathcal{H})$ is a corresponding linear combination of positive semidefinite functions (see Theorem 2.22). In the finite dimensional case $\mathcal{H} = \mathbb{R}^n$, one has the inclusion $\mathcal{S}(\mathbb{R}^n) \subset \mathcal{F}(\mathbb{R}^n)$.

For the applications that will follow, it is convenient to introduce a particular definition of finite dimensional Fresnel integrals, which differs from the general definition 5.1 for the presence of a normalization constant.

Definition 5.3. A function $f : \mathbb{R}^n \to \mathbb{C}$ is called Fresnel integrable if for each $\phi \in \mathcal{S}(\mathbb{R}^n)$ such that $\phi(0) = 1$ the limit

$$\lim_{\epsilon \to 0} (2\pi i \hbar)^{-n/2} \int e^{\frac{i}{2\hbar} \langle x, x \rangle} f(x) \phi(\epsilon x) dx \tag{5.8}$$

exists and is independent of ϕ. In this case the limit is called the Fresnel integral of f and denoted by

$$\widetilde{\int} e^{\frac{i}{2\hbar} \langle x, x \rangle} f(x) dx. \tag{5.9}$$

The factor $(2\pi i \hbar)^{-n/2}$ in the definition of the Fresnel integral plays the role of a normalization constant. Indeed if $f = 1$, it is easy to verify that

$$\widetilde{\int} e^{\frac{i}{2\hbar} \langle x, x \rangle} f(x) dx = 1.$$

The description of the full class of Fresnel integrable functions is still an open problem, but the following theorem shows that it includes $\mathcal{F}(\mathbb{R}^n)$.

Theorem 5.2. *Let* $L : \mathbb{R}^n \to \mathbb{R}^n$ *be a self-adjoint linear operator such that* $I - L$ *is invertible* (I *being the identity operator on* $\mathbb{R}^n$) *and let* $f \in \mathcal{F}(\mathbb{R}^n)$. *Then the Fresnel integral of the function*

$$x \mapsto e^{-\frac{i}{2\hbar} \langle x, Lx \rangle} f(x), \qquad x \in \mathbb{R}^n$$

is well defined and is given by the following Parseval-type equality:

$$\widetilde{\int} e^{\frac{i}{2\hbar} \langle x, (I-L)x \rangle} f(x) dx$$

$$= e^{-\frac{\pi i}{2} \operatorname{Ind}(I-L)} |\det(I - L)|^{-1/2} \int_{\mathbb{R}^n} e^{-\frac{i\hbar}{2} \langle x, (I-L)^{-1}x \rangle} d\mu_f(x), \tag{5.10}$$

where $\operatorname{Ind}(I - L)$ *is the index of the operator* $I - L$, *that is the number of negative eigenvalues.*

Proof. We present here the proof first appeared in [129].

Let $\phi \in \mathcal{S}(\mathbb{R}^n)$ and $\epsilon > 0$ and let us consider the integral

$$(2\pi i \hbar)^{-n/2} \int_{\mathbb{R}^n} e^{\frac{i}{2\hbar} \langle x, (I-L)x \rangle} f(x) \phi(\epsilon x) dx \tag{5.11}$$

We claim that for any $f \in \mathcal{F}(\mathbb{R}^n)$ and $g \in \mathcal{S}(\mathbb{R}^n)$, the following holds:

$$(2\pi i\hbar)^{-n/2} \int_{\mathbb{R}^n} e^{\frac{i}{2\hbar}\langle x,(I-L)x\rangle} f(x)g(x)dx$$

$$= e^{-\frac{\pi i}{2}\mathrm{Ind}(I-L)} |\det(I-L)|^{-1/2}$$

$$\times \int_{\mathbb{R}^n \times \mathbb{R}^n} e^{-\frac{i\hbar}{2}\langle x,(I-L)^{-1}x\rangle} \tilde{g}(x-y)d\mu_f(y)dx. \qquad (5.12)$$

By Eq. (5.12) the integral (5.11) is equal to

$$(2\pi i\hbar)^{-n/2} \int_{\mathbb{R}^n} e^{\frac{i}{2\hbar}\langle x,(I-L)x\rangle} f(x)\phi(\epsilon x)dx$$

$$= e^{-\frac{\pi i}{2}\mathrm{Ind}(I-L)} |\det(I-L)|^{-1/2} \int_{\mathbb{R}^n \times \mathbb{R}^n} e^{-\frac{i\hbar}{2}\langle x,(I-L)^{-1}x\rangle} \epsilon^{-n} \tilde{\phi}(\frac{x-y}{\epsilon}) d\mu_f(y)dx$$

$$= e^{-\frac{\pi i}{2}\mathrm{Ind}(I-L)} |\det(I-L)|^{-1/2} \int_{\mathbb{R}^n \times \mathbb{R}^n} e^{-\frac{i\hbar}{2}\langle y+\epsilon x,(I-L)^{-1}(y+\epsilon x)\rangle} \tilde{\phi}(x)d\mu_f(y)dx$$

By letting $\epsilon \downarrow 0$, taking into account that $\int \tilde{\phi}(x)dx = 1$, we get Eq. (5.10).

Let us now prove Eq. (5.12). First of all, let us assume that $f = 1$.

(1) If $I - L$ is positive definite, by a change of variable Eq. (5.12) is equivalent to

$$(2\pi i\hbar)^{-n/2} \int_{\mathbb{R}^n} e^{\frac{i}{2\hbar}\langle x,x\rangle} g(x)dx = \int_{\mathbb{R}^n} e^{-\frac{i\hbar}{2}\langle x,x\rangle} \tilde{g}(x)dx.$$

Indeed, as $(2\pi i\hbar)^{-n/2}e^{\frac{i}{2\hbar}\langle x,x\rangle}$ is a bounded continuous function, it has a Fourier transform in the sense of tempered distributions

$$(2\pi i\hbar)^{-n/2} \int_{\mathbb{R}^n} e^{\frac{i}{2\hbar}\langle x,x\rangle} e^{i\langle x,y\rangle} dx = e^{-\frac{i\hbar}{2}\langle y,y\rangle}.$$

(2) If $I - L$ is negative definite, $\mathrm{Ind}(I - L) = n$, the result follows by replacing $\hbar$ with $-\hbar$ and from the fact that

$$(-2\pi i\hbar)^{-n/2} = (2\pi i\hbar)^{-n/2}e^{-\frac{\pi i n}{2}}.$$

(3) For general $I - L$, it is possible to decompose $\mathbb{R}^n$ as the product of the positive and negative eigenspaces, i.e.

$$\mathbb{R}^n = E^+ \times E^-,$$

and analogously $I - L$ is the product of its positive and negative part

$$I - L = (I - L)^+ \times (I - L)^-,$$

where $(I - L)^+ : E^+ \to E^+$ and $(I - L)^- : E^+ \to E^-$, $\dim(E^-) = \text{Ind}(I - L)$. The result follows from the points (1) and (2) in the case of a function $g : E^+ \times E^- \to \mathbb{C}$ of the form

$$g(x_+, x_-) = g_+(x_+)g_-(x_-), \qquad g_+ : E^+ \to \mathbb{C}, \quad g_- : E^- \to \mathbb{C}.$$
(5.13)

The same can be deduced for g that are finite linear combinations of factorisable functions of the form (5.13). For general g it is sufficient to note that $\mathcal{S}(E^+) \times \mathcal{S}(E^-)$ is dense in $\mathcal{S}(\mathbb{R}^n)$ and the result follows by a continuity argument.

Let us now prove Eq. (5.12) for $f \in \mathcal{F}(\mathbb{R}^n)$ and $g \in \mathcal{S}(\mathbb{R}^n)$. Let us denote with I_{rs}, with $r + s = n$, the linear operator on $\mathbb{R}^n$ having r eigenvalues equal to -1 and s eigenvalues equal to $+1$. By substituting in the left-hand side of Eq. (5.12) $f(x) = \int e^{i\langle x,y\rangle} d\mu_f(y)$ and by applying Fubini theorem, we get:

$$(2\pi i\hbar)^{-n/2} \int_{\mathbb{R}^n} e^{\frac{i}{2\hbar}\langle x, I_{rs}x\rangle} f(x)g(x)dx$$

$$= (2\pi i\hbar)^{-n/2} \int_{\mathbb{R}^n} d\mu_f(y) \int_{\mathbb{R}^n} e^{\frac{i}{2\hbar}\langle x, I_{rs}x\rangle} e^{i\langle x,y\rangle} g(x)dx.$$

By a change of variable the latter is equal to

$$(2\pi i\hbar)^{-n/2} \int_{\mathbb{R}^n} d\mu_f(y) e^{-\frac{i\hbar}{2}\langle y, I_{rs}y\rangle} \int_{\mathbb{R}^n} e^{\frac{i}{2\hbar}\langle x, I_{rs}x\rangle} g(x - I_{rs}\hbar y)dx$$

$$= e^{-\frac{i\pi r}{2}} \int_{\mathbb{R}^n} d\mu_f(y) \int_{\mathbb{R}^n} \tilde{g}(x - y) e^{-\frac{i\hbar}{2}\langle x, I_{rs}x\rangle} dx.$$

By a change of scale the latter equality is equivalent to

$$(2\pi i\hbar)^{-n/2} \int_{\mathbb{R}^n} e^{\frac{i}{2\hbar}\langle x,(I-L)x\rangle} f(x)g(x)dx$$

$$= e^{-\frac{\pi i}{2}\text{Ind}(I-L)} |\det(I-L)|^{-1/2} \int_{\mathbb{R}^n \times \mathbb{R}^n} e^{-\frac{i\hbar}{2}\langle x,(I-L)^{-1}x\rangle} \tilde{g}(x-y) d\mu_f(y)dx. \qquad \square$$

5.3 Generalized Fresnel integrals

In the present section, following [31], we shall generalize the result of the previous section to a larger class of phase functions Φ, in particular those given by an even polynomial $P(x)$ in the variables $x_1, \ldots, x_n$:

$$P(x) = A_{2M}(x, \ldots, x) + A_{2M-1}(x, \ldots, x) + \cdots + A_1(x) + A_0, \qquad (5.14)$$

where A_k are k_{th}-order covariant tensors on $\mathbb{R}^n$:

$$A_k : \underbrace{\mathbb{R}^n \times \mathbb{R}^n \times \ldots \times \mathbb{R}^n}_{k-times} \to \mathbb{R}$$

and the leading term, namely $A_{2M}(x, \ldots, x)$, is a $2M_{th}$-order completely symmetric positive covariant tensor on $\mathbb{R}^n$.

By Theorem 5.1 the *generalized Fresnel integral*

$$\int_{\mathbb{R}^n}^{o} e^{\frac{i}{\hbar} P(x)} f(x) dx \qquad (5.15)$$

is well defined for P of the form (5.14) and f belonging to the class of symbols (see Definition 5.2).

In the case where $\hbar \in \mathbb{C}$ with $Im(\hbar) < 0$ and if Φ is of the form (5.14), then the generalized Fresnel integral (5.15) also exists, even in Lebesgue sense, as an analytic function in $\hbar$, as easily seen by the fact that the integrand is bounded by $|f| \exp(\frac{Im(\hbar)}{|\hbar|^2} \Phi)$.

The key tool for the generalization of Parseval-type equality (5.10) is an estimate of the Fourier transform of the function $x \mapsto e^{\frac{i}{\hbar} P(x)}$, $x \in \mathbb{R}^n$.

Lemma 5.1. *Let $P : \mathbb{R}^n \to \mathbb{R}$ be given by (5.14) and let $\hbar \in \mathbb{C}$, with $Im(\hbar) \leq 0$. Then the Fourier transform of the distribution $e^{\frac{i}{\hbar} P(x)}$:*

$$\tilde{F}(k) = \int_{\mathbb{R}^n} e^{ik \cdot x} e^{\frac{i}{\hbar} P(x)} dx, \qquad \hbar \in \mathbb{R} \setminus \{0\} \qquad (5.16)$$

is an entire bounded function and admits in the case where $\hbar \in \mathbb{R}$ the following representation:

$$\tilde{F}(k) = e^{in\pi/4M} \int_{\mathbb{R}^n} e^{ie^{i\pi/4M} k \cdot x} e^{\frac{i}{\hbar} P(e^{i\pi/4M} x)} dx, \qquad \hbar > 0, \qquad (5.17)$$

or

$$\tilde{F}(k) = e^{-in\pi/4M} \int_{\mathbb{R}^n} e^{ie^{-i\pi/4M} k \cdot x} e^{\frac{i}{\hbar} P(e^{-i\pi/4M} x)} dx, \qquad \hbar < 0. \qquad (5.18)$$

Remark 5.2. The integral on the r.h.s. of (5.17) is absolutely convergent as

$$e^{\frac{i}{\hbar} P(e^{i\pi/4M} x)} = e^{-\frac{1}{\hbar} A_{2M}(x, \ldots, x)} e^{\frac{i}{\hbar}(A_{2M-1}(e^{i\pi/4M} x, \ldots, e^{i\pi/4M} x) + \cdots + A_1(xe^{i\pi/4M}) + A_0)}.$$

A similar calculation shows the absolute convergence of the integral on the r.h.s. of (5.18).

Proof (of Lemma 5.1). The proof is divided into three steps.

(1) Let us prove first of all formulas (5.17) and (5.18) by using the analyticity of $e^{ikz + \frac{i}{\hbar}P(z)}$, $z \in \mathbb{C}$, and a change of integration contour. Let us denote D the following region of the complex plane:

$$D \subset \mathbb{C}, \qquad D \equiv \{z \in \mathbb{C} \mid Im(z) < 0\}$$

and assume that $\hbar$ is a complex variable belonging to the region $\bar{D} \setminus \{0\}$. Further, let us introduce the polar coordinates in $\mathbb{R}^n$:

$$\int_{\mathbb{R}^n} e^{ik \cdot x} e^{\frac{i}{\hbar}P(x)} dx$$
$$= \int_{S_{n-1}} \left(\int_0^{+\infty} e^{i|k|r f_k(\phi_1, \dots, \phi_{n-1})} e^{\frac{i}{\hbar}\mathcal{P}_{(\phi_1, \dots, \phi_{n-1})}(r)} r^{n-1} dr \right) d\Omega_{n-1}$$

$$(5.19)$$

where instead of n Cartesian coordinates we use $n-1$ angular coordinates $(\phi_1, \dots, \phi_{n-1})$ and the variable $r = |x|$. S_{n-1} denotes the $(n-1)$-dimensional spherical surface, $d\Omega_{n-1}$ is the measure on it, $\mathcal{P}_{(\phi_1, \dots, \phi_{n-1})}(r)$ is a $2M_{th}$ order polynomial in the variable r with coefficients depending on the $n-1$ angular variables $(\phi_1, \dots, \phi_{n-1})$, namely:

$$P(x) = r^{2M} A_{2M}\left(\frac{x}{|x|}, \dots, \frac{x}{|x|}\right) + r^{2M-1} A_{2M-1}\left(\frac{x}{|x|}, \dots, \frac{x}{|x|}\right)$$
$$+ \dots + r A_1\left(\frac{x}{|x|}\right) + A_0$$
$$= a_{2M}(\phi_1, \dots, \phi_{n-1})r^{2M} + a_{2M-1}(\phi_1, \dots, \phi_{n-1})r^{2M-1}$$
$$| \dots + a_1(\phi_1, \dots, \phi_{n-1})r + a_0$$
$$= \mathcal{P}_{(\phi_1, \dots, \phi_{n-1})}(r)$$

where $a_{2M}(\phi_1, \dots, \phi_{n-1}) > 0$ for all $(\phi_1, \dots, \phi_{n-1}) \in S_{n-1}$. The function $f_k : S_{n-1} \to [-1, 1]$ is defined by

$$\frac{k}{|k|} \frac{x}{|x|} = f_k(\phi_1, \dots, \phi_{n-1})$$

We can now focus on the integral

$$\int_0^{+\infty} e^{i|k|r f_k(\phi_1, \dots, \phi_{n-1})} e^{\frac{i}{\hbar}\mathcal{P}_{(\phi_1, \dots, \phi_{n-1})}(r)} r^{n-1} dr, \qquad (5.20)$$

which can be interpreted as the Fourier transform of the distribution on the real line

$$F(r) = \Theta(r) r^{n-1} e^{\frac{i}{\hbar}\mathcal{P}_{(\phi_1, \dots, \phi_{n-1})}(r)},$$

with $\Theta(r) = 1$ for $r \geq 0$ and $\Theta(r) = 0$ for $r < 0$. Let us introduce the notation $k' \equiv |k| f_k(\phi_1, \dots, \phi_{n-1})$, $a_j \equiv a_j(\phi_1, \dots, \phi_{n-1})$, $j = 0, \dots, 2M$, $P'(r) = \sum_{j=0}^{2M} a_j r^j$ and $\hbar \in \mathbb{C}$, $\hbar = |\hbar| e^{i\phi}$, with $-\pi \leq \phi \leq 0$.

Let us consider the complex plane and set $z = \rho e^{i\theta}$. If $Im(\hbar) < 0$ the integral (5.20) is absolutely convergent, while if $\hbar \in \mathbb{R} \setminus \{0\}$ it needs a regularization. If $\hbar \in \mathbb{R}$, $\hbar > 0$ we have

$$\int_0^{+\infty} e^{ik'r} e^{\frac{i}{\hbar}P'(r)} r^{n-1} dr = \lim_{\epsilon \downarrow 0} \int_{z=\rho e^{i\epsilon}} e^{ik'z} e^{\frac{i}{\hbar}P'(z)} z^{n-1} dz$$

while if $\hbar < 0$

$$\int_0^{+\infty} e^{ik'r} e^{\frac{i}{\hbar}P'(r)} r^{n-1} dr = \lim_{\epsilon \downarrow 0} \int_{z=\rho e^{-i\epsilon}} e^{ik'z} e^{\frac{i}{\hbar}P'(z)} z^{n-1} dz \quad (5.21)$$

We deal first of all with the case $\hbar \in \mathbb{R}$, $\hbar > 0$ (the case $\hbar < 0$ can be handled in a completely similar way). Let

$$\gamma_1(R) = \{z = \rho e^{i\theta} \in \mathbb{C} \mid 0 \leq \rho \leq R, \quad \theta = \epsilon\},$$

$$\gamma_2(R) = \{z = \rho e^{i\theta} \in \mathbb{C} \mid \rho = R, \quad \epsilon \leq \theta \leq \pi/4M\},$$

$$\gamma_3(R) = \{z = \rho e^{i\theta} \in \mathbb{C} \mid 0 \leq \rho \leq R, \quad \theta = \pi/4M\}.$$

From the analyticity of the integrand and the Cauchy theorem we have

$$\int_{\gamma_1(R) \cup \gamma_2(R) \cup \gamma_3(R)} e^{ik'z} e^{\frac{i}{\hbar}P'(z)} z^{n-1} dz = 0.$$

In particular:

$$\left| \int_{\gamma_2(R)} e^{ik'z} e^{\frac{i}{\hbar}P'(z)} z^{n-1} dz \right| = R^n \left| \int_\epsilon^{\pi/4M} e^{ik'Re^{i\theta}} e^{\frac{i}{\hbar}P'(Re^{i\theta})} e^{in\theta} d\theta \right|$$

$$\leq R^n \int_\epsilon^{\pi/4M} e^{-k'R\sin(\theta)} e^{-\frac{1}{\hbar} \sum_{j=1}^{2M} a_j R^j \sin(j\theta)} d\theta$$

$$\leq R^n \int_\epsilon^{\pi/4M} e^{-k''R\theta} e^{-a_{2M} \frac{4M}{\hbar\pi} R^{2M}\theta} e^{-\sum_{j=1}^{2M-1} a_j' R^j \theta} d\theta,$$

where k'', a_k' for $k = 1, \ldots, 2M - 1$ are suitable constants. We have used the fact that if $\alpha \in [0, \pi/2]$ then $\frac{2}{\pi}\alpha \leq \sin(\alpha) \leq \alpha$. The latter integral can be explicitly computed and gives:

$$R^n \left(\frac{e^{-\epsilon(a_{2M} \frac{4M}{\hbar\pi} R^{2M} + k''R + \sum_{j=1}^{2M-1} a_j' R^j)} - e^{-\frac{\pi}{4M}(a_{2M} \frac{4M}{\hbar\pi} R^{2M} + k''R + \sum_{j=1}^{2M-1} a_j' R^j)}}{a_{2M} \frac{4M}{\hbar\pi} R^{2M} + k''R + \sum_{j=1}^{2M-1} a_j' R^j} \right),$$

which converges to 0 as $R \to \infty$. We get

$$\int_{z=\rho e^{i\epsilon}} e^{ik'z} e^{\frac{i}{\hbar}P'(z)} z^{n-1} dz = \int_{z=\rho e^{i(\pi/4M)}} e^{ik'z} e^{\frac{i}{\hbar}P'(z)} z^{n-1} dz.$$

By taking the limit as $\epsilon \downarrow 0$ of both sides one gets:

$$\int_0^{+\infty} e^{ik'r} e^{\frac{i}{\hbar}P'(r)} r^{n-1} dr = e^{in\pi/4M} \int_0^{+\infty} e^{ik\rho e^{i\pi/4M}} e^{\frac{i}{\hbar}P'(re^{i\pi/4M})} \rho^{n-1} d\rho.$$

By substituting into (5.19) we obtain the final result:

$$\tilde{F}(k) = \int_{\mathbb{R}^n} e^{ik \cdot x} e^{\frac{i}{\hbar}P(x)} dx = e^{in\pi/4M} \int_{\mathbb{R}^n} e^{ie^{i\pi/4M}k \cdot x} e^{\frac{i}{\hbar}P(e^{i\pi/4M}x)} dx.$$

(5.22)

In the case $\hbar < 0$ an analogous reasoning gives:

$$\tilde{F}(k) = \int_{\mathbb{R}^n} e^{ik \cdot x} e^{\frac{i}{\hbar}P(x)} dx = e^{-in\pi/4M} \int_{\mathbb{R}^n} e^{ie^{-i\pi/4M}k \cdot x} e^{\frac{i}{\hbar}P(e^{-i\pi/4M}x)} dx.$$

(5.23)

(2) The analyticity of $\tilde{F}(k)$ is trivial in the case $Im(\hbar) < 0$, and follows from Eqs. (5.22) and (5.23) when $\hbar \in \mathbb{R} \setminus \{0\}$.

(3) Let us now prove that $\tilde{F}$

$$\tilde{F}(k) = \int_{\mathbb{R}^n} e^{ikx} e^{\frac{i}{\hbar}P(x)} dx$$

is bounded as a function of k by studying its asymptotic behavior as $|k| \to \infty$.

Let us focus on the case $\hbar \in \mathbb{R} \setminus \{0\}$ (in the case $Im(\hbar) < 0$ $|\tilde{F}|$ is trivially bounded by

$$\int_{\mathbb{R}^n} |e^{\frac{i}{\hbar}P(x)}| dx = \int_{\mathbb{R}^n} e^{\frac{Im(\hbar)}{|\hbar|^2}P(x)} dx < +\infty).$$

Let us assume for notational simplicity that $\hbar = 1$, the general case can be handled in a completely similar way. In order to study $\int_{\mathbb{R}^N} e^{ikx} e^{iP(x)} dx$ one has to introduce a suitable regularization. Chosen $\psi \in \mathcal{S}(\mathbb{R}^n)$, such that $\psi(0) = 1$ we have

$$e^{iP(x)}\psi(\epsilon x) \to e^{iP(x)}, \qquad \text{in } \mathcal{S}'(\mathbb{R}^n) \text{ as } \epsilon \to 0,$$

$$\tilde{F}(k) = \lim_{\epsilon \to 0} \int_{\mathbb{R}^n} e^{ikx} e^{iP(x)} \psi(\epsilon x) dx.$$

Let us consider first of all the case $n = 1$ and $P(x) = x^{2M}/2M$. The unique real stationary point of the phase function $\Phi(x) = kx + x^{2M}/2M$ is $c_k = (-k)^{\frac{1}{2M-1}}$. Let χ_1 be a positive C^∞ function such that $\chi_1(x) = 1$ if $|x - c_k| \leq 1/2$, $\chi_1(x) = 0$ if $|x - c_k| \geq 1$ and $0 \leq \chi_1(x) \leq 1$ if $1/2 \leq |x - c_k| \leq 1$. Let $\chi_0 \equiv 1 - \chi_1$. Then $\tilde{F}(k) = I_1(k) + I_0(k)$, where

$$I_0(k) = \lim_{\epsilon \to 0} \int e^{ikx} e^{ix^{2M}/2M} \chi_0(x) \psi(\epsilon x) dx,$$

$$I_1(k) = \int e^{ikx} e^{ix^{2M}/2M} \chi_1(x) dx.$$

For the study of the boundedness of $|\tilde{F}(k)|$ as $|k| \to \infty$ it is enough to look at I_0, since one has, by the choice of χ_1, that $|I_1| \le 2$. By repeating the same reasoning used in the proof of Theorem 5.1, I_0 can be computed by means of Stokes formula:

$$\lim_{\epsilon \to 0} \int e^{ikx} e^{ix^{2M}/2M} \chi_0(x) \psi(\epsilon x) dx = i \lim_{\epsilon \to 0} \epsilon \int e^{ikx} e^{i\frac{x^{2M}}{2M}} \frac{\chi_0(x) \psi'(\epsilon x)}{k + x^{2M-1}} dx$$

$$+ i \lim_{\epsilon \to 0} \int e^{ikx} e^{ix^{2M}/2M} \frac{d}{dx} \Big(\frac{\chi_0(x)}{k + x^{2M-1}} \Big) \psi(\epsilon x) dx.$$

Both integrals are absolutely convergent and, by dominated convergence, we can take the limit $\epsilon \to 0$, so that

$$I_0(k) = i \int e^{ikx} e^{ix^{2M}/2M} \frac{d}{dx} \Big(\frac{\chi_0(x)}{k + x^{2M-1}} \Big) dx$$

$$= i \int e^{ikx} e^{ix^{2M}/2M} \Big(\frac{\chi_0'(x)}{k + x^{2M-1}} \Big) dx$$

$$- i \int e^{ikx} e^{ix^{2M}/2M} \Big(\frac{(2M-1)\chi_0(x) x^{2M-2}}{(k + x^{2M-1})^2} \Big) dx.$$

Thus,

$$|I_0(k)| \le \sup |\chi_0'| \int_{c_k-1}^{c_k-1/2} \Big| \frac{1}{k + x^{2M-1}} \Big| dx + \sup |\chi_0'| \int_{c_k+1/2}^{c_k+1} \Big| \frac{1}{k + x^{2M-1}} \Big| dx$$

$$+ (2M-1) \int_{-\infty}^{c_k-1/2} \Big| \frac{x^{2M-2}}{(k + x^{2M-1})^2} \Big| dx + (2M-1) \int_{c_k+1/2}^{+\infty} \Big| \frac{x^{2M-2}}{(k + x^{2M-1})^2} \Big| dx$$

By a change of variables it is possible to see that both integrals remain bounded as $|k| \to \infty$ (see [31] for more details) By such considerations we can deduce that $|\tilde{F}(k)|$ is bounded as $|k| \to \infty$.

A similar reasoning holds also in the case $n = 1$ and $P(x) = \sum_{i=1}^{2M} a_i x^i$ is a generic polynomial. Indeed for $|k|$ sufficiently large the derivative of the phase function $\Phi'(x) = k + P'(x)$ has only one simple real root, denoted by c_k. One can repeat the same reasoning valid for the case $P(x) = x^{2M}/2M$ and prove that for $|k| \to \infty$ one has $|\int e^{ikx + iP(x)} dx| \le C$ (where C is a function of the coefficients a_i of P at most with polynomial growth).

The general case $\mathbb{R}^n$ can also be essentially reduced to the one-dimensional case. Indeed let use consider a generic vector $k \in \mathbb{R}^n$,

$k = |k|u_1$, and study the behavior of $\tilde{F}(k)$ as $|k| \to \infty$. By choosing as orthonormal base $u_1, \ldots, u_n$ of $\mathbb{R}^n$, where $u_1 = k/|k|$, we have

$$\tilde{F}(k) = \lim_{\epsilon \to 0} \int_{\mathbb{R}^{n-1}} e^{iQ(x_2,\ldots,x_n)} \psi(\epsilon x_2) \cdots \psi(\epsilon x_n)$$
$$\left(\int_{\mathbb{R}} e^{i|k||x_1|} e^{iP_{x_2,\ldots,x_n}(x_1)} \psi(\epsilon x_1) dx_1 \right) dx_2 \ldots dx_n, \quad (5.24)$$

where $\psi \in \mathcal{S}(\mathbb{R})$, $\psi(0) = 1$; $x_i = x \cdot u_i$, $P_{x_2,\ldots,x_n}(x_1)$ is the polynomial in the variable x_1 with coefficients depending on powers of the remaining $n-1$ variables $x_2, \ldots, x_n$, obtained by considering in the initial polynomial $P(x_1, x_2, \ldots, x_n)$ all the terms containing some power of x_1. The polynomial Q in the $n-1$ variables $x_2, \ldots, x_n$ is given by $P(x_1, x_2, \ldots, x_n) - P_{x_2,\ldots,x_n}(x_1)$.

Let us set

$$I^\epsilon(k, x_2, \ldots, x_n) \equiv \int_{\mathbb{R}} e^{i|k||x_1|} e^{iP_{x_2,\ldots,x_n}(x_1)} \psi(\epsilon x_1) dx_1.$$

By the previous considerations we know that, for each $\epsilon \geq 0$, $|I^\epsilon(k, x_2, \ldots, x_n)|$ is bounded by a function of $G(x_2, \ldots, x_n)$ of polynomial growth. By the same reasoning as in the proof of Theorem 5.1 we can deduce that the oscillatory integral (5.24) is a well defined bounded function of k. $\qquad\square$

Remark 5.3. A representation similar to (5.17) holds also in the more general case $\hbar \in \mathbb{C}$, $Im(\hbar) < 0$, $\hbar \neq 0$. By setting $\hbar \equiv |\hbar| e^{i\phi}$, $\phi \in [-\pi, 0]$ one has:

$$\tilde{F}(k) = \int_{\mathbb{R}^n} e^{ik \cdot x} e^{\frac{i}{\hbar} P(x)} dx$$
$$= e^{in(\pi/4M + \phi/2M)} \int_{\mathbb{R}^n} e^{ie^{i(\pi/4M + \phi/2M)} k \cdot x} e^{\frac{i}{\hbar} P(e^{i(\pi/4M + \phi/2M)} x)} dx. \quad (5.25)$$

By mimicking the proof of Eq. (5.17), one can prove in the case $\hbar > 0$ the following result (a similar one holds also in the case $\hbar < 0$):

Theorem 5.3. *Let us denote by Λ_M the subset of the complex plane*

$$\Lambda_M = \{ z \in \mathbb{C} \mid 0 < \arg(z) < \pi/4M \} \subset \mathbb{C}, \quad (5.26)$$

and let $\bar{\Lambda}_M$ be its closure. Let $f : \mathbb{R}^n \to \mathbb{C}$ be a Borel function defined for all y of the form $y = \lambda x$, where $\lambda \in \bar{\Lambda}_M$ and $x \in \mathbb{R}^n$, with the following properties:

(1) the function $\lambda \mapsto f(\lambda x)$ is analytic in Λ_M and continuous in $\bar{\Lambda}_M$ for each $x \in \mathbb{R}^n$, $|x| = 1$,

(2) for all $x \in \mathbb{R}^n$ and all $\theta \in (0, \pi/4M)$

$$|f(e^{i\theta}x)| \le AG(x),$$

where $A \in \mathbb{R}$ and $G : \mathbb{R}^n \to \mathbb{R}$ is a positive function satisfying bound (a) or (b) respectively:

(a) if P is as in the general case defined by (5.14)

$$G(x) \le e^{B|x|^{2M-1}}, \quad B > 0$$

(b) if P is homogeneous, i.e. $P(x) = A_{2M}(x, \ldots, x)$,

$$G(x) \le e^{\frac{\sin(2M\theta)}{\hbar} A_{2M}(x,x,\ldots,x)} g(|x|),$$

where $g(t) = O(t^{-(n+\delta)})$, $\delta > 0$, as $t \to \infty$.

Then the limit of regularized integrals:

$$\lim_{\epsilon \downarrow 0} \int e^{\frac{i}{\hbar}P(xe^{i\epsilon})} f(xe^{i\epsilon}) dx, \quad 0 < \epsilon < \pi/4M, \quad \hbar > 0$$

is given by:

$$e^{in\pi/4M} \int_{\mathbb{R}^n} e^{\frac{i}{\hbar}P(e^{i\pi/4M}x)} f(e^{i\pi/4M}x) dx \tag{5.27}$$

The latter integral is absolutely convergent and it is understood in Lebesgue sense.

The class of functions satisfying conditions (1) and (2) in Theorem 5.3 includes for instance the polynomials of any degree and the exponentials. In the case $f \in S_\lambda^N$ for some N, λ, one is tempted to interpret expression (5.27) as an explicit formula for the evaluation of the generalized Fresnel integral $\int e^{\frac{i}{\hbar}P(x)} f(x) dx$, $\hbar > 0$, whose existence is assured by Theorem 5.1. This is, however, not necessarily true for all $f \in S_\lambda^\infty$ satisfying (1) and (2). Indeed the Definition 5.1 of oscillatory integral requires that the limit of the sequence of regularized integrals exists and is independent on the regularization. The identity

$$\lim_{\epsilon \to 0} \int_{\mathbb{R}^n} e^{\frac{i}{\hbar}P(x)} f(x)\psi(\epsilon x) dx = e^{in\pi/4M} \int_{\mathbb{R}^n} e^{\frac{i}{\hbar}P(e^{i\pi/4M}x)} f(e^{i\pi/4Mx}) dx,$$

can be proved only by choosing regularizing functions $\psi \in \mathcal{S}(\mathbb{R}^n)$ with $\psi(0) = 1$ and ψ in the class Σ consisting of all $\psi \in \mathcal{S}(\mathbb{R}^n)$ which satisfy assumption (1) of Theorem 5.3 and are such that $|\psi(e^{i\theta}x)|$ is bounded as $|x| \to \infty$ for each $\theta \in (0, \pi/4M)$. In fact we will prove that expression (5.27) coincides with the oscillatory integral (5.15), i.e. one can take $\Sigma = \mathcal{S}(\mathbb{R}^N)$, by imposing stronger assumptions on the function f. First of all we show that the representation (5.17) (resp. (5.18)) for the Fourier transform of $e^{\frac{i}{\hbar}P(x)}$ allows a generalization of equation (5.10). Let us denote by $\bar{D} \subset \mathbb{C}$ the lower semiplane in the complex plane

$$\bar{D} \equiv \{z \in \mathbb{C} \mid Im(z) \leq 0\} \tag{5.28}$$

Theorem 5.4. *Let* $f \in \mathcal{F}(\mathbb{R}^n)$, $f = \hat{\mu}_f$. *Then the generalized Fresnel integral*

$$I(f) \equiv \int_{\mathbb{R}^n} e^{\frac{i}{\hbar}P(x)} f(x) dx, \qquad \hbar \in \bar{D} \setminus \{0\}$$

is well defined and it is given by the formula of Parseval's type:

$$\int_{\mathbb{R}^n} e^{\frac{i}{\hbar}P(x)} f(x) dx = \int_{\mathbb{R}^n} \tilde{F}(k) d\mu_f(k), \tag{5.29}$$

where $\tilde{F}(k)$ *is given by (5.25) (see Lemma 5.1 and remark 5.3)*

$$\tilde{F}(k) = \int_{\mathbb{R}^n} e^{ikx} e^{\frac{i}{\hbar}P(x)} dx.$$

The integral on the r.h.s. of (5.29) is absolutely convergent (hence it can be understood in Lebesgue sense).

Proof. Let us choose a test function $\psi \in \mathcal{S}(\mathbb{R}^n)$, such that $\psi(0) = 1$ and let us compute the limit

$$I(f) \equiv \lim_{\epsilon \downarrow 0} \int_{\mathbb{R}^n} e^{\frac{i}{\hbar}P(x)} \psi(\epsilon x) f(x) dx.$$

By hypothesis $f(x) = \int_{\mathbb{R}^n} e^{ikx} d\mu_f(k)$, $x \in \mathbb{R}^N$, and substituting in the previous expression we get:

$$I(f) = \lim_{\epsilon \downarrow 0} \int_{\mathbb{R}^n} e^{\frac{i}{\hbar}P(x)} \psi(\epsilon x) \left(\int_{\mathbb{R}^n} e^{ikx} d\mu_f(k) \right) dx.$$

By Fubini theorem (which applies for any $\epsilon > 0$ since the integrand is bounded by $|\psi(\epsilon x)|$ which is dx-integrable, and μ_f is a bounded measure) the r.h.s. is

$$= \lim_{\epsilon \downarrow 0} \int_{\mathbb{R}^n} \left(\int_{\mathbb{R}^n} e^{\frac{i}{\hbar}P(x)} \psi(\epsilon x) e^{ikx} dx \right) d\mu_f(k)$$

$$= \frac{1}{(2\pi)^n} \lim_{\epsilon \downarrow 0} \int_{\mathbb{R}^n} \int_{\mathbb{R}^n} \tilde{F}(k - \alpha\epsilon) \tilde{\psi}(\alpha) d\alpha d\mu_f(k) \tag{5.30}$$

(here we have used the fact that the integral with respect to x is the Fourier transform of $e^{\frac{iP(x)}{\hbar}} \psi(\epsilon x)$ and the inverse Fourier transform of a product is a convolution). Now we can pass to the limit using the Lebesgue dominated convergence theorem and get the desired result:

$$\lim_{\epsilon \downarrow 0} \int_{\mathbb{R}^n} e^{\frac{i}{\hbar} P(x)} \psi(\epsilon x) f(x) dx = \int_{\mathbb{R}^n} \tilde{F}(k) d\mu_f(k),$$

where we have used the identity $\int \tilde{\psi}(\alpha) d\alpha = (2\pi)^n \psi(0)$ and Lemma 5.1, which assures the boundedness of $\tilde{F}(k)$.   □

Corollary 5.1. *Let* $\hbar = |\hbar| e^{i\phi}$, $\phi \in [-\pi, 0]$, $\hbar \neq 0$, $f \in \mathcal{F}(\mathbb{R}^n)$, $f = \hat{\mu}_f$ *be such that* $\forall x \in \mathbb{R}^n$

$$\int_{\mathbb{R}^n} e^{-kx \sin(\pi/4M + \phi/2M)} d|\mu_f|(k) \leq AG(x), \qquad (5.31)$$

where $A \in \mathbb{R}$ *and* $G : \mathbb{R}^n \to \mathbb{R}$ *is a positive function satisfying bound (1) or (2) respectively:*

(1) if P is defined by (5.14),

$$G(x) \leq e^{B|x|^{2M-1}}, \quad B > 0$$

(2) if P is homogeneous, i.e. $P(x) = A_{2M}(x, \ldots, x)$:

$$G(x) \leq e^{\frac{1}{\hbar} A_{2M}(x,x,\ldots,x)} g(|x|),$$

where $g(t) = O(t^{-(n+\delta)})$, $\delta > 0$, as $t \to \infty$.

Then f extends to an analytic function on $\mathbb{C}^n$ and its generalized Fresnel integral (5.15) is well defined and it is given by

$$\int_{\mathbb{R}^n} e^{\frac{i}{\hbar} P(x)} f(x) dx$$

$$= e^{in(\pi/4M + \phi/2M)} \int_{\mathbb{R}^n} e^{\frac{i}{\hbar} P(e^{i(\pi/4M + \phi/2M)} x)} f(e^{i(\pi/4M + \phi/2M)} x) dx.$$

Proof.   By bound (5.31) it follows that the Laplace transform $f^L : \mathbb{C}^n \to \mathbb{C}$, $f^L(z) = \int_{\mathbb{R}^n} e^{kz} d\mu_f(k)$, of μ_f is a well defined entire function such that, for $x \in \mathbb{R}^n$, $f^L(ix) = f(x)$. By Theorem 5.4 the generalized Fresnel integral can be computed by means of the Parseval type equality

$$\int_{\mathbb{R}^n} e^{\frac{i}{\hbar} P(x)} f(x) dx = \int_{\mathbb{R}^n} \tilde{F}(k) d\mu_f(k)$$

$$= e^{in(\pi/4M + \phi/2M)} \int_{\mathbb{R}^n} \left(\int_{\mathbb{R}^n} e^{ikx e^{i(\pi/4M + \phi/2M)}} e^{\frac{i}{\hbar} P(e^{i(\pi/4M + \phi/2M)} x)} dx \right) d\mu_f(k)$$

By Fubini theorem, which applies given the assumptions on the measure μ_f, this is equal to

$$e^{in(\pi/4M+\phi/2M)} \int_{\mathbb{R}^n} e^{\frac{i}{\hbar}P(e^{i(\pi/4M+\phi/2M)}x)} \int_{\mathbb{R}^n} e^{ikxe^{i(\pi/4M+\phi/2M)}} d\mu_f(k)dx$$

$$= e^{in(\pi/4M+\phi/2M)} \int_{\mathbb{R}^n} e^{\frac{i}{\hbar}P(e^{i(\pi/4M+\phi/2M)}x)} f^L(ie^{i(\pi/4M+\phi/2M)}x)dx$$

$$= e^{in(\pi/4M+\phi/2M)} \int_{\mathbb{R}^n} e^{\frac{i}{\hbar}P(e^{i(\pi/4M+\phi/2M)}x)} f(e^{i(\pi/4M+\phi/2M)}x)dx$$

and the conclusion follows. $\qquad\square$

5.4 Infinite dimensional oscillatory integrals

Definition 5.3 of the (finite dimensional) Fresnel integral can be generalized to the case the integration is performed on an infinite dimensional real separable Hilbert space $(\mathcal{H}, \langle\ ,\ \rangle)$. An *infinite dimensional oscillatory integral* is defined as the limit of a sequence of finite dimensional approximations. This approach was proposed in [129] and further developed in [7] in connection with the study of an infinite dimensional version of the stationary phase method. It is also related to previous work by K. Ito [195, 196] and S. Albeverio and R. Høegh-Krohn [17, 18]

Definition 5.4. A Borel measurable function $f : \mathcal{H} \to \mathbb{C}$ is called Fresnel integrable if for each sequence $\{P_n\}_{n\in\mathbb{N}}$ of projectors onto n-dimensional subspaces of $\mathcal{H}$, such that

- $P_n \leq P_{n+1}$ (i.e. $P_n(\mathcal{H}) \subset P_{n+1}(\mathcal{H})$),
- $P_n \to I$ strongly as $n \to \infty$ (I being the identity operator in $\mathcal{H}$),

the finite dimensional approximations of the oscillatory integral of f, suitably normalized

$$\int_{P_n\mathcal{H}}^{o} e^{\frac{i}{2\hbar}\|P_n x\|^2} f(P_n x)d(P_n x)\Big(\int_{P_n\mathcal{H}}^{o} e^{\frac{i}{2\hbar}\|P_n x\|^2}d(P_n x)\Big)^{-1},$$

are well defined (in the sense of Definition 5.1) and the limit

$$\lim_{n\to\infty} \int_{P_n\mathcal{H}}^{o} e^{\frac{i}{2\hbar}\|P_n x\|^2} f(P_n x)d(P_n x)\Big(\int_{P_n\mathcal{H}}^{o} e^{\frac{i}{2\hbar}\|P_n x\|^2}d(P_n x)\Big)^{-1},$$

exists and is independent on the sequence $\{P_n\}$.

In this case the limit is called the *infinite dimensional oscillatory integral of* f and is denoted

$$\widetilde{\int_{\mathcal{H}}} e^{\frac{i}{2\hbar}\|x\|^2} f(x)dx.$$

Analogously to Definition 5.3, also in Definition 5.4 a normalization constant

$$\left(\int_{P_n \mathcal{H}}^{\circ} e^{\frac{i}{2\hbar} \|P_n x\|^2} d(P_n x) \right)^{-1} = (2\pi i \hbar)^{-n/2}$$

is present. Indeed if $f : \mathcal{H} \to \mathbb{C}$ is the identity function, i.e. $f(x) = 1$ $\forall x \in \mathcal{H}$, it is simple to verify that

$$\widetilde{\int_{\mathcal{H}}} e^{\frac{i}{2\hbar} \|x\|^2} f(x) dx = 1.$$

In addition, the presence of the normalization constant makes Definition 5.4 the direct generalization of Definition 5.3. Indeed if $f : \mathcal{H} \to \mathbb{C}$ is a finite based function depending on a finite number of variables, i.e. if $f(x) = f(P_n x)$ for a finite dimensional projection operator P_n in $\mathcal{H}$, it is possible to see that the infinite dimensional oscillatory integral of f on $\mathcal{H}$ coincides with the finite dimensional Fresnel integral on the finite dimensional subspace $P_n(\mathcal{H})$ of the restriction f_n of the function f on $P_n(\mathcal{H})$:

$$f_n : P_n \mathcal{H} \to \mathbb{C}, \qquad f_n(x) := f(P_n x), \qquad x \in P_n(\mathcal{H}),$$

$$\widetilde{\int_{\mathcal{H}}} e^{\frac{i}{2\hbar} \|x\|^2} f(x) dx = \widetilde{\int_{P_n(\mathcal{H})}} e^{\frac{i}{2\hbar} \|x\|^2} f_n(x) dx.$$

The "concrete" description of the class of all Fresnel integrable functions is still an open problem of harmonic analysis, even when $dim(\mathcal{H}) < \infty$ (as already discussed in Sections 5.1–5.3). However, analogously to the finite dimensional case, it is possible to prove that this class includes $\mathcal{F}(\mathcal{H})$, the Banach algebra of functions that are Fourier transform of complex Borel measures on $\mathcal{H}$, and a Parseval-type formula analogous to Eq. (5.10) holds. In order that all the terms on the right-hand side of Eq. (5.10) make sense even in an infinite dimensional setting, we have to impose some conditions to the operator L.

In the following we shall consider a self-adjoint trace class operator $L : \mathcal{H} \to \mathcal{H}$, such that $I - L$ is invertible. As L is compact, it has a complete set of eigenvectors, with eigenvalues of finite multiplicity and with 0 as their only possible limit point. As $I - L$ is invertible by assumption, the index of $(I - L)$, i.e. the number of negatives eigenvalues, counted with their multiplicity, is finite. The following lemma is taken from [129].

Lemma 5.2. *Let $\{P_n\}_{n \in \mathbb{N}}$ be a sequence of projectors onto n-dimensional subspaces of $\mathcal{H}$, convergent strongly to the identity when $n \to \infty$. For any compact self-adjoint operator L the following holds:*

$$\lim_{n \to \infty} \text{Ind}(I - P_n L P_n) = \text{Ind}(I - L). \tag{5.32}$$

Proof. Let $\mathrm{Ind}(I - L) = p$ and let $\lambda_1, \ldots \lambda_p$ be the eigenvalues of L that are greater than 1, with corresponding orthonormal vectors $e_1, \ldots, e_p$. We have

$$\langle e_i, (I - L)e_i \rangle = 1 - \lambda_i < 0,$$

and by the strong convergence of the sequence $\{P_n\}_{n \in \mathbb{N}}$ to the identity,

$$\lim_n \langle P_n e_i, (I - L)P_n e_i \rangle = 1 - \lambda_i.$$

In particular for sufficiently large n we have

$$\langle P_n e_i, (I - L)P_n e_i \rangle < 0.$$

As $\langle P_n e_i, (I_n - P_n L P_n)P_n e_i \rangle = \langle P_n e_i, (I - L)P_n e_i \rangle$ (I_n being the identity operator on $P_n \mathcal{H}$), it is possible to conclude that the vectors $P_n e_1, \ldots, P_n e_p$ lie in a subspace of $P_n \mathcal{H}$ in which $(I_n - P_n L P_n)$ is negative definite. Since $P_n e_1, \ldots, P_n e_p$ are linearly independent for sufficiently large n, as

$$\lim_n \langle P_n e_i, P_n e_j \rangle = \delta_{ij},$$

we can conclude that

$$\underline{\lim}_n \mathrm{Ind}(I_n - P_n L P_n) \geq \mathrm{Ind}(I - L).$$

In order to prove the converse inequality, it is sufficient to note that, if V_n is the subspace of $P_n \mathcal{H}$ spanned by the negative eigenvectors of $I_n - P_n L P_n$, then for $v \in V_n$ we have

$$\langle v, (I - L)v \rangle = \langle P_n v, (I - L)P_n v \rangle = \langle v, (I_n - P_n L P_n)v \rangle < 0.$$

This gives, for sufficiently large n,

$$\mathrm{Ind}(I - L) \geq \mathrm{Ind}(I_n - P_n L P_n). \qquad \square$$

For a self-adjoint trace class operator $L : \mathcal{H} \to \mathcal{H}$, it is possible to define the Fredholm determinant of the operator $I - L$ [301], given by the infinite product of its eigenvalues

$$\Pi_n (1 - \lambda_n).$$

It is denoted with $\det(I - L)$ and given by

$$\det(I - L) = |\det(I - L)|e^{-\pi i \, \mathrm{Ind} \, (I-L)},$$

where $|\det(I - L)|$ is its absolute value and $\mathrm{Ind}((I - L))$ is the number of negative eigenvalues of the operator $(I - L)$, counted with their multiplicity.

We can now state the fundamental theorem of this section.

Theorem 5.5. *Let $L : \mathcal{H} \to \mathcal{H}$ be a self-adjoint trace class operator such that $(I - L)$ is invertible (I being the identity operator in $\mathcal{H}$). Let us assume that $f \in \mathcal{F}(\mathcal{H})$. Then the function $g : \mathcal{H} \to \mathbb{C}$ given by*

$$g(x) = e^{-\frac{i}{2\hbar}\langle x, Lx \rangle} f(x), \qquad x \in \mathcal{H}$$

is Fresnel integrable and the corresponding infinite dimensional oscillatory integral is given by the following Parseval-type formula:

$$\widetilde{\int_{\mathcal{H}}} e^{\frac{i}{2\hbar}\langle x, (I-L)x \rangle} f(x)dx = (\det(I - L))^{-1/2} \int_{\mathcal{H}} e^{-\frac{i\hbar}{2}\langle x, (I-L)^{-1}x \rangle} d\mu_f(x) \tag{5.33}$$

where

$$\det(I - L) = |\det(I - L)|e^{-\pi i \, \mathrm{Ind} \, (I-L)}$$

is the Fredholm determinant of the operator $(I - L)$.

Proof. Given a sequence $\{P_n\}_{n \in \mathbb{N}}$ of projectors onto n-dimensional subspaces of $\mathcal{H}$, such that $P_n \leq P_{n+1}$ and $P_n \to I$ strongly as $n \to \infty$ (I being the identity operator in $\mathcal{H}$), the finite dimensional approximations of the infinite dimensional oscillatory integral $\widetilde{\int_{\mathcal{H}}} e^{\frac{i}{2\hbar}\langle x, (I-L)x \rangle} f(x)dx$ are equal to

$$(2\pi i \hbar)^{-\dim P_n \mathcal{H}} \int_{P_n \mathcal{H}}^{\circ} e^{\frac{i}{2\hbar}\|x\|^2} e^{-\frac{i}{2\hbar}\langle x, L_n x \rangle} f_n(x)dx, \tag{5.34}$$

where $f_n : P_n \mathcal{H} \to \mathbb{C}$ is given by

$$f_n(x) := f(P_n x), \qquad x \in P_n \mathcal{H}.$$

The function f_n belongs to $\mathcal{F}(P_n \mathcal{H})$, indeed for $x \in P_n \mathcal{H}$ we have:

$$f_n(x) = \int_{\mathcal{H}} e^{i\langle y, P_n x \rangle} d\mu_f(y) = \int_{\mathcal{H}} e^{i\langle P_n y, P_n x \rangle} d\mu_f(y) = \int_{P_n \mathcal{H}} e^{i\langle y, x \rangle} d\mu_{f,n}(y),$$

where $\mu_{f,n}$ is the image measure of μ under the action of $P_n : \mathcal{H} \to P_n \mathcal{H}$, while $L_n : P_n \mathcal{H} \to P_n \mathcal{H}$ is the operator on $P_n \mathcal{H}$ given by $L_n := P_n L P_n$. As $(I - L)$ is invertible, it is easy to see that for n sufficiently large the operator $(I_n - L_n)$ on $P_n \mathcal{H}$ (I_n being the identity operator in $P_n \mathcal{H}$) is invertible. By Parseval's formula (5.10) the expression (5.34) is equal to

$$(\det(I_n - L_n))^{-1/2} \int_{P_n \mathcal{H}} e^{-\frac{i\hbar}{2}\langle x, (I_n-L_n)^{-1}x \rangle} d\mu_{f,n}(x)$$

$$= (\det(I_n - L_n))^{-1/2} \int_{\mathcal{H}} e^{-\frac{i\hbar}{2}\langle P_n x, (I-L_n)^{-1}P_n x \rangle} d\mu_f(x) \tag{5.35}$$

By letting $n \to \infty$, $L_n \to L$ in trace norm and expression (5.35) converges to the right-hand side of (5.33). $\qquad\square$

A fundamental property of infinite dimensional oscillatory integrals is the covariance under translations of vectors belonging to $\mathcal{H}$, more precisely the following holds [129]:

Theorem 5.6. *Let $a \in \mathcal{H}$ and $f \in \mathcal{F}(\mathcal{H})$. Let us define $f_a : \mathcal{H} \to \mathbb{C}$ by $f_a(x) := f(x + a)$. Then the function $g : \mathcal{H} \to \mathbb{C}$, given by*

$$g(x) := e^{\frac{i}{\hbar}\langle a, x \rangle} f_a(x), \qquad x \in \mathcal{H} \tag{5.36}$$

is Fresnel integrable and the corresponding infinite dimensional oscillatory integral is given by

$$\widetilde{\int_{\mathcal{H}}} e^{\frac{i}{2\hbar}\langle x, x \rangle} e^{\frac{i}{\hbar}\langle a, x \rangle} f_a(x) dx = e^{-\frac{i}{2\hbar}\|a\|^2} \widetilde{\int_{\mathcal{H}}} e^{\frac{i}{2\hbar}\langle x, x \rangle} f(x) dx. \tag{5.37}$$

Proof. It is easy to verify that the function g belongs to $\mathcal{F}(\mathcal{H})$, indeed it is the Fourier transform of the measure μ_g, whose action on a Borel bounded function $h : \mathcal{H} \to \mathbb{C}$ is given by

$$\int_{\mathcal{H}} h(x) d\mu_g(x) = \int_{\mathcal{H} \times \mathcal{H}} h(x + y) \delta_{a/\hbar}(x) e^{i\langle y, a \rangle} d\mu_f(y).$$

By Theorem 5.5, we have:

$$\widetilde{\int_{\mathcal{H}}} e^{\frac{i}{2\hbar}\langle x, x \rangle} e^{\frac{i}{\hbar}\langle a, x \rangle} f_a(x) dx$$

$$= \int_{\mathcal{H} \times \mathcal{H}} e^{-\frac{i\hbar}{2}\langle x+y, x+y \rangle} \delta_{a/\hbar}(x) e^{i\langle y, a \rangle} d\mu_f(y)$$

$$= e^{-\frac{i}{2\hbar}\|a\|^2} \int_{\mathcal{H}} e^{-\frac{i}{2\hbar}\|a\|^2} \widetilde{\int_{\mathcal{H}}} e^{-\frac{i\hbar}{2}\langle y, y \rangle} d\mu_f(y)$$

$$= e^{-\frac{i}{2\hbar}\|a\|^2} \widetilde{\int_{\mathcal{H}}} e^{\frac{i}{2\hbar}\langle x, x \rangle} f(x) dx. \qquad \square$$

The Parseval type equality (5.33) allows also the proof of the following Fubini theorem [7]:

Theorem 5.7. *Let $T : \mathcal{H} \to \mathcal{H}$, $T := I - L$ with L self-adjoint trace class, a linear invertible operator. Let $\mathcal{H} = Y + Z$ be the direct sum of two closed subspaces, with Z being finite dimensional. Assume that*

$$\langle Ty, z \rangle = 0 \qquad \forall y \in Y, z \in Z.$$

Let

$$T_1 y = (P_Y \circ T)(y), \qquad y \in Y,$$

$$T_2 z = (P_Z \circ T)(z), \qquad z \in Z,$$

where P_Y and P_Z are respectively orthogonal projections onto Y and Z. Assume that both T_1 and T_2 are isomorphisms of Y and Z respectively. Then, if $f \in \mathcal{F}(\mathcal{H})$:

$$\widetilde{\int_{\mathcal{H}}} e^{\frac{i}{2\hbar}\langle x, Tx \rangle} f(x) dx = C_T \widetilde{\int_Z} e^{\frac{i}{2\hbar}\langle z, T_2 z \rangle} \widetilde{\int_Y} e^{\frac{i}{2\hbar}\langle y, T_1 y \rangle} f(y+z) dy dz, \quad (5.38)$$

with

$$C_T = (\det T)^{-1/2} (\det T_1)^{1/2} (\det T_2)^{1/2}$$

Proof. For the proof of Equation (5.38) it is convenient to introduce a different notation. Let us define the bilinear form $((\ , \ ))$ on $\mathcal{H} \times \mathcal{H}$:

$$((x_1, x_2)) = \langle x_1, T x_2 \rangle.$$

Since $f \in \mathcal{F}(\mathcal{H})$, there exists a complex Borel measure $\mu \in \mathcal{M}(\mathcal{H})$ such that

$$f(x) = \int_{\mathcal{H}} e^{i \langle Tx, \xi \rangle} d\mu(\xi) = \int_{\mathcal{H}} e^{i((x,\xi))} d\mu(\xi).$$

The measure μ is given by

$$\int_{\mathcal{H}} h(x) d\mu(x) = \int_{\mathcal{H}} h(T^{-1}x) d\mu_f(x), \qquad f(x) = \int_{\mathcal{H}} e^{i \langle x, \xi \rangle} d\mu_f(\xi).$$

By Parseval type equality (5.33), we have

$$\widetilde{\int_{\mathcal{H}}} e^{\frac{i}{2\hbar}\langle x, Tx \rangle} f(x) dx = (\det T)^{-1/2} \int_{\mathcal{H}} e^{-\frac{i\hbar}{2}((x,x))} d\mu(x).$$

By considering $y + z, \eta + \xi \in Y \oplus Z$, we have that the function $f_z : Y \to \mathbb{C}$, given by

$$f_z(y) = f(y + z)$$

is of the form

$$f_z(y) = \int_Y e^{i((y,\eta))} d\mu_z(\eta),$$

indeed

$$\begin{aligned}
f(y+z) &= \int_{\mathcal{H}} e^{i \langle T(y+z), \eta + \zeta \rangle} d\mu(\eta + \zeta) \\
&= \int_{\mathcal{H}} e^{i \langle Ty, \eta \rangle} e^{i \langle Tz, \zeta \rangle} d\mu(\eta + \zeta) \\
&= \int_Y e^{i \langle Ty, \eta \rangle} d\mu_z(\eta) = \int_Y e^{i((y,\eta))} d\mu_z(\eta).
\end{aligned}$$

The measure μ_z is defined by

$$\int_Y h(\eta)d\mu_z(\eta) = \int_{\mathcal{H}} h(\eta)e^{i\langle Tz,\zeta\rangle}d\mu(\eta + \zeta),$$

for any Borel bounded function $h : Y \to \mathbb{C}$.

Since

$$((y,y)) = \langle y, T_1 y\rangle, \quad ((z,z)) = \langle z, T_2 z\rangle, \qquad y \in Y, z \in Z,$$

and

$$\widetilde{\int_Y} e^{\frac{i}{2\hbar}\langle y,T_1 y\rangle} f(y + z)dy = (\det T_1)^{-1/2}\int_Y e^{-\frac{i\hbar}{2}((y,y))}d\mu_z(y), \qquad (5.39)$$

we have that the function $I : Z \to \mathbb{C}$ given by

$$I(z) := \widetilde{\int_Y} e^{\frac{i}{2\hbar}\langle y,T_1 y\rangle} f(y + z)dy$$

is of the form $I(z) = \int_Z e^{i((z,\zeta))}d\tilde{\mu}(\zeta)$, with $\tilde{\mu} \in \mathcal{M}(Z)$. Indeed

$$(\det T_1)^{-1/2}\int_Y e^{-\frac{i\hbar}{2}((y,y))}d\mu_z(y)$$

$$= (\det T_1)^{-1/2}\int_{\mathcal{H}} e^{-\frac{i\hbar}{2}((\eta,\eta))}e^{i((z,\zeta))}d\mu(\eta + \zeta)$$

$$= \int_Z e^{i((z,\zeta))}d\tilde{\mu}(\zeta)$$

with $\tilde{\mu}$ defined by

$$\int_Z h(z)\tilde{\mu}(z) - (\det T_1)^{-1/2}\int_{\mathcal{H}} h(z)e^{-\frac{i\hbar}{2}((\eta,\eta))}d\mu(\eta + \zeta),$$

for any Borel bounded function $h : Z \to \mathbb{C}$.

Analogously to Eq. (5.39) we have

$$\widetilde{\int_Z} e^{\frac{i}{2\hbar}\langle z,T_2 z\rangle} I(z)dz = (\det T_2)^{-1/2}\int_Z e^{-\frac{i\hbar}{2}((z,z))}d\tilde{\mu}(z),$$

and the final result is a consequence of the following chain of equalities

$$\widetilde{\int_Z} e^{\frac{i}{2\hbar}\langle z,T_2 z\rangle} \widetilde{\int_Y} e^{\frac{i}{2\hbar}\langle y,T_1 y\rangle} f(y + z)dydz$$

$$= (\det T_2)^{-1/2}(\det T_1)^{-1/2}\int_{\mathcal{H}} e^{-\frac{i\hbar}{2}((\zeta,\zeta))}e^{-\frac{i\hbar}{2}((\eta,\eta))}d\mu(\eta + \zeta)$$

$$= (\det T_2)^{-1/2}(\det T_1)^{-1/2}\int_{\mathcal{H}} e^{-\frac{i\hbar}{2}((\zeta+\eta,\zeta+\eta))}d\mu(\eta + \zeta)$$

$$= (\det T_2)^{-1/2}(\det T_1)^{-1/2}(\det T)^{1/2}\widetilde{\int_{\mathcal{H}}} e^{\frac{i}{2\hbar}\langle x,Tx\rangle} f(x)dx.$$

$\square$

As we have already remarked above, the normalization constant in the Definition 5.4 plays a crucial role. Other alternative definitions of infinite dimensional oscillatory integrals can be considered. They can be obtained by introducing in the finite dimensional approximations different normalization constants.

Given a a self-adjoint invertible operator B on $\mathcal{H}$ (we do not impose any assumption on its trace), we can consider the definition of the *normalized infinite dimensional oscillatory integral with respect to B*.

Definition 5.5. A Borel function $f : \mathcal{H} \to \mathbb{C}$ is called $\mathcal{F}_B^{\hbar}$ integrable if for each sequence $\{P_n\}_{n \in \mathbb{N}}$ of projectors onto n-dimensional subspaces of $\mathcal{H}$, such that $P_n \leq P_{n+1}$ and $P_n \to I$ strongly as $n \to \infty$ (I being the identity operator in $\mathcal{H}$) the finite dimensional approximations

$$\widetilde{\int_{P_n \mathcal{H}}} e^{\frac{i}{2\hbar}\langle P_n x, B P_n x\rangle} f(P_n x) d(P_n x),$$

are well defined (in the sense of Definition 2.3) and the limit

$$\lim_{n \to \infty} (\det P_n B P_n)^{\frac{1}{2}} \widetilde{\int_{P_n \mathcal{H}}} e^{\frac{i}{2\hbar}\langle P_n x, B P_n x\rangle} f(P_n x) d(P_n x) \qquad (5.40)$$

exists and is independent on the sequence $\{P_n\}$.

In this case the limit is called the normalized oscillatory integral of f with respect to B and is denoted by

$$\widetilde{\int_{\mathcal{H}}^{B}} e^{\frac{i}{2\hbar}\langle x, B x\rangle} f(x) dx.$$

Again, given a function $f \in \mathcal{F}(\mathcal{H})$, it is possible to prove that f is $\mathcal{F}_B^{\hbar}$ integrable and the corresponding normalized infinite dimensional oscillatory integral can be computed by means of a formula similar to (5.33):

Theorem 5.8. *Let us assume that $f \in \mathcal{F}(\mathcal{H})$. Then f is $\mathcal{F}_B^{\hbar}$ integrable and the corresponding normalized oscillatory integral is given by the following Parseval-type formula:*

$$\widetilde{\int_{\mathcal{H}}^{B}} e^{\frac{i}{2\hbar}\langle x, B x\rangle} f(x) dx = \int_{\mathcal{H}} e^{-\frac{i\hbar}{2}\langle x, B^{-1} x\rangle} d\mu_f(x). \qquad (5.41)$$

Proof. The proof is completely similar to the proof of Theorem 5.5 and we left the details to the reader.   □

Remark 5.4. Formula (5.41) has already been discussed in the first part of [18].

The difference between Definitions 5.4 and 5.5 is the normalization constant. In fact Definition 5.5 can be seen as a generalization of Definition 5.4, which can be obtained by setting $B = I$, the identity operator.

The integral $\widetilde{\int_{\mathcal{H}}^{B}} e^{\frac{i}{2\hbar} \langle x, Bx \rangle} f(x) dx$ is called "normalized" because if we substitute into Eq. (5.41) the function $f = 1$, we have

$$\widetilde{\int_{\mathcal{H}}^{B}} e^{\frac{i}{2\hbar} \langle x, Bx \rangle} f(x) dx = 1.$$

The importance of the normalization constant in the finite dimensional approximations is highlighted by Theorems 5.5 and 5.8. Indeed Theorem 5.8 makes sense even if the operator $L := I - B$ is not trace class (in that case the Fredholm determinant $\det(I - B)$ cannot be defined).

In fact it is possible to introduce different normalization constants in the finite dimensional approximations and the properties of the corresponding infinite dimensional oscillatory integrals are related to the trace properties of the operator associated to the quadratic part of the phase function [9]. More precisely, for any $p \in \mathbb{N}$, let us consider the Schatten class $\mathcal{T}_p(\mathcal{H})$ of bounded linear operators L in $\mathcal{H}$ such that

$$\|L\|_p = (\mathrm{Tr}(L^+ L)^{p/2})^{1/p}$$

is finite. $(\mathcal{T}_p(\mathcal{H}), \| \cdot \|_p)$ is a Banach space (see [301]). For any $p \in \mathbb{N}$, $p \geq 2$ and $L \in \mathcal{T}_p(\mathcal{H})$ one defines the regularized Fredholm determinant $\det_{(p)} : I + \mathcal{T}_p(\mathcal{H}) \to \mathbb{R}$:

$$\det_{(p)}(I + L) = \det \left((I + L) \exp \sum_{j=1}^{p-1} \frac{(-1)^j}{j} L^j \right), \qquad L \in \mathcal{T}_p(\mathcal{H}),$$

where det denotes the usual Fredholm determinant, which is well defined as it is possible to prove that the operator $(I + L) \exp \sum_{j=1}^{p-1} \frac{(-1)^j}{j} L^j - I$ is trace class [301]. In particular $\det_{(2)}$ is called Carleman determinant.

For $p \in \mathbb{N}$, $p \geq 2$, $L \in \mathcal{T}_1(\mathcal{H})$, let us define the normalized quadratic form on $\mathcal{H}$:

$$N_p(L)(x) = (x, Lx) - i\hbar \mathrm{Tr} \sum_{j=1}^{p-1} \frac{L^j}{j}, \qquad x \in \mathcal{H}. \qquad (5.42)$$

Again, for $p \in \mathbb{N}$, $p \geq 2$, let us define the *class p normalized oscillatory integral*:

Definition 5.6. Let $p \in \mathbb{N}$, $p \geq 2$, L a bounded linear operator in $\mathcal{H}$, $f : \mathcal{H} \to \mathbb{C}$ a Borel measurable function. The class p normalized oscillatory integral of the function f with respect to the operator L is well defined if for each sequence $\{P_n\}_{n\in\mathbb{N}}$ of projectors onto n-dimensional subspaces of $\mathcal{H}$, such that $P_n \leq P_{n+1}$ and $P_n \to I$ strongly as $n \to \infty$ (I being the identity operator in $\mathcal{H}$) the finite dimensional approximations

$$\widetilde{\int_{P_n\mathcal{H}}^{\,\circ}} e^{\frac{i}{2\hbar}\|x\|^2} e^{-\frac{i}{2\hbar}N_p(P_nLP_n)(P_nx)} f(P_nx)d(P_nx), \tag{5.43}$$

are well defined and the limit

$$\lim_{n\to\infty} \widetilde{\int_{P_n\mathcal{H}}^{\,\circ}} e^{\frac{i}{2\hbar}\|x\|^2} e^{-\frac{i}{2\hbar}N_p(P_nLP_n)(P_nx)} f(P_nx)d(P_nx) \tag{5.44}$$

exists and is independent on the sequence $\{P_n\}$.

In this case the limit is denoted by

$$\mathcal{I}_{p,L}(f) = \widetilde{\int_{\mathcal{H}}^{\,p}} e^{\frac{i}{2\hbar}\|x\|^2} e^{-\frac{i}{2\hbar}\langle x, Lx\rangle} f(x)dx.$$

If L is not a trace class operator, then the quadratic form (5.42) is not well defined. Nevertheless expression (5.43) still makes sense thanks to the fact that all the functions under the integral are restricted to finite dimensional subspaces. Moreover the limit (5.44) can make sense, as the following result shows [9].

Theorem 5.9. *Let us assume that $f \in \mathcal{F}(\mathcal{H})$ and let L be a self-adjoint operator such that $L \in \mathcal{T}_p(\mathcal{H})$ and $\det_{(p)}(I - L) \neq 0$. Then the class-p normalized oscillatory integral of the function f with respect to the operator L exists and is given by the following Parseval-type formula:*

$$\widetilde{\int_{\mathcal{H}}^{\,p}} e^{\frac{i}{2\hbar}\|x\|^2} e^{-\frac{i}{2\hbar}\langle x, Lx\rangle} f(x)dx = [\det_{(p)}(I - L)]^{-1/2} \int_{\mathcal{H}} e^{-\frac{i\hbar}{2}\langle x, (I-L)^{-1}x\rangle} d\mu_f(x). \tag{5.45}$$

For the detailed proof of this result we refer to [9].

5.5 Polynomial phase functions

As we have seen in the previous section, the possible generalizations of the definition of finite dimensional oscillatory integrals to an infinite dimensional Hilbert space $\mathcal{H}$

$$\widetilde{\int_{\mathcal{H}}} e^{\frac{i}{\hbar}\Phi(x)} dx$$

concerns only quadratic phase functions $\Phi : \mathcal{H} \to \mathbb{C}$ of the form $\Phi(x) = \langle x, x \rangle / 2$ or, more generally $\Phi(x) = \langle x, Bx \rangle / 2$, with $B : \mathcal{H} \to \mathcal{H}$ a linear operator satisfying suitable assumptions.

As one can understand by a careful reading of Theorem 5.5 (and analogous Theorems 5.8 and 5.9), the key tool allowing the extension of the results of Section 5.2 is the Parseval type equality (5.10). Indeed, given a function $f \in \mathcal{F}(\mathbb{R}^n)$, its Fresnel integral is given by

$$\widetilde{\int} e^{\frac{i}{2\hbar} \langle x, (I-L)x \rangle} f(x) dx = (\det(I - L))^{-1/2} \int_{\mathbb{R}^n} e^{-\frac{i\hbar}{2} \langle x, (I-L)^{-1}x \rangle} d\mu_f(x),$$

(5.46)

with $L : \mathbb{R}^n \to \mathbb{R}^n$ self-adjoint and $I - L$ invertible.

Equation (5.46) admits a generalization on an infinite dimensional Hilbert space $\mathcal{H}$ because, even if the left-hand side loses any meaning when $n \to \infty$, the right-hand side can have a well defined meaning also when $\mathbb{R}^n$ is replaced with $\mathcal{H}$, provided the operator L satisfies suitable assumptions (such as, e.g., the existence of the inverse as well as of the Fredholm determinant of $I - L$).

In the case the quadratic phase function is replaced with an higher degree polynomial function, as in Section 5.3, the situation becomes more involved. Theorem 5.4 generalizes a Parseval type equality to the case of oscillatory integrals of the form:

$$\int_{\mathbb{R}^n}^{o} e^{\frac{i}{\hbar} P(x)} f(x) dx, \qquad f \in \mathcal{F}(\mathbb{R}^n),$$

where $\Gamma : \mathbb{R}^n \to \mathbb{R}$ is a $2M$ degree polynomial with positive leading coefficient.[1] More precisely the following formula holds:

$$\int_{\mathbb{R}^n} e^{\frac{i}{\hbar} P(x)} f(x) dx = \int_{\mathbb{R}^n} \tilde{F}(k) d\mu_f(k),$$

(5.47)

with the function $\tilde{F} : \mathbb{R}^n \to \mathbb{C}$ is a smooth function, represented by an absolutely convergent integral

$$\tilde{F}(k) = e^{in\pi/4M} \int_{\mathbb{R}^n} e^{ie^{i\pi/4M}kx} e^{\frac{i}{\hbar} P(e^{i\pi/4M}x)} dx.$$

(5.48)

The generalization of formulae (5.47) and (5.48) to the infinite dimensional case is not straightforward, since neither of them makes sense in the limit $n \to \infty$.

The quadratic phase functions are much simpler to deal with. Indeed, when the polynomial P is of the form

$$P(x) = \frac{1}{2} \langle x, x \rangle,$$

---

[1]An analogous result is obtained in the case where the leading coefficient is negative.

the function $\tilde{F}$, i.e. the Fourier transform of the imaginary exponential of the phase function, can be explicitly computed:

$$\tilde{F}(k) = (2\pi i\hbar)^{-n/2} \int_{\mathbb{R}^n} e^{ikx} e^{\frac{i}{2\hbar}\langle x,x\rangle} dx = e^{-\frac{i\hbar}{2}\langle x,x\rangle}. \qquad (5.49)$$

If the left-hand side of Eq. (5.49) loses any meaning in the limit $n \to \infty$ because of the non-convergence of the constant $(2\pi i\hbar)^{-n/2}$ and the non-existence of the Lebesgue measure dx, the right-hand side is still meaningful, even when x belongs to an infinite dimensional Hilbert space.

In the following we shall present the case of a polynomial phase function with quartic growth, where an infinite dimensional generalization of Parseval type equality (5.47) is allowed.

Let us deal first of all with the finite dimensional case, i.e. $dim(\mathcal{H}) = n$. Let $A : \mathcal{H} \times \mathcal{H} \times \mathcal{H} \times \mathcal{H} \to \mathbb{R}$ be a completely symmetric and positive fourth order covariant tensor on $\mathcal{H}$. After the introduction of an orthonormal basis in $\mathcal{H}$, the elements $x \in \mathcal{H}$ can be identified with $n-$ple of real numbers, i.e. $x = (x_1, \dots, x_n)$, and the action of the tensor A on the 4-ple (x, x, x, x) is represented by an homogeneous fourth order polynomial in the variables $x_1, \dots, x_n$:

$$P(x) = A(x, x, x, x) = \sum_{j,k,l,m} a_{j,k,l,m} x_j x_k x_l x_m \qquad (5.50)$$

with $a_{j,k,l,m} \in \mathbb{R}$.

We are going to define the following generalized Fresnel integral:

$$\widetilde{\int} e^{\frac{i}{2\hbar} x \cdot (I-B)x} e^{\frac{-i\lambda}{\hbar} P(x)} f(x) dx \qquad (5.51)$$

where I, B are $n \times n$ matrices, I being the identity, $\lambda \in \mathbb{R}$, $f \in \mathcal{F}(\mathbb{R}^n)$ and $\hbar > 0$.

Lemma 5.3. *Let $P : \mathbb{R}^n \to \mathbb{R}$ be given by (5.50). Then the Fourier transform of the distribution $\frac{e^{\frac{i}{2\hbar} x \cdot (I-B)x}}{(2\pi i\hbar)^{n/2}} e^{\frac{-i\lambda}{\hbar} P(x)}$:*

$$\tilde{F}(k) = \int_{\mathbb{R}^n} e^{ik \cdot x} \frac{e^{\frac{i}{2\hbar} x \cdot (I-B)x}}{(2\pi i\hbar)^{n/2}} e^{\frac{-i\lambda}{\hbar} P(x)} dx \qquad (5.52)$$

is a bounded complex-valued entire function on $\mathbb{R}^n$ admitting, if A is strictly positive, the following representations

$$\tilde{F}(k) = \begin{cases} e^{in\pi/8} \int_{\mathbb{R}^n} e^{ie^{i\pi/8} k \cdot x} \dfrac{e^{\frac{ie^{i\pi/4}}{2\hbar} x \cdot (I-B)x}}{(2\pi i\hbar)^{N/2}} e^{\frac{\lambda}{\hbar} P(x)} dx & \lambda < 0, \\[4mm] e^{-in\pi/8} \int_{\mathbb{R}^n} e^{ie^{-i\pi/8} k \cdot x} \dfrac{e^{\frac{ie^{-i\pi/4}}{2\hbar} x \cdot (I-B)x}}{(2\pi i\hbar)^{n/2}} e^{-\frac{\lambda}{\hbar} P(x)} dx & \lambda > 0. \end{cases}$$

$$(5.53)$$

Moreover, for general $A \geq 0$, if $\lambda \leq 0$ and $(I - B)$ is symmetric strictly positive then $\tilde{F}(k)$ can also be represented by

$$\tilde{F}(k) = \int_{\mathbb{R}^n} e^{ie^{i\pi/4}k \cdot x} \frac{e^{-\frac{1}{2\hbar}x \cdot (I-B)x}}{(2\pi\hbar)^{n/2}} e^{\frac{i\lambda}{\hbar}P(x)} dx = \mathbb{E}[e^{ie^{i\pi/4}k \cdot x} e^{\frac{i\lambda}{\hbar}P(x)} e^{\frac{1}{2\hbar}x \cdot Bx}],$$

(5.54)

where $\mathbb{E}$ denotes the expectation value with respect to the centered Gaussian measure on $\mathbb{R}^n$ with covariance operator $\hbar I$.

Proof. Representation (5.53) and the boundedness of $\tilde{F}$ follow from Lemma 5.1 , where a more general case is handled. From the representations (5.53) and (5.54) the analyticity of $\tilde{F}(k)$, $k \in \mathbb{C}$ follows immediately.

Let us now prove representation (5.54) for the Fourier transform

$$\tilde{F}(k) = \int_{\mathbb{R}^n} e^{ik \cdot x} \frac{e^{\frac{i}{2\hbar}x \cdot (I-B)x}}{(2\pi i\hbar)^{n/2}} e^{\frac{-i\lambda}{\hbar}P(x)} dx,$$

by mimicking the same procedure used in the proof of Lemma 5.1. Without loss of generality we can assume that the quadratic form $x \cdot (I-B)x$ is equal to $x \cdot x$, as it can always be reduced to this form by a change of coordinates.

Let us compute the n–dimensional integral defining $\tilde{F}(k)$ by introducing the polar coordinates in $\mathbb{R}^n$:

$$\int_{\mathbb{R}^n} e^{ik \cdot x} \frac{e^{\frac{i}{2\hbar}x \cdot x}}{(2\pi i\hbar)^{n/2}} e^{\frac{-i\lambda}{\hbar}P(x)} dx$$

$$= \int_{S_{n-1}} \left(\int_0^{+\infty} e^{i|k|rf(\phi_1,\dots,\phi_{n-1})} \frac{e^{\frac{i}{2\hbar}r^2}}{(2\pi i\hbar)^{n/2}} e^{\frac{-i\lambda}{\hbar}P(r)} r^{n-1} dr \right) d\Omega_{n-1} \quad (5.55)$$

where instead of n Cartesian coordinates we use $n - 1$ angular coordinates $(\phi_1, \dots, \phi_{n-1})$ and the variable $r = |x|$. S_{n-1} denotes the $(n - 1)$-dimensional spherical surface, $d\Omega_{n-1}$ is the Haar measure on it, $f(\phi_1, \dots, \phi_{n-1}) = (k \cdot x)/|k|r$, $P(r)$ is a fourth order polynomial in the variable r with coefficients depending on the $n - 1$ angular variables $(\phi_1, \dots, \phi_{n-1})$, namely:

$$P(r) = r^4 A\left(\frac{x}{|x|}, \frac{x}{|x|}, \frac{x}{|x|}, \frac{x}{|x|}\right) = r^4 a(\phi_1, \dots, \phi_{n-1}), \quad (5.56)$$

where $a(\phi_1, \dots, \phi_{n-1}) > 0$ for all $(\phi_1, \dots, \phi_{n-1}) \in S_{n-1}$. Let us focus on the integral

$$\int_0^{+\infty} e^{i|k|rf(\phi_1,\dots,\phi_{n-1})} \frac{e^{\frac{i}{2\hbar}r^2}}{(2\pi i\hbar)^{n/2}} e^{\frac{-i\lambda}{\hbar}P(r)} r^{n-1} dr.$$

This can be interpreted as the Fourier transform of the distribution on the real line

$$F(r) = \theta(r) r^{n-1} \frac{e^{\frac{i}{2\hbar} r^2}}{(2\pi i \hbar)^{n/2}} e^{\frac{-i\lambda}{\hbar} P(r)},$$

with $\theta(r) = 1$ if $r \geq 0$ and $\theta(r) = 0$ otherwise, $\lambda < 0$ and $P(r) = ar^4$, $a > 0$:

$$\int_0^{+\infty} e^{ikr} \frac{e^{\frac{i}{2\hbar} r^2}}{(2\pi i \hbar)^{n/2}} e^{\frac{-i\lambda}{\hbar} P(r)} r^{n-1} dr. \tag{5.57}$$

Let us consider the complex plane and set $z = re^{i\theta}$. We have

$$\int_0^{+\infty} e^{ikr} \frac{e^{\frac{i}{2\hbar} r^2}}{(2\pi i \hbar)^{n/2}} e^{\frac{-i\lambda}{\hbar} P(r)} r^{n-1} dr$$

$$= \lim_{\epsilon \downarrow 0} \int_{z = \rho e^{i\epsilon}} e^{ikz} \frac{e^{\frac{i}{2\hbar} z^2}}{(2\pi i \hbar)^{n/2}} e^{\frac{-i\lambda}{\hbar} P(z)} z^{n-1} dz$$

$$= \lim_{\epsilon \downarrow 0} \lim_{R \to +\infty} \int_0^R e^{ik\rho e^{i\epsilon}} \frac{e^{\frac{i}{2\hbar} \rho^2 e^{2i\epsilon}}}{(2\pi i \hbar)^{n/2}} e^{\frac{-i\lambda}{\hbar} P(\rho e^{i\epsilon})} \rho^{n-1} e^{ni\epsilon} d\rho$$

Given:

$$\gamma_1(R) = \{ z \in \mathbb{C} \mid 0 \leq \rho \leq R, \quad \theta = \epsilon \}$$
$$\gamma_2(R) = \{ z \in \mathbb{C} \mid \rho = R, \quad \epsilon \leq \theta \leq \pi/4 - \epsilon \}$$
$$\gamma_3(R) = \{ z \in \mathbb{C} \mid 0 \leq \rho \leq R, \quad \theta = \pi/4 - \epsilon \},$$

with $\epsilon > 0$ small, from the analyticity of the integrand and the Cauchy theorem we have

$$\int_{\gamma_1(R) \cup \gamma_2(R) \cup \gamma_3(R)} e^{ikz} \frac{e^{\frac{i}{2\hbar} z^2}}{(2\pi i \hbar)^{n/2}} e^{\frac{-i\lambda}{\hbar} P(z)} z^{n-1} dz = 0.$$

In particular:

$$\left| \int_{\gamma_2(R)} e^{ikz} \frac{e^{\frac{i}{2\hbar}z^2}}{(2\pi i\hbar)^{n/2}} e^{\frac{-i\lambda}{\hbar}P(z)} z^{n-1} dz \right|$$

$$= R^n \left| \int_\epsilon^{\pi/4-\epsilon} e^{ikRe^{i\theta}} \frac{e^{\frac{ie^{i2\theta}}{2\hbar}R^2}}{(2\pi i\hbar)^{n/2}} e^{\frac{-i\lambda}{\hbar}P(Re^{i\theta})} e^{in\theta} d\theta \right|$$

$$\leq R^n \int_\epsilon^{\pi/4-\epsilon} e^{-kR\sin(\theta)} \frac{e^{-\frac{\sin(2\theta)}{2\hbar}R^2}}{(2\pi\hbar)^{n/2}} e^{\frac{\lambda}{\hbar}(aR^4\sin(4\theta))} d\theta$$

$$\leq R^n \int_\epsilon^{\pi/8} e^{-k'R\theta} \frac{e^{\frac{-2\theta}{\pi\hbar}R^2}}{(2\pi\hbar)^{n/2}} e^{\frac{\lambda}{\hbar}(aR^4\frac{8}{\pi})\theta} d\theta$$

$$+ R^n e^{\frac{\lambda}{\hbar}2aR^4} \int_{\pi/8}^{\pi/4-\epsilon} e^{-k'R\theta} \frac{e^{\frac{-2\theta}{\pi\hbar}R^2}}{(2\pi\hbar)^{n/2}} e^{\frac{\lambda}{\hbar}(-aR^4\frac{8}{\pi})\theta} d\theta$$

$$= \frac{R^n}{(2\pi\hbar)^{n/2}} \left\{ \left(\frac{e^{(\frac{8a\lambda}{\pi\hbar}R^4 - \frac{2}{\pi\hbar}R^2 - k'R)\pi/8} - e^{(\frac{8a\lambda}{\pi\hbar}R^4 - \frac{2}{\pi\hbar}R^2 - k'R)\epsilon}}{(\frac{8a\lambda}{\pi\hbar}R^4 - \frac{2}{\pi\hbar}R^2 - k'R)} \right) \right.$$

$$\left. + \left(\frac{e^{\frac{8ea\lambda}{\pi\hbar}R^4} e^{(-\frac{2}{\pi\hbar}R^2 - k'R)(\pi/4-\epsilon)} - e^{\frac{a\lambda}{\hbar}R^4} e^{(-\frac{2}{\pi\hbar})R^2 - k'R)\pi/8}}{(\frac{-8a\lambda}{\pi\hbar}R^4 - \frac{2}{\pi\hbar}R^2 + -k'R)} \right) \right\}, \quad (5.58)$$

where $k' \in \mathbb{R}$ is a suitable constant. We have used the fact that if $\alpha \in [0, \pi/2]$ then $\frac{2}{\pi}\alpha \leq \sin(\alpha) \leq \alpha$, while if $\alpha \in [\pi/2, \pi]$ then $\sin(\alpha) \geq 2 - \frac{2}{\pi}\alpha$. From the last line one can deduce that

$$\left| \int_{\gamma_2(R)} e^{ikz} \frac{e^{\frac{i}{2\hbar}z^2}}{(2\pi i\hbar)^{n/2}} e^{\frac{-i\lambda}{\hbar}P(z)} z^{n-1} dz \right| \to 0, \qquad R \to \infty,$$

hence the following holds

$$\int_{z=\rho e^{i\epsilon}} e^{ikz} \frac{e^{\frac{i}{2\hbar}z^2}}{(2\pi i\hbar)^{n/2}} e^{\frac{-i\lambda}{\hbar}P(z)} z^{n-1} dz$$

$$= \int_{z=\rho e^{i(\pi/4-\epsilon)}} e^{ikz} \frac{e^{\frac{i}{2\hbar}z^2}}{(2\pi i\hbar)^{n/2}} e^{\frac{-i\lambda}{\hbar}P(z)} z^{n-1} dz.$$

By taking the limit as $\epsilon \downarrow 0$ of both sides one gets:

$$\int_0^{+\infty} e^{ikr} \frac{e^{\frac{i}{2\hbar}r^2}}{(2\pi i\hbar)^{n/2}} e^{\frac{-i\lambda}{\hbar}P(r)} r^{n-1} dr$$

$$= \int_0^{+\infty} e^{ik\rho e^{i\pi/4}} \frac{e^{\frac{-\rho^2}{2\hbar}}}{(2\pi\hbar)^{n/2}} e^{\frac{-i\lambda}{\hbar}P(\rho e^{i\pi/4})} \rho^{n-1} d\rho \quad (5.59)$$

By substituting into (5.55) we get the final result:

$$\int_{\mathbb{R}^n} e^{ik\cdot x} \frac{e^{\frac{i}{2\hbar}x\cdot x}}{(2\pi i\hbar)^{n/2}} e^{\frac{-i\lambda}{\hbar} P(x)} dx$$

$$= \int_{\mathbb{R}^n} e^{ie^{i\pi/4}k\cdot x} \frac{e^{\frac{-x\cdot x}{2\hbar}}}{(2\pi\hbar)^{n/2}} e^{\frac{-i\lambda}{\hbar} P(xe^{i\pi/4})} dx$$

$$= \mathbb{E}[e^{ie^{i\pi/4}k\cdot x} e^{\frac{-i\lambda}{\hbar} P(xe^{i\pi/4})}]. \qquad \square$$

Remark 5.5. A careful reading of this proof shows that the second part of the statement, that is representation (5.54), is valid if and only if the degree of P is 4, but cannot be generalized to polynomial functions of higher (even) degree. In fact the proof is based on the analyticity of the integrand and on a deformation of the contour of integration into a region of the complex plane in which the real part of the leading term of the polynomial, that is $Re(-i\lambda az^4)$, is negative, where $\lambda < 0$, $a > 0$. By setting $z = \rho e^{i\theta}$ one can immediately verify that this condition is satisfied if and only if $0 \le \theta \le \pi/4$. By considering a polynomial of higher even degree $2M$ this condition becomes $0 \le \theta \le \pi/2M$ and if $M > 2$ the angle $\theta = \pi/4$ is no longer included. This angle is fundamental as the oscillatory function $\frac{e^{\frac{i}{2\hbar}z^2}}{(2\pi i\hbar)^{1/2}}$ evaluated in $z = \rho e^{i\pi/4}$ gives $e^{-i\pi/4}\frac{e^{\frac{-\rho^2}{2\hbar}}}{(2\pi\hbar)^{1/2}}$, that is the density of the normal distribution with mean zero and variance $\hbar^2$, multiplied by the factor $e^{-i\pi/4}$. These considerations also show the necessity of considering $\lambda \le 0$.

Remark 5.6. We note that to have $\lambda = 0$ is equivalent to take $P = 0$. In this case one has immediately:

$$\int_{\mathbb{R}^n} e^{ik\cdot x} \frac{e^{\frac{i}{2\hbar}x\cdot x}}{(2\pi i\hbar)^{n/2}} dx$$

$$= \int_{\mathbb{R}^n} e^{ik\cdot xe^{i\pi/4}} \frac{e^{\frac{-x\cdot x}{2\hbar}}}{(2\pi\hbar)^{n/2}} dx = \mathbb{E}[e^{ik\cdot xe^{i\pi/4}}] = e^{\frac{-i\hbar}{2}k\cdot k}. \quad (5.60)$$

These results can now be applied to the definition of a generalized Fresnel integral of the following form

$$I(f) \equiv \widetilde{\int} e^{\frac{i}{2\hbar}x\cdot(I-B)x} e^{\frac{-i\lambda}{\hbar}P(x)} f(x)dx. \qquad (5.61)$$

Theorem 5.10 ("Parseval equality"). *Let $f \in \mathcal{F}(\mathbb{R}^n)$, $f = \hat{\mu}_f$. Then the generalized Fresnel integral (5.61) is well defined and it is given by*

$$\widetilde{\int} e^{\frac{i}{2\hbar}x\cdot(I-B)x} e^{\frac{-i\lambda}{\hbar}P(x)} f(x)dx = \int \tilde{F}(k)d\mu_f(k), \qquad (5.62)$$

where $\tilde{F}(k)$ is given by Eq. (5.53) if A in (5.50) is strictly positive, or by Eq. (5.54) if $A \geq 0$, $\lambda \leq 0$ and $(I - B)$ is symmetric strictly positive. The integral on the r.h.s. of (5.62) is absolutely convergent (hence it can be understood in Lebesgue sense).

The proof is completely analogous to the proof of Theorem 5.4.

Corollary 5.2. *Let $(I - B)$ be symmetric and strictly positive, $\lambda \leq 0$ and $f \in \mathcal{F}(\mathbb{R}^n)$, $f = \hat{\mu}_f$, such that $\forall x \in \mathbb{R}^n$ the integral $\int e^{-\frac{\sqrt{2}}{2}kx}|\mu_f|(dk)$ is convergent and the positive function $g : \mathbb{R}^n \to \mathbb{R}$, defined by*

$$g(x) = e^{\frac{1}{2\hbar}x \cdot Bx} \int e^{-\frac{\sqrt{2}}{2}kx}d|\mu_f|(k), \qquad x \in \mathbb{R}^n$$

is summable with respect to the centered Gaussian measure on $\mathbb{R}^n$ with covariance $\hbar I$.

Then f extends to an analytic function on $\mathbb{C}^n$ and the corresponding generalized Fresnel integral is well defined and it is given by

$$\widetilde{\int_{\mathbb{R}^n}} \frac{e^{\frac{i}{2\hbar}x \cdot (I-B)x}}{(2\pi i\hbar)^{n/2}} e^{\frac{-i\lambda}{\hbar}P(x)} f(x)dx = \mathbb{E}[e^{\frac{i\lambda}{\hbar}P(x)}e^{\frac{1}{2\hbar}x \cdot Bx}f(e^{i\pi/4}x)]. \qquad (5.63)$$

Proof. By the assumption on the measure μ_f, it follows that its Laplace transform $f^L : \mathbb{C}^n \to \mathbb{C}$,

$$f^L(z) = \int_{\mathbb{R}^n} e^{kz}d\mu_f(k),$$

is a well defined entire function such that $f^L(ix) = f(x)$, $x \in \mathbb{R}^n$. By Theorem 5.10 the generalized Fresnel integral can be computed by means of Parseval equality

$$\widetilde{\int_{\mathbb{R}^n}} \frac{e^{\frac{i}{2\hbar}x \cdot (I-B)x}}{(2\pi i\hbar)^{n/2}} e^{\frac{-i\lambda}{\hbar}P(x)} f(x)dx = \int_{\mathbb{R}^n} \tilde{F}(k)d\mu_f(k)$$

$$= \int_{\mathbb{R}^n} \mathbb{E}[e^{ikxe^{i\pi/4}}e^{\frac{1}{2\hbar}x \cdot Bx}e^{\frac{i\lambda}{\hbar}P(x)}]d\mu_f(k).$$

By Fubini theorem, which applies given the assumptions on the measure μ_f, this is equal to

$$\mathbb{E}[e^{\frac{1}{2\hbar}x \cdot Bx}e^{\frac{i\lambda}{\hbar}P(x)} \int_{\mathbb{R}^n} e^{ikxe^{i\pi/4}}d\mu_f(k)] = \mathbb{E}[e^{\frac{1}{2\hbar}x \cdot Bx}e^{\frac{i\lambda}{\hbar}P(x)}f^L(ie^{i\pi/4}x)]$$

$$= \mathbb{E}[e^{\frac{1}{2\hbar}x \cdot Bx}e^{\frac{i\lambda}{\hbar}P(x)}f(e^{i\pi/4}x)] \qquad (5.64)$$

and the conclusion follows. $\qquad \square$

Remark 5.7. The latter theorem shows that, under suitable assumptions on the function f, the generalized Fresnel integral (5.61) can be explicitly computed by means of a Gaussian integral. By mimicking the proof of Lemma 5.3 one can be tempted to generalize Eq. (5.63) to a larger class of functions that are analytic in a suitable region of $\mathbb{C}^n$ but do not belong to $\mathcal{F}(\mathbb{R}^n)$. In fact this is not possible, as the Definition 5.1 of oscillatory integral requires that the limit of the sequence of regularized integrals exists and is independent of the regularization. Let us consider the subset of the complex plane

$$\Lambda = \{z \in \mathbb{C} \mid 0 < \arg(z) < \pi/4\} \subset \mathbb{C}, \tag{5.65}$$

and let $\bar{\Lambda}$ be its closure. The identity

$$\lim_{\epsilon \to 0} \int \int_{\mathbb{R}^n} \frac{e^{\frac{i}{2\hbar} x \cdot (I-B)x}}{(2\pi i \hbar)^{n/2}} e^{\frac{-i\lambda}{\hbar} P(x)} f(x)\psi(\epsilon x)dx = \mathbb{E}[e^{\frac{1}{2\hbar} x \cdot Bx} e^{\frac{i\lambda}{\hbar} P(x)} f(e^{i\pi/4}x)]$$

(with $(I - B)$ symmetric strictly positive and $\lambda \leq 0$) can only be proved by choosing a regularizing function $\psi \in \mathcal{S}$, $\psi(0) = 1$, such that the function $z \mapsto \psi(zx)$ is analytic for $z \in \Lambda$ and continuous for $z \in \bar{\Lambda}$ for each $x \in \mathbb{R}^n$. Moreover one has to assume that $|\psi(e^{i\theta}x)|$ is bounded as $|x| \to \infty$ for each $\theta \in (0, \pi/4)$.

We can now extend these results to the case the generalized Fresnel integral is defined on a real separable infinite dimensional Hilbert space $(\mathcal{H}, \langle\, , \,\rangle, \|\ \|)$.

Let us consider the *abstract Wiener space* built on $\mathcal{H}$ (see [169, 237] and Section 3.7). In particular, let ν be the finitely additive cylinder measure on $\mathcal{H}$, defined by its characteristic functional $\hat{\nu}(x) = e^{-\frac{\hbar}{2}\|x\|^2}$. Let $|\ |$ be a measurable norm on $\mathcal{H}$, that is a norm $|\ |$ such that for every $\epsilon > 0$ there exist a finite-dimensional projection $P_\epsilon : \mathcal{H} \to \mathcal{H}$, such that for all $P \perp P_\epsilon$ one has

$$\nu(\{x \in \mathcal{H} \mid |P(x)| > \epsilon\}) < \epsilon,$$

where P and P_ϵ are called orthogonal ($P \perp P_\epsilon$) if their ranges are orthogonal in $(\mathcal{H}, \langle\, , \,\rangle)$. In fact this implies that $|\ |$ is weaker than $\|\ \|$ and, denoted by $\mathcal{B}$ the completion of $\mathcal{H}$ in the $|\ |$-norm and by i the continuous inclusion of $\mathcal{H}$ in $\mathcal{B}$, one can prove that $\mu \equiv \nu \circ i^{-1}$ is a countably additive Gaussian measure on the Borel subsets of $\mathcal{B}$ [169]. The triple $(i, \mathcal{H}, \mathcal{B})$ is called abstract Wiener space.

Given $y \in \mathcal{B}^*$ one can easily verify that the restriction of y to $\mathcal{H}$ is continuous on $\mathcal{H}$, so that one can identify $\mathcal{B}^*$ as a subset of $\mathcal{H}$. Moreover

$\mathcal{B}^*$ is dense in $\mathcal{H}$ and we have the dense continuous inclusions $\mathcal{B}^* \subset \mathcal{H} \subset \mathcal{B}$. Each element $y \in \mathcal{B}^*$ can be regarded as a random variable $n(y)$ on $(\mathcal{B}, \mu)$. A direct computation shows that $n(y)$ is normally distributed, with covariance $|y|^2$ (see Theorem 3.14 and Eq. (3.58)). The density of $\mathcal{B}^*$ in $\mathcal{H}$ and Eq. (3.58) allow the extension of the map $n : \mathcal{B}^* \to L^2(\mathcal{B}, \mu)$ to the whole space $\mathcal{H}$ (the extended map will be denoted with the same symbol with an abuse of notation).

Given an orthogonal projection P in $\mathcal{H}$, with

$$P(x) = \sum_{i=1}^{n} \langle e_i, x \rangle e_i$$

for some orthonormal $e_1, \ldots, e_n \in \mathcal{H}$, the stochastic extension $\tilde{P}$ of P on $\mathcal{B}$ is defined by

$$\tilde{P}(\,\cdot\,) = \sum_{i=1}^{n} n(e_i)(\,\cdot\,)e_i.$$

Analogously, given a function $f : \mathcal{H} \to \mathcal{B}_1$, where $(\mathcal{B}_1, |\ |_{\mathcal{B}_1})$ is another real separable Banach space, the stochastic extension $\tilde{f}$ of f to $\mathcal{B}$ exists if the functions $f \circ \tilde{P} : \mathcal{B} \to \mathcal{B}_1$ converges to $\tilde{f}$ in probability with respect to μ as P converges strongly to the identity in $\mathcal{H}$.

Let us consider now a completely symmetric positive covariant tensor operator $A : \mathcal{H} \times \mathcal{H} \times \mathcal{H} \times \mathcal{H} \to \mathbb{R}$ on $\mathcal{H}$ and let us assume that the map $V : \mathcal{H} \to \mathbb{R}^+$,

$$x \mapsto V(x) \equiv A(x, x, x, x), \qquad x \in \mathcal{H},$$

is continuous in the measurable norm $|\ |$. As a consequence V is continuous in the Hilbert norm $\|\ \|$, moreover it can be extended by continuity to a random variable $\bar{V}$ on $\mathcal{B}$, with $\bar{V}|_{\mathcal{H}} = V$. By Theorem 3.16, the stochastic extension $\tilde{V}$ of V exists and coincides with $\bar{V} : \mathcal{B} \to \mathbb{R}$ μ−a.e. Moreover for any increasing sequence of n−dimensional projectors P_n in $\mathcal{H}$, the family of bounded random variables on $(\mathcal{B}, \mu)$

$$e^{i\frac{\lambda}{\hbar} V \circ \tilde{P}_n(\,\cdot\,)} \equiv e^{i\frac{\lambda}{\hbar} V^n(\,\cdot\,)}$$

converges μ−a.e. to

$$e^{i\frac{\lambda}{\hbar} \bar{V}(\,\cdot\,)}.$$

Furthermore for any $h \in \mathcal{H}$ the sequence of random variables

$$\sum_{i=1}^{n} h_i n(e_i), \qquad h_i = \langle e_i, h \rangle$$

converges in $L^2(\mathcal{B}, \mu)$, and by subsequences almost everywhere, to the random variable $n(h)$. Analogously, given a self-adjoint trace class operator $B : \mathcal{H} \to \mathcal{H}$, the quadratic form on $\mathcal{H} \times \mathcal{H}$:

$$x \in \mathcal{H} \mapsto \langle x, Bx \rangle$$

can be extended to a random variable on $\mathcal{B}$, denoted again by $\langle \, \cdot \, , B \, \cdot \, \rangle$ (see Section 3.7).

Let us assume that the largest eigenvalue of B is strictly less than 1 (or, in other words, that the operator $(I - B)$ is strictly positive) and let $y \in \mathcal{H}$. Then one can prove (see Theorem 3.19) that the sequence of random variables $f_n : \mathcal{B} \to \mathbb{R}$

$$f_n(\, \cdot \,) = e^{\sum_{i=1}^n y_i n(e_i)(\, \cdot \,)} e^{\frac{1}{2h} \sum_{i=1}^n b_i ([n(e_i)(\, \cdot \,)]^2},$$

where $y_i = \langle y, e_i \rangle$, converges μ−a.e. as n goes to ∞ to the random variable

$$f(\, \cdot \,) = e^{n(y)(\, \cdot \,)} e^{\frac{1}{2h} \langle \, \cdot \, , B \, \cdot \, \rangle}$$

and that

$$\int f_n d\mu \to \int f d\mu = (\det(I - B))^{-1/2} e^{\frac{h}{2} \langle y, (I-B)^{-1} y \rangle}. \tag{5.66}$$

In this setting it is possible to prove the following result:

Lemma 5.4. *Let $B : \mathcal{H} \to \mathcal{H}$ be a self adjoint and trace class operator such that $I - B$ is strictly positive, let $k \in \mathcal{H}$ and $\lambda \le 0$. Then for any increasing sequence P_n of projectors onto n-dimensional subspaces of $\mathcal{H}$ such that $P_n \uparrow I$ strongly as $n \to \infty$, the following sequence of finite dimensional integrals:*

$$F_n(k) \equiv (2\pi i \hbar)^{-n/2} \int_{P_n \mathcal{H}} e^{i \langle P_n k, P_n x \rangle} e^{\frac{i}{2\hbar} \langle P_n x, (I-B) P_n x \rangle} e^{-i \frac{\lambda}{\hbar} V(P_n x)} d(P_n x)$$

converges, as $n \to \infty$, to the Gaussian integral on $\mathcal{B}$:

$$F(k) \equiv \mathbb{E}[e^{in(k)(\, \cdot \,)} e^{i\pi/4} e^{\frac{1}{2\hbar} \langle \, \cdot \, , B \, \cdot \, \rangle} e^{i \frac{\lambda}{\hbar} \bar{V}(\, \cdot \,)}] \tag{5.67}$$

($\mathbb{E}$ *being the expectation with respect to μ on $\mathcal{B}$).*

Proof. By Lemma 5.3 and Eq. (5.54) one has

$$(2\pi i \hbar)^{-n/2} \int_{P_n \mathcal{H}} e^{i \langle P_n k, P_n x \rangle} e^{\frac{i}{2\hbar} \langle P_n x, (I-B) P_n x \rangle} e^{-i \frac{\lambda}{\hbar} V(P_n x)} d(P_n x)$$

$$= (2\pi \hbar)^{-n/2} \int_{P_n \mathcal{H}} e^{i \langle P_n k, P_n x \rangle e^{i\pi/4}} e^{-\frac{1}{2\hbar} \langle P_n x, P_n x \rangle} e^{\frac{1}{2\hbar} \langle P_n x, B P_n x \rangle} e^{i \frac{\lambda}{\hbar} V(P_n x)} d(P_n x).$$

Let us introduce an orthonormal base $\{e_i\}$ of $\mathcal{H}$ such that P_n is the projector onto the span of the first n vectors. Each element $P_n x \in P_n \mathcal{H}$ can be represented as an n−ple of real numbers $(x_1, \ldots, x_n)$, where $x_i = \langle x, e_i \rangle$. The latter integral can be written in the following form:

$$(2\pi\hbar)^{-n/2} \int_{\mathbb{R}^n} e^{i \sum_{i=1}^n k_i x_i e^{i\pi/4}} e^{-\frac{1}{2\hbar} \sum_{i=1}^n x_i^2} e^{\frac{1}{2\hbar} \sum_{i,j=1}^n B_{ij} x_i x_j}$$

$$e^{i\frac{\lambda}{\hbar} \sum_{ij,k,h=1}^n A_{ijkh} x_i x_j x_k x_h} dx_1 \ldots dx_n,$$

where $B_{ij} = \langle e_i, B e_j \rangle$ and $A_{ijkh} = A(e_i, e_j, e_k, e_h)$.
On the other hand, this coincides with the Gaussian integral on $(\mathcal{B}, \mu)$:

$$\mathbb{E}[e^{i \sum_{i=1}^n k_i n(e_i)(\,\cdot\,) e^{i\pi/4}} e^{\frac{1}{2\hbar} \sum_{i,j=1}^n \langle e_i, B e_j \rangle n(e_i)(\,\cdot\,) n(e_j)(\,\cdot\,)} e^{\frac{\lambda}{\hbar} V \circ \tilde{P}_n(\,\cdot\,)}].$$

By Lebesgue's dominated convergence theorem (which holds because of the assumption on the strict positivity of the operator $I - B$) this converges as $n \to \infty$ to

$$\mathbb{E}[e^{in(k)(\,\cdot\,) e^{i\pi/4}} e^{\frac{1}{2\hbar} \langle \,\cdot\,, B \,\cdot\, \rangle} e^{i\frac{\lambda}{\hbar} \bar{V}(\,\cdot\,)}]$$

and the conclusion follows. □

The above result allows the following generalization of Theorem 5.10 to the infinite dimensional case [28, 26].

Theorem 5.11. *Let $B : \mathcal{H} \to \mathcal{H}$ be self-adjoint trace class, $(I - B)$ strictly positive, $\lambda \le 0$ and $f \in \mathcal{F}(\mathcal{H})$, $f \equiv \hat{\mu}_f$, and let us assume that the bounded variation measure μ_f satisfies the following condition:*

$$\int_{\mathcal{H}} e^{\frac{\hbar}{4} \langle k, (I-B)^{-1} k \rangle} |\mu_f|(dk) < +\infty. \tag{5.68}$$

Then the infinite dimensional oscillatory integral

$$\widetilde{\int_{\mathcal{H}}} e^{\frac{i}{2\hbar} \langle x, (I-B)x \rangle} e^{-i\frac{\lambda}{\hbar} A(x,x,x,x)} f(x) dx \tag{5.69}$$

exists and is given by

$$\int_{\mathcal{H}} \mathbb{E}[e^{in(k)(\,\cdot\,) e^{i\pi/4}} e^{\frac{1}{2\hbar} \langle \,\cdot\,, B \,\cdot\, \rangle} e^{i\frac{\lambda}{\hbar} \bar{V}(\,\cdot\,)}] \mu_f(dk).$$

Proof. By definition, choosing an increasing sequence of finite dimensional projectors P_n on $\mathcal{H}$, with $P_n \uparrow I$ strongly as $n \to \infty$, the oscillatory integral (5.69) is given by

$$\lim_{n \to \infty} (2\pi i\hbar)^{-n/2} \int_{P_n \mathcal{H}} e^{\frac{i}{2\hbar} \langle P_n x, (I-B) P_n x \rangle} e^{-i\frac{\lambda}{\hbar} A(P_n x, P_n x, P_n x, P_n x)} f(P_n x) dP_n x. \tag{5.70}$$

Let $f^n : P_n\mathcal{H} \to \mathbb{C}$ be the function defined by

$$f^n(y) \equiv f(y), \qquad y \in P_n\mathcal{H}.$$

One can easily verify that $f^n \in \mathcal{F}(P_n\mathcal{H})$, $f^n = \hat{\mu}_f^n$, where μ_f^n is the bounded variation measure on $P_n\mathcal{H}$ defined by

$$\mu_f^n(I) = \mu_f(P_n^{-1}I),$$

I being a Borel subset of $P_n\mathcal{H}$, indeed:

$$f^n(y) = f(y) = \int_{\mathcal{H}} e^{i\langle y,k\rangle} \mu_f(dk)$$

$$= \int_{\mathcal{H}} e^{i\langle P_n y, P_n k\rangle} \mu_f(dk) = \int_{P_n\mathcal{H}} e^{i\langle y, P_n k\rangle} \mu_f^n(dP_n k), \quad (5.71)$$

where $y = P_n y$. By Theorem 5.10 the limit (5.70) is equal to

$$\lim_{n\to\infty} \int_{P_n\mathcal{H}} G_n(P_n k)\mu_f^n(dP_n k), \qquad (5.72)$$

where $G_n : P_n\mathcal{H} \to \mathbb{C}$ is given by:

$$G_n(P_n k) = (2\pi\hbar)^{-n/2} \int_{P_n\mathcal{H}} e^{i\langle P_n k, P_n x\rangle} e^{i\pi/4} e^{-\frac{1}{2\hbar}\langle P_n x, (I-B)P_n x\rangle}$$

$$e^{i\frac{\lambda}{\hbar} A(P_n x, P_n x, P_n x, P_n x)} dP_n x.$$

This, on the other hand (see the proof of Lemma 5.4) is equal to

$$\mathbb{E}[e^{in(P_n k)(\,\cdot\,)} e^{i\pi/4} e^{\frac{1}{2\hbar}\sum_{i,j=1}^{n} B_{ij} n(e_i)(\,\cdot\,)n(e_j)(\,\cdot\,)} e^{i\frac{\lambda}{\hbar} V^n(\,\cdot\,)}],$$

where $V^n = V \circ \tilde{P}_n$. By substituting the latter expression into (5.72) we have

$$\lim_{n\to\infty} \int_{P_n\mathcal{H}} \mathbb{E}[e^{in(P_n k)(\,\cdot\,)} e^{i\pi/4} e^{\frac{1}{2\hbar}\sum_{i,j=1}^{n} B_{ij} n(e_i)(\,\cdot\,)n(e_j)(\,\cdot\,)} e^{i\frac{\lambda}{\hbar} V^n(\,\cdot\,)}]\mu_f^n(dP_n k)$$

$$= \lim_{n\to\infty} \int_{\mathcal{H}} \mathbb{E}[e^{in(P_n k)(\,\cdot\,)} e^{i\pi/4} e^{\frac{1}{2\hbar}\sum_{i,j=1}^{n} B_{ij} n(e_i)(\,\cdot\,)n(e_j)(\,\cdot\,)} e^{i\frac{\lambda}{\hbar} V^n(\,\cdot\,)}]\mu_f(dk)$$

$$= \lim_{n\to\infty} \int_{\mathcal{H}} F_n(k)\mu_f(dk).$$

By Lemma 5.4 and the dominated convergence theorem, applicable to the integral with respect to μ_f, due to assumption (5.68), we then get

$$\int_{\mathcal{H}} F(k)\mu_f(dk) = \int_{\mathcal{H}} \mathbb{E}[e^{in(k)(\,\cdot\,)} e^{i\pi/4} e^{\frac{1}{2\hbar}\langle\,\cdot\,,B\,\cdot\,\rangle} e^{i\frac{\lambda}{\hbar}\bar{V}(\,\cdot\,)}]\mu_f(dk),$$

and the conclusion follows. $\qquad\square$

Also Corollary 5.2 can be generalized to the infinite dimensional case. Indeed due to the assumption (5.68) the function f on the real Hilbert space $\mathcal{H}$ can be extended to those vectors $y \in \mathcal{H}^{\mathbb{C}}$ in the complex Hilbert space $\mathcal{H}^{\mathbb{C}}$ of the form $y = zx$, $x \in \mathcal{H}$, $z \in \mathbb{C}$, as the integral

$$\int_{\mathcal{H}} e^{iz\langle x,k\rangle} \mu_f(dk)$$

is absolutely convergent. Moreover the latter can be uniquely extended to a random variable on $\mathcal{B}$, denoted again by f, by

$$f^z(\,\cdot\,) \equiv f(z\,\cdot\,) \equiv \int_{\mathcal{H}} e^{izn(k)(\,\cdot\,)} \mu_f(dk). \tag{5.73}$$

Moreover the random variable

$$e^{\frac{1}{2\hbar}\langle\,\cdot\,,B\,\cdot\,\rangle} f^z(\,\cdot\,)$$

belongs to $L^1(\mathcal{B}, \mu)$ if $Im(z)^2 \leq 1/2$.

Theorem 5.12. *Let $B : \mathcal{H} \to \mathcal{H}$ be self-adjoint trace class, $I - B$ strictly positive, $\lambda \leq 0$ and $f \in \mathcal{F}(\mathcal{H})$ be the Fourier transform of a bounded variation measure μ_f satisfying assumption (5.68).*

Then the infinite dimensional oscillatory integral (5.69) is well defined and it is given by:

$$\widetilde{\int_{\mathcal{H}}} e^{\frac{i}{2\hbar}\langle x,(I-B)x\rangle} e^{-i\frac{\lambda}{\hbar}A(x,x,x,x)} f(x)dx = \mathbb{E}[e^{i\frac{\lambda}{\hbar}\bar{V}(\,\cdot\,)} e^{\frac{1}{2\hbar}\langle\,\cdot\,,B\,\cdot\,\rangle} f(e^{i\pi/4}\,\cdot\,)] \tag{5.74}$$

Proof. By Theorem 5.11 the infinite dimensional oscillatory integral (5.69) can be computed by means of the Parseval-type formula:

$$\widetilde{\int_{\mathcal{H}}} e^{\frac{i}{2\hbar}\langle x,(I-B)x\rangle} e^{-i\frac{\lambda}{\hbar}A(x,x,x,x)} f(x)dx$$

$$= \int_{\mathcal{H}} \mathbb{E}[e^{in(k)(\,\cdot\,)e^{i\pi/4}} e^{\frac{1}{2\hbar}\langle\,\cdot\,,B\,\cdot\,\rangle} e^{i\frac{\lambda}{\hbar}\bar{V}(\,\cdot\,)}] \mu_f(dk). \tag{5.75}$$

By Fubini theorem, which can be applied under the assumption (5.68), the integral on the r.h.s. of (5.75) is equal to

$$\mathbb{E}[e^{i\frac{\lambda}{\hbar}\bar{V}(\,\cdot\,)} e^{\frac{1}{2\hbar}\langle\,\cdot\,,B\,\cdot\,\rangle} \int_{\mathcal{H}} e^{in(k)(\,\cdot\,)e^{i\pi/4}} \mu_f(dk)]$$

$$= \mathbb{E}[e^{i\frac{\lambda}{\hbar}\bar{V}(\,\cdot\,)} e^{\frac{1}{2\hbar}\langle\,\cdot\,,B\,\cdot\,\rangle} f e^{i\pi/4}(\,\cdot\,)] = \mathbb{E}[e^{i\frac{\lambda}{\hbar}\bar{V}(\,\cdot\,)} e^{\frac{1}{2\hbar}\langle\,\cdot\,,B\,\cdot\,\rangle} f(e^{i\pi/4}\,\cdot\,)].$$

The integral on the right-hand side is absolutely convergent as $|e^{i\frac{\lambda}{\hbar}\bar{V}}| = 1$ and $e^{\frac{1}{2\hbar}\langle\,\cdot\,,B\,\cdot\,\rangle} f e^{i\pi/4} \in L^1(\mathcal{B}, \mu)$ as $Im(e^{i\pi/4}) = 1/\sqrt{2}$ (see Fernique's theorem in [237]). $\qquad\square$

Remark 5.8. In the simpler case $\lambda = 0$, under the above assumptions on the function f and the operator B, the infinite dimensional oscillatory integral (given by (5.74) with $V = 0$) can also be explicitly computed by means of the absolutely convergent integrals:

$$\widetilde{\int_{\mathcal{H}}} e^{\frac{i}{2\hbar}\langle x,(I-B)x\rangle} f(x)dx = \frac{1}{\sqrt{\det(I-B)}} \int_{\mathcal{H}} e^{-\frac{i\hbar}{2}\langle k,(I-B)^{-1}k\rangle} \mu_f(dk).$$

$$(5.76)$$

In fact, by means of different methods (see Section 5.4), Eq. (5.76) can be proved even without the assumption on the positivity of the operator $(I - B)$ (it is sufficient that $(I - B)$ is invertible).

Remark 5.9. So far we have proved, under suitable assumptions on the function $f : \mathcal{H} \to \mathbb{C}$ and the operator B, that if $\lambda \le 0$ the infinite dimensional generalized Fresnel integral (5.69)

$$I^F(\lambda) \equiv \widetilde{\int_{\mathcal{H}}} e^{\frac{i}{2\hbar}\langle x,(I-B)x\rangle} e^{-i\frac{\lambda}{\hbar}A(x,x,x,x)} f(x)dx$$

on the Hilbert space $\mathcal{H}$ is exactly equal to a Gaussian integral on $\mathcal{B}$:

$$I^G(\lambda) \equiv \int_{\mathcal{H}} \mathbb{E}[e^{in(k)(\,\cdot\,)}e^{i\pi/4}e^{\frac{1}{2\hbar}\langle\,\cdot\,,B\,\cdot\,\rangle}e^{i\frac{\lambda}{\hbar}\bar{V}(\,\cdot\,)}]\mu_f(dk)$$

(Theorem 5.11), and to

$$I^A(\lambda) \equiv \mathbb{E}[e^{i\frac{\lambda}{\hbar}\bar{V}(\,\cdot\,)}e^{\frac{1}{2\hbar}\langle\,\cdot\,,B\,\cdot\,\rangle}f(e^{i\pi/4}\,\cdot\,)]$$

(Theorem 5.12). One can easily verify that I^G and I^A are analytic functions of the complex variable λ in the region of the complex λ plane $\{Im(\lambda) > 0\}$, while they are continuous in $\{Im(\lambda) = 0\}$ and coincide with I^F in $\{Im(\lambda) = 0, Re(\lambda) \le 0\}$.

The generalization of these techniques to infinite dimensional oscillatory integrals with polynomial phase function of higher degree is not straightforward. Some partial result concerning particular complex phase function of higher arbitrary degree has been obtained in [35].

Chapter 6

Feynman path integrals and the Schrödinger equation

The present chapter is devoted to the application of infinite dimensional oscillatory integrals to the construction of the Feynman path integral representation of the solution of the Schrödinger equation with different potentials.

6.1 The anharmonic oscillator with a bounded anharmonic potential

Let us consider the Schrödinger equation in $L^2(\mathbb{R}^d)$

$$i\hbar\frac{\partial}{\partial t}\psi = H\psi, \tag{6.1}$$

with initial datum $\psi_{|t=0} = \psi_0 \in L^2(\mathbb{R}^d)$ and quantum mechanical Hamiltonian, given on the smooth vectors $\psi \in \mathcal{S}(\mathbb{R}^d)$ by

$$H\psi(x) = -\frac{\hbar^2}{2m}\Delta\psi(x) + \frac{1}{2}x\Omega^2 x\psi(x) + V(x)\psi(x), \qquad x \in \mathbb{R}^d,$$

where $m > 0$ is the mass, $\Omega^2 \geq 0$ is a positive $d \times d$ matrix, and V is a bounded continuous real function on $\mathbb{R}^d$. In the following we shall put for notational simplicity $m \equiv 1$, but the whole discussion can be repeated for arbitrary values of the mass parameter.

Let us denote by H_0 the harmonic oscillator Hamiltonian, given on $\psi \in \mathcal{S}(\mathbb{R}^d)$ by

$$H_0\psi(x) = -\frac{\hbar^2}{2}\Delta\psi(x) + \frac{1}{2}x\Omega^2 x\psi(x), \qquad x \in \mathbb{R}^d.$$

Both H_0 and H are self-adjoint operators on $L^2(\mathbb{R}^d)$, on the natural domain of definition of H_0, which can be easily described by writing the vectors $\psi \in L^2(\mathbb{R}^d)$ as linear combination of Hermite functions (see [280, 281]).

In particular, both H and H_0 generate unitary groups in $L^2(\mathbb{R}^d)$, denoted by $U(t) = e^{-\frac{i}{\hbar}Ht}$ and $U_0(t) = e^{-\frac{i}{\hbar}H_0t}$. The solution of the Schrödinger Eq. (6.1) with initial datum $\psi_{|t=0} = \psi_0 \in L^2(\mathbb{R}^d)$, is given by

$$\psi(t,x) = (e^{-\frac{i}{\hbar}Ht}\psi_0)(x). \tag{6.2}$$

In the following we shall show how the Feynman path integral representation for the solution (6.2)

$$\psi(t,x) = \int_{\gamma(t)=x} e^{\frac{i}{\hbar}S_t(\gamma)}\psi_0(\gamma(0))d\gamma, \tag{6.3}$$

can be mathematically realized in terms of a well defined infinite dimensional oscillatory integral on a suitable Hilbert space of paths:

$$\psi(t,x) = \int_{\gamma(t)=0} e^{\frac{i}{\hbar}\int_0^t \left(\dot{\gamma}(s)^2/2 - \gamma(s)\Omega^2\gamma(s)/2 - V(\gamma(s)+x)\right)ds}\psi_0(\gamma(0)+x)d\gamma, \tag{6.4}$$

where we performed the formal change of variable $\gamma \mapsto \gamma + x$ in order to deal with the homogeneous condition $\gamma(t) = 0$, which is preserved by linear combinations.

Let us consider the *Cameron-Martin space* $\mathcal{H}_t$, that is the Sobolev space of absolutely continuous functions $\gamma : [0, t] \to \mathbb{R}^d$, such that $\gamma(t) = 0$, and with square integrable weak derivative $\dot{\gamma}$:

$$\int_0^t |\dot{\gamma}(s)|^2 ds < \infty,$$

endowed with the inner product

$$\langle \gamma_1, \gamma_2 \rangle = \int_0^t \dot{\gamma}_1(s) \cdot \dot{\gamma}_2(s)ds.$$

(This space was already introduced in (1.15) in the introduction). From a physical point of view, $\mathcal{H}_t$ represents a space of Feynman path with finite kinetic energy.

Let us consider the linear operator $L : \mathcal{H}_t \to \mathcal{H}_t$ given by

$$(L\gamma)(s) := \int_s^t ds' \int_0^{s'} (\Omega^2\gamma)(s'')ds''. \tag{6.5}$$

One can easily verify that

$$\langle \gamma_1, L\gamma_2 \rangle = \int_0^t \gamma_1(s)\Omega^2\gamma_2(s)ds,$$

and conclude that L is self-adjoint and positive on $\mathcal{H}_t$:

$$\langle \gamma, L\gamma \rangle = \int_0^t \gamma(s)\Omega^2\gamma(s)ds \geq 0, \qquad \forall \gamma \in \mathcal{H}_t.$$

Let us now investigate the conditions for the existence of the inverse of the operator $I - L$.

Lemma 6.1. *Let Ω_j, with $j = 1, \ldots, d$, be the eigenvalues of the matrix Ω. If*

$$t \neq \left(n + \frac{1}{2}\right)\frac{\pi}{\Omega_j}, \qquad n \in \mathbb{N}, \, j = 1, \ldots, d,$$

then the operator $I - L$ is invertible and its inverse is given by:

$$(I - L)^{-1}\gamma(s) = \gamma(s) - \Omega \int_s^t \sin[\Omega(s - s')]\gamma(s')ds'$$

$$+ \sin[\Omega(t - s)] \int_0^t (\cos\Omega t)^{-1}\Omega\cos(\Omega s')\gamma(s')ds'. \quad (6.6)$$

Proof. By Eq. (6.5), given a vector $\eta \in \mathcal{H}_t$, the vector $\gamma \in \mathcal{H}_t$ is of the form $\gamma = (I - L)^{-1}\eta$ if it satisfies the integral equation:

$$\eta(s) = \gamma(s) - \int_s^t ds' \int_0^{s'} (\Omega^2\gamma)(s'')ds''. \quad (6.7)$$

If η is sufficiently smooth, by differentiating twice, (6.7) is equivalent to the differential equation:

$$\ddot{\gamma}(s) + \Omega^2\gamma(s) = \ddot{\eta}(s)$$

with conditions $\gamma(t) = 0$ and $\dot{\gamma}(0) = \dot{\eta}(0)$. By standard technique the solution is given by

$$(I - L)^{-1}\eta(s) = \eta(s) - \Omega \int_s^t \sin[\Omega(s - s')]\eta(s')ds'$$

$$+ \sin[\Omega(t - s)] \int_0^t (\cos\Omega t)^{-1}\Omega\cos(\Omega s')\eta(s')ds'. \quad (6.8)$$

It is easy to verify that the latter formula is valid for general $\eta \in \mathcal{H}_t$, and the conclusion follows. □

The following lemma gives a characterization of the spectrum of L [129].

Lemma 6.2. *Under the assumptions of Lemma 6.1, the self-adjoint operator $L : \mathcal{H}_t \to \mathcal{H}_t$ given by Eq. (6.5) is trace class and*

$$\mathrm{Ind}(I - L) = \sum_{j=1}^d \left[\frac{\Omega_j t}{\pi} + \frac{1}{2}\right],$$

where $[\cdot]$ denotes the integer part and $\Omega_1, \ldots, \Omega_d$ are the eigenvalues of the matrix Ω, counted with their multiplicity.
Moreover the Fredholm determinant of $I - L$ is given by:

$$\det(I - L) = \det(\cos(\Omega t)).$$

Proof. We follow here the method by Elworthy and Truman [129]. Let us assume, without loss of generality, that the matrix Ω is diagonal.

Because of the self-adjointness and the positivity of L, we look for eigenvalues of the form p^2, with $p \in \mathbb{R}$. The real positive number p^2 is an eigenvalue of L, if there exists a $\gamma \in \mathcal{H}_t$ such that

$$L\gamma = p^2\gamma,$$

$$\int_s^t ds' \int_0^{s'} (\Omega^2\gamma)(s'')ds'' = p^2\gamma(s).$$

By differentiating twice, we obtain the differential equation

$$p^2\ddot{\gamma}(s) + \Omega^2\gamma(s) = 0,$$

with the conditions $\gamma(t) = 0$ and $\dot{\gamma}(0) = 0$. The solutions are given by the vectors

$$\gamma(s) = (\gamma^1(s), \dots, \gamma^d(s)),$$

where $\gamma^j(s) = A_j \sin[\Omega_j(s - \phi_j)/p]$, for constant $A_j, \phi_j, j = 1, \dots, d$, satisfying the conditions:

$$\Omega_j\phi_j/p = \left(m_j + \frac{1}{2}\right)\pi, \qquad m_j \in \mathbb{Z},$$

$$\Omega_j t/p = \left(n_j + \frac{1}{2}\right)\pi, \qquad n_j \in \mathbb{Z}.$$

The only possible values for p are of the form

$$p = \frac{\Omega_j t}{\left(n_j + \frac{1}{2}\right)\pi}, \qquad n_j \in \mathbb{Z}.$$

Since $\sum_{n=0}^{\infty}(n+\frac{1}{2})^{-2} < \infty$, one can conclude that the operator $L : \mathcal{H}_t \to \mathcal{H}_t$ is trace class.

For the calculation of the index of the operator $I - L$, we have to take into account the eigenvalues of the operator L of the form

$$p^2 = \frac{\Omega_j t}{\left(n_j + \frac{1}{2}\right)\pi},$$

with $p > 1$, that is

$$n_j = 0, 1, \dots, \left\lfloor \frac{\Omega_j t}{\pi} - \frac{1}{2}\right\rfloor,$$

$\lfloor\ \rfloor$ denoting the integer part. They correspond to the following eigenvectors

$$\gamma_{n_j}(s) = \left(0, 0, \dots, 0, \cos\left(\left(n_j + \frac{1}{2}\right)\frac{s\pi}{t}\right), 0, \dots, 0\right),$$

where the jth entry being non-zero.

The result follows from the equality:

$$\cos(x) = \prod_{n=0}^{\infty} \left(1 - \frac{x^2}{\left(n + \frac{1}{2}\right)^2\pi^2}\right). \qquad \square$$

The next lemma gives sufficient conditions for the definition of the infinite dimensional oscillatory integral (6.4) and the application of Theorem 5.5.

Lemma 6.3. *Let $\psi_0, V \in \mathcal{F}(\mathbb{R}^d)$. Then the function $f : \mathcal{H}_t \to \mathbb{C}$, given by*

$$f(\gamma) = e^{-\frac{i}{\hbar} \int_0^t V(\gamma(s)+x)ds} \psi_0(\gamma(0) + x), \qquad \gamma \in \mathcal{H}_t$$

belongs to $\mathcal{F}(\mathcal{H}_t)$.

Proof. We give here the proof in the case where $d = 1$ in order to simplify the notation, but the whole reasoning can be easily extended to the case of general dimension d.

Let us denote by γ_τ, $\tau \in [0,t]$, the vector in $\mathcal{H}$ defined by

$$\gamma_\tau(s) = t - \tau \vee s, \qquad s \in [0,t].$$

One can easily verify that for an $\gamma \in \mathcal{H}_t$

$$\langle \gamma_\tau, \gamma \rangle = \gamma(\tau).$$

If $\psi_0 = \hat{\mu}_0$, with $\mu_0 \in \mathcal{M}(\mathbb{R})$, we have:

$$\psi_0(\gamma(0) + x) = \int_{\mathbb{R}} e^{ik(\gamma(0)+x)} d\mu_0(k)$$

$$= \int_{\mathbb{R}} e^{i\langle \gamma, k\gamma_0 \rangle} e^{ikx} d\mu_0(k) \tag{6.9}$$

$$= \int_{\mathbb{R}} \int_{\mathcal{H}} e^{i\langle \gamma, \eta \rangle} \delta_{k\gamma_0}(d\eta) e^{ikx} d\mu_0(k). \tag{6.10}$$

Analogously, if $V \in \mathcal{F}(\mathbb{R})$, $V = \hat{\mu}_V$, with $\mu_V \in \mathcal{M}(\mathbb{R})$, we have

$$\int_0^t V(\gamma(s) + x) ds = \int_0^t \int_{\mathbb{R}} e^{ik(\gamma(s)+x)} d\mu_v(k) ds$$

$$= \int_0^t \int_{\mathbb{R}} \int_{\mathcal{H}} e^{i\langle \gamma, \eta \rangle} \delta_{k\gamma_s}(d\eta) e^{ikx} d\mu_v(k) ds. \tag{6.11}$$

It follows that both functions on $\mathcal{H}_t$, i.e. $\gamma \mapsto \psi_0(\gamma(0) + x)$ and $\int_0^t V(\gamma(s) + x) ds$, belong to $\mathcal{F}(\mathcal{H}_t)$. As $\mathcal{F}(\mathcal{H}_t)$ is a Banach algebra by pointwise multiplications (see Section 5.2), it is possible to conclude that also the function

$$\gamma \mapsto e^{-\frac{i}{\hbar} \int_0^t V(\gamma(s)+x)ds} \psi_0(\gamma(0) + x), \qquad \gamma \in \mathcal{H}_t,$$

is an element of $\mathcal{F}(\mathcal{H}_t)$. $\qquad\square$

The Parseval type equality for infinite dimensional oscillatory integrals (Theorem 5.5) and the previous lemmas allow us to define and compute the Feynman path integral representation for the solution of the Schrödinger equation with the harmonic oscillator Hamiltonian:

$$\begin{cases} i\hbar \frac{\partial}{\partial t}\psi(t,x) = -\frac{\hbar^2}{2}\Delta\psi(t,x) + \frac{1}{2}x\Omega^2 x\psi(t,x) \\ \psi(0,x) = \psi_0(x) \end{cases} \tag{6.12}$$

that is

$$\psi(t,x) = \int_{\gamma(t)=0} e^{\frac{i}{2\hbar}\int_0^t \left(\dot{\gamma}(s)^2 - (\gamma(s)+x)\Omega^2(\gamma(s)+x)\right)ds}\psi_0(\gamma(0)+x)d\gamma, \tag{6.13}$$

Theorem 6.1. *Let $\psi_0 \in \mathcal{S}(\mathbb{R}^d)$ and $t \neq \left(n+\frac{1}{2}\right)\frac{\pi}{\Omega_j}$, $n \in \mathbb{N}$. Then the solution of the Schrödinger Eq. (6.12) is given by the infinite dimensional oscillatory integral on the Cameron-Martin space*

$$\psi(t,x) = \widetilde{\int_{\mathcal{H}_t}} e^{\frac{i}{2\hbar}\int_0^t \dot{\gamma}(s)^2 ds}e^{-\frac{i}{2\hbar}\int_0^t(\gamma(s)+x)\Omega^2(\gamma(s)+x)ds}\psi_0(\gamma(0)+x)d\gamma$$

$$= \widetilde{\int_{\mathcal{H}_t}} e^{\frac{i}{2\hbar}\langle\gamma,(I-L)\gamma\rangle}e^{-\frac{i}{2\hbar}x\Omega^2 xt}e^{-\frac{i}{\hbar}\int_0^t \gamma(s)\Omega^2 xds}\psi_0(\gamma(0)+x)d\gamma \tag{6.14}$$

Proof. We give here the proof in the case where $d = 1$ in order to simplify the notation, but the whole reasoning can be easily extended to the case of general dimension d.

First of all we can see that the function on $f : \mathcal{H}_t \to \mathbb{C}$ given by

$$f(\gamma) := e^{-\frac{i}{2\hbar}\int_0^t(\gamma(s)+x)\Omega^2(\gamma(s)+x)ds}\psi_0(\gamma(0)+x), \qquad \gamma \in \mathcal{H}_t$$

is Fresnel integrable. Indeed it is of the form

$$f(\gamma) = e^{-\frac{i}{2\hbar}\langle\gamma,L\gamma\rangle}g(\gamma), \qquad \gamma \in \mathcal{H}_t,$$

with $L : \mathcal{H}_t \to \mathcal{H}_t$ is the self-adjoint operator (6.5), that by Lemmas 6.1 and 6.2 is trace class with $I - L$ being invertible, and $g : \mathcal{H}_t \to \mathbb{C}$, given by

$$g(\gamma) = e^{-\frac{i}{2\hbar}x\Omega^2 xt}e^{-\frac{i}{\hbar}\int_0^t \gamma(s)\Omega^2 xds}\psi_0(\gamma(0)+x), \qquad \gamma \in \mathcal{H}_t$$

belongs to $\mathcal{F}(\mathcal{H}_t)$. Indeed, as $\mathcal{S}(\mathbb{R}^d) \subset \mathcal{F}(\mathbb{R}^d)$, by Lemma 6.3 the function $\gamma \mapsto \psi_0(\gamma(0)+x)$ belongs to $\mathcal{F}(\mathcal{H}_t)$. Moreover the function $\gamma \mapsto e^{-\frac{i}{\hbar}\int_0^t \gamma(s)\Omega^2 xds}$ is of the form

$$e^{-\frac{i}{\hbar}\int_0^t \gamma(s)\Omega^2 xds} = \int_{\mathcal{H}_t} e^{i\langle\gamma,\eta\rangle}\delta_{\eta_x}(d\eta), \qquad \gamma \in \mathcal{H}_t,$$

where η_x is the vector of $\mathcal{H}_t$ given by

$$\eta_x(s) = \frac{\Omega^2 x}{\hbar}\left(\frac{s^2}{2} - \frac{t^2}{2}\right), \qquad s \in [0,t].$$

By Theorem 5.5 the infinite dimensional oscillatory integral

$$\widetilde{\int_{\mathcal{H}_t}} e^{\frac{i}{2\hbar}\langle\gamma,(I-L)\gamma\rangle} e^{-\frac{i}{2\hbar}x\Omega^2 xt} e^{-\frac{i}{\hbar}\int_0^t \gamma(s)\Omega^2 x ds} \psi_0(\gamma(0)+x)d\gamma$$

is well defined and is given by

$$\det(I-L)^{-1/2}\int_{\mathcal{H}_t} e^{-\frac{i\hbar}{2}\langle\gamma,(I-L)^{-1}\gamma\rangle}d\mu_g(\gamma),$$

with $g = \hat{\mu}_g$. By Lemmas 6.1 and 6.2, and some calculations we have that the latter expressing is equal to:

$$\sqrt{\frac{\Omega}{2\pi i \sin(\Omega t)}} \int_{\mathbb{R}} e^{\frac{i}{2\hbar}\left(x\Omega\cot(\Omega t)x+y\Omega\cot(\Omega t)y-2x\Omega\sin(\Omega t)^{-1}y\right)} \psi_0(y)dy, \quad (6.15)$$

that is the solution of the Schrödinger Eq. (6.12) with initial datum $\psi_0 \in \mathcal{S}(\mathbb{R})$. $\qquad\square$

Remark 6.1. By diagonalizing the symmetric matrix Ω, Eq. (6.15) can be generalized to arbitrary dimension d and becomes

$$\sqrt{\det\left(\frac{\Omega}{2\pi i \sin(\Omega t)}\right)} \int_{\mathbb{R}^d} e^{\frac{i}{2\hbar}\left(x\Omega\cot(\Omega t)x+y\Omega\cot(\Omega t)y-2x\Omega\sin(\Omega t)^{-1}y\right)} \psi_0(y)dy.$$

We can now state the main result of the present section. the proof is taken by [129] and is based on a technique present in [302].

Theorem 6.2. *Let $\psi_0, V \in \mathcal{F}(\mathbb{R}^d)$ and $t \neq \left(n+\frac{1}{2}\right)\frac{\pi}{\Omega_j}$, $n \in \mathbb{N}$. Then the solution of the Schrödinger equation*

$$\begin{cases} i\hbar\frac{\partial}{\partial t}\psi(t,x) = -\frac{\hbar^2}{2}\Delta\psi(t,x) + \frac{1}{2}x\Omega^2 x\psi(t,x) + V(x)\psi(t,x) \\ \psi(0,x) = \psi_0(x) \end{cases}$$

is given by the infinite dimensional oscillatory integral on the Cameron-Martin space

$$\widetilde{\int_{\mathcal{H}_t}} e^{\frac{i}{2\hbar}\int_0^t \dot\gamma(s)^2 ds} e^{-\frac{i}{2\hbar}\int_0^t (\gamma(s)+x)\Omega^2(\gamma(s)+x)ds} e^{-\frac{i}{\hbar}\int_0^t V(\gamma(s)+x)ds} \psi_0(\gamma(0)+x)d\gamma$$

$$= \widetilde{\int_{\mathcal{H}_t}} e^{\frac{i}{2\hbar}\langle\gamma,(I-L)\gamma\rangle} e^{-\frac{i}{2\hbar}x\Omega^2 xt} e^{-\frac{i}{\hbar}\int_0^t \gamma(s)\Omega^2 x ds} e^{-\frac{i}{\hbar}\int_0^t V(\gamma(s)+x)ds} \psi_0(\gamma(0)+x)d\gamma.$$

$$(6.16)$$

Proof. By Lemma 6.3 and by repeating the reasoning in the proofs of Theorem 6.1, the function on $\mathcal{H}_t$

$$\gamma \mapsto e^{-\frac{i}{2\hbar}\int_0^t (\gamma(s)+x)\Omega^2(\gamma(s)+x)ds} e^{-\frac{i}{\hbar}\int_0^t V(\gamma(s)+x)ds} \psi_0(\gamma(0)+x), \quad \gamma \in \mathcal{H}_t$$

is Fresnel integrable.

For $u \in [0,t]$ let $\mu_u(V,x)$, $\nu_u^t(V,x)$, $\lambda_u^t(x)$ be the measures on $\mathcal{H}_t$, whose Fourier transforms when evaluated at $\gamma \in \mathcal{H}_t$ are respectively $V(x+\gamma(u))$, $e^{-\frac{i}{\hbar}\int_u^t V(x+\gamma(s))ds}$, and $e^{-\frac{i}{\hbar}\int_u^t x\Omega^2\gamma(s)ds}$. We shall use the following notation

$$\mu_u \equiv \mu_u(V,x),$$
$$\nu_u^t \equiv \nu_u^t(V,x), \tag{6.17}$$
$$\lambda_u^t \equiv \lambda_u^t(x). \tag{6.18}$$

Moreover, for $\psi_0 \in \mathcal{F}(\mathbb{R}^d)$, we shall denote with μ_{ψ_0} the measure on $\mathcal{H}_t$ whose Fourier transform is the function $\gamma \mapsto \psi_0(\gamma(0)+x)$.

If $\{\mu_u : a \le u \le b\}$ is a family in $\mathcal{M}(\mathcal{H}_t)$, we shall let $\int_a^b \mu_u du$ denote the measure on $\mathcal{H}_t$ defined by

$$f \mapsto \int_a^b \int_{\mathcal{H}_t} f(\gamma)\mu_u(d\gamma)du, \qquad f \in C_0(\mathcal{H}_t)$$

whenever it exists. Since for any continuous path γ we have

$$\exp\left(-\frac{i}{\hbar}\int_0^t V(\gamma(s)+x)ds\right)$$
$$= 1 - \frac{i}{\hbar}\int_0^t V(\gamma(u)+x)\exp\left(-\frac{i}{\hbar}\int_u^t V(\gamma(s)+x)ds\right)du,$$

we get

$$\nu_0^t = \delta_0 - \frac{i}{\hbar}\int_0^t (\mu_u * \nu_u^t)du, \tag{6.19}$$

where δ_0 is the Dirac measure at $0 \in \mathcal{H}_t$.

We set for $t > 0$ and $x \in \mathbb{R}^d$

$$U(t)\psi_0(x) = \widetilde{\int_{\mathcal{H}_t}} e^{\frac{i}{2\hbar}\int_0^t |\dot{\gamma}(s)|^2 ds - \frac{i}{2\hbar}\int_0^t (\gamma(s)+x)\Omega^2(\gamma(s)+x)ds}$$
$$e^{-\frac{i}{\hbar}\int_0^t V(\gamma(s)+x)ds} \psi_0(\gamma(0)+x)d\gamma, \tag{6.20}$$

and

$$U_0(t)\psi_0(x) = \widetilde{\int_{\mathcal{H}_t}} e^{\frac{i}{2\hbar}\int_0^t |\dot{\gamma}(s)|^2 ds - \frac{i}{2\hbar}\int_0^t (\gamma(s)+x)\Omega^2(\gamma(s)+x)ds}$$
$$\psi_0(\gamma(0)+x)d\gamma. \tag{6.21}$$

By Parseval-type equality (Theorem 5.5), we have

$$U(t)\psi_0(x) = e^{-i\frac{t}{2\hbar}x\Omega^2 x}(\det(I-L))^{-1/2}\int_{\mathcal{H}_t} e^{-\frac{i\hbar}{2}\langle\gamma,(I-L)^{-1}\gamma\rangle}\lambda_0^t*\nu_0^t*\mu_{\psi_0}(d\gamma).$$

By applying Eq. (6.19) we obtain:

$$U(t)\psi_0(x) = C(t)\int_{\mathcal{H}_t} e^{-\frac{i\hbar}{2}\langle\gamma,(I-L)^{-1}\gamma\rangle}\lambda_0^t * \mu_{\psi_0}(d\gamma)$$

$$-\frac{i}{\hbar}C(t)\int_0^t\int_{\mathcal{H}_t} e^{-\frac{i\hbar}{2}\langle\gamma,(I-L)^{-1}\gamma\rangle}\lambda_0^t * \mu_u * \nu_u^t * \mu_{\psi_0}(d\gamma)du,$$

where $C(t) = e^{-i\frac{t}{2\hbar}x\Omega^2 x}(\det(I - L))^{-1/2}$. By applying the Parseval-type equality (5.33) in the other direction we get

$$U(t)\psi_0(x) = U_0(t)\psi_0(x) - \frac{i}{\hbar}e^{-i\frac{t}{2\hbar}x\Omega^2 x}$$

$$\int_0^t\widetilde{\int_{\mathcal{H}_t}} e^{\frac{i}{2\hbar}\int_0^t|\dot{\gamma}(s)|^2 ds - \frac{i}{2\hbar}\int_0^t\gamma(s)\Omega^2\gamma(s)ds}e^{-\frac{i}{\hbar}\int_0^t x\Omega^2\gamma(s)ds}$$

$$V(\gamma(u) + x)e^{-\frac{i}{\hbar}\int_u^t V(\gamma(s)+x)ds}\psi_0(\gamma(0) + x)d\gamma du \quad (6.22)$$

Denoting by $\mathcal{H}_{r,s}$ the Cameron-Martin space of paths $\gamma : [r, s] \to \mathbb{R}^d$, we have $\mathcal{H}_t \equiv \mathcal{H}_{0,t} = \mathcal{H}_{0,u} \oplus \mathcal{H}_{u,t}$, indeed each $\gamma \in \mathcal{H}_t$ can uniquely be associated to a couple (γ_1, γ_2), with $\gamma_1 \in \mathcal{H}_{0,u}$ and $\gamma_2 \in \mathcal{H}_{u,t}$, $\gamma(s) = \gamma_2(s)$ for $s \in [u, t]$ and $\gamma(s) = \gamma_1(s) + \gamma_2(u)$ for $s \in [0, u)$. By means of these notations and by Fubini's theorem for oscillatory integrals (see Theorem 5.7), Eq. (6.22) can be written in the following form:

$$U(t)\psi_0(x) = U_0(t)\psi_0(x) - \frac{i}{\hbar}\int_0^t\widetilde{\int_{\mathcal{H}_{u,t}}} e^{\frac{i}{2\hbar}\int_u^t|\dot{\gamma}_2(s)|^2 ds}$$

$$e^{-\frac{i}{2\hbar}\int_u^t(\gamma_2(s)+x)\Omega^2(\gamma_2(s)+x)ds}e^{-\frac{i}{\hbar}\int_u^t V(\gamma_2(s)+x)ds}V(\gamma_2(u) + x)$$

$$\left(\widetilde{\int_{\mathcal{H}_{0,u}}} e^{\frac{i}{2\hbar}\int_0^u|\dot{\gamma}_1(s)|^2 ds - \frac{i}{2\hbar}\int_0^u(\gamma_1(s)+\gamma_2(u)+x)\Omega^2(\gamma_1(s)+\gamma_2(u)+x)ds}\right.$$

$$\left.\psi_0(\gamma_1(0) + \gamma_2(u) + x)d\gamma_1\right)d\gamma_2 du,$$

and by Eqs. (6.20) and (6.21) the latter is equal to

$$U(t)\psi_0(x) = U_0(t)\psi_0(x) - \frac{i}{\hbar}\int_0^t U(t - u)VU_0(u)\psi_0(x)du. \quad (6.23)$$

For $\psi \in \mathcal{S}(\mathbb{R})$, by Theorem 6.1, the function $U_0(t)\psi_0(x)$ is the vector $e^{-\frac{i}{\hbar}H_0 t}\psi_0 \in L^2(\mathbb{R})$ evaluated at the point x. By Eq. (6.23), the operator $U(t)$ on $L^2(\mathbb{R})$ is obtained as the solution of an integral equation, whose iterative solution is the convergent Dyson series [281] for $e^{-\frac{i}{\hbar}Ht}$ (see also [129, 18]). $\qquad\square$

It is interesting to remark that, by choosing in the sequential definition of the infinite dimensional oscillatory integral

$$
\widetilde{\int_{\mathcal{H}_t}} e^{\frac{i}{2\hbar}\int_0^t |\dot{\gamma}(s)|^2 ds - \frac{i}{\hbar}\int_0^t V(\gamma(s)+x)ds} \psi_0(\gamma(0)+x)d\gamma
$$

$$
= \lim_{n\to\infty} \frac{\int_{P_n\mathcal{H}_t}^o e^{\frac{i}{2\hbar}\int_0^t |\dot{\gamma}_n(s)|^2 ds - \frac{i}{\hbar}\int_0^t V(\gamma_n(s)+x)ds} \psi_0(\gamma_n(0)+x)d\gamma_n}{\int_{P_n\mathcal{H}_t}^o e^{\frac{i}{2\hbar}\int_0^t |\dot{\gamma}_n(s)|^2 ds} d\gamma_n}, \quad (6.24)
$$

(with $\gamma_n := P_n\gamma$) a suitable sequence of finite dimensional projection operators $\{P_n\}_{n\in\mathbb{N}}$, it is possible to recover in the right-hand side of Eq. (6.24), an analogous of the Feynman formula (1.9) derived by the Trotter product formula, i.e.

$$
e^{-\frac{i}{\hbar}Ht}\psi_0(x) = \lim_{n\to\infty} \left(\frac{2\pi i\hbar t}{mn}\right)^{-\frac{nd}{2}} \int_{\mathbb{R}^{nd}} e^{\frac{i}{\hbar}\sum_{j=1}^n \left(\frac{m}{2}\frac{(x_j-x_{j-1})^2}{(t/n)^2} - V(x_j)\right)\frac{t}{n}}
$$

$$
\psi_0(x_0)dx_0\ldots dx_{n-1} \quad (6.25)
$$

Let us consider indeed the *polygonal path approximation* [324]. For any $n \in \mathbb{N}$ and $\gamma \in \mathcal{H}_t$, let $P_n\gamma$ be the piecewise linear polygonal approximation of γ, that is:

$$
P_n\gamma(s) = \gamma_j + \left(s - \frac{jt}{n}\right)(\gamma_{j+1} - \gamma_j)\frac{n}{t}, \quad \frac{jt}{n} \le s \le \frac{(j+1)t}{n},
$$

where $j = 0, \ldots, n-1$, $\gamma_j := \gamma(jt/n)$. Clearly $P_n^2 = P_n$, moreover for any $\gamma, \eta \in \mathcal{H}_t$

$$
\langle \eta, P_n\gamma \rangle = \sum_{j=0}^{n-1} (\eta_{j+1} - \eta_j)(\gamma_{j+1} - \gamma_j)\frac{n}{t} = \langle P_n\eta, \gamma \rangle,
$$

and we can conclude that the operators P_n are orthogonal projections. Moreover the following holds:

Lemma 6.4. *The sequence of operators $\{P_n\}_{n\in\mathbb{N}}$ converges to the identity* $I : \mathcal{H}_t \to \mathcal{H}_t$ *in the strong operator topology* .

Proof. [324] Let us denote with $U \subset \mathcal{H}_t$ the set

$$
U := \{\gamma \in \mathcal{H}_t, \mid \lim_{n\to\infty} \|P_n\gamma - \gamma\| = 0\},
$$

and let us prove that $U = \mathcal{H}_t$ by showing that it is a closed subspace of $\mathcal{H}_t$ containing its basis functions.

U is a subspace of $\mathcal{H}_t$ because P_n is linear, indeed given $\eta, \gamma \in U$, $\alpha, \beta \in \mathbb{R}$

$$\|P_n(\alpha\eta + \beta\gamma) - (\alpha\eta + \beta\gamma)\| = \|\alpha(P_n\eta - \eta) + \beta(P_n\gamma - \gamma)\|$$
$$\leq |\alpha|\|P_n\eta - \eta\| + |\beta|\|P_n\gamma - \gamma\| \to 0, \qquad n \to \infty$$

U is closed, indeed given a sequence $\{\gamma_n\} \subset U$, with $\|\gamma_n - \gamma\| \to 0$, the limit vector γ belongs to U. Indeed

$$\|P_n\gamma - \gamma\| \leq \|P_n(\gamma - \gamma_m)\| + \|\gamma - \gamma_m\| + \|P_n\gamma_m - \gamma_m\| \qquad (6.26)$$
$$\leq 2\|\gamma - \gamma_m\| + \|P_n\gamma_m - \gamma_m\| \qquad (6.27)$$

Given $\epsilon > 0$, $\exists N_\epsilon > 0$ such that $\|\gamma - \gamma_m\| < \epsilon/4$ for $m = N_\epsilon$. There exists also $N(m, \epsilon)$, such that $\|P_n\gamma_m - \gamma_m\| < \epsilon/2$ for $n > N(m, \epsilon)$, and consequently $\|P_n\gamma - \gamma\| < \epsilon$.

The inclusion of a basis of $\mathcal{H}_t$ in U can be proved in the following way [321].

Given $\gamma \in \mathcal{H}_t$, let $\alpha_0, \alpha_n, \beta_n$, with $\sum_{n=1}^{\infty}(\alpha_n^2 + \beta_n^2) < \infty$, be the Fourier coefficients of $\dot{\gamma} \in L^2([0, t])$ and let $\dot{\gamma}_N$, γ_N be the corresponding partial sums, i.e.:

$$\dot{\gamma}_N(s) = \alpha_0 + \sum_{n=1}^{N} \alpha_n \cos\frac{2\pi n s}{t} + \sum_{n=1}^{N} \beta_n \sin\frac{2\pi n s}{t}, \qquad s \in [0, t],$$

$$\gamma_N(s) = \alpha_0(s-t) + \sum_{n=1}^{N} \frac{\alpha_n t}{2\pi n} \sin\frac{2\pi n s}{t} + \sum_{n=1}^{N} \frac{\beta_n t}{2\pi n}\left(1 - \cos\frac{2\pi n s}{t}\right), \qquad s \in [0, t].$$

One has

$$\|\gamma_N - \gamma\| = \|\dot{\gamma}_N - \dot{\gamma}\|_{L^2([0,t])} \to 0, \qquad \text{as } N \to \infty,$$

and it is quite simple to see that $\gamma_N \in U$ for each N $\qquad\square$

6.2 Time dependent potentials

The results of the previous section can be generalized to the case the potential in the Schrödinger equation depends explicitly on the time variable t [30].

Let us consider first of all a linearly forced harmonic oscillator. Let us assume that the quantum mechanical Hamiltonian is given on the smooth vectors $\psi \in C_0(\mathbb{R}^d)$ by

$$H\psi(x) = -\frac{\hbar^2}{2}\Delta\psi(x) + V(t, x)\psi(x), \quad V(t, x) = \frac{1}{2}x\Omega^2 x + f(t)\cdot x, \quad x \in \mathbb{R}^d,$$
$$(6.28)$$

where Ω is a positive symmetric constant $d \times d$ matrix with eigenvalues Ω_j, $j = 1 \ldots d$, and $f : I \subset \mathbb{R} \to \mathbb{R}^d$ is a continuous function (I being a closed interval).

This potential is particularly interesting from a physical point of view as it is used in simple models for a large class of processes, as the vibration-relaxation of a diatomic molecule in gas kinetics and the interaction of a particle with the field oscillators in quantum electrodynamics. Feynman calculated heuristically the Green function for the Schrödinger equation associated to (6.28) in his famous paper on the path integral formulation of quantum mechanics [138]. Our aim is to give meaning to the Feynman path integral representation of the solution of Schrödinger equation

$$i\hbar \frac{d}{dt}\psi = H\psi,$$

with H given by Eq. (6.28):

$$\psi(t, x) = \int_{\gamma(t)=x} e^{\frac{i}{2\hbar}\int_0^t |\dot\gamma(s)|^2 ds - \frac{i}{2\hbar}\int_0^t \gamma(s)\Omega^2\gamma(s)ds - \frac{i}{\hbar}\int_0^t f(s)\cdot\gamma(s)ds} \psi_0(\gamma(0))d\gamma$$

(6.29)

in terms of a well defined infinite dimensional oscillatory integral on the Cameron-Martin space $\mathcal{H}_t$. We recall that a similar result has been obtained in the case $d = 1$ by means of the white-noise approach [134].

Let $L : \mathcal{H}_t \to \mathcal{H}_t$ be the symmetric operator on $\mathcal{H}_t$ given by Eq. (6.5). By Lemmas 6.1 and 6.2, if

$$t \neq \left(n + \frac{1}{2}\right)\frac{\pi}{\Omega_j}, \qquad n \in \mathbb{N}, \; j = 1, \ldots, d,$$

then L is trace class and $(I - L)$ is invertible and its inverse is given by Eq. (6.6).

Let w, v be the vectors in $\mathcal{H}_t$ defined by

$$w(s) \equiv \frac{\Omega^2 x}{2\hbar}(s^2 - t^2), \tag{6.30}$$

$$v(s) \equiv \frac{1}{\hbar}\int_t^s \int_0^{s'} f(s'')ds''ds' \qquad s \in [0, t]. \tag{6.31}$$

With these notations heuristic expression (6.29) can be written in terms of a well defined infinite dimensional oscillatory integrals, i.e.

$$\psi(t, x) = \int_{\gamma(t)=0} e^{\frac{i}{2\hbar}\int_0^t |\dot\gamma(s)|^2 ds - \frac{i}{2\hbar}\int_0^t (\gamma(s)+x)\Omega^2(\gamma(s)+x)ds}$$

$$e^{-\frac{i}{\hbar}\int_0^t f(s)\cdot(\gamma(s)+x)ds}\psi_0(\gamma(0) + x)d\gamma = e^{-i\frac{t}{2\hbar}x\Omega^2 x}e^{-i\frac{x}{\hbar}\cdot\int_0^t f(s)ds}$$

$$\widetilde{\int_{\mathcal{H}_t}} e^{\frac{i}{2\hbar}\langle\gamma,(I-L)\gamma\rangle}e^{i\langle v,\gamma\rangle}e^{i\langle w,\gamma\rangle}\psi_0(\gamma(0) + x)d\gamma. \tag{6.32}$$

Under the assumption that $\psi_0 \in \mathcal{F}(\mathbb{R}^d)$, by Lemma 6.3 the functional on $\mathcal{H}_t$ given by

$$\gamma \mapsto \psi_0(\gamma(0) + x), \qquad \gamma \in \mathcal{H}_t$$

belongs to $\mathcal{F}(\mathcal{H}_t)$. Indeed if $\psi_0(x) = \int_{\mathbb{R}^d} e^{ik \cdot x} d\mu_0(k)$, then

$$\psi_0(\gamma(0) + x) = \int_{\mathcal{H}_t} e^{i\langle \eta, \gamma \rangle} \mu_{\psi_0}(d\eta),$$

where

$$\int_{\mathcal{H}_t} f(\gamma) d\mu_{\psi_0}(\gamma) = \int_{\mathbb{R}^d} e^{ik \cdot x} f(k\gamma_0) d\mu_0(k) \tag{6.33}$$

and, for any $k \in \mathbb{R}^d$, $k\gamma_0$ is the element in $\mathcal{H}_t$ such that $\langle k\gamma_0, \gamma \rangle = k \cdot \gamma(0)$, that is

$$k\gamma_0(s) = k(t - s), \quad s \in [0, t].$$

In this case the functional on $\mathcal{H}_t$ defined by

$$\gamma \mapsto e^{i\langle v, \gamma \rangle} e^{i\langle w, \gamma \rangle} \psi_0(\gamma(0) + x), \qquad \gamma \in \mathcal{H}_t,$$

belongs to $\mathcal{F}(\mathcal{H}_t)$ and the infinite dimensional oscillatory integral on the r.h.s. of Eq. (6.32) on $\mathcal{H}_t$ can be explicitly computed by means of the Parseval-type equality (Theorem 5.5):

$$\widetilde{\int}_{\mathcal{H}_t} e^{\frac{i}{2\hbar}\langle \gamma, (I-L)\gamma \rangle} e^{i\langle v, \gamma \rangle} e^{i\langle w, \gamma \rangle} \psi_0(\gamma(0) + x) d\gamma$$

$$= (\det(I - L))^{-1/2} \int_{\mathcal{H}_t} e^{-\frac{i\hbar}{2}\langle \gamma, (I-L)^{-1}\gamma \rangle} \delta_v * \delta_w * \mu_{\psi_0}(d\gamma)$$

and we have

$$\psi(t, x) = \frac{e^{-i\frac{t}{2\hbar} x \Omega^2 x} e^{-i\frac{x}{\hbar} \cdot \int_0^t f(s) ds}}{\sqrt{\det(\cos(\Omega t))}} \int_{\mathbb{R}^d} e^{ik \cdot x}$$

$$e^{-\frac{i\hbar}{2}\langle (v+w+k\gamma_0), (I-L)^{-1}(v+w+k\gamma_0) \rangle} d\mu_0(k).$$

If $\psi_0 \in \mathcal{S}(\mathbb{R}^d)$, we can proceed further and compute explicitly the Green function $G(0, t, x, y)$ for the Schrödinger equation:

$$\psi(t, x) = \int_{\mathbb{R}^d} G(0, t, x, y) \psi_0(y) dy,$$

where

$$G(0, t, x, y) = (2\pi i\hbar)^{-d/2} \sqrt{\det\left(\frac{\Omega}{\sin(\Omega t)}\right)}$$

$$e^{\frac{i\Omega \sin(\Omega t)^{-1}}{2\hbar}(x\cos(\Omega t)x + y\cos(\Omega t)y - 2xy)}$$

$$e^{-\frac{i}{\hbar}x\sin(\Omega t)^{-1}\int_0^t \sin(\Omega s)f(s)ds - \frac{i}{\hbar}y(\int_0^t \cos(\Omega s)f(s)ds - \cos(\Omega t)\sin(\Omega t)^{-1}\int_0^t \sin(\Omega s)f(s)ds)}$$

$$e^{\frac{i}{\hbar}\Omega^{-1}(\frac{1}{2}\cos(\Omega t)\sin(\Omega t)^{-1}(\int_0^t \sin(\Omega s)f(s)ds)^2 - \int_0^t \sin(\Omega s)f(s)ds \int_0^t \cos(\Omega s)f(s)ds)}$$

$$e^{\frac{i}{\hbar}\Omega^{-1}\int_0^t \cos(\Omega s)f(s)\int_s^t \sin(\Omega s')f(s')ds'ds}. \tag{6.34}$$

One can easily verify by a direct computation that (6.34) is the Green's function for the Schrödinger equation with the time dependent Hamiltonian (6.28).

Remark 6.2. Our result can be obtained even if the initial assumption on the continuity of the function $f : [0, t] \to \mathbb{R}^d$ is weakened, in fact it is sufficient to assume that the function v in (6.30) belongs to $\mathcal{H}_t$, that is $\int_0^t |\int_0^s f(s')ds'|^2 ds < \infty$.

Let us consider now the Schrödinger equation with an harmonic-oscillator Hamiltonian with a time-dependent frequency:

$$H\psi(x) = -\frac{\hbar^2}{2}\Delta\psi(x) + \frac{1}{2}x\Omega^2(t)x\psi(x), \qquad x \in \mathbb{R}^d, \psi \in C_0^2(\mathbb{R}^d), \quad (6.35)$$

where $\Omega : [0, t] \to L(\mathbb{R}^d, \mathbb{R}^d)$ is a continuous map from the time interval $[0, t]$ to the space of symmetric positive $d \times d$ matrices.

This problem has been analyzed by several authors (see for instance [272, 226] and references therein) as an approximated description for the vibration of complex physical systems, as well as an exact model for some physical phenomena, as the motion of an ion in a Paul trap, quantum mechanical description of highly cooled ions, the emergence of non-classical optical states of light owing to a time-dependent dielectric constant, or even in cosmology for the study of a three-dimensional isotropic harmonic oscillator in a spatially flat universe such that $g_{ij} = R(t)\delta_{ij}$, with $R(t)$ being the scale factor at time t and $i, j = 1, 2, 3$ the indexes relative to the spatial coordinates.

If $d = 1$ it is possible to solve the Schrödinger equation with Hamiltonian (6.35) (and also the corresponding classical equation of motion) by adopting a suitable transformation of the time and space variables which allows to map the solution of the time-independent harmonic oscillator to the solution of the time-dependent one (see [283, 287, 286] and references therein).

Let us consider the classical equation of motion for the time-dependent harmonic oscillator (6.35)

$$\ddot{u}(s) + \Omega^2(s)u(s) = 0. \tag{6.36}$$

Let u_1 and u_2 be two independent solutions of (6.36) such that $u_1(0) = \dot{u}_2(0) = 0$ and $u_2(0) = \dot{u}_1(0) = 1$. Then it easy to prove that the Wronskian $w(u_1, u_2) = u_1\dot{u}_2 - \dot{u}_1 u_2$ is the constant function $w = 1$. Let us define the function $\xi := u_1^2 + u_2^2$. It is possible to prove that $\xi(s) > 0 \ \forall s$ and it satisfies the following differential equation:

$$2\xi\ddot{\xi} - \dot{\xi}^2 + 4\xi^2 - 4 = 0.$$

Moreover the function $\eta : [0, \infty] \to \mathbb{R}$, given by

$$\eta(s) = \int_0^s \xi(\tau)^{-1} d\tau$$

is well defined and strictly increasing. It is possible to verify that

$$u(s) = \xi(s)^{1/2}(A\cos(\eta(s)) + B\sin(\eta(s))) \tag{6.37}$$

is the general solution of the classical equation of motion (6.36). In other words, by rescaling the time variable $s \mapsto \eta(s)$ and the space variable $x \mapsto \xi^{-1/2}x$ it is possible to map the solution of the equation of motion for the time-independent harmonic oscillator $\ddot{u}(s) + u(s) = 0$ into the solution of (6.36). By another point of view, it is possible to find (see, for instance, [218] for more details) a general canonical transformation $(x, p, t) \mapsto (X, P, \tau)$, given by

$$\begin{cases} X = \xi(t)^{-1/2}x \\ \frac{d\tau(t)}{dt} = \xi(t)^{-1} \\ P = \frac{dX}{d\tau} = (\xi^{1/2}\dot{x} - \frac{1}{2}\xi^{-1/2}\dot{\xi}x). \end{cases} \tag{6.38}$$

The Hamiltonian in the new variables is time independent

$$H(X, P, \tau) = \frac{1}{2}(P^2 + X^2).$$

The generating function of the transformation $(x, p, t) \mapsto (X, P, \tau)$ is given by $F(x, P, t) = \xi(t)^{-1/2}xP + \frac{\xi(t)^{-1}\dot{\xi}}{4}x^2$ and the transformation is given more explicitly as

$$\begin{cases} p = \frac{\partial}{\partial x}F(x, P, t) \\ X = \frac{\partial}{\partial P}F(x, P, t) \\ H(X, P; \tau)\dot{\tau} = H(x, p; t) + \frac{\partial}{\partial t}F(x, P, t). \end{cases} \tag{6.39}$$

A similar result holds also in the quantum case. In fact by considering the Schrödinger equations for the time-independent and time-dependent harmonic oscillator respectively,

$$\left(i\hbar\frac{\partial}{\partial t} + \frac{\hbar^2}{2}\Delta - \frac{1}{2}x^2\right)\phi(t,x) = 0, \tag{6.40}$$

$$\left(i\hbar\frac{\partial}{\partial t} + \frac{\hbar^2}{2}\Delta - \frac{1}{2}\Omega^2(t)x^2\right)\psi(t,x) = 0, \tag{6.41}$$

where $\phi(t,x)$ and $\psi(t,x)$ are continuously differentiable with respect to t and twice continuously differentiable with respect to x, it is possible to prove the following [283]:

Theorem 6.3. *Let $\phi(t,x)$ be a solution of (6.40). Then*

$$\psi(t,x) = \xi(t)^{-1/4}\exp[i\dot{\xi}(t)x^2/4\hbar\xi(t)]\phi(\eta(t),\xi(t)^{-1/2}x)$$

is a solution of (6.41).

In an analogous way, by denoting with $K_{TI}(t,0;x,y)$ and $K_{TD}(t,0;x,y)$ the Green functions for the Schrödinger Eqs. (6.40) and (6.41) respectively, it is possible to prove that the following holds:

$$K_{TD}(t,0;x,y) = \xi(t)^{-1/4}\exp[i\dot{\xi}(t)x^2/4\hbar\xi(t)]K_{TI}(\eta(t),0;\xi(t)^{-1/2}x,y). \tag{6.42}$$

It is interesting to note that the "correction term"

$$\xi(t)^{-1/4}\exp[i\dot{\xi}(t)x^2/4\hbar\xi(t)]$$

in Theorem 6.3 can be interpreted in terms of the classical canonical transformation (6.39) (see [218] for more details).

We can give a rigorous mathematical meaning to the Feynman path integral representation of the solution of Eq. (6.41) by means of an well defined infinite dimensional oscillatory integral on the Cameron-Martin space $\mathcal{H}_t$ and prove formula (6.42). A similar result has been obtained in the framework of the white-noise approach [170].

Let us consider the following linear operator $L : \mathcal{H}_t \to \mathcal{H}_t$:

$$(L\gamma)(s) = -\int_t^s \int_0^r \Omega^2(u)\gamma(u)\,du\,dr, \qquad \gamma \in \mathcal{H}_t.$$

One can easily verify that L is self-adjoint and positive, as for any $\gamma_1, \gamma_2 \in \mathcal{H}_t$ one has:

$$\langle \gamma_1, L\gamma_2 \rangle = \int_0^t \gamma_1(s)\Omega^2(s)\gamma_2(s)\,ds.$$

By using formula (6.37), it is possible to prove that, if $t \neq \eta^{-1}(\pi/2+n\pi)$, $n \in \mathbb{N}$, the operator $I - L$ is invertible and its inverse is given by

$$
(I - L)^{-1}\gamma(s) = \Big[- \frac{\sin(\eta(t))}{\cos(\eta(t))} \big(\int_0^t \xi(s')^{1/2} \cos(\eta(s'))\ddot{\gamma}(s')ds'
$$
$$
+ \dot{\gamma}(0) \big) - \int_t^s \xi(s')^{1/2} \sin(\eta(s'))\ddot{\gamma}(s')ds' \Big] \xi(s)^{1/2} \cos(\eta(s))
$$
$$
+ \Big[\int_0^t \xi(s')^{1/2} \cos(\eta(s'))\ddot{\gamma}(s')ds' + \dot{\gamma}(0)
$$
$$
+ \int_t^s \xi(s')^{1/2} \cos(\eta(s'))\ddot{\gamma}(s')ds' \Big] \xi(s)^{1/2} \sin(\eta(s)), \quad s \in [0,t].
$$
$$(6.43)$$

The Fredholm determinant of the operator $I - L$ can be computed by exploiting the general relation between infinite dimensional determinants of the form

$$
\det(I + \epsilon L), \qquad \epsilon \in \mathbb{C}, L \text{ is of trace class,}
$$

and finite dimensional determinants associated with the solution of a certain Sturm-Liouville problem [5, 286]. According to [5], by using the fact that $v(s) = L\gamma(s)$ is the unique solution of the problem:

$$
\begin{cases} \ddot{v}(s) = -\Omega^2(s)\gamma(s), & s \in (0,t), \\ \dot{v}(0) = 0, \quad v(t) = 0 \end{cases}
\tag{6.44}
$$

and by using the ellipticity of the problem (6.44), it is possible to prove that the range of L is contained in $H^3((0,t);\mathbb{R})$, the Sobolev space of functions belonging to $L^2((0,t);\mathbb{R})$, whose derivatives up to order 3 belong also to $L^2((0,t);\mathbb{R})$, hence L is a trace class operator. Moreover by considering the solution K_ϵ of the initial value problem

$$
\begin{cases} \ddot{K}_\epsilon(s) + \epsilon\Omega^2(s)K_\epsilon(s) = 0, \\ \dot{K}_\epsilon(0) = 0, \quad K_\epsilon(0) = 1 \end{cases}
\tag{6.45}
$$

we obtain

$$
K_\epsilon(t) = \det(I - \epsilon L).
$$

By substituting $\epsilon = 1$ in (6.45) and by using formula (6.37) for the general solution of the differential Eq. (6.36) we eventually get:

$$
\det(I - L) = \xi(t)^{1/2} \cos(\eta(t)).
\tag{6.46}
$$

Let us consider now the vectors $\gamma_0, w \in \mathcal{H}_t$ given by

$$
\gamma_0(s) = t - s,
$$
$$
w(s) = \frac{x}{\hbar} \int_t^s \int_0^u \Omega^2(r)drdu, \qquad s \in [0,t].
\tag{6.47}
$$

With the notations introduced so far and by assuming that the initial vector ψ_0 belongs to $\mathcal{F}(\mathbb{R})$, so that $\psi_0 = \hat{\mu}_0$, the heuristic Feynman path integral representation for the solution of the Schrödinger equation with the time-dependent Hamiltonian (6.35)

$$\psi(t, x) = \int_{\{\gamma | \gamma(t) = 0\}} e^{\frac{i}{2\hbar} \int_0^t \dot{\gamma}^2(s)ds - \frac{i}{2\hbar} \int_0^t \Omega^2(s)(\gamma(s)+x)^2 ds} \psi_0(\gamma(0) + x) D\gamma$$

can be rigorously realized as the infinite dimensional oscillatory integral on the Cameron-Martin space $\mathcal{H}_t$:

$$\psi(t, x) = e^{-\frac{ix^2}{2\hbar} \int_0^t \Omega^2(s)ds} I_t, \qquad I_t = \widetilde{\int_{\mathcal{H}_t}} e^{\frac{i}{2\hbar} \langle \gamma, (I-L)\gamma \rangle} e^{i\langle w, \gamma \rangle} \hat{\mu}_{\psi_0}(\gamma) d\gamma,$$

where μ_{ψ_0} is given by formula (6.33). By Parseval-type equality I_t can be explicitly computed and one has:

$$I_t = \det(I - L)^{-1/2} \int_{\mathcal{H}_t} e^{\frac{i}{2\hbar} \langle \gamma, (I-L)^{-1}\gamma \rangle} \delta_w * \mu_{\psi_0}(d\gamma). \tag{6.48}$$

By assuming $\psi_0 \in \mathcal{S}(\mathbb{R})$, we can proceed further and compute explicitly the Green function of the problem, that is

$$\psi(t, x) = \int_{\mathbb{R}} K_{TD}(t, 0; x, y) \psi_0(y) dy.$$

By substituting in (6.48) formulae (6.43) and (6.46), and performing a simple calculation we obtain:

$$K_{TD}(t, 0; x, y) = \xi(t)^{-1/4} e^{\frac{ix^2}{4\hbar} \xi(t)^{-1} \dot{\xi}(t)} \frac{e^{\frac{i}{2\hbar} (\frac{\cos(\eta(t))}{\sin(\eta(t))})(\xi(t)^{-1} x^2 + y^2) - \frac{2\xi(t)^{-1/2} xy}{\sin(\eta(t))}}}{(2\pi i\hbar \sin(\eta(t)))^{1/2}},$$

and by recalling the well known formula for the Green function $K_{TI}(t, 0; x, y)$ of the Schrödinger equation with a time-independent harmonic oscillator Hamiltonian (see, e.g., [302]):

$$K_{TI}(t, 0; x, y) = \frac{e^{\frac{i}{2\hbar} (\frac{\cos(t)}{\sin(t)}(x^2 + y^2) - \frac{2xy}{\sin(t)})}}{(2\pi i\hbar \sin(t))^{1/2}},$$

one can verify directly formula (6.42).

Remark 6.3. The case where $d > 1$ is more complicated. In fact neither a transformation formula analogous to (6.38) exists in general, nor a formula analogous to (6.42) relating the Green function of the Schrödinger equation with a time-dependent harmonic oscillator potential with the Green function of the Schrödinger equation with a time-independent harmonic oscillator potential (see for instance [287, 286] for some partial results in this direction).

Let us finally consider a quantum mechanical Hamiltonian of the following form:

$$H\psi(x) = H_0\psi(x) + V(t,x)\psi(x), \qquad \psi \in C_0^2(\mathbb{R}^d), \qquad (6.49)$$

where H_0 is of the type (6.28) or (6.35) and $V : [0,t] \times \mathbb{R}^d \to \mathbb{R}$ satisfies the following assumptions:

(1) for each $s \in [0,t]$, the application $V(s, \cdot) : \mathbb{R}^d \to \mathbb{R}$ belongs to $\mathcal{F}(\mathbb{R}^d)$, i.e. $V(s,x) = \int_{\mathbb{R}^d} e^{ikx} \mu_s(dk)$, $\mu_s \in \mathcal{M}(\mathbb{R}^d)$;
(2) the application $s \in [0,t] \mapsto \mu_s \in \mathcal{M}(\mathbb{R}^d)$ is continuous in the norm $\| \cdot \|$ of the Banach space $\mathcal{M}(\mathbb{R}^d)$.

We remark that condition (1) implies that for each $s \in [0,t]$, the function $V(s, \cdot) : \mathbb{R}^d \to \mathbb{R}$ is bounded. Moreover by condition (2) one can easily verify that the application $s \in [0,t] \mapsto V(s,\cdot) \in C(\mathbb{R}^d) \cap L^\infty(\mathbb{R}^d)$ in continuous in the sup-norm.

Under these assumptions it is possible to prove that the application on $\mathcal{H}_t$, given by

$$\gamma \mapsto \int_0^t V(s, \gamma(s) + x)ds, \qquad \gamma \in \mathcal{H}_t$$

belongs to $\mathcal{F}(\mathcal{H}_t)$. More precisely it is the Fourier transform of the complex bounded variation measure μ_v on the Cameron-Martin space $\mathcal{H}_t$ defined by:

$$\mu_v(I) = \int_0^t \int_{\mathbb{R}^d} e^{ikx} \chi_I(k\gamma_s)\mu_s(dk)ds, \qquad I \in \mathcal{B}(\mathcal{H}_t),$$

where χ_I is the characteristic function of the Borel set $I \subset \mathcal{H}_t$ and, for any $k \in \mathbb{R}^d$, $k\gamma_s$ is the element in $\mathcal{H}_t$ given by

$$k\gamma_s(s') = k(t-s), \qquad s' \leq s$$
$$k\gamma_s(s') = k(t-s'), \qquad s' > s. \qquad (6.50)$$

As a consequence also the application

$$\gamma \mapsto e^{-\frac{i}{\hbar} \int_0^t V(s,\gamma(s)+x)ds}, \qquad \gamma \in \mathcal{H}_t,$$

belongs to $\mathcal{F}(\mathcal{H}_t)$ (let us denote by ν_v the bounded variation measure on $\mathcal{H}_t$ associated to it) and the infinite dimensional oscillatory integral associated with the Cameron-Martin space $\mathcal{H}_t$ giving the rigorous mathematical realization of the Feynman path integral representation of the solution of the Schrödinger equation with time dependent Hamiltonian (6.49) is well defined. In the following we give some details of the case where the "free

Hamiltonian" H_0 is given by (6.28), but the same reasoning can be repeated in the case where H_0 is given by (6.35). One has:

$$\psi(t,x) = \int_{\gamma(t)=0} e^{\frac{i}{2\hbar}\int_0^t |\dot{\gamma}(s)|^2 ds - \frac{i}{2\hbar}\int_0^t (\gamma(s)+x)\Omega^2(\gamma(s)+x)ds} e^{-\frac{i}{\hbar}\int_0^t f(s)\cdot(\gamma(s)+x)ds}$$

$$e^{-\frac{i}{\hbar}\int_0^t V(s,\gamma(s)+x)ds}\psi_0(\gamma(0)+x)d\gamma = e^{-i\frac{t}{2\hbar}x\Omega^2 x}e^{-i\frac{x}{\hbar}\cdot\int_0^t f(s)ds}I_t, \qquad (6.51)$$

where

$$I_t \equiv \int_{\mathcal{H}_t} e^{\frac{i}{2\hbar}\langle\gamma,(I-L)\gamma\rangle}e^{i\langle v,\gamma\rangle}e^{i\langle w,\gamma\rangle}e^{-\frac{i}{\hbar}\int_0^t V(s,\gamma(s)+x)ds}\psi_0(\gamma(0)+x)d\gamma$$

$$\qquad (6.52)$$

is well defined and can be explicitly computed using the Parseval type equality (Theorem 5.5):

$$I_t = (\det(I-L))^{-1/2}\int_{\mathcal{H}_t} e^{-\frac{i\hbar}{2}\langle\gamma,(I-L)^{-1}\gamma\rangle}\delta_v * \delta_w * \nu_v * \mu_{\psi_0}(d\gamma). \quad (6.53)$$

The detailed proof that the right-hand side of (6.51) is the solution of the Schrödinger equation with Hamiltonian (6.49) is completely similar to the proof of Theorem 6.2 and we refer to [30] for more details.

6.3   Phase space Feynman path integrals

Let us recall that Feynman's original aim was to give a Lagrangian formulation of quantum mechanics. On the other hand an Hamiltonian formulation could be preferable from many points of view. For instance the discussion of the approach from quantum mechanics to classical mechanics, i.e the study of the behavior of physical quantities taking into account that $\hbar$ is small, is more natural in an Hamiltonian setting (see for instance [7, 248, 247, 250] and Section 7.4 for a discussion of this behavior). In other words the "phase space" rather then the "configuration space" is the natural framework of classical mechanics.

As a consequence one is tempted to propose a "phase space Feynman path integral" representation for the solution of the Schrödinger equation, that is an heuristic formula of the following form:

$$\psi(t,x) = \text{const} \int_{q(t)=x} e^{\frac{i}{\hbar}S(q,p)}\psi_0(q(0))dqdp. \qquad (6.54)$$

Here the integral is meant on the space of paths $q(s), p(s)$, $s \in [0,t]$, in the phase space of the system, where $q(s)_{s\in[0,t]}$ is the path in configuration space and $p(s)_{s \in [0,t]}$ is the path in momentum space, and S is the action functional in the Hamiltonian formulation:

$$S(q,p) = \int_0^t (\dot{q}(s)p(s) - H(q(s),p(s)))ds,$$

(H being the classical Hamiltonian of the system).

An approach of phase space Feynman path integrals via analytic continuation of "phase space Wiener integrals" has been presented by I. Daubechies and J. Klauder [104, 105, 102, 103]. Analytic continuation was also used in other "path space" approaches, see [265, 203, 87] and references therein.

Analogously to Feynman Lagrangian formula (1.6), a formal derivation of its Hamiltonian version (6.54) can be given by means of Lie-Trotter product formula [320, 91, 92]. Let us first recall here an abstract version of it, which will be also used in the definition of the quantum dynamics.

Lemma 6.5. *Let A and B be self-adjoint operators in a Hilbert space $\mathcal{H}$ and let $A + B$ be essentially self-adjoint on $D(A) \cap D(B)$. Then*

$$s - \lim_{n \to \infty} (e^{itA/n}e^{itB/n})^n = e^{i(A+B)t}, \qquad t \in \mathbb{R} \qquad (6.55)$$

Here $s - \lim$ is the strong operator limit.[1] For a proof and a discussion of this lemma see e.g. [92, 280].

Let us now define the quantum Hamiltonian operator H on $L^2(\mathbb{R}^d)$ associated to a classical potential V depending both on position and on momentum in the following way

$$H = -\frac{\hbar^2}{2m}\Delta_x + V_1(x) + V_2(p).$$

The operator V_1 is defined as a self-adjoint operator on $L^2(\mathbb{R}^d)$, with its natural domain as a multiplication operator. V_2 is the operator in $L^2(\mathbb{R}^d)$ with domain

$$D(V_2(p)) = \{\psi \in L^2(\mathbb{R}^d) \mid \alpha \to V_2(\alpha)\hat{\psi}(\alpha) \in L^2(\mathbb{R}^d)\},$$

where $\hat{\psi}$ is the Fourier transform of ψ. It coincides with the operator defined by functional calculus as $V_2(p)$, with p denoting the self-adjoint operator in $\mathcal{H}$ given by $p := -i\hbar\nabla_x$, with its natural definition domain. The operator

---

[1] A sequence $(A_n)_{n \in \mathbb{N}}$ of linear operators $A_n : D \subseteq \mathcal{H} \to \mathcal{H}$ with a common domain D in a Hilbert space $(\mathcal{H}, \|\cdot\|)$ converges strongly to an operator A is for each $\psi \in D$, one has $\lim_{n \to \infty} \|A_n\psi - A\psi\| = 0$.

V is then the sum, as a self-adjoint operator in $L^2(\mathbb{R}^d)$, of the self-adjoint operators V_1 and V_2. We assume that the functions V_1 and V_2 are such that the corresponding operators have a common dense domain of essentially self-adjointness D. This is the case, e.g., when $V_1 \in L^2(\mathbb{R}^d) + L^\infty(\mathbb{R}^d)$, V_2 is bounded measurable, and $D = C_0^\infty(\mathbb{R}^d)$ or $D = \mathcal{S}(\mathbb{R}^d)$.

In order to define the quantum dynamics by applying Lemma 6.5, let us assume that V_1 and V_2 are such that the operators $-\frac{\hbar^2}{2m}\Delta + V_2$ and $-\frac{\hbar^2}{2m}\Delta + V_1 + V_2$ are essentially self-adjoint on D. We denote by H the closure of the latter operator. H (which we also write simply as $-\frac{\hbar^2}{2m}\Delta + V_1 + V_2$), is then the quantum Hamiltonian.

Let $U(t)_{t\in\mathbb{R}}$ be the one-parameter group of unitary operators on $L^2(\mathbb{R}^d)$ generated by the self-adjoint operator $H/\hbar$:

$$U(t) = e^{-\frac{it}{\hbar}H} = e^{-\frac{it(p^2/2m+V)}{\hbar}}.$$

Given an initial vector $\psi_0 \in L^2(\mathbb{R}^d)$, the solution of the Cauchy problem

$$\begin{cases} \frac{d}{dt}\psi = -\frac{i}{\hbar}H\psi \\ \psi(0,x) = \psi_0(x) \end{cases} \tag{6.56}$$

is given by $\psi(t) = e^{-\frac{it(p^2/2m+V)}{\hbar}}\psi_0$.

By Lemma 6.5 we have

$$e^{-\frac{it(p^2/2m+V)}{\hbar}} = s-\lim_{n\to\infty}\left(e^{-\frac{i\epsilon(p^2/2m+V_2)}{\hbar}}e^{-\frac{i\epsilon(V_1)}{\hbar}}\right)^n, \qquad \epsilon \equiv \frac{t}{n}$$

$$\psi(t) = e^{-\frac{it(p^2/2m+V)}{\hbar}}\psi_0 = \lim_{n\to\infty}\left(e^{-\frac{i\epsilon(p^2/2m+V_2)}{\hbar}}e^{-\frac{i\epsilon V_1}{\hbar}}\right)^n\psi_0,$$

(see e.g. [91, 92, 281] for related uses of the Lie-Trotter formula).

Let us consider a smooth vector $\psi_0 \in C_0^\infty(\mathbb{R}^d)$. By shifting from the position representation to the momentum representation and vice versa and by assuming that V_1 and V_2 are continuous, we can write in the strong $L^2(\mathbb{R}^d)$-sense, for all $t > 0$:

$$\psi(t,x) = \lim_{n\to\infty}\int_{\mathbb{R}^d} e^{-\frac{i\epsilon(p_{n-1}^2/2m+V_2(p_{n-1}))}{\hbar}}.$$

$$\left(e^{-\frac{i\epsilon V_1}{\hbar}}\left(e^{-\frac{i\epsilon(p^2/2m+V_2)}{\hbar}}e^{-\frac{i\epsilon(V_1)}{\hbar}}\right)^{n-1}\psi_0\right)(p_{n-1})\frac{e^{i\frac{xp_{n-1}}{\hbar}}}{(2\pi\hbar)^{d/2}}dp_{n-1}$$

$$= \lim_{n\to\infty}\int_{\mathbb{R}^{2d}} e^{-\frac{i\epsilon(p_{n-1}^2/2m+V_2(p_{n-1}))}{\hbar}}e^{-\frac{i\epsilon V_1(x_{n-1})}{\hbar}}.$$

$$\left(\left(e^{-\frac{i\epsilon(p^2/2m+V_2)}{\hbar}}e^{-\frac{i\epsilon(V_1)}{\hbar}}\right)^{n-1}\psi_0\right)(x_{n-1})\frac{e^{i\frac{xp_{n-1}}{\hbar}}}{(2\pi\hbar)^{d/2}}\frac{e^{-i\frac{x_{n-1}p_{n-1}}{\hbar}}}{(2\pi\hbar)^{d/2}}dp_{n-1}dx_{n-1}$$

$$= \lim_{n \to \infty} \left(\frac{1}{\sqrt{2\pi\hbar}} \right)^{2nd} \cdot$$

$$\int_{\mathbb{R}^{2nd}} e^{-\frac{i\epsilon}{\hbar} \sum_{j=0}^{n-1} \left(\frac{p_j^2}{2m} + V_1(x_j) + V_2(p_j) - p_j \frac{(x_{j+1} - x_j)}{\epsilon} \right)} \psi_0(x_0) \prod_{j=0}^{n-1} dp_j dx_j,$$
(6.57)

where[2] $x_n \equiv x$.

Now the latter expression suggests the following heuristic formula for the limit

$$\psi(t,x) = \text{const} \int_{q(t)=x} e^{\frac{i}{\hbar} \left(\int_0^t p(s)\dot{q}(s) - H(q(s),p(s))ds \right)} \psi_0(q(0)) dq dp, \quad (6.58)$$

which can be seen phase space-Hamiltonian version of Feynman's path integral (1.6). The aim of the present section is the rigorous mathematical realization of the heuristic formula (6.58) in terms of a well defined infinite dimensional oscillatory integral and to prove that, under suitable assumptions on the initial datum ψ_0 and on the classical potential V, it gives a representation of the solution of the Schrödinger equation.

Let us consider first of all Eq. (6.58) in the particular case of the free particle, namely when the Hamiltonian is just the kinetic energy: $H = p^2/2m$. In this case we have heuristically

$$\psi(t,x) = \text{const} \int_{q(t)=x} e^{\frac{i}{\hbar} \int_0^t (p(s)\dot{q}(s) - p(s)^2/2m)ds} \psi_0(q(0)) dq dp \quad (6.59)$$

From now on we will assume for notational simplicity that $m = 1$, but the whole discussion can be generalized to arbitrary m.

Following [325, 326], let us introduce the Hilbert space $\mathcal{H}_t \times \mathcal{L}_t$, namely the space of paths in the d-dimensional phase space $(q(s), p(s))_{s \in [0,t]}$, such that the path $(q(s))_{s \in [0,t]}$ belongs to the Cameron Martin space $\mathcal{H}_t$, while the path in the momentum space $(p(s))_{s \in [0,t]}$ belongs to $\mathcal{L}_t = L^2([0,t], \mathbb{R}^d)$. $\mathcal{H}_t \times \mathcal{L}_t$ is an Hilbert space with the natural inner product

$$\langle q, p; Q, P \rangle = \int_0^t \dot{q}(s)\dot{Q}(s)ds + \int_0^t p(s)P(s)ds.$$

Let us introduce the following bilinear form:

$$[q, p; Q, P] =$$

$$\int_0^t \dot{q}(s)P(s)ds + \int_0^t p(s)\dot{Q}(s)ds - \int_0^t p(s)P(s)ds = \langle q, p; A(Q,P) \rangle,$$

---

[2]The integrals in Eq. (6.57) are to be understood as limits as $\Lambda \uparrow \mathbb{R}^d$, $n \to \infty$ in the $L^2(\mathbb{R}^{2nd})$ sense of the corresponding integrals over Λ^{2nd}, with Λ bounded (see [265]). Formula (6.57) holds first as a strong L^2-limit, but then (possibly by subsequences) also for Lebesgue almost everywhere in $\mathbb{R}^d$. It also follows from this that Eq. (6.57) gives the solution to the Cauchy problem (6.56).

where A is the following operator in $\mathcal{H}_t \times \mathcal{L}_t$:

$$A(Q,P)(s) = \left(\int_t^s P(u)du, \dot{Q}(s) - P(s) \right). \qquad (6.60)$$

A is densely defined, e.g. on $C^1([0,t]; \mathbb{R}^d) \times C^1([0,t]; \mathbb{R}^d)$.

Moreover A is invertible with inverse given by

$$A^{-1}(Q,P)(s) = \left(\int_t^s P(u)du + Q(s), \dot{Q}(s) \right) \qquad (6.61)$$

(on the range of A).

Let $\psi_0 \in \mathcal{F}(\mathbb{R}^d)$, it is easy to see that the functional on $\mathcal{H}_t \times \mathcal{L}_t$ given by

$$(q,p) \mapsto \psi_0(q(0) + x), \qquad (q,p) \in \mathcal{H}_t \times \mathcal{L}_t,$$

belongs to $\mathcal{F}(\mathcal{H}_t \times \mathcal{L}_t)$:

$$\psi_0(q(0) + x) = \int_{\mathcal{H}_t \times \mathcal{L}_t} e^{i\langle q,p;Q,P \rangle} d\mu_0(Q,P).$$

Now expression (6.58) can be realized rigorously as the normalized oscillatory integral with respect to the operator A (in the sense of Definition 5.5) on the Hilbert space $\mathcal{H}_t \times \mathcal{L}_t$:

$$\widetilde{\int_{\mathcal{H}_t \times \mathcal{L}_t}^A} e^{\frac{i}{2\hbar} \langle q,p;A(q,p) \rangle} \psi_0(q(0) + x) dq dp, \qquad (6.62)$$

and by Theorem 5.8 the normalized oscillatory integral is well defined and can be computed by means of Parseval type equality (5.41):

$$\widetilde{\int_{\mathcal{H}_t \times \mathcal{L}_t}^A} e^{\frac{i}{2\hbar} \langle q,p;A(q,p) \rangle} \psi_0(q(0)+x) dq dp = \int_{\mathcal{H}_t \times \mathcal{L}_t} e^{-\frac{i\hbar}{2\hbar} \langle q,p;A^{-1}(q,p) \rangle} d\mu_0(q,p).$$

By choosing a suitable sequence of finite dimensional projection operators $\{P_n\}$ on $\mathcal{H}_t \times \mathcal{L}_t$, it is possible to recover in the definition of the normalized oscillatory integral the expression (6.57), obtained by means of the Lie-Trotter product formula.

Let us consider a sequence of partitions π_n of the interval $[0,t]$ into n subintervals of amplitude $\epsilon \equiv t/n$:

$$t_0 = 0, t_1 = \epsilon, \ldots, t_i = i\epsilon, \ldots, t_n = n\epsilon = t.$$

To each π_n we associate a projector $P_n : \mathcal{H}_t \times \mathcal{L}_t \to \mathcal{H}_t \times \mathcal{L}_t$ onto a finite dimensional subspace of $\mathcal{H}_t \times \mathcal{L}_t$, namely the subspace of polygonal paths.

In other words each projector P_n acts on a phase space path $(q, p) \in \mathcal{H}_t \times \mathcal{L}_t$ in the following way:

$$P_n(q, p)(s) =$$

$$\left(\sum_{i=1}^{n} \chi_{[t_{i-1}, t_i]}(s) \left(q(t_{i-1}) + \frac{(q(t_i) - q(t_{i-1}))}{t_i - t_{i-1}} (s - t_{i-1}) \right), \sum_{i=1}^{n} \chi_{[t_{i-1}, t_i]}(s) p_i \right),$$

where

$$p_i = \frac{\int_{t_{i-1}}^{t_i} p(s) ds}{t_i - t_{i-1}} = \frac{1}{\epsilon} \int_{t_{i-1}}^{t_i} p(s) ds.$$

Analogously to Lemma 6.4, it is possible to prove the following.

Lemma 6.6. *For each $n \in \mathbb{N}$, P_n is a projector in $\mathcal{H}_t \times \mathcal{L}_t$. Moreover for $n \to \infty$, $P_n \to I$ as a strong operator limit.*

Proof.

- P_n is symmetric, indeed for all $(Q, P) \in \mathcal{H}_t \times \mathcal{L}_t$ and all $(q, p) \in \mathcal{H}_t \times \mathcal{L}_t$

$$\langle Q, P; P_n(q, p) \rangle = \int_0^t \dot{Q}(s) \sum_{i=1}^{n} \chi_{[t_{i-1}, t_i]}(s) \frac{(q(t_i) - q(t_{i-1}))}{t_i - t_{i-1}} ds$$

$$+ \int_0^t P(s) \sum_{i=1}^{n} \chi_{[t_{i-1}, t_i]}(s) p_i ds = \sum_{i=1}^{n} \frac{(q(t_i) - q(t_{i-1})(Q(t_i) - Q(t_{i-1}))}{t_i - t_{i-1}}$$

$$+ \sum_{i=1}^{n} \frac{\int_{t_{i-1}}^{t_i} p(s) ds \int_{t_{i-1}}^{t_i} P(s) ds}{t_i - t_{i-1}} = \langle P_n(Q, P); q, p \rangle \qquad (6.63)$$

- $P_n^2 = P_n$, indeed

$$P_n^2(q, p)(s) =$$

$$\left(\sum_{i=1}^{n} \chi_{[t_{i-1}, t_i]}(s) \left(q(t_{i-1}) + \frac{(q(t_i) - q(t_{i-1}))}{t_i - t_{i-1}} (s - t_{i-1}) \right), \sum_{i=1}^{n} \chi_{[t_{i-1}, t_i]}(s) p_i \right)$$

$$= P_n(q, p)(s).$$

- $\forall (q, p) \in \mathcal{H}_t \times \mathcal{L}_t, \| P_n(q, p) - (q, p) \| \to 0$ as $n \to \infty$:
  Let us consider the subset $\mathcal{K} \subseteq \mathcal{H}_t \times \mathcal{L}_t$,

$$\mathcal{K} = \{ (q, p) \in \mathcal{H}_t \times \mathcal{L}_t \ : \ \| P_n(q, p) - (q, p) \| \to 0, \ n \to \infty \}.$$

It is enough to prove that the closure of $\mathcal{K}$ is $\mathcal{H}_t \times \mathcal{L}_t$. To prove this, it is sufficient to show that $\mathcal{K}$ is a closed subspace of $\mathcal{H}_t \times \mathcal{L}_t$ and contains a dense subset of $\mathcal{H}_t \times \mathcal{L}_t$. This follows from the density of the piecewise linear paths in $\mathcal{H}_t$ (see Lemma 6.4 and [324]) and the density of the piecewise constant paths in $\mathcal{L}_t$. $\qquad \square$

By means of these results and under the assumption that $\psi_0 \in \mathcal{F}(\mathbb{R}^d)$, it is possible to prove that the infinite dimensional oscillatory integral (6.62) coincides with the limit (6.57), which can be taken for Lebesgue almost every $x \in \mathbb{R}^d$.

Theorem 6.4. *Let the function* $(q,p) \to \psi_0(x + q(0))$, $\psi_0 \in L^2(\mathbb{R}^d)$, *be Fresnel integrable[3] with respect to* A *(with* A *defined by (6.60)). Then the phase space Feynman path integral, namely the limit*

$$\lim_{n \to \infty} \frac{\sqrt{\det(P_n A P_n)}}{(2\pi i \hbar)^{nd}} \int_{P_n(\mathcal{H}_t \times \mathcal{L}_t)} e^{\frac{i}{2\hbar} \langle P_n(q,p), A P_n(q,p) \rangle} \psi_0(x + q(0)) dP_n(q,p)$$

$$(6.64)$$

coincides with the limit (6.57), namely with the solution of the Schrödinger equation with a free Hamiltonian.

Proof. The result follows by direct computation, by showing that the two limits (6.57) and (6.64) coincide. Indeed (6.64) is a pointwise limit by hypothesis. On the other hand (6.57) is a limit in the L_2 sense, hence, passing if necessary to a subsequence, it is also a pointwise limit. □

Remark 6.4. The latter result is equivalent to the "traditional" formulation of the Feynman path integral in the configuration space. Indeed it can be obtained by means of Fubini theorem and an integration with respect to the momentum variables:

$$\lim_{n \to \infty} \left(\frac{1}{\sqrt{2\pi\hbar}} \right)^{2nd} \int_{\mathbb{R}^{2nd}} e^{-\frac{i\epsilon}{\hbar} \sum_{j=0}^{n-1} \left(\frac{p_j^2}{2m} - p_j \frac{(x_{j+1} - x_j)}{\epsilon} \right)} \psi_0(x_0) \prod_{j=0}^{n-1} dp_j dx_j$$

$$= \lim_{n \to \infty} \left(\frac{1}{\sqrt{2\pi i \hbar}} \right)^{nd} \int_{\mathbb{R}^{nd}} e^{-\frac{i\epsilon}{\hbar} \sum_{j=0}^{n-1} m \frac{(x_{j+1} - x_j)^2}{2\epsilon^2}} \psi_0(x_0) \prod_{j=0}^{n-1} dx_j. \quad (6.65)$$

The latter expression yields the Feynman functional on the configuration space, i.e. heuristically

$$\text{const} \int e^{\int_0^t \mathcal{L}(q(s), \dot{q}(s)) ds} dq,$$

($\mathcal{L}$ being the classical Lagrangian density).

Even if the integration with respect to the momentum variables might seem to be superfluous, it is very useful when we introduce a potential

---

[3]This condition is satisfied if for instance $\psi_0 \in \mathcal{F}(\mathbb{R}^d)$.

depending explicitly on the momentum variables, as the following theorem shows.

Theorem 6.5. *Let us consider a semibounded potential V depending explicitly on the momentum: $V = V(p)$ and the corresponding quantum mechanical Hamiltonian $H = -\frac{\hbar^2}{2}\Delta + V(p)$. Let us assume H is an essentially self-adjoint operator on $L_2(\mathbb{R}^d)$. Let the functional on the Hilbert space $\mathcal{H}_t \times \mathcal{L}_t$, given by*

$$(q,p) \mapsto e^{-\frac{i}{\hbar}\int_0^t V(p(s))ds}\psi_0(x + q(0)), \qquad (q,p) \in \mathcal{H}_t \times \mathcal{L}_t$$

be Fresnel integrable with respect to the operator A, with A defined by (6.60). Then the solution to the Schrödinger equation

$$\begin{cases} \frac{d}{dt}\psi = -\frac{i}{\hbar}H\psi \\ \psi(0,x) = \psi_0(x), \qquad \psi_0 \in \mathcal{S}(\mathbb{R}^d) \end{cases} \tag{6.66}$$

is given by the phase space path integral

$$\lim_{n\to\infty} (2\pi i\hbar)^{-nd}(\det(P_n A P_n))^{1/2} \int_{P_n(\mathcal{H}_t \times \mathcal{L}_t)} e^{\frac{i}{2\hbar}\langle P_n(q,p), A P_n(q,p)\rangle}$$

$$e^{-\frac{i}{\hbar}\int_0^t V(P_n(p(s)))ds}\psi_0(x + q(0))dP_n(q,p).$$

Proof. We can proceed in a completely analogous way as in the proof of Theorem 6.4, therefore we shall omit the details. □

We can now handle the case of a classical potential V depending both on position Q and on momentum P of the form $V = V(Q,P) = V_1(Q) + V_2(P)$ (the general case presents problems due to the non-commutativity of the quantized expression of Q and P). For a different approach with more general Hamiltonians see [304]).

Let us consider an initial wave function $\psi_0 \in \mathcal{F}(\mathbb{R}^d)$ and let us assume that $V_1 \in \mathcal{F}(\mathbb{R}^d)$ and the function on $\mathcal{L}_t$ given by

$$p(s)_{s\in[0,t]} \to e^{-\frac{i}{\hbar}\int_0^t V_2(p(s))ds}$$

belongs to $\mathcal{F}(\mathcal{L}_t)$. Then it is easy to see that the functional $f : \mathcal{H}_t \times \mathcal{L}_t \to \mathbb{C}$

$$f(q,p) = \psi_0(x + q(0))e^{-\frac{i}{\hbar}\int_0^t V(q(s)+x,p(s))ds},$$

belongs to $\mathcal{F}(\mathcal{H}_t \times \mathcal{L}_t)$, i.e. $f = \hat{\mu}_f$, $\mu_f \in \mathcal{M}(\mathcal{H}_t \times \mathcal{L}_t)$. Indeed $f(q,p)$ is the product of two functions: the first, say f_1, depends only on the first variable q, while the second f_2 depends only on the variable p, more precisely

$$f_1(q) = \psi_0(x + q(0))e^{-\frac{i}{\hbar}\int_0^t V_1(q(s)+x)ds}, \qquad f_2(p) = e^{-\frac{i}{\hbar}\int_0^t V_2(p(s))ds}.$$

Under the given hypothesis on V_1 and ψ_0, f_1 belongs to $\mathcal{F}(\mathcal{H}_t)$ (see Lemma 6.3). For f_2 one must pay more attention: indeed the same proof given for f_1 does not work, as f_2 is defined on a different Hilbert space and we have to require explicitly that $e^{-\frac{i}{\hbar}\int_0^t V_2(p(s))ds} \in \mathcal{F}(\mathcal{L}_t)$. This holds for instance if V_2 is linear, i.e. $V_2(p) = a \cdot p$, $a \in \mathbb{R}^d$, or if the function $p(s)_{s\in[0,t]} \to \int_0^t V_2(p(s))ds \in \mathcal{F}(\mathcal{L}_t)$.

Now if $f_1 = \hat{\mu}_{f_1} \in \mathcal{F}(\mathcal{H}_t)$, f_1 can be extended to a function, denoted again by f_1, in $\mathcal{F}(\mathcal{H}_t \times \mathcal{L}_t)$: it is the Fourier transform of the product measure on $\mathcal{H}_t \times \mathcal{L}_t$ of $\mu_{f_1}(dq)$ and $\delta_0(dp)$. The same holds for $f_2 = \hat{\mu}_{f_2}$:
$$f_2 = (\widehat{\delta_0(dq)\mu_{f_2}(dp)}).$$
Finally, as $\mathcal{F}(\mathcal{H}_t \times \mathcal{L}_t)$ is a Banach algebra, the product of two elements $f_1 f_2$ is again an element of $\mathcal{F}(\mathcal{H}_t \times \mathcal{L}_t)$: more precisely it is the Fourier transform of the convolution of the two measures in $\mathcal{M}(\mathcal{H}_t \times \mathcal{L}_t)$ corresponding to f_1 and f_2 respectively.

By applying Theorem 5.8, the phase space Feynman path integral of the function f is well defined and can be computed in terms of a Parseval type equality:

$$\widetilde{\int_{\mathcal{H}_t \times \mathcal{L}_t}^A} e^{\frac{i}{2\hbar}\langle (q,p), A(q,p)\rangle} f(q,p)dqdp = \int_{\mathcal{H}_t \times \mathcal{L}_t} e^{-\frac{i\hbar}{2}\langle (q,p), A^{-1}(q,p)\rangle} d\mu_f(q,p).$$

The next theorem shows that the above oscillatory integral

$$\widetilde{\int_{\mathcal{H}_t \times \mathcal{L}_t}^A} e^{\frac{i}{2\hbar}\langle (q,p), A(q,p)\rangle} e^{-\frac{i}{\hbar}\int_0^t V_1(q(s)+x)ds} e^{-\frac{i}{\hbar}\int_0^t V_2(p(s))ds} \psi_0(x + q(0))dqdp$$

$$(6.67)$$

gives the solution to the Schrödinger Eq. (6.56).

Theorem 6.6. *Let us consider the following Hamiltonian*

$$H(Q; P) = \frac{P^2}{2} + V_1(Q) + V_2(P)$$

in $L^2(\mathbb{R}^d)$ and the corresponding Schrödinger equation

$$\begin{cases} \frac{d}{dt}\psi = -\frac{i}{\hbar}H\psi \\ \psi(0, x) = \psi_0(x), \quad x \in \mathbb{R}^d. \end{cases}$$

Let us suppose that $V_1, \psi_0 \in \mathcal{F}(\mathbb{R}^d)$ and $\int_0^t V_2(p(s))ds \in \mathcal{F}(\mathcal{L}_t)$. Then the solution to the Cauchy problem (6.56) is given by the phase space Feynman path integral (6.67).

Proof.   As in the proof of Theorem 6.2, we follow the technique proposed by Elworthy and Truman in [129].

For $0 \leq u \leq t$ let $\mu_u(V_1, x) \equiv \mu_u$, $\nu_u^t(V_1, x) \equiv \nu_u^t$, $\eta_u^t(V_2) \equiv \eta_u^t$ and $\mu_0(\psi)$ be the measures on $\mathcal{H}_t \times \mathcal{L}_t$, whose Fourier transforms when evaluated at $(q, p) \in \mathcal{H}_t \times \mathcal{L}_t$ are $V_1(x + q(u))$, $\exp\left(-\frac{i}{\hbar}\int_u^t V_1(x + q(s))ds\right)$, $\exp\left(-\frac{i}{\hbar}\int_u^t V_2(p(s))ds\right)$ and $\psi_0(q(0) + x)$.

We set

$$U(t)\psi_0(x) = \int_{\mathcal{H}_t \times \mathcal{L}_t}^{\overset{A}{\frown}} e^{\frac{i}{2\hbar}\langle q,p; A(q,p)\rangle} e^{-\frac{i}{\hbar}\int_0^t (V_1(q(s)+x)+V_2(p(s)))ds} \psi_0(q(0) + x)dqdp$$

and

$$U_0(t)\psi_0(x) = \int_{\mathcal{H}_t \times \mathcal{L}_t}^{\overset{A}{\frown}} e^{\frac{i}{2\hbar}\langle q,p; A(q,p)\rangle} e^{-\frac{i}{\hbar}\int_0^t V_2(p(s))ds} \psi_0(q(0) + x)dqdp.$$

By Theorem 5.8 we have:

$$U(t)\psi_0(x) = \int_{\mathcal{H}_t \times \mathcal{L}_t} e^{\frac{-i\hbar}{2}\langle q,p; A^{-1}(q,p)\rangle} (\eta_0^t * \nu_0^t * \mu_0(\psi))(dqdp). \tag{6.68}$$

Now, if $\{\mu_u : a \leq u \leq t\}$ is a family in $\mathcal{M}(\mathcal{H}_t \times \mathcal{L}_t)$, we shall let $\int_a^b \mu_u du$ denote the measure on $\mathcal{H}_t \times \mathcal{L}_t$ given by :

$$f \rightarrow \int_a^b \int_{\mathcal{H}_t \times \mathcal{L}_t} f(q, p)d\mu_u(q, p)du, \qquad f \in C_0(\mathcal{H}_t \times \mathcal{L}_t),$$

whenever it exists.

Since for any continuous path q we have

$$\exp\left(-\frac{i}{\hbar}\int_0^t V_1(q(s)+x)ds\right) = 1 - \frac{i}{\hbar}\int_0^t V_1(q(u)+x)\exp\left(-\frac{i}{\hbar}\int_u^t V_1(q(s)+x)ds\right)du$$

the following relation holds

$$\nu_0^t = \delta_0 - \frac{i}{\hbar}\int_0^t (\mu_u * \nu_u^t)du \tag{6.69}$$

where δ_0 is the Dirac measure at $0 \in \mathcal{H}_t \times \mathcal{L}_t$.

Applying this relation to (6.68) we obtain:

$$U(t)\psi_0(x) = \int_{\mathcal{H}_t \times \mathcal{L}_t} e^{\frac{-i\hbar}{2}\langle q,p; A^{-1}(q,p)\rangle} (\eta_0^t * \mu_0(\psi))(dqdp)$$

$$-\frac{i}{\hbar}\int_0^t \int_{\mathcal{H}_t \times \mathcal{L}_t} e^{\frac{-i\hbar}{2}\langle q,p; A^{-1}(q,p)\rangle} (\eta_0^t * \mu_u(V_1, x) * \nu_u^t * \mu_0(\psi))(dqdp)du$$

$$= U_0(t)\psi_0(x) - \frac{i}{\hbar}\int_0^t \int_{\mathcal{H}_t \times \mathcal{L}_t}^{\overset{A}{\frown}} e^{\frac{i}{2\hbar}\langle q,p; A(q,p)\rangle} e^{-\frac{i}{\hbar}\int_u^t V_1(q(s)+x)ds}$$

$$e^{-\frac{i}{\hbar}\int_0^t V_2(p(s)))ds} V_1(q(u) + x)\psi_0(q(0) + x)dqdpdu.$$

The conclusion follows by repeating the same procedure, as in the proof of Theorem 6.2. □

6.4   Quartic potential

The examples we have seen so far have shown that infinite dimensional oscillatory integrals are a flexible tool and provide a rigorous mathematical realization for a large class of Feynman path integrals representations. However all the examples have a common problem, that is the restriction on the class of classical potentials V which can be handled. Indeed, in order to apply Theorem 5.5 (or Theorems 5.8 and 5.9), we have to assume that the potential V describing the dynamics of the quantum particle is of the type "harmonic oscillator plus a bounded perturbation which is the Fourier transform of a measure". This situation is rather unsatisfactory from a physical point of view, as this class of potentials does not include several interesting unbounded ones. There are some extension of the theory to unbounded potentials which are Laplace transform of measures, such as those with exponential growth at infinity [10, 236, 20], but even this class does not include the functions with a generic polynomial growth at infinity. In fact the problem for such polynomial potentials is not simple, as it has been proved [334] that in one dimension, if the potential is time independent and super-quadratic in the sense that

$$V(x) \geq C(1 + |x|)^{2+\epsilon}, \qquad x \to \infty,$$

where $C > 0$ and $\epsilon > 0$, then, as a function of (t, x, y), the fundamental solution $G(t, 0, x, y)$ of the time dependent Schrödinger equation is nowhere C^1. In other words, we do not have even the hope to define in terms of a functional integral an expression of the form

$$G(0, x; t, y) = \int e^{\frac{i}{\hbar} S_t(\gamma)} D\gamma, ,$$

depending from the variables (t, x, y), that can be differentiated in order to prove that it is a fundamental solution for the Schrödinger equation.

Even if the problem is not only technical, but rather fundamental, some interesting results can be obtained in the case where the potential V has a quartic polynomial growth at infinity, by means of the results of Section 5.5.

Let us consider the Schrödinger equation in $L^2(\mathbb{R}^d)$

$$\begin{cases} i\hbar \frac{\partial}{\partial t} \psi(t, x) = -\frac{\hbar^2}{2} \Delta \psi(t, x) + V(x) \psi(t, x) \\ \psi(0, x) = \psi_0(x) \end{cases} \tag{6.70}$$

where the potential V is of the following form:

$$V(x) = \frac{1}{2} x \Omega^2 x + \lambda |x|^4, \qquad x \in \mathbb{R}^d, \tag{6.71}$$

where Ω^2 is a positive symmetric $d \times d$ matrix, and $\lambda \in \mathbb{R}$ is a real constant. More generally, we can also consider a polynomial potential of the following form

$$V(x) = \frac{1}{2}x\Omega^2 x + \lambda C(x,x,x,x), \qquad x \in \mathbb{R}^d, \tag{6.72}$$

where C is a completely symmetric positive fourth order covariant tensor on $\mathbb{R}^d$. In the following we shall focus on expression (6.71) because of notational simplicity, but all the reasoning can be repeated in the general case of Eq. (6.72).

In the case where λ is positive, the quantum mechanical Hamiltonian $H : D(H) \subset L^2(\mathbb{R}^d) \to L^2(\mathbb{R}^d)$ given on vector $\psi \in C_0^\infty(\mathbb{R}^d)$ by

$$H\psi(x) = -\frac{\Delta}{2}\psi(x) + V(x)\psi(x) \tag{6.73}$$

is essentially self-adjoint [281] and determines uniquely the quantum dynamics. In the case where λ is strictly negative, i.e. when the potential V represents a quartic (double well) polynomial potentials unbounded from below, the quantum Hamiltonian is not essentially self-adjoint as one can deduce by a limit point argument (see [281], Theorem X.9) and the quantum evolution is not uniquely determined.

In the following we shall present the results of [28, 26, 251] and show that the infinite dimensional oscillatory integrals with polynomial phase function studied in Section 5.5 can provide a mathematical realization for the Feynman path integral representation for the weak solution of the Schrödinger Eq. (6.70) with potential V given by Eq. (6.71), i.e. the matrix elements $\langle \phi, e^{-\frac{i}{\hbar}Ht}\psi_0 \rangle$, $\phi, \psi_0 \in L^2(\mathbb{R}^d)$:

$$\int_{\mathbb{R}^d} \bar{\phi}(x) \int_{\gamma(t)=x} e^{\frac{i}{2\hbar}\int_0^t \dot{\gamma}^2(s)ds - \frac{i}{\hbar}\int_0^t \gamma(s)\Omega^2\gamma(s)ds - \frac{i\lambda}{\hbar}\int_0^t |\gamma(s)|^4 ds} \psi_0(\gamma(0))d\gamma dx.$$
$$\tag{6.74}$$

Let us consider the Cameron-Martin space[4] $\mathcal{H}_t$, that is the Hilbert space of absolutely continuous paths $\gamma : [0,t] \to \mathbb{R}^d$, with $\gamma(0) = 0$ and inner product

$$\langle \gamma_1, \gamma_2 \rangle = \int_0^t \dot{\gamma}_1(s)\dot{\gamma}_2(s)ds.$$

The cylindrical Gaussian measure on $\mathcal{H}_t$ with covariance operator the identity extends to a σ-additive measure on the Wiener space

---

[4]With an abuse of notation we call here Cameron-Martin space the space of paths γ belonging to the Sobolev space $H_{(t)}^{1,2}([0,t], \mathbb{R}^d)$ such that $\gamma(0) = 0$, while in the previous sections with the same name we denoted the space of paths $\gamma \in H_{(t)}^{1,2}([0,t], \mathbb{R}^d)$ such that $\gamma(t) = 0$.

$C_t = \{\omega \in C([0, t]; \mathbb{R}^d) \mid \gamma(0) = 0\}$: the Wiener measure W. $(i, \mathcal{H}_t, C_t)$ is an abstract Wiener space (see Section 3.7 for the definition and the main properties of abstract Wiener spaces).

Let us consider moreover the Hilbert space $\mathcal{H} = \mathbb{R}^d \times \mathcal{H}_t$ and the Banach space $\mathcal{B} = \mathbb{R}^d \times C_t$ endowed with the product measure $N(dx) \times W(d\omega)$, N being the Gaussian measure on $\mathbb{R}^d$ with covariance equal to the $d \times d$ identity matrix. $(i, \mathcal{H}, \mathcal{B})$ is an abstract Wiener space.

Let us consider the operator $B : \mathcal{H} \to \mathcal{H}$ given by:

$$(x, \gamma) \mapsto (y, \eta) = B(x, \gamma), \qquad (x, \gamma) \in \mathbb{R}^d \times \mathcal{H}_t \qquad (6.75)$$

$$y = t\Omega^2 x + \Omega^2 \int_0^t \gamma(s)ds,$$

$$\eta(s) = \Omega^2 x \left(ts - \frac{s^2}{2}\right) - \int_0^s \int_t^u \Omega^2 \gamma(r)drdu, \qquad s \in [0, t]. \qquad (6.76)$$

It is easy to verify that $B : \mathcal{H} \to \mathcal{H}$ is positive and symmetric, indeed

$$\langle (y, \eta), B(x, \gamma) \rangle = \int_0^t (\eta(s) + y)\Omega^2(\gamma(s) + x)ds, \qquad (y, \eta), (x, \gamma) \in \mathcal{H}.$$

Let us introduce the fourth order tensor operator $A : \mathcal{H} \times \mathcal{H} \times \mathcal{H} \times \mathcal{H} \to \mathbb{R}$ given by:

$$A((x_1, \gamma_1), (x_2, \gamma_2), (x_3, \gamma_3), (x_4, \gamma_4))$$
$$= \int_0^t (\gamma_1(s) + x_1)(\gamma_2(s) + x_2)(\gamma_3(s) + x_3)(\gamma_4(s) + x_4)ds, \qquad (6.77)$$

and the homogeneous fourth order polynomial function $V_4 : \mathcal{H} \to \mathbb{R}$ given by

$$V_4(x, \gamma) = \lambda A((x, \gamma), (x, \gamma), (x, \gamma), (x, \gamma)) = \lambda \int_0^t |\gamma(s) + x|^4 ds.$$

Given two vectors $\phi, \psi_0 \in L^2(\mathbb{R}^d) \cap \mathcal{F}(\mathbb{R}^d)$, let us consider the function $f : \mathcal{H} \to \mathbb{C}$ given by

$$f(x, \gamma) = (2\pi i\hbar)^{d/2} e^{-\frac{i}{2\hbar}|x|^2} \bar{\phi}(x)\psi_0(\gamma(t) + x), \qquad (x, \gamma) \in \mathbb{R}^d \times \mathcal{H}_t. \qquad (6.78)$$

By means of these notations, and by imposing suitable condition on the vectors ϕ, ψ_0 as well as to the time variable t, expression (6.74) can be realized as the following infinite dimensional oscillatory integral on $\mathcal{H}$:

$$\widetilde{\int_{\mathcal{H}}} e^{\frac{i}{2\hbar}(|x|^2 + |\gamma|^2)} e^{-\frac{i}{2\hbar}\langle (x, \gamma), B(x, \gamma) \rangle} e^{-\frac{i}{\hbar}V_4(x, \gamma)} f(x, \gamma)dxd\gamma. \qquad (6.79)$$

In the following we will denote by Ω_i, $i = 1, \ldots, d$, the eigenvalues of the matrix Ω.

Lemma 6.7. *The operator $B : \mathcal{H}_t \to \mathcal{H}_t$ given by Eq. (6.75) is trace class. Moreover, if for each $i = 1, \ldots, d$ the following inequalities are satisfied*

$$\Omega_i t < \frac{\pi}{2}, \qquad 1 - \Omega_i \tan(\Omega_i t) > 0. \tag{6.80}$$

then the operator $(I - B)$ is strictly positive.

Proof. Let us study the spectrum of the self-adjoint operator B on $\mathcal{H}$ given by (6.75). In order to avoid the use of too many indexes we will assume $d = 1$, but our reasoning remains valid also in the case $d > 1$. A positive real number c_l and a vector $(x_l, \gamma_l) \in \mathcal{H}$ are respectively an eigenvalue and an eigenvector of B if and only if:

$$\begin{cases} t\Omega^2 x_l + \Omega^2 \int_0^t \gamma_l(s)ds = c_l x_l \\ \Omega^2 x_l(ts - \frac{s^2}{2}) - \int_0^s \int_t^s \Omega^2 \gamma_l(r)drdu = c_l \gamma_l(s). \end{cases}$$

By differentiating twice, the vector $(x_l, \gamma_l) \in \mathcal{H}$ is a solution of the following system:

$$\begin{cases} t\Omega^2 x_l + \Omega^2 \int_0^t \gamma_l(s)ds = c_l x_l \\ c_l \ddot{\gamma}_l(s) + \Omega^2 \gamma_l(s) = -\Omega^2 x_l \\ \gamma_l(0) = 0 \\ \dot{\gamma}_l(t) = 0. \end{cases}$$

By a direct calculation one can verify that the latter system indeed admits a (unique) solution if and only if c_l satisfies the following equation

$$\frac{\Omega}{\sqrt{c_l}} \tan \frac{\Omega t}{\sqrt{c_l}} = 1.$$

A graphical representation of the position of the solutions shows that the operator B is trace class. Moreover if the conditions (6.80) are fulfilled the maximum eigenvalue of B is strictly less than 1, so that $(I - B)$ is positive definite. $\qquad\square$

Lemma 6.8. *Let $\phi, \psi_0 \in L^2(\mathbb{R}^d) \cap \mathcal{F}(\mathbb{R}^d)$. Let μ_0 be the complex bounded variation measure on $\mathbb{R}^d$ such that $\hat{\mu}_0 = \psi_0$. Let μ_ϕ be the complex bounded variation measure on $\mathbb{R}^d$ such that $\hat{\mu}_\phi(x) = (2\pi i\hbar)^{d/2} e^{-\frac{i}{2\hbar}|x|^2} \hat{\phi}(x)$.*
Let us assume that t satisfies inequalities (6.80) and that the measures μ_0, μ_ϕ satisfy the following assumption:

$$\int_{\mathbb{R}^d} \int_{\mathbb{R}^d} e^{\frac{\hbar}{4}(y - \cos(\Omega t)^{-1} x)(1 - \Omega \tan(\Omega t))^{-1}(y - \cos(\Omega t)^{-1} x)}$$

$$e^{\frac{\hbar}{4} x \Omega^{-1} \tan(\Omega t) x} |\mu_0|(dx)|\mu_\phi|(dy) < \infty. \tag{6.81}$$

Then the function $f : \mathcal{H} \to \mathbb{C}$, given by (6.78) is the Fourier transform of a bounded variation measure μ_f on $\mathcal{H}$ satisfying

$$\int_{\mathcal{H}} e^{\frac{\hbar}{4}\langle (y,\eta),(I-B)^{-1}(y,\eta) \rangle} |\mu_f|(dyd\eta) < \infty, \tag{6.82}$$

(B being given by (6.75)).

Proof. By the assumptions on ϕ, one can easily verify that the function $(2\pi i\hbar)^{d/2} e^{-\frac{i}{2\hbar}|x|^2} \bar{\phi}(x)$ is the Fourier transform of the bounded variation measure on $\mathbb{R}^d \times \mathcal{H}_t$, which is the product measure $\mu_\phi(dx) \times \delta_0(d\gamma)$, where $\delta_0(d\gamma)$ is the measure on $\mathcal{H}_t$ concentrated on the vector $0 \in \mathcal{H}_t$. Analogously the function $(x,\gamma) \mapsto \psi_0(\gamma(t)+x)$ is the Fourier transform of the bounded variation measure μ_ψ on $\mathbb{R}^d \times \mathcal{H}_t$ given by

$$\int_{\mathbb{R}^d \times \mathcal{H}_t} f(x,\gamma)\mu_\psi(dxd\gamma) = \int_{\mathbb{R}^d \times \mathcal{H}_t} f(x,x\gamma)\delta_{\gamma_t}(d\gamma)\mu_0(dx),$$

where γ_t is the vector in $\mathcal{H}_t$ given by

$$\gamma_t(s) = s, \qquad s \in [0,t].$$

As $\mathcal{F}(\mathbb{R}^d \times \mathcal{H}_t)$ is a Banach algebra, the product

$$f(x,\gamma) := (2\pi i\hbar)^{d/2} e^{-\frac{i}{2\hbar}|x|^2} \bar{\phi}(x)\psi(\gamma(t)+x)$$

still belongs to $\mathcal{F}(\mathbb{R}^d \times \mathcal{H}_t)$, in fact it is the Fourier transform of the convolution $\mu_f \equiv (\mu_\phi \times \delta_0) * \mu_\psi$.

Let us now prove that the measure μ_f satisfies assumptions (5.68) of Theorem 5.11, that is (6.82), if μ_0 and μ_ϕ satisfy (6.81).

By Theorem 3.20, we have

$$e^{\frac{\hbar}{4}\langle (y,\eta),(I-B)^{-1}(y,\eta) \rangle} = \sqrt{\det(I-B)} F(y/\sqrt{2}, \eta/\sqrt{2}),$$

where the function $F : \mathcal{H} \to \mathbb{C}$ is given by the Gaussian integral

$$F(y,\eta) = \int_{\mathbb{R}^d \times C_t} e^{\sqrt{\hbar}xy + \sqrt{\hbar}n(\eta)(\omega)} e^{\frac{1}{2}\langle (x,\omega),B(x,\omega) \rangle} N(dx)W(d\omega).$$

By a direct computation and by Fubini theorem, F is equal to

$$F(y,\eta) = (2\pi)^{-\frac{d}{2}} \int_{\mathbb{R}^d} e^{\sqrt{\hbar}xy} e^{-x\frac{(I-t\Omega^2)}{2}x} \left(\int_{C_t} e^{\sqrt{\hbar}n(\eta)(\omega)} e^{x \int_0^t \Omega^2 \omega(s)ds} \right.$$

$$\left. e^{\frac{1}{2}\int_0^t \omega(s)\Omega^2 \omega(s)ds} W(d\omega) \right) dx$$

$$= (2\pi)^{-\frac{d}{2}} \int_{\mathbb{R}^d} e^{\sqrt{\hbar}xy} e^{-x\frac{(I-t\Omega^2)}{2}x} \left(\int_{C_t} e^{\sqrt{\hbar}n(\eta)(\omega)} e^{n(v_x)(\omega)} e^{\frac{1}{2}\langle \omega,L\omega \rangle} W(d\omega) \right) dx, \tag{6.83}$$

where $L : \mathcal{H}_t \to \mathcal{H}_t$ is the operator given by

$$L\gamma(s) = -\int_0^s \int_t^{s'} \Omega^2 \gamma(s'')ds''ds'$$

and $v_x \in \mathcal{H}_t$ is the vector given by

$$v_x(s) = \Omega^2 x\left(ts - \frac{s^2}{2}\right).$$

One can easily verify that L is symmetric and trace class. Indeed by denoting by α^2, γ_α respectively the eigenvalues and the eigenvectors of the operator L, we have

$$\alpha^2 \ddot{\gamma}_\alpha(s) + \Omega^2 \gamma_\alpha(s) = 0, \qquad s \in [0, t],$$

with the conditions

$$\gamma_\alpha(0) = 0, \quad \dot{\gamma}_\alpha(t) = 0.$$

Without loss of generality we can assume Ω^2 is diagonal with eigenvalues Ω_i^2, $i = 1, \ldots d$. The components $\gamma_{\alpha,i}$, $i = 1, \ldots d$, of the eigenvector γ_α corresponding to the eigenvalue α^2 are equal to

$$\gamma_{\alpha,i}(s) = A_i \sin \frac{\Omega_i s}{\alpha}, \qquad s \in [0, t].$$

By imposing the condition $\dot{\gamma}(t) = 0$, we have

$$\Omega_i t / \alpha = \pi/2 + n_i \pi, \qquad n_i \in \mathbb{Z}.$$

The possible α^2 are of the form

$$\alpha^2 = \frac{\Omega_i^2 t^2}{\left(n_i + \frac{1}{2}\right)^2 \pi^2}, \qquad n_i \in \mathbb{Z}.$$

It follows that the operator $I - L$ is positive definite if and only if $\Omega_i t < \pi/2$ for all $i = 1, \ldots d$. Moreover the Fredholm determinant of L can easily be computed by means of the equality $\cos x = \prod \left(1 - \frac{x^2}{\pi^2(n+1/2)^2}\right)$ and it is equal to $\det \cos \Omega t$.

By Theorem 3.20 the function $G : \mathbb{R}^d \to \mathbb{R}$ given by

$$G(x) = \int_{C_t} e^{\sqrt{\hbar} n(\eta)(\omega) + n(v_x)(\omega)} e^{\frac{1}{2}\langle \omega, L\omega \rangle} W(d\omega)$$

is equal to

$$\frac{1}{\sqrt{\det \cos \Omega t}} e^{\frac{1}{2}\langle \sqrt{\hbar}\eta + v_x, (I-L)^{-1}(\sqrt{\hbar}\eta + v_x)\rangle},$$

where $(I - L)^{-1}\gamma$, for $\gamma \in \mathcal{H}_t$ sufficiently regular, is given by

$$(I - L)^{-1}\gamma(s) = \Omega^{-1} \int_0^s \sin[\Omega(s - s')]\ddot{\gamma}(s')ds' + \frac{\sin(\Omega s)}{\Omega \cos(\Omega t)} \Big[\dot{\gamma}(t) +$$

$$- \int_0^t \cos[\Omega(t - s')]\ddot{\gamma}(s')ds' \Big].$$

Moreover by direct computation we see that

$$G(x) = \frac{1}{\sqrt{\det \cos \Omega t}} e^{\frac{1}{2}\langle \sqrt{\hbar}\eta, (I-L)^{-1}\sqrt{\hbar}\eta \rangle} e^{\frac{1}{2}x(-t\Omega^2 + \Omega \tan \Omega t)x} e^{\langle v_x, (I-L)^{-1}\sqrt{\hbar}\eta \rangle}.$$

By inserting this into (6.83), we have

$$F(y, \eta) = \frac{(2\pi)^{-d/2}}{\sqrt{\det \cos \Omega t}} e^{\frac{1}{2}\langle \sqrt{\hbar}\eta, (I-L)^{-1}\sqrt{\hbar}\eta \rangle}$$

$$\int_{\mathbb{R}^d} e^{\sqrt{\hbar}xy} e^{-\frac{1}{2}x(I-\Omega \tan \Omega t)x} e^{\langle v_x, (I-L)^{-1}\sqrt{\hbar}\eta \rangle} dx. \quad (6.84)$$

By imposing Eq. (6.82), that is the condition

$$\int_{\mathcal{H}} F(y/\sqrt{2}, \eta/\sqrt{2}) d|\mu_f|(y, \eta) < \infty,$$

we get

$$\int_{\mathcal{H} \times \mathcal{H}} F((x + y)/\sqrt{2}, (\gamma + \eta)/\sqrt{2}) d|(\mu_\phi \times \delta_0)|(y, \eta) d|\mu_\psi|(x, \gamma)$$

$$= \int_{\mathbb{R}^d \times \mathbb{R}^d} F((x + y)/\sqrt{2}, (x\gamma_t)/\sqrt{2}) d|\mu_\phi|(y) d|\mu_0|(x) < \infty. \quad (6.85)$$

By substituting into Eq. (6.84) we have

$$F((x + y)/\sqrt{2}, (x\gamma_t)/\sqrt{2}) = \frac{(2\pi)^{-d/2}}{\sqrt{\det \cos \Omega t}} e^{\frac{\hbar}{4}\langle x\gamma_t, (I-L)^{-1}x\gamma_t \rangle}$$

$$\int_{\mathbb{R}^d} e^{\sqrt{\frac{\hbar}{2}}z(x+y)} e^{-\frac{1}{2}z(I-\Omega \tan \Omega t)z} e^{\sqrt{\frac{\hbar}{2}}\langle v_z, (I-L)^{-1}x\gamma_t \rangle} dz$$

$$= \frac{(2\pi)^{-d/2}}{\sqrt{\det \cos \Omega t}} e^{\frac{\hbar}{4}x\frac{\sin(\Omega t)}{\Omega \cos(\Omega t)}x} \int_{\mathbb{R}^d} e^{\sqrt{\frac{\hbar}{2}}z(y-\cos(\Omega t)^{-1}x)} e^{-\frac{1}{2}z(I-\Omega \tan \Omega t)z} dz$$

$$= \frac{(2\pi)^{-d/2}}{\sqrt{\det(\cos \Omega t - \Omega \sin \Omega t)}} e^{\frac{\hbar}{4}x\frac{\sin(\Omega t)}{\Omega \cos(\Omega t)}x} e^{\frac{\hbar}{4}(y-\cos(\Omega t)^{-1}x)(I-\Omega \tan \Omega t)^{-1}(y-\cos(\Omega t)^{-1}x)}.$$

By substituting in Eq. (6.85), we obtain condition (6.81). $\qquad \square$

Theorem 6.7. *Let us assume that $\lambda < 0$ and that the time variable t satisfies conditions (6.80). Let $\phi, \psi_0 \in L^2(\mathbb{R}^d) \cap \mathcal{F}(\mathbb{R}^d)$, with $\hat{\mu}_0 = \psi_0$ and $\hat{\mu}_\phi(x) = (2\pi i\hbar)^{d/2} e^{-\frac{i}{2\hbar}|x|^2} \bar\phi(x)$. Assume in addition that the measures $\mu_0, \mu_\phi \in \mathcal{M}(\mathbb{R}^d)$ satisfy the assumption (6.81).*

Then the function $f : \mathcal{H} \to \mathbb{C}$ given by (6.78) is the Fourier transform of a bounded variation measure μ_f on $\mathcal{H}$ satisfying (6.82) and the infinite dimensional oscillatory integral (6.79) is well defined and is given by:

$$
\int_{\mathbb{R}^d \times H_t} \left(\int_{\mathbb{R}^d \times C_t} e^{ie^{i\pi/4}(x \cdot y + \sqrt{\hbar} n(\gamma)(\omega))} e^{\frac{1}{2\hbar} \int_0^t (\sqrt{\hbar}\omega(s) + x)\Omega^2(\sqrt{\hbar}\omega(s) + x) ds} \right.
$$
$$
\left. e^{-i\frac{\lambda}{\hbar} \int_0^t |\sqrt{\hbar}\omega(s) + x|^4 ds} W(d\omega) \frac{e^{-\frac{|x|^2}{2\hbar}}}{(2\pi\hbar)^{d/2}} dx \right) \mu_f(dy d\gamma). \qquad (6.86)
$$

This is also equal to

$$
(i)^{d/2} \int_{\mathbb{R}^d \times C_t} e^{\frac{1}{2\hbar} \int_0^t (\sqrt{\hbar}\omega(s) + x)\Omega^2(\sqrt{\hbar}\omega(s) + x) ds}
$$
$$
e^{-i\frac{\lambda}{\hbar} \int_0^t |\sqrt{\hbar}\omega(s) + x|^4 ds} \bar\phi(e^{i\pi/4} x) \psi_0(e^{i\pi/4}\sqrt{\hbar}\omega(t) + e^{i\pi/4}x) W(d\omega) dx. \qquad (6.87)
$$

Proof. By Lemma 6.7, the operator $B : \mathcal{H} \to \mathcal{H}$ given by (6.75) is bounded symmetric and trace class. Moreover if assumptions (6.80) are satisfied, $I - B$ is positive definite.

By Lemma 6.8 the function $f : \mathcal{H} \to \mathbb{C}$, given by (6.78) is the Fourier transform of a bounded variation measure μ_f on $\mathcal{H}$ satisfying assumptions (5.68) of Theorem 5.11, that is Eq. (6.82).

A direct computation shows that the function $V_4 : \mathcal{H} \to \mathbb{R}$,

$$
V_4(x, \gamma) = \int_0^t |\gamma(s) + x|^4 ds, \qquad (x, \gamma) \in \mathcal{H},
$$

is continuous in the norm of the Banach space B and extends to a function $\bar{V}_4$ on it.

By applying Theorems 5.11 and 5.12, the conclusion follows. $\qquad \square$

Remark 6.5. The class of states $\phi, \psi_0 \in L^2(\mathbb{R}^d) \cap \mathcal{F}(\mathbb{R}^d)$ satisfying assumption (6.81) is sufficiently rich. Indeed both ϕ and ψ_0 can be chosen in two dense subsets of the Hilbert space $L^2(\mathbb{R}^d)$.

More precisely one can take for instance $\psi_0, \phi \in \mathcal{S}(\mathbb{R}^d)$ of the form

$$
|\hat{\psi}_0(x)| = P(x) e^{-\frac{\alpha\hbar}{2} x^2}, \qquad |\hat{\phi}(x)| = Q(x) e^{-\frac{\beta\hbar}{2} x^2}, \qquad x \in \mathbb{R}^d,
$$

and with $\alpha, \beta \in \mathbb{R}^+$ and with P, Q arbitrary polynomials. Moreover α and β have to satisfy the following conditions, for all $i = 1, \ldots, d$:

$$\begin{cases} \beta - (1 - \Omega_i \tan \Omega_i t)^{-1} > 0 \\ \alpha - \frac{(1-\Omega_i \tan \Omega_i t)^{-1}}{\cos^2 \Omega_i t} + \frac{\sin \Omega_i t}{\Omega \cos \Omega_i t} > 0 \\ \left(\beta - (1 - \Omega_i \tan \Omega_i t)^{-1}\right)\left(\alpha - \frac{(1-\Omega_i \tan \Omega_i t)^{-1}}{\cos^2 \Omega_i t} + \frac{\sin \Omega_i t}{\Omega \cos \Omega_i t}\right) \\ \quad - \left(\frac{(1-\Omega_i \tan \Omega_i t)^{-1}}{\cos^2 \Omega_i t}\right)^2 > 0. \end{cases} \tag{6.88}$$

By the density of the finite linear combinations of Hermite functions in $L^2(\mathbb{R}^d)$ (see for instance [280, 281]), it is easy to see that the vectors of the above form are dense in $L^2(\mathbb{R}^d)$.

Remark 6.6. In [29, 32] the above result has been generalized to the case the quartic potential is explicitly time-dependent.

Let us consider the (Wiener Gaussian) integrals (6.86) and (6.87). According to Theorem 6.7, for $\lambda < 0$ they are equal to the infinite dimensional oscillatory integral (6.79).

By the considerations in Remark 5.9 the absolutely convergent integrals (6.86) and (6.87) are analytic functions of the complex variable λ if $Im(\lambda) > 0$, continuous in $Im(\lambda) = 0$. The following theorem shows that they solve the Schrödinger Eq. (6.70) with potential given by Eq. (6.71) in a weak sense (see also [36] for further comments).

Theorem 6.8. *Let $\lambda \in \mathbb{R}$. Let $t \in \mathbb{R}^+$ satisfying conditions (6.80). Let $\phi, \psi_0 \in L^2(\mathbb{R}^d) \cap \mathcal{F}(\mathbb{R}^d)$ satisfying the assumption (6.81).*
Then the Gaussian integral $I_t(\phi, \psi_0)$, given by

$$I_t(\phi, \psi_0) := (i)^{d/2} \int_{\mathbb{R}^d \times C_t} e^{\frac{1}{2\hbar} \int_0^t (\sqrt{\hbar}\omega(s)+x)\Omega^2(\sqrt{\hbar}\omega(s)+x)ds}$$
$$e^{-i\frac{\lambda}{\hbar}\int_0^t |\sqrt{\hbar}\omega(s)+x|^4 ds} \bar{\phi}(e^{i\pi/4}x)\psi_0(e^{i\pi/4}\sqrt{\hbar}\omega(t) + e^{i\pi/4}x)W(d\omega)dx,$$

is a quadratic form in the variables (ϕ, ψ_0) and it satisfies the Schrödinger Eq. (6.70) in the following weak sense, namely

$$I_0(\phi, \psi) = \langle \phi, \psi \rangle, \tag{6.89}$$

$$i\hbar \frac{d}{dt} I_t(\phi, \psi_0) = I_t(\phi, H\psi_0) = I_t(H\phi, \psi_0), \tag{6.90}$$

(H being given on the smooth vector $\phi \in C_0^2(\mathbb{R}^d)$ by (6.73)).

Remark 6.7. The result of Theorem 6.8 does not depend on the sign of the coupling constant λ.

Proof. Equation (6.90) follows by an application of Ito's formula (see for instance [219]). Equation (6.89), that is

$$(i)^{d/2} \int_{\mathbb{R}^d \times C_t} \bar{\phi}(e^{i\pi/4}x)\psi_0(e^{i\pi/4}x)W(d\omega)dx$$

$$= (i)^{d/2} \int_{\mathbb{R}^d} \bar{\phi}(e^{i\pi/4}x)\psi_0(e^{i\pi/4}x)dx \qquad (6.91)$$

$$= \int_{\mathbb{R}^d} \bar{\phi}(x)\psi_0(x)dx \qquad (6.92)$$

follows by the analyticity of the functions ϕ, ψ_0 and by a rotation of the integration contour in the complex plane. □

The particular case where the coupling constant λ is negative is rather interesting. Indeed in this case the potential (6.71) is unbounded from below, and the quantum Hamiltonian

$$H = -\frac{\Delta}{2m} + V$$

is not essentially self-adjoint as one can deduce by a limit point argument (see [281], Theorem X.9). In this case the quantum evolution is not uniquely determined. Nelson [265] was the first mathematician proposing Feynman path integrals as a tool defining the quantum dynamics in the case of not essentially self-adjoint Hamiltonians. In [265], by means of a generalized Trotter product formula and an analytic continuation technique, a strongly continuous contraction semigroup

$$U(t) : L^2(\mathbb{R}^d) \to L^2(\mathbb{R}^d), \qquad t \geq 0$$

is constructed and, given a $\psi \in L^2(\mathbb{R}^d)$, the vector $\psi(t) \equiv U(t)\psi$ satisfies the Schrödinger equation in a distributional way, i.e. for any $\phi \in L^2(\mathbb{R}^d)$ sufficiently regular, one has

$$i\hbar\frac{d}{dt}\langle \phi, \psi(t) \rangle = \langle H\phi, \psi(t) \rangle.$$

Even if the starting point of Nelson's derivation is a Wiener integral representation of the solution of an heat equation with imaginary potential, the evolution operators $U(t)$ are defined in an abstract way by means of a limiting procedure and, in general, a path integral representation for its matrix elements $\langle \phi, U(t)\psi \rangle$ cannot be defined (even for very regular vectors $\phi, \psi \in L^2(\mathbb{R}^d)$). A technical problem of Nelson's result, directly connected with the method of the proof (i.e. the application of the Fatou-Privaloff theorem) is a restriction to the allowed values of the mass parameter m, which cannot belong to a certain set N of Lebesgue measure 0.

In the particular case of quartic potential however, following [251], we can show that both problems of Nelson's paper can be overcome. Indeed we can prove that the infinite dimensional oscillatory integral with polynomial phase function (6.79) studied in Theorem 6.7 provides a Feynman path integral representation for the matrix elements $\langle \phi, e^{-\frac{i}{\hbar}Ht}\psi \rangle$ of the evolution operator defined by means of Nelson's method.

This result provides a link between two different approaches to the mathematical definition of Feynman path integral (the analytic continuation approach and the infinite dimensional oscillatory integral approach).

Theorem 6.9. *Let $\lambda \in \mathbb{R}$, $\lambda < 0$. Let $t \in \mathbb{R}^+$ satisfying conditions* (6.80). *Let $\phi, \psi_0 \in L^2(\mathbb{R}^d) \cap \mathcal{F}(\mathbb{R}^d)$ satisfying the assumption* (6.81).

Then the infinite dimensional oscillatory integral (6.79) *is equal to the Gaussian integral $I_t(\phi, \psi_0)$, given by Eq.* (6.87), *and the latter is equal to the inner product $\langle \phi, U(t)\psi_0 \rangle$, with $U(t)$, $t \geq 0$, strongly continuous contraction semigroup and*

$$i\hbar \frac{d}{dt}\langle \phi, U(t)\psi_0 \rangle = \langle H\phi, U(t)\psi_0 \rangle,$$

(H being given on the smooth vector $\phi \in C_0^2(\mathbb{R}^d)$ by (6.73)).

Proof. Let us consider the heat equation with complex potential

$$\begin{cases} -\hbar \frac{\partial}{\partial t}\psi(t,x) = -\frac{\hbar^2}{2m}\Delta\psi(t,x) + iV(x)\psi(t,x) \\ \psi(0,x) = \psi_0(x) \end{cases} \tag{6.93}$$

with V given by (6.71). For any $\psi_0 \in L_2(\mathbb{R}^d)$, the integral

$$U_m(t)\psi_0(x) \equiv \int_{C_t} e^{i\frac{\lambda}{\hbar}\int_0^t |\sqrt{\hbar/m}\,\omega(s)+x|^4 ds}$$

$$e^{-\frac{i}{2\hbar}\int_0^t (\sqrt{\hbar/m}\,\omega(s)+x)\Omega^2(\sqrt{\hbar}\,\omega(s)+x)ds}\psi_0(\sqrt{\hbar/m}\,\omega(t)+x)W(d\omega) \tag{6.94}$$

is convergent and defines a contraction operator $U_m(t)$, as

$$|U_m(t)\psi_0(x)| \leq \int_{C_t} |\psi_0(\sqrt{\hbar/m}\omega(t)+x)|W(d\omega)$$

$$\leq (2\pi t\hbar/m)^{-d/2}\int_{\mathbb{R}^d} e^{-\frac{m}{2t\hbar}|x-y|^2}\psi_0(y)dy = K_m(t)|\psi_0|(x), \tag{6.95}$$

where $K_m(t)$ is the heat semigroup $K_m(t) = e^{\frac{t\hbar}{2m}\Delta}$. By writing the cylindrical approximations of the Wiener integral (6.94) one has

$$U_m(t)\psi_0(x) = \lim_{n\to\infty}(K_m(t/n)M_V(t/n))^n\psi_0(x), \tag{6.96}$$

where $M_V(t)$ is the group given by the multiplication operator $M_V(t) = e^{-it\hbar V}$. By mimicking Nelson's argument [265] one can see that the limit (6.96) can be taken in L^2-sense, it defines a strongly continuous contraction semigroup $U_m(t)$ and for any $\psi_0 \in C_0^2(\mathbb{R}^d)$ the generator A_m is given by

$$A_m \psi_0 = \lim_{t \to 0} \frac{1}{t}(U_m(t)\psi_0 - \psi_0) = \left(\frac{\hbar}{2m}\Delta - \frac{i}{\hbar}V\right)\psi_0. \qquad (6.97)$$

As Δ is a negative operator, for any $t \geq 0$ $K_m(t)$ is an holomorphic operator-valued function of m in the half plane $Re(m) > 0$. It follows that for any $\psi_0 \in L^2$ and for any $n \in \mathbb{N}$, the expression

$$F_n(m) := \big(K_m(t/n)M_V(t/n)\big)^n \psi_0$$

defines an L^2-valued function holomorphic in the half plane $Re(m) > 0$ and continuous on $Re(m) \geq 0$. Since the sequence of functions $\{F_n\}_{n \in \mathbb{N}}$ is uniformly bounded on $Re(m) \geq 0$ by $\|\psi_0\|$ and converges for $m > 0$, by Vitali's theorem it converges on the whole domain $Re(m) > 0$ and the limit

$$\lim_{n \to \infty} (K_m(t/n)M_V(t/n))^n \psi_0 \equiv U_m(t)\psi_0$$

defines an holomorphic L^2-valued function $U_m(t)\psi_0$ on $Re(m) > 0$. By analytic continuation, one can prove that $U_m(t)$ is a strongly continuous contraction semigroup whose generator is given on vectors $\psi_0 \in C_0^2(\mathbb{R}^d)$ by Eq. (6.97).

The purely quantum mechanical - Schrödinger case is obtained for m in (6.128) purely imaginary, i.e. $m = -i$. Let us consider ϕ, ψ_0 satisfying the assumptions of the theorem. For $m > 0$, the inner product $\langle \phi, U_m(t)\psi \rangle$ is given by

$$\int_{\mathbb{R}^d \times C_t} e^{-\frac{i}{2\hbar} \int_0^t (\sqrt{\hbar/m}\,\omega(s)+x)\Omega^2(\sqrt{\hbar/m}\,\omega(s)+x)ds}$$
$$e^{i\frac{\lambda}{\hbar} \int_0^t |\sqrt{\hbar/m}\,\omega(s)+x|^4 ds} \psi_0(\sqrt{\hbar/m}\,\omega(t) + x)\bar\phi(x)W(d\omega)dx. \qquad (6.98)$$

By a change of variable $x \mapsto x/\sqrt{m}$ the latter becomes

$$\langle \phi, U_m(t)\psi_0 \rangle = m^{-d/2} \int_{\mathbb{R}^d \times C_t} e^{-\frac{i}{2\hbar m} \int_0^t (\sqrt{\hbar}\,\omega(s)+x)\Omega^2(\sqrt{\hbar}\,\omega(s)+x)ds}$$
$$e^{i\frac{\lambda}{m^2\hbar} \int_0^t |\sqrt{\hbar}\,\omega(s)+x|^4 ds} \psi_0(\sqrt{\hbar/m}\,\omega(t) + x/\sqrt{m})\bar\phi(x/\sqrt{m})W(d\omega)dx. \qquad (6.99)$$

By assumptions (6.80) and (6.81), the right-hand side of (6.99) is an holomorphic function of m in the domain $\{Re(m) > 0\} \cap \{Im(m) < 0\}$ and continuous on the boundary. On the other hand, by previous considerations, the matrix element $\langle \phi, U_m(t)\psi_0 \rangle$ is an holomorphic function of m in the domain $\{Re(m) > 0\}$ and coincides with the functional integral (6.99) on the half line $m > 0$. By the uniqueness of analytic continuation, both sides of (6.99) coincides on the domain $\{Re(m) > 0\} \cap \{Im(m) < 0\}$. In particular there exists the limit

$$\lim_{m \to -i} \langle \phi, U_m(t)\psi_0 \rangle$$

and, by bounded convergence theorem, it is equal to (6.87).   □

Remark 6.8. The results of Theorems 6.7, 6.8 and 6.9 hold also in the case the potential V is of the form:

$$V(x) = \lambda|x|^4 - \frac{1}{2}x\Omega^2 x, \qquad x \in \mathbb{R}^d$$

with $\lambda \in \mathbb{R}$ and Ω^2 a positive symmetric $d \times d$ matrix. In this case conditions (6.80) can be dropped and condition (6.81) has to be replaced by

$$\int_{\mathbb{R}^d} \int_{\mathbb{R}^d} e^{\frac{\hbar}{4}(y - \cosh(\Omega t)^{-1} x)(1 + \Omega \tanh(\Omega t))^{-1}(y - \cosh(\Omega t)^{-1} x)}$$

$$e^{-\frac{\hbar}{4}x\Omega^{-1} \tanh(\Omega t)x} |\mu_0|(dx)|\mu_\phi|(dy) < \infty.$$

6.5   Magnetic field

Let us consider the Schrödinger equation for a non-relativistic quantum particle moving under the influence of a magnetic field $\mathbf{B}$ associated to a vector potential $\mathbf{a}$

$$i\hbar \frac{\partial}{\partial t}\psi(t, x) = \frac{1}{2}\left(-i\hbar\nabla - \lambda\mathbf{a}(x)\right)^2 \psi(t, x), \tag{6.100}$$

where $\mathbf{a}(x) \in \mathbb{R}^d$, $\mathbf{B} = \mathrm{rot}\, \mathbf{a}$, and $\lambda \in \mathbb{R}$ plays the role of a coupling constant. As mentioned in the introduction, the traditional construction of the heuristic Feynman path integral representation for the solution of Eq. (6.100)

$$\psi(t, x) = \int_{\gamma(t)=x} e^{\frac{i}{2\hbar}\int_0^t |\dot\gamma(s)|^2 ds + \frac{i}{\hbar}\int_0^t \lambda\mathbf{a}(\gamma(s))\cdot\dot\gamma(s)ds}\psi_0(\gamma(0))d\gamma, \tag{6.101}$$

based on the time-slicing approximation

$$\int e^{\frac{i}{2\hbar}\sum_i \frac{|\gamma(t_{i+1}) - \gamma(t_i)|^2}{t_{i+1} - t_i} + \frac{i}{\hbar}\sum_i \lambda\mathbf{a}(\gamma(\tilde{t}_i))\cdot(\gamma(t_{i+1}) - \gamma(t_i))}$$

$$\psi_0(\gamma(0)) \prod_i \frac{d\gamma(t_i)}{(2\pi i\hbar(t_{i+1} - t_i))^{1/2}},$$

(with $0 = t_0 < t_1 < t_2 < \cdots < t_n = t$ and $\tilde{t}_i \in [t_i, t_{i+1}]$, $i = 0, \ldots, n-1$) presents some ambiguities since different choices of the point $\tilde{t}_i \in [t_i, t_{i+1}]$ lead to different results. The correct choice relies on the so-called mid-point rule which requires that in the formula above the vector potential $\mathbf{a}$ is evaluated at the point $\tilde{t}_i \equiv \frac{\gamma(t_{i+1}) + \gamma(t_i)}{2}$. In the Euclidean version of Feynman formula, namely the Feynman-Kac-Itô formula [302] for the solution of the corresponding heat equation in a magnetic field, this procedure yields the Wiener integral representation

$$u(t, x) = \mathbb{E}\left[u(x + \omega(t)) e^{-i \int_0^t \lambda \mathbf{a}(\omega(s) + x) \circ d\omega(s)} \right], \qquad (6.102)$$

where $\circ d\omega(s)$ denotes the Stratonovich stochastic integral [194] and $\mathbb{E}$ is the expectation with respect to Wiener measure. In fact other choices for the point $\tilde{t}_i \in [t_i, t_{i+1}]$ would lead to different stochastic integrals but, as pointed out in [302, 296], the *mid point rule*, or, equivalently, the Stratonovich stochastic integration, is the only one yielding the *gauge invariance* of representation formulas (6.101) and (6.102).

In the case where $\mathbf{B}$ is constant and the vector potential satisfies the Coulomb gauge, then Feynman path integral representation for the fundamental solution $G(t, x, y)$ of Eq. (6.100) is rigorously defined in terms of infinite dimensional oscillatory integrals [7, 9].

The case of a rather general magnetic field, without any requirement on the gauge choice, has been recently studied in [11]. In the following we shall assume that the components a_j, $j = 1, 2, 3$, of the vector potential $\mathbf{a}$ are real valued functions and belong to the space $\mathcal{F}_c(\mathbb{R}^3)$ of Fourier transforms of complex Borel measures μ_j on $\mathbb{R}^3$ with compact support, i.e. they are functions of the form

$$a_j(x) = \int_{\mathbb{R}^3} e^{iyx} d\mu_j(y), \qquad x \in \mathbb{R}^3, \quad j = 1, 2, 3, \qquad (6.103)$$

with μ_j of compact support. In particular, this condition implies that any function a_j has an analytic continuation to a function on $\mathbb{C}^3$, denoted (with an abuse of notation) with the same symbol. Under this assumption, it is possible to prove (see, e.g. [270, 302, 101, 74]) that the quantum Hamiltonian operator H given on smooth compactly supported vector $\psi \in C_c^\infty(\mathbb{R}^3)$ by

$$H\psi = \frac{1}{2}(-i\hbar\nabla - \lambda\mathbf{a}(x))^2 \psi, \qquad \psi \in C_c^\infty(\mathbb{R}^3)$$

is positive, symmetric and, $\mathbf{a}$ being bounded, has a unique self-adjoint extension $H : D(H) \subset L^2(\mathbb{R}^3) \to L^2(\mathbb{R}^3)$, with domain

$$D(H) = \left\{ \psi \in L^2(\mathbb{R}^3) : \int_{\mathbb{R}^3} |y|^4 |\hat{\psi}(y)|^2 dk < +\infty \right\}, \qquad (6.104)$$

where $\hat{\psi} \in L^2(\mathbb{R}^3)$ denotes the Fourier transform of the vector $\psi \in L^2(\mathbb{R}^3)$ and $|y|$ is the norm of the vector $y \in \mathbb{R}^3$.

By Stone's theorem, H generates a one-parameter group $U(t) = e^{-\frac{i}{\hbar}Ht}$, $t \in \mathbb{R}$, of unitary operators on $L^2(\mathbb{R}^3)$, solving the Schrödinger equation in the following sense:

$$i\hbar\partial_t U(t)\psi_0 = HU(t)\psi_0, \qquad \psi_0 \in D(H) \qquad (6.105)$$

(where the derivation on the left is a strong one in $L^2(\mathbb{R}^3)$).

We shall assume, without loss of generality, that the vector potential **a** satisfies the Coulomb gauge, namely that div $\mathbf{a} = 0$. In this case the Hamiltonian operator H can be written as $H = H_0 + W$, where H_0 is the free Hamiltonian, namely the operator $H_0 = -\frac{\hbar^2}{2}\Delta$ on $D(H_0) = D(H)$ (defined by (6.104)) and $W = \lambda A + \lambda^2 B$, where $A = i\hbar\mathbf{a} \cdot \nabla$ and B is the multiplication operator associated to the function $\frac{1}{2}|\mathbf{a}^2|$ (both well defined on $D(H_0)$). Under the assumption that the initial datum $\psi_0 \in L^2(\mathbb{R}^3)$ has a compactly supported Fourier transform, i.e. $\psi_0 \in \mathcal{F}_c(\mathbb{R}^3) \cap L^2(\mathbb{R}^3)$, the Dyson series expansion (in powers of the coupling constant λ for the vector $U(t)\psi_0$ has a finite radius of convergence as the following theorem shows.

Theorem 6.10. *Let us assume that the Fourier transform $\hat{\psi}_0$ of the vector $\psi_0 \in L^2(\mathbb{R}^3)$ has a compact support included in the ball $B_\rho \equiv \{x \subset \mathbb{R}^3 : |x| < \rho\}$ and that the component a_j, $j = 1,\ldots,3$, of the vector potentials **a** are of the form (6.103), with μ_j, $j = 1,\ldots,3$, bounded Borel measures with support contained in the ball B_R, for some $R \in \mathbb{R}^+$. Then the expansion in powers of the coupling constant λ for the vector $U(t)\psi_0$, namely $U(t)\psi_0 = \sum_m \lambda^m \phi_m(t)$, with*

$$\phi_m(t) = \sum_{(n,k)\in\mathbb{N}^2 :\, 2n-k=m,\, k\leq n} \left(-\frac{i}{\hbar}\right)^n \phi_{n,k}$$

and $\phi_{n,k}$ given by (6.108) below, converges in $L^2(\mathbb{R}^3)$ for $|\lambda| < \lambda^*$, with

$$\lambda^* = \left(\frac{2\alpha^2 t}{\hbar}\left(2r^2 t\hbar + 1\right)\right)^{-1/2}, \qquad (6.106)$$

where $\alpha = \sup_{x\in\mathbb{R}^3}|\mathbf{a}(x)|$ and $r = \max\{\rho, R\}$.

We start by proving the following lemma.

Lemma 6.9. *Under the assumptions of Theorem 6.10, the following holds:*

$$\|U_0(t_1)O_1 U_0(t_2)O_2 \cdots U_0(t_n)O_n U_0(t_{n+1})\psi_0\|$$

$$\leq \hbar^k \alpha^k \left(\frac{\alpha^2}{2}\right)^{n-k} \prod_{j=0}^{k-1}(\rho + 2R(n-k) + jR)\|\psi_0\|,$$

where $U_0(t) = e^{-\frac{i}{\hbar}H_0 t}$, $O_j \in \{A, B\}$, $j = 1, \ldots, n$, $k = \#\{j : O_j = A\}$, with $A = i\hbar \mathbf{a} \cdot \nabla$, $B = \frac{1}{2}|\mathbf{a}|^2$ and $\alpha = \sup_{x \in \mathbb{R}^3} |\mathbf{a}(x)|$.

Proof. Let $\phi \in L^2(\mathbb{R}^3)$ be a function whose Fourier transform $\hat{\phi}$ has support in a ball B_ρ centered at the origin with radius $\rho \in \mathbb{R}^+$. Then the following holds.

- For any $t \in \mathbb{R}$, the vector $U_0(t)\phi \in L^2(\mathbb{R}^3)$ has a Fourier transform with support contained in B_ρ. Indeed $\widehat{U_0(t)\phi}$ is simply given by

$$\widehat{U_0(t)\phi}(y) = e^{-\frac{i}{\hbar}|y|^2 t}\hat{\phi}(y), \qquad y \in \mathbb{R}^3.$$

- The vector $B\phi \in L^2(\mathbb{R}^3)$ has a Fourier transform with support contained in $B_{\rho+2R}$. Indeed, under the assumptions on the components a_j of the vector field $\mathbf{a}$, the function $x \mapsto |\mathbf{a}|^2$ is Fourier transform of a Borel measure $\mu_{\mathbf{a}^2}$ on $\mathbb{R}^3$ with support contained in the ball B_{2R}, with $\mu_{\mathbf{a}^2} = \sum_{j=1}^{3} \mu_j * \mu_j$, where the symbol $\mu * \nu$ stands for the convolution of the measures μ and ν. It is simple to verify that if the support of the measure μ_j is contained in B_R, then the support of the convolution $\mu_j * \mu_j$ is contained in B_{2R}. Correspondingly, the Fourier transform of $B\phi$ is given by

$$\widehat{B\phi}(y) = \frac{1}{2}\int_{\mathbb{R}^3} \hat{\phi}(y - y')d\mu_{\mathbf{a}^2}(y'), \qquad y \in \mathbb{R}^3,$$

  and its support is contained in $B_{\rho+2R}$.
- The norm of the vector $B\phi$ is bounded by

$$\|B\phi\| \leq \frac{1}{2}\|\mathbf{a}^2\|_\infty\|\phi\|,$$

  where $\|\mathbf{a}^2\|_\infty = \sup_{x \in \mathbb{R}^3} |\mathbf{a}(x)|^2$, which is finite by the assumptions on the components a_j, $j = 1, \ldots, 3$.
- The vector $A\phi \in L^2(\mathbb{R}^3)$ (with A as in Lemma 6.9) has a Fourier transform with support contained in $B_{\rho+R}$, given by

$$\widehat{A\phi}(y) = -\hbar\sum_{j=1}^{3}\int_{\mathbb{R}^3}(y_j - y_j')\hat{\phi}(y - y')d\mu_j(y'), \qquad k \in \mathbb{R}^3.$$

  Moreover, the norm of $A\phi$ satisfies the following bound

$$\|A\phi\| \leq \hbar\sqrt{\|\mathbf{a}^2\|_\infty}\,\rho\,\|\phi\|.$$

Now it is straightforward to verify that, if $\#\{j : O_j = A\} = k$:

$$\|U_0(t_1)O_1U_0(t_2)O_2\cdots U_0(t_n)O_nU_0(t_{n+1})\psi_0\|$$

$$\leq \|U_0(t_1)AU_0(t_2)\cdots U_0(t_k)AU_0(t_{k+1})B\cdots U_0(t_n)BU_0(t_{n+1})\psi_0\|$$

$$\leq \hbar^k\alpha^k\left(\frac{\alpha^2}{2}\right)^{n-k}\prod_{j=0}^{k-1}(\rho+2R(n-k)+jR)\|\psi_0\|.$$

$\square$

Proof (of Theorem 6.10). By the classical Dyson-Phillips expansion for the vector $U(t)\psi_0$ [270], we have

$$U(t)\psi = \sum_{n=0}^{\infty}\left(-\frac{i}{\hbar}\right)^n\int_{\Delta_n(t)}e^{-\frac{i}{\hbar}H_0(t-s_n)}We^{-\frac{i}{\hbar}H_0(s_n-s_{n-1})}$$

$$\cdots We^{-\frac{i}{\hbar}H_0(s_2-s_1)}We^{-\frac{i}{\hbar}H_0s_1}\psi_0ds_1\ldots ds_n$$

$$= \sum_{n=0}^{\infty}\left(-\frac{i}{\hbar}\right)^n\int_{\Delta_n(t)}U_0(t-s_n)(\lambda A+\lambda^2B)U_0(s_n-s_{n-1})$$

$$\cdots(\lambda A+\lambda^2B)U_0(s_1)\psi_0ds_1\ldots ds_n,$$

where $\Delta_n(t)\subset\mathbb{R}^n$ is the n-dimensional simplex defined as $\Delta_n(t) = \{(s_1,\ldots,s_n)\in\mathbb{R}^n : 0\leq s_1\leq\cdots\leq s_n\leq t\}$ and $W = \lambda A+\lambda^2B$. The dependence on the coupling constant λ can be made explicit as

$$U(t)\psi_0 = \sum_{n=0}^{\infty}\left(-\frac{i}{\hbar}\right)^n\sum_{k=0}^{n}\lambda^k(\lambda^2)^{n-k}\phi_{n,k}, \qquad (6.107)$$

where the term $\phi_{n,k}\in L^2(\mathbb{R}^3)$ is a sum of $\binom{n}{k}$ terms of the form

$$\int_{\Delta_n(t)}U_0(t-s_n)O_1U_0(s_n-s_{n-1})O_2\cdots O_nU_0(s_1)\psi_0ds_1\ldots ds_n,$$

where we recall that $O_j = A, B$, $j = 1,\ldots,n$ and $\#\{j : O_j = A\} = k$. More precisely:

$$\phi_{n,k} = \sum_{E\in C_k^n}\int_{\Delta_n(t)}U_0(t-s_n)O_1^EU_0(s_n-s_{n-1})O_2^E\cdots O_n^EU_0(s_1)\psi_0ds_1\ldots ds_n,$$

$$(6.108)$$

where the sum is taken over the set C_k^n of all possible subsets $E\subset\{1,\ldots,n\}$ with k elements and the map $O^E : \{1,\ldots,n\}\to\{A,B\}$ is defined as $O_i^E := A$ if $i\in E$ and $O_i^E := B$ if $i\notin E$.

By Lemma 6.9 we have

$$\|\phi_{n,k}\| \leq \binom{n}{k}\frac{t^n}{n!}\hbar^k\alpha^k\left(\frac{\alpha^2}{2}\right)^{n-k}\prod_{j=0}^{k-1}(\rho+2R(n-k)+jR)\|\psi_0\|. \quad (6.109)$$

In particular, by setting $r := \max\{\rho,R\}$, we obtain:

$$\|\phi_{n,k}\| \leq \frac{t^n}{(n-k)!}\hbar^k\alpha^{2n-k}\left(\frac{1}{2}\right)^{n-k}r^k\binom{2n-k}{k}\|\psi_0\|. \quad (6.110)$$

Now, the sum appearing in (6.107) can be written as

$$U(t)\psi_0 = \sum_{m=0}^{\infty} \lambda^m \sum_{(n,k)\in\mathbb{N}^2\,:\,2n-k=m,\,k\leq n} \left(-\frac{i}{\hbar}\right)^n \phi_{n,k}$$

$$= \sum_m \lambda^m \phi_m, \qquad (6.111)$$

where

$$\phi_m = \left(-\frac{i}{\hbar}\right)^{m/2} \sum_{h=0}^{m/2} \left(-\frac{i}{\hbar}\right)^h \phi_{h+\frac{m}{2},2h}, \qquad m \text{ even};$$

$$\phi_m = \left(-\frac{i}{\hbar}\right)^{(m+1)/2} \sum_{h=0}^{(m-1)/2} \left(-\frac{i}{\hbar}\right)^h \phi_{h+\frac{m+1}{2},2h+1}, \qquad m \text{ odd}.$$

$$(6.112)$$

By estimate (6.110) we have for m even, namely $m = 2M$, $M \in \mathbb{N}$:

$$\|\phi_{2M}\| \leq \left(\frac{\alpha^2 t}{2\hbar}\right)^M \sum_{h=0}^{M} \left(2r^2 t\hbar\right)^h \frac{1}{(M-h)!} \binom{2M}{2h} \|\psi_0\|$$

$$\leq \left(\frac{\alpha^2 t}{2\hbar}\right)^M \left(2r^2 t\hbar + 1\right)^M \max_{h\in\{0,\dots,M\}} \frac{h!}{M!} \binom{2M}{2h} \|\psi_0\|$$

$$\leq \left(\frac{2\alpha^2 t}{\hbar} \left(2r^2 t\hbar + 1\right)\right)^M \|\psi_0\|;$$

analogously, for m odd, namely $m = 2M + 1$, $M \in \mathbb{N}$ we get

$$\|\phi_{2M+1}\| \leq 2rt\alpha\hbar \left(\frac{2\alpha^2 t}{\hbar} \left(2r^2 t\hbar + 1\right)\right)^M \|\psi_0\|.$$

Hence,

the series (6.111) converges in $L^2(\mathbb{R}^d)$ for $|\lambda| < \left(\frac{2\alpha^2 t}{\hbar} \left(2r^2 t\hbar + 1\right)\right)^{-1/2}$.

$\square$

Remark 6.9. Since the Hamiltonian operator H is self-adjoint and positive, as pointed out at the beginning of this section, it generates an analytic semigroup. Hence, for $z \in \mathbb{C}$ belonging to the closure $\bar{D}$ of the open sector $D \subset \mathbb{C}$ of the complex plane defined as

$$D = \{z \in \mathbb{C} : \mathrm{Re}(z) > 0\},$$

it is possible to define the operator $V(z) = e^{-zH}$ yielding for $z = \frac{i}{\hbar}t$ and $t \in \mathbb{R}$ the Schrödinger group and for $z \in \mathbb{R}^+$ the heat semigroup. In both cases, under the assumptions of Theorem 6.10 the perturbative Dyson expansion for the vector $V(z)\psi_0$ has a positive radius of convergence (depending on $|z|$). Indeed, if $z \in D$, the Dyson expansion can be written as

$$e^{-zH}\psi_0 = \sum_{n=0}^{\infty}(-z)^n \int_{\Delta_n} e^{-zH_0(1-s_n)}We^{-zH_0(s_n-s_{n-1})}\cdots$$
$$We^{-zH_0(s_2-s_1)}We^{-zH_0 s_1}\psi_0 ds_1 \ldots ds_n \quad (6.113)$$

with $\Delta_n \equiv \Delta_n(1) = \{(s_1,\ldots,s_n) \in \mathbb{R}^n : 0 \le s_1 \le \cdots \le s_n \le 1\}$ and $W = \lambda A + \lambda^2 B$. By collecting in the sum (6.113) all the terms associated to the same power of the coupling constant λ, we get

$$e^{-zH}\psi_0 = \sum_m \lambda^m \phi_m(z) \quad (6.114)$$

where $\phi_m(z) = \sum_{(n,k)\in\mathbb{N}^2 : 2n-k=m, k\le n}(-z)^n\phi_{n,k}(z)$ with

$$\phi_{n,k}(z) = \sum_{E\in C_k^n}\int_{\Delta_n} e^{-zH_0(t-s_n)}O_1^E e^{-zH_0(s_n-s_{n-1})}O_2^E \cdots$$
$$O_n^E e^{-zH_0 s_1}\psi_0 ds_1 \ldots ds_n, \quad (6.115)$$

where, analogously to Eq. (6.108), the sum is taken over the set C_k^n of all possible subsets $E \subset \{1,\ldots,n\}$ with k elements and the map $O^E : \{1,\ldots,n\} \to \{A,B\}$ is defined as $O_i^E := A$ if $i \in E$ and $O_i^E := B$ if $i \notin E$.

By repeating the arguments in the proof of Theorem 6.10, it is now easy to verify that the expansion (6.114) converges in $L^2(\mathbb{R}^3)$ for $|\lambda| < \lambda^*(z)$, with

$$\lambda^*(z) = \left(2\alpha^2|z|\left(2r^2\hbar^2|z| + 1\right)\right)^{-1/2}, \qquad z \in D. \quad (6.116)$$

Remark 6.10. The results of Theorem 6.10 and Remark 6.9 can easily be extended to the case where a (bounded) scalar potential V is added to the Hamiltonian. Indeed, let us consider the following

$$H\psi(x) = \frac{1}{2}\left(-i\hbar\nabla - \lambda\mathbf{a}(x)\right)^2\psi(x) + \lambda V(x)\psi(x), \qquad \psi \in C_0^\infty(\mathbb{R}^3),$$
$$(6.117)$$

where $V \in \mathcal{F}_c(\mathbb{R}^3)$, i.e. $V : \mathbb{R}^3 \to \mathbb{R}$ a function of the form:

$$V(x) = \int_{\mathbb{R}^3} e^{ixy}d\mu_V(y), \qquad x \in \mathbb{R}^3, \quad (6.118)$$

with μ_V complex Borel measure with support contained in the ball B_R. Under the assumptions of Theorem 6.10, Lemma 6.9 still holds. In particular, by setting $A = i\hbar\mathbf{a}\cdot\nabla + V$, $B = \frac{1}{2}|\mathbf{a}|^2$, $\alpha = |\mathbf{a}|_\infty$ and $\tilde{\alpha} = 2\max\{\hbar|\mathbf{a}|_\infty, |V|_\infty\}$, we get:

$$\|U_0(t_1)O_1U_0(t_2)O_2\cdots U_0(t_n)O_nU_0(t_{n+1})\psi_0\|$$

$$\leq \tilde{\alpha}^k\left(\frac{\alpha^2}{2}\right)^{n-k}\prod_{j=0}^{k-1}(\rho + 2R(n-k) + jR)\|\psi_0\|, \quad (6.119)$$

where $U_0(t) = e^{-\frac{i}{\hbar}H_0t}$, $O_j \in \{A, B\}$, $j = 1,\ldots,n$, $k = \#\{j\colon O_j = A\}$. By using (6.119) it is now possible to repeat the proof of Theorem 6.10, obtaining the convergence in $L^2(\mathbb{R}^3)$ of the perturbative Dyson expansion for the vector $e^{-\frac{i}{\hbar}Ht}\psi_0$ for $\lambda < \tilde{\lambda}$, where

$$\tilde{\lambda} = \left(\frac{2\tilde{\alpha}^2t}{\hbar}\left(\frac{2r^2t}{\hbar} + 1\right)\right)^{-1/2}. \quad (6.120)$$

Let us consider now the Cameron-Martin space $\mathcal{H}_t$ and the associated abstract Wiener space $(\mathcal{H}_t, C_t, i)$ (see Example 3.7 in Section 3.7). For fixed $n \in \mathbb{N}$, let $H_n \subset \mathcal{H}_t$ be the finite dimensional subspace of piecewise linear paths of the form

$$\gamma(s) = \sum_{k=1}^{n}\chi_{[t_{k-1},t_k]}(s)\left(\gamma(t_{k-1}) + \frac{\gamma(t_k) - \gamma(t_{k-1})}{t_k - t_{k-1}}(s - t_{k-1})\right), \quad (6.121)$$

where $s \in [0, t]$, $t_k = \frac{kt}{n}$ and $k = 0,\ldots,n$, and let $P_n : \mathcal{H}_t \to \mathcal{H}_t$ the projector operator onto H_n, whose action on a generic vector $\gamma \in \mathcal{H}_t$ is given by the right-hand side of (6.121). Correspondingly, let $\{\tilde{P}_n\}_n$ be the sequence of random variables $\tilde{P}_n : C_t \to \mathcal{H}_t$ given by

$$\tilde{P}_n(\omega)(s) = \sum_{k=1}^{n}\chi_{[t_{k-1},t_k]}(s)\left(\omega(t_{k-1}) + \frac{\omega(t_k) - \omega(t_{k-1})}{t_k - t_{k-1}}(s - t_{k-1})\right), \quad s \in [0, t],$$

$$(6.122)$$

with $\omega \in C_t$ and $t_k = kt/n$, $k = 1,\ldots,n$ as above. For future use, it is convenient to introduce the shortened notation $\tilde{P}_n(\omega)(s) \equiv \omega_n(s)$. Actually the map $s \mapsto \omega_n(s)$, $s \in [0, t]$, can be regarded as a piecewise smooth approximation of the Brownian path $\omega \in C_t$.

The following technical lemma provides a convergence result for a sequence of random variable constructed out of the vector potential $\mathbf{a}$ and the sequence $\{\tilde{P}_n\}$.

Lemma 6.10. *Let* $\mathbf{a}$ *be a three dimensional vector field fulfilling the assumptions of Theorem 6.10. Let* $\{f_n\}$ *be the sequence of random variables* $f_n : C_t \to \mathbb{C}$ *defined by*

$$f_n(\omega) = \int_0^t \mathbf{a}\left(\sqrt{i\hbar}\omega_n(s)\right)\cdot\dot{\omega}_n(s)ds,$$

where $\omega_n(s) \equiv P_n(\omega)(s)$ and $P_n(\omega)$ is defined by the right-hand side of
(6.122).

$$\omega_n(s) = \sum_{k=1}^{n} \chi_{[t_{k-1}, t_k]}(s) \left(\omega(t_{k-1}) + \frac{\omega(t_k) - \omega(t_{k-1})}{t_k - t_{k-1}} (s - t_{k-1}) \right).$$

Then for any $p \in \mathbb{N}$, $1 \leq p \leq \infty$, f_n converges, as $n \to \infty$, in $L^p(C_t, \mathbb{P})$ to the random variable f defined as the Stratonovich stochastic integral

$$f(\omega) = \int_0^t \mathbf{a}(\sqrt{i\hbar}\omega(s)) \circ d\omega(s).$$

where $\circ d\omega(s)$ denotes the Stratonovich integral.

For the proof we refer to the original paper [11]. For definition and properties of the Stratonovich stochastic integral see, e.g., [194].

It is convenient to introduce now the definition of *Feynman map*, originally proposed by A. Truman [323]. It is defined as a linear functional M_F whose action on functions $f : \mathcal{H}_t \to \mathbb{C}$ is given by

$$M_F(f) = \lim_{n \to \infty} \frac{\int_{P_n \mathcal{H}}^o e^{i \frac{\|P_n \gamma\|^2}{2\hbar}} f(P_n \gamma) d(P_n \gamma)}{\int_{P_n \mathcal{H}}^o e^{i \frac{\|P_n \gamma\|^2}{2\hbar}} d(P_n \gamma)}, \tag{6.123}$$

whenever the limit on the right-hand side exists, where $\{P_n\}$ is the sequence of projectors operators onto the spaces H_n of piecewise linear paths. It is important to remark that M_F is not an infinite dimensional oscillatory integral on $\mathcal{H}_t$, the latter being described by Definition 5.4 requiring the existence of the limit on the right-hand side of (6.123) for *any* increasing sequence of projectors $\{P_n\}$ and the *independence* of the limit of $\{P_n\}$.

Lemma 6.11. *Let the vector field $\mathbf{a}$ and the function $\psi_0 \in L^2(\mathbb{R}^3)$ satisfy the assumptions of Theorem 6.10. Then for any $m \geq 0$, the Feynman map of the function $g_m^x : \mathcal{H}_t \to \mathbb{C}$ defined as*

$$g_m^x(\gamma) := \psi_0(\gamma(t) + x) \left(\int_0^t \mathbf{a}(\gamma(s) + x) \cdot \dot{\gamma}(s) ds \right)^m \tag{6.124}$$

is given by

$$M_F(g_m^x) = \int_{C_t} \left(\sqrt{i\hbar} \int_0^t \mathbf{a} \left(\sqrt{i\hbar}\omega(s) + x \right) \circ d\omega(s) \right)^m \psi_0 \left(\sqrt{i\hbar}\omega(t) + x \right) dW(\omega),$$
$$\tag{6.125}$$

where W stands for the Wiener measure on $(C_t, \mathcal{B}(C_t))$ and $\int_0^t \mathbf{a}(\sqrt{i\hbar}\omega(s) + x) \circ d\omega(s)$ denotes the Stratonovich stochastic integral.

For the proof we refer to [11].

Theorem 6.11. *Under the assumption of Theorem 6.10, the solution of the Schrödinger equation with magnetic field*

$$i\hbar\partial_t\psi(t) = H\psi(t,x), \quad \psi(0,x) = \psi_0(x), \quad H = \frac{1}{2}(-i\hbar\nabla - \lambda\mathbf{a}(x))^2$$

can be expressed by the perturbative Dyson series expansion as

$$e^{-\frac{i}{\hbar}Ht}\psi_0 = \sum_{m=0}^{\infty} \lambda^m \psi_m(t),$$

where the vector ψ_m can be expressed by the Feynman map $M_F(g_m^x)$ of the function g_m^x defined by (6.124) and it is equal to the Wiener integral (6.125):

$$\psi_m(t,x) = \frac{1}{m!}\left(-\frac{i}{\hbar}\right)^m M_F(g_m^x)$$

$$= \frac{1}{m!}\left(-\frac{i}{\hbar}\right)^m \mathbb{E}\Big[\left(\sqrt{i\hbar}\int_0^t \mathbf{a}\left(\sqrt{i\hbar}\omega(s) + x\right)\circ d\omega(s)\right)^m$$

$$\psi_0\left(\sqrt{i\hbar}\omega(t) + x\right)\Big]. \quad (6.126)$$

The expansion is convergent in $L^2(\mathbb{R}^3)$ for $\lambda \in \mathbb{C}$, with $|\lambda| < \lambda^*$, λ^* given by (6.106). The integral under the expectation is to be understood as a Stratonovich stochastic integral.

Proof. By Theorem 6.10 for $|\lambda| < \lambda^*$ the vector $\psi(t) = e^{-\frac{i}{\hbar}Ht}\psi_0$ in $L^2(\mathbb{R}^3)$ is given by the convergent power series expansion (6.111). Hence, we are left to prove that for any $m \in \mathbb{N}$ the term ϕ_m in (6.112) is equal to ψ_m as given in (6.126).

By Remark 6.9, the Hamiltonian operator H generates an analytic semigroup $e^{-zH} : L^2(\mathbb{R}^3) \to L^2(\mathbb{R}^3)$, where $z \in \mathbb{C}$, $Re(z) \geq 0$, with a convergent Dyson expansion of the form $e^{-zH}\psi_0 = \sum_m \lambda^m \phi_m(z)$ with a radius of convergence $\lambda^*(z)$ depending on $|z|$ (see Eq. (6.116)). In particular, for $z \in \mathbb{R}^+$, namely $z = \frac{t}{\hbar}$, the family of operators $T(t) = e^{-\frac{t}{\hbar}H}$, $t \in \mathbb{R}^+$, yields the heat semigroup generated by H. In this case, by Feynman-Kac-Itô formula (6.102) the vector $e^{-\frac{t}{\hbar}H}\psi_0$ is given by the Wiener integral [302]

$$e^{-\frac{t}{\hbar}H}\psi_0(x) = \mathbb{E}\left[\psi_0(\sqrt{\hbar}\omega(t) + x)e^{-\frac{i\lambda}{\hbar}\int_0^t \mathbf{a}(\sqrt{\hbar}\omega(s)+x)\circ d\omega(s)}\right]. \quad (6.127)$$

For any $\phi \in L^2(\mathbb{R}^3)$ the inner product $\langle\phi, e^{-zH}\psi_0\rangle$ is an analytic function of $z \in D$, $D = \{z \in \mathbb{C}, Re(z) \geq 0\}$, continuous in the closure $\bar{D}$ of D and admitting the expansions

$$\langle\phi, e^{-zH}\psi_0\rangle = \sum_{m=0}^{\infty} \lambda^m \langle\phi, \phi_m(z)\rangle.$$

By formula (6.115) each term $\langle \phi, \phi_m(z) \rangle$ is an analytic function of $z \in D$, continuous in the closure $\bar{D}$ and for $z = t/\hbar$, $t \in \mathbb{R}^+$, by formula (6.127) it is equal to

$$
\langle \phi, \phi_m(t/\hbar) \rangle
$$
$$
= \left(-\frac{i}{\hbar} \right)^m \int_{\mathbb{R}^3} \bar{\phi}(x) \mathbb{E} \left[\psi_0(\sqrt{\hbar}\omega(t) + x) \left(\sqrt{\hbar} \int_0^t \mathbf{a}(\sqrt{\hbar}\omega(s) + x) \circ d\omega(s) \right)^m \right] dx.
$$
(6.128)

By replacing t with $t\xi$, with $\xi \in \mathbb{R}^+$, the expression above assumes the following form:

$$
\left(-\frac{i}{\hbar} \right)^m \int_{\mathbb{R}^3} \bar{\phi}(x) \mathbb{E} \left[\psi_0(\sqrt{\hbar}\omega(t\xi) + x) \left(\sqrt{\hbar} \int_0^{t\xi} \mathbf{a}(\sqrt{\hbar}\omega(s) + x) \circ d\omega(s) \right)^m \right] dx,
$$

and by a change of variable we obtain

$$
\langle \phi, \phi_m(t\xi/\hbar) \rangle
$$
$$
= \left(-\frac{i}{\hbar} \right)^m \int_{\mathbb{R}^3} \bar{\phi}(x) \mathbb{E} \left[\psi_0(\sqrt{\hbar\xi}\omega(t) + x) \left(\sqrt{\xi\hbar} \int_0^t \mathbf{a}(\sqrt{\hbar\xi}\omega(s) + x) \circ d\omega(s) \right)^m \right] dx.
$$
(6.129)

By the discussion above, both the right-hand side and the left-hand side of (6.129) are analytic for $\xi \in D$ and continuous for $\xi \in \bar{D}$, by setting $\xi \equiv i$ we obtain the required equality, namely:

$$
\langle \phi, \phi_m(it/\hbar) \rangle =
$$
$$
\left(-\frac{i}{\hbar} \right)^m \int_{\mathbb{R}^3} \bar{\phi}(x) \mathbb{E} \left[\psi_0(\sqrt{i\hbar}\omega(t) + x) \left(\sqrt{i\hbar} \int_0^t \mathbf{a}(\sqrt{i\hbar}\omega(s) + x) \circ d\omega(s) \right)^m \right] dx.
$$

$\square$

Actually Theorem 6.11 can be generalized to the case where a scalar potential V is added to the right-hand side of (1.12), see [11] for a discussion of this problem.

Theorem 6.11 provides a convergent constructive expansion for the Feynman path integral representation for the solution of the Schrödinger equation with magnetic field. It relies upon a particular class of finite dimensional approximations, namely the ones related to piecewise linear path. On the other hand, as remarked above, the Feynman map is not an infinite dimensional oscillatory integral whose very definition requires the existence of a unique limit of finite dimensional approximations independently of the particular sequence of projection operators $\{P_n\}$ converging strongly to the identity in $\mathcal{H}_t$. This property is fundamental since it characterizes the infinite dimensional oscillatory integral as an intrinsically well defined functional, whose value does not depend on the calculation procedure. In the final part of this chapter we tackle this issue and show that in the case where the Schrödinger equation contains a magnetic field term, the requirement of the independence of the limit of the sequence of finite dimensional approximations leads to the introduction of a natural renormalization term. This result is a further development of a similar one obtained in [9], the latter being only valid in the Coulomb gauge div $\mathbf{a} = 0$. On the contrary, the results presented below (Theorem 6.13 and Corollary 6.1) provide a gauge-independent construction of the renormalization term as well as of the Feynman path integral, yielding a rigorous construction of the solution of the Schrödinger equation with linear vector potential $\mathbf{a}$.

Let $\mathbf{a} : \mathbb{R}^3 \to \mathbb{R}^3$ be a linear vector potential corresponding to a constant magnetic field $\mathbf{B} = \text{rot}\,\mathbf{a}$. More precisely, we assume that the vector field $\mathbf{a}$ is given on a vector $(x_1, x_2, x_3) \in \mathbb{R}^3$ by

$$\mathbf{a}(x_1, x_2, x_3) = (\alpha_1^1 x_1 + \alpha_2^1 x_2 + \alpha_3^1 x_3, \alpha_1^2 x_1 + \alpha_2^2 x_2 + \alpha_3^2 x_3, \alpha_1^3 x_1 + \alpha_2^3 x_2 + \alpha_3^3 x_3),$$
(6.130)

where $\alpha_j^i \in \mathbb{R}$, $i, j = 1, \ldots, 3$ are real constants. We are going to study the Fresnel integrability (in the sense of Definition 5.4) of the function $f : \mathcal{H}_t \to \mathbb{C}$, defined on the Cameron-Martin space $\mathcal{H}_t$ as

$$f(\gamma) := e^{-\frac{i}{\hbar} \int_0^t \mathbf{a}(\gamma(s)) \cdot \dot{\gamma}(s) ds}, \qquad \gamma \in \mathcal{H}_t.$$
(6.131)

For any sequence $\{P_n\}_n$ of projectors onto n-dimensional subspaces of $\mathcal{H}_t$, such that $P_n \leq P_{n+1}$ and $P_n \to \mathbb{I}$ strongly as $n \to \infty$, we have to study the limit of the sequence of finite dimensional oscillatory integrals

$$\lim_{n \to \infty} (2\pi i \hbar)^{-n/2} \int_{P_n \mathcal{H}_t}^o e^{i \frac{\|P_n \gamma\|^2}{2\hbar}} f(P_n \gamma) d(P_n \gamma).$$

As we shall see, the limit above cannot be independent of the sequence $\{P_n\}_n$. In fact it is necessary to renormalize the term $f(P_n \gamma) \equiv e^{-\frac{i}{\hbar} g(P_n \gamma)}$ by replacing the exponent $g(P_n \gamma) = \int_0^t \mathbf{a}(P_n \gamma(s)) \cdot \dot{P}_n \gamma(s) ds$ by $g(P_n \gamma) - r_n$, where r_n is a suitable constant depending on the projector P_n as well as on the magnetic field $\mathbf{B}$.

First of all, let us consider the linear operator $G : \mathcal{H}_t \to \mathcal{H}_t$ defined by

$$G(\gamma)(s) := \int_0^s \mathbf{a}(\gamma(r)) dr, \qquad \gamma \in \mathcal{H}_t, \ s \in [0, t],$$
(6.132)

in such a way that the function $f : \mathcal{H}_t \to \mathbb{C}$ can be written as $f(\gamma) = e^{-\frac{i}{\hbar}\langle G(\gamma),\gamma \rangle}$, i.e. the function $g : \mathcal{H}_t \to \mathbb{C}$, with $g(\gamma) = \langle G(\gamma), \gamma \rangle$ can be represented as the quadratic form associated to G. Note that G is bounded in $\mathcal{H}_t$ due to our assumptions on $\mathbf{a}$. The following lemma provides some properties of G.

Lemma 6.12. *The operator* $G : \mathcal{H}_t \to \mathcal{H}_t$ *is Hilbert-Schmidt . The eigenvalues of the positive symmetric operator* $G^\dagger G$, *are given by* $\lambda_{m,j} = \frac{4a_j t^2}{\pi^2(1+2m)^2}$, *with* $j = 1, 2, 3$, $m \in \mathbb{N}$ *and* $a_j \in \mathbb{R}^+$ *eigenvalues of the matrix* (6.134) *below.*

Proof.   Let us consider the positive symmetric operator $L \equiv G^\dagger G : \mathcal{H}_t \to \mathcal{H}_t$ ($\dagger$ standing for the adjoint), whose matrix elements are given by

$$\langle \eta, L\gamma \rangle = \langle G\eta, G\gamma \rangle = \int_0^t \eta(s) A\gamma(s)^T dt, \qquad \eta, \gamma \in \mathcal{H}_t, \qquad (6.133)$$

where A is the 3×3 symmetric matrix with real elements given by

$$A_{ij} = \sum_{l=1}^3 \alpha_i^l \alpha_j^l, \qquad i, j = 1, ..., 3. \qquad (6.134)$$

Hence, for $\gamma \in \mathcal{H}_t$ the vector $L(\gamma) \in \mathcal{H}_t$ is given by

$$L(\gamma)(s)^T = - \int_0^s \int_t^r A\gamma(\tau)^T d\tau dr,$$

T stands for transpose. L is a compact operator in $\mathcal{H}_t$ and has a discrete spectrum. Indeed, by introducing in $\mathbb{R}^3$ an orthonormal basis $\{\hat{u}_1, \hat{u}_2, \hat{u}_3\}$ of eigenvectors of the symmetric matrix A, with corresponding eigenvalues $a_1, a_2, a_3 \in \mathbb{R}^+$, the eigenvectors $\{\gamma_m\}$ of L can be represented as linear combination of $\hat{u}_1$, $\hat{u}_2$ and $\hat{u}_3$ by $\gamma_m = \eta_{m,1}\hat{u}_1 + \eta_{m,2}\hat{u}_2 + \eta_{m,3}\hat{u}_3$, with $\eta_{m,j} : [0,t] \to \mathbb{R}$. Recalling the form of the scalar product in $\mathcal{H}_t$, for the expression (6.133) we get that the components $\{\eta_{m,j}\}$ of the eigenvectors (with eigenvalues $\lambda_{m,j}$) are the solutions of

$$\begin{cases} \lambda_{m,j}\ddot{\eta}_{m,j} + a_j\eta_{m,j} = 0 \\ \dot{\eta}_{m,j}(t) = 0 \qquad\qquad j = 1, 2, 3, \\ \eta_{m,j}(0) = 0 \end{cases}$$

with

$$\lambda_{m,j} = \frac{4a_j t^2}{\pi^2(1+2m)^2}, \qquad m \in \mathbb{N}, \quad j = 1, 2, 3. \qquad (6.135)$$

$\square$

Remark 6.11. From (6.135) we see easily that the operator G is not trace class on $\mathcal{H}_t$.

The next lemma shows that the finite dimensional approximations of the infinite dimensional oscillatory integral of the function (6.131) can be computed in terms of suitable Wiener integrals. This is an important technical result since it paves the way for the application of classical result of analysis in Wiener spaces to the problem of Feynman integration.

Lemma 6.13. *Let* $\mathbf{a}$ *be the linear vector potential* (6.130) *and let* $\psi_0 \in L^2(\mathbb{R}^3)$ *be such that its Fourier transform* $\hat{\psi}_0$ *has compact support. Let* $\{e_j\}_j$ *be an orthonormal basis of the Cameron-Martin space* $\mathcal{H}_t$ *and let* P_n *be the projection operator onto the span of the first* n *vectors. Define the function* $g^x_{\exp} : \mathcal{H}_t \to \mathbb{C}$ *by*

$$g^x_{\exp}(\gamma) = \psi_0(\gamma(t) + x)\exp\left(-\frac{i}{\hbar}\int_0^t \mathbf{a}(\gamma(s) + x)\cdot\dot{\gamma}(s)ds\right), \qquad \gamma \in \mathcal{H}_t,$$

and let $\bar{a} \in \mathbb{R}^+$ *be the constant defined as* $\bar{a} = \max_{j=1,2,3}\{a_j\}$, *where* a_j, $j = 1,2,3$ *are the eigenvalues of the (positive semidefinite) matrix* A *defined in* (6.134). *Then, for fixed* n *and for*

$$t < t^* := \frac{\pi}{4\sqrt{\bar{a}}}, \tag{6.136}$$

the finite dimensional oscillatory integral

$$(2\pi i\hbar)^{-n/2}\int_{P_n\mathcal{H}_t}^o e^{\frac{i}{2\hbar}\|P_n\gamma\|^2}g^x_{\exp}(P_n\gamma)dP_n\gamma$$

is equal to the Wiener integral:

$$\int_{P_n\mathcal{H}_t}^o e^{\frac{i}{2\hbar}\|\gamma\|^2}g^x_{\exp}(\gamma)d\gamma = \mathbb{E}\left[\psi_0(\sqrt{i\hbar}\omega_n(t) + x)e^{-\frac{i}{\hbar}\sqrt{i\hbar}\int_0^t \mathbf{a}(\sqrt{i\hbar}\omega_n(s)+x)\cdot\dot{\omega}_n(s)ds}\right],$$

(6.137)

where $\omega_n := \tilde{P}_n(\omega)$ *is defined by* (6.122).

Proof. By definition we have, for fixed n, by setting $\gamma_n \equiv P_n\gamma$:

$$(2\pi i\hbar)^{-n/2}\int_{P_n\mathcal{H}_t}^o e^{i\frac{\|\gamma_n\|^2}{2\hbar}}g^x_{exp}(\gamma_n)d\gamma_n$$

$$= (2\pi i\hbar)^{-n/2}\int_{P_n\mathcal{H}_t}^o e^{\frac{i}{2\hbar}\|\gamma_n\|^2}e^{-\frac{i}{\hbar}\int_0^t \mathbf{a}(\gamma_n(s)+x)\dot{\gamma}_n ds}\psi_0(\gamma_n(t) + x)d\gamma_n$$

$$= (2\pi i\hbar)^{-n/2}\int_{P_n\mathcal{H}_t}^o e^{\frac{i}{2\hbar}\langle\gamma_n,(\mathbb{I}-2G)\gamma_n\rangle}e^{-\frac{i}{\hbar}\mathbf{a}(x)\cdot\gamma_n(t)}\psi_0(\gamma_n(t) + x)d\gamma_n.$$

Let us consider the function $F \to \mathbb{R}^+ \to \mathbb{C}$ defined by

$$F(z) = (2\pi i\hbar)^{-n/2} z^n \int_{P_n \mathcal{H}_t}^{o} e^{\frac{iz^2}{2\hbar}\langle\gamma_n,(\mathbb{I}-2G)\gamma_n\rangle} e^{-z\frac{i}{\hbar}\mathbf{a}(x)\cdot\gamma_n(t)} \psi_0(z\gamma_n(t) + x) d\gamma_n.$$

(6.138)

By the classical change of variable formula, for $z \in \mathbb{R}^+$ the function F is a constant equal to the finite dimensional oscillatory integral above. In fact, if $t < t^*$, with t^* given by (6.136), F can be extended to an analytic function defined on the open sector $D_{\pi/2} = \{z \in \mathbb{C} : z = |z|e^{i\theta}, \theta \in (0, \pi/2), |z| > 0\}$ of the complex plane. Indeed, for any $\gamma \in \mathcal{H}_t$, if condition (6.136) is fulfilled, we have

$$\langle\gamma, (\mathbb{I} - 2G)\gamma\rangle \geq \epsilon\|\gamma\|^2,$$

where $\epsilon > 0$ is given by $\epsilon = 1 - \frac{2t\sqrt{\bar{a}}}{\pi}$. Indeed:

$$\langle\gamma, (\mathbb{I} - 2G)\gamma\rangle = \langle\gamma, \mathbb{I}\gamma\rangle - \langle\gamma, 2G\gamma\rangle$$
$$= \langle\gamma, \gamma\rangle - \langle\gamma, 2|G|U\gamma\rangle,$$

where, by the polar decomposition formula, $G = |G|U$, with $|G| = \sqrt{G^\dagger G}$ and U a unitary operator. Furthermore

$$|\langle\gamma, 2|G|U\gamma\rangle| \leq 2\|\gamma\|\|U\gamma\|\||G\|| \leq 2\|\gamma\|^2 \sup_m \tilde{\lambda}_m,$$

where $\||G\||$ denotes the operator norm of the positive operator $|G|$, while $\{\tilde{\lambda}_m\}_m$ are its eigenvalues, namely $\tilde{\lambda}_m = \sqrt{\lambda_{m,j}}$, with $\lambda_{m,j}$ given by (6.135). Hence, we get

$$|\langle\gamma, 2|G|U\gamma\rangle| \leq \frac{4t\sqrt{\bar{a}}}{\pi}\|\gamma\|^2,$$

hence, for $t < t^*$, we have, using the Fourier transform $\hat{\psi}_0$ of ψ_0, the following bound on the oscillatory integral in (6.138):

$$\int_{P_n \mathcal{H}_t}^{o} \left|e^{z^2 \frac{i}{2\hbar}\langle\gamma_n,(\mathbb{I}-2G)\gamma_n\rangle} e^{-z\frac{i}{\hbar}\mathbf{a}(x)\cdot\gamma_n(t)} \psi_0(z\gamma_n(t) + x)\right| d\gamma_n$$
$$\leq \int_{P_n \mathcal{H}_t}^{o} \int_{\mathbb{R}^3} e^{-\sin(2\theta)\frac{|z|^2}{2\hbar}\epsilon\|\gamma\|^2 + \sin\theta\frac{|z|}{\hbar}\mathbf{a}(x)\cdot\gamma_n(t) - k|z|\gamma_n(t)} \frac{|\hat{\psi}_0(k)|}{(2\pi)^3} dk d\gamma_n < \infty,$$

where the convergence of the integral in the second line is assured by the conditions $\theta \in (0, \pi/2)$, $\epsilon = 1 - \frac{2t\sqrt{\bar{a}}}{\pi} > 0$ and $\hat{\psi}_0$ is compactly supported. Hence, by applying Fubini and Morera's theorems, it is simple to check that the function $F : \bar{D}_{\pi/2} \to \mathbb{C}$ is analytic on $D_{\pi/2}$ and continuous up to $\mathbb{R}^+$. Since by the classical change of variables formula the value of $F(z)$ does not depend on $|z|$, i.e. F is constant along rays $\{z \in D_{\pi/2} : z = |z|e^{i\theta}, |z| \in \mathbb{R}^+\}$, by analyticity F is constant on $D_{\pi/2}$ and by continuity up to $\mathbb{R}^+$ we obtain in particular, $F(1) = F(\sqrt{\hbar}e^{i\pi/4})$, namely:

$$(2\pi i\hbar)^{-n/2} \int_{P_n \mathcal{H}_t}^{o} e^{\frac{i}{2\hbar}\|\gamma_n\|^2} e^{-\frac{i}{\hbar}\int_0^t \mathbf{a}(\gamma_n(s)+x)\dot{\gamma}_n ds} \psi_0(\gamma_n(t) + x) d\gamma_n$$

$$= (2\pi)^{-n/2} \int_{P_n \mathcal{H}_t}^{o} e^{-\frac{1}{2}\|\gamma_n\|^2} e^{-\frac{i\sqrt{i}}{\sqrt{\hbar}}\int_0^t \mathbf{a}(\sqrt{i\hbar}\gamma_n(s)+x)\dot{\gamma}_n ds} \psi_0(\sqrt{i\hbar}\gamma_n(t) + x) d\gamma_n$$

and the last line is equal to the r.h.s. of (6.137), namely to:

$$\mathbb{E}\left[\psi_0\left(\sqrt{i\hbar}\omega_n(t) + x\right) e^{-\frac{i}{\hbar}\sqrt{i\hbar}\int_0^t \mathbf{a}(\sqrt{i\hbar}\omega_n(s)+x)\cdot\dot{\omega}_n(s)ds}\right]. \qquad \square$$

The next step is the study of the convergence of the Wiener integrals on the r.h.s. of (6.137) which can be written as

$$\mathbb{E}\left[\psi_0\left(\sqrt{i\hbar}\omega_n(t) + x\right) e^{-\frac{i}{\hbar}\sqrt{i\hbar}\mathbf{a}(x)\omega_n(t)} e^{g_n(\omega)}\right],$$

where, given an orthonormal basis $\{e_n\}$ of $\mathcal{H}_t$, the random variables $g_n : C_t \to \mathbb{C}$ are defined by

$$g_n(\omega) := \int_0^t \mathbf{a}(\omega_n(s)) \cdot \dot{\omega}_n(s)ds, \qquad \omega \in C_t. \tag{6.139}$$

Further, let us consider the sequence $\{\mathfrak{g}_n\}$ of real random variables on $(C_t, \mathcal{B}(C_t), \mathbb{P})$ defined as

$$\mathfrak{g}_n(\omega) := \int_0^t \mathbf{a}(\omega(t)) \cdot \dot{\omega}_n(t)dt, \qquad \omega \in C_t, \tag{6.140}$$

and the linear operator $\mathfrak{G} : C_t \to \mathcal{H}_t$ defined by

$$\mathfrak{G}(\omega)(s) = \int_0^s \mathbf{a}(\omega(r))dr, \qquad \omega \in C_t, \ s \in [0, t]. \tag{6.141}$$

In fact, the functions $\{\mathfrak{g}_n\}$ and $\{g_n\}$ can be represented by the inner products:

$$\mathfrak{g}_n(\omega) = \langle \mathfrak{G}(\omega), \tilde{P}_n(\omega)\rangle, \qquad g_n(\omega) = \langle \mathfrak{G}(\tilde{P}_n(\omega)), \tilde{P}_n(\omega)\rangle. \tag{6.142}$$

The following lemma shows that the sequences $\{\mathfrak{g}_n\}$ and $\{g_n\}$ converge to the same limit in $L^2(C_t, W)$.

Lemma 6.14. *Let $\mathfrak{G} : C_t \to \mathcal{H}_t$ be a linear operator such that its restriction $\mathfrak{G}_{\mathcal{H}_t}$ on $\mathcal{H}_t$ is Hilbert-Schmidt. Let $\{P_n\}_n$ be a sequence of finite dimensional projection operators in $\mathcal{H}_t$ converging strongly to the identity. Then the sequences of random variables $\{\mathfrak{g}_n\}$ and $\{\mathfrak{g}'_n\}$ on C_t defined as*

$$\mathfrak{g}_n(\omega) = \langle \mathfrak{G}(\omega), \tilde{P}_n(\omega)\rangle, \qquad \omega \in C_t,$$

$$\mathfrak{g}'_n(\omega) = \langle \mathfrak{G}(\tilde{P}_n(\omega)), \tilde{P}_n(\omega)\rangle, \qquad \omega \in C_t,$$

satisfy

$$\lim_{n\to\infty} \mathbb{E}[|\mathfrak{g}_n - \mathfrak{g}'_n|^2] = 0. \tag{6.143}$$

Proof.

$$\mathbb{E}[|\mathfrak{g}_n - \mathfrak{g}'_n|^2] = \int |\langle \mathfrak{G}(\omega) - G(\tilde{P}_n(\omega)), \tilde{P}_n(\omega) \rangle|^2 d\mathbb{P}(\omega)$$

$$= \int |\langle \mathfrak{G}(\omega - \tilde{P}_n(\omega)), \tilde{P}_n(\omega) \rangle|^2 d\mathbb{P}(\omega)$$

$$= \int |\langle \mathfrak{G}(\sum_{j=n+1}^{\infty} e_j n_{e_j}(\omega)), \sum_{i=1}^{n} e_i n_{e_i}(\omega) \rangle|^2 d\mathbb{P}(\omega)$$

$$= \sum_{j,j'=n+1}^{\infty} \sum_{i,i'=1}^{n} \langle \mathfrak{G}e_j, e_i \rangle \langle \mathfrak{G}e_{j'}, e_{i'} \rangle \mathbb{E}[n_{e_j} n_{e_{j'}} n_{e_i} n_{e_{i'}}]$$

$$= \sum_{j=n+1}^{\infty} \sum_{i=1}^{n} (\langle \mathfrak{G}e_j, e_i \rangle)^2$$

$$= \sum_{j=n+1}^{\infty} \langle P_n \mathfrak{G}e_j, P_n \mathfrak{G}e_j \rangle,$$

where in the third step we have applied Itô-Nisio theorem. By using the assumption that $\mathfrak{G}_{\mathcal{H}_t}$ is an Hilbert-Schmidt operator we obtain (6.143). □

It is now convenient to introduce the definition of $\mathcal{H}$-differentiable function, following, e.g., [278]:

Definition 6.1. A function $\mathfrak{G} : C_t \to C_t$ with $\mathfrak{G}(C_t) \subset \mathcal{H}_t$ is said to be $\mathcal{H}_t$-*differentiable* if for any $\omega \in C_t$ the function $\mathfrak{G}_\omega : \mathcal{H}_t \to \mathcal{H}_t$ defined as $\mathfrak{G}_\omega(\gamma) = \mathfrak{G}(\omega + \gamma)$, $\gamma \in \mathcal{H}_t$, is Fréchet differentiable at the origin in $\mathcal{H}_t$. Its Fréchet derivative, namely the linear operator $\mathfrak{D}\mathfrak{G}_\omega(0) \in L(\mathcal{H}_t; \mathcal{H}_t)$, will be denoted with the symbol $\mathfrak{D}\mathfrak{G}(\omega)$ and called the $\mathcal{H}_t$-*derivative of* $\mathfrak{G}$ *at* ω.

The following result is a direct consequence of Lemmas 4.2 and 4.3 in [278].

Theorem 6.12. *Let* $\mathfrak{G} : C_t \to C_t$, *with* $\mathfrak{G}(C_t) \subset \mathcal{H}_t$, *be a* $\mathcal{H}_t$-*differentiable map such that for any* $\omega \in C_t$ *the* $\mathcal{H}_t$-*derivative* $\mathfrak{D}\mathfrak{G}(\omega) \in L(\mathcal{H}_t, \mathcal{H}_t)$ *is an Hilbert-Schmidt operator. Let us assume furthermore that the maps* $\|\mathfrak{G}\| : C_t \to \mathbb{R}$ *and* $\|\mathfrak{D}\mathfrak{G}\|_2 : C_t \to \mathbb{R}$, *where* $\|\mathfrak{D}\mathfrak{G}(\omega)\|_2$ *denotes the Hilbert-Schmidt norm of* $\mathfrak{D}\mathfrak{G}(\omega)$, *belong to* $L^2(C_t, \mathbb{P})$. *Let* $\{e_i\}$ *be an orthonormal basis of* $\mathcal{H}_t$ *and let* $\{P_n\}_n$ *and* $\{\tilde{P}_n\}_n$ *be the sequence of finite dimensional projectors on the span of* $e_1, \ldots, e_n$ *and their stochastic extensions to* C_t *respectively. Then the sequence of random variables* $\{\mathfrak{h}_n\}$ *defined as*

$$\mathfrak{h}_n(\omega) := \langle \mathfrak{G}(\omega), \tilde{P}_n(\omega) \rangle - \text{Tr}(P_n \mathfrak{D}\mathfrak{G}(\omega)), \qquad \omega \in C_t,$$

converges in $L^2(C_t, \mathbb{P})$ *and the limit does not depend on the basis* $\{e_i\}$, $i = 1, \ldots, n$.

The previous theorem applied to our particular case provides actually a no-go result on the convergence of the sequence of random variables $\{g_n\}$ given by (6.139). Indeed, since in our particular case $\mathfrak{D}\mathfrak{G}(\omega) = G$ and by Remark 6.11 the operator $G : \mathcal{H}_t \to \mathcal{H}_t$ is not trace class, the sequence of real numbers $\mathrm{Tr}(P_n\mathfrak{D}\mathfrak{G}(\omega)) \equiv \mathrm{Tr}(P_n G)$ does not in general converge independently on the choice of finite dimensional approximations $\{P_n\}_n$. Hence, by Theorem 6.12 and Lemma 6.14 neither the random variables $\{\mathfrak{g}_n\}$ nor $\{g_n\}$ admit a well defined limit in $L^2(C_t, \mathbb{P})$ independent of the sequence $\{P_n\}_n$ of finite dimensional projectors. The following theorem provides a suitable renormalization term, namely a sequence $\{r_n\}$ of real numbers such that the renormalized random variables $h_n : C_t \to \mathbb{R}$ given by $h_n(\omega) := g_n(\omega) - r_n$ converge in $L^p(C_t, \mathbb{P})$ for all $p \geq 1$ and in probability to the Stratonovich stochastic integral $h(\omega) = \int_0^t \mathbf{a}(\omega(s)) \circ d\omega(s)$.

Theorem 6.13. *Let* $\mathbf{a} : \mathbb{R}^3 \to \mathbb{R}^3$ *be a linear vector field, let* $\{e_k\}$ *be an orthonormal basis of* $\mathcal{H}_t$ *and let* $\{P_n\}_n$ *and* $\{\tilde{P}_n\}_n$ *be the sequence of finite dimensional projectors on the span of* $e_1, \ldots, e_n$ *in* $\mathcal{H}_t$ *and their stochastic extensions to* C_t *respectively. Then by setting*

$$r_n := \mathbf{B} \cdot \frac{1}{2} \sum_{k=1}^n \int_0^t e_k(s) \wedge \dot{e}_k(s) ds \qquad (6.144)$$

with $\mathbf{B} = \mathrm{rot}\,\mathbf{a}$*, the sequence of random variables* $h_n : C_t \to \mathbb{R}$ *defined as*

$$h_n(\omega) := \int_0^t \mathbf{a}(\omega_n(s)) \cdot \dot{\omega}_n(s) ds - r_n, \quad \omega \in C_t,$$

where $\omega_n := \tilde{P}_n(\omega)$*, converges in* $L^2(C_t, \mathbb{P})$*, independently of* $\{P_n\}_n$*, to*

$$\int_0^t \mathbf{a}(\omega(s)) \circ d\omega(s).$$

Proof. Let us set

$$X_n(\omega) = \int_0^t \mathbf{a}(\omega_n(s)) \cdot \dot{\omega}_n(s) ds = \int_0^t a_1(\omega_n(s)) \dot{\omega}_{n,1}(s) \, ds$$
$$+ \int_0^t a_2(\omega_n(s)) \dot{\omega}_{n,2}(s) ds + \int_0^t a_3(\omega_n(s)) \dot{\omega}_{n,3}(s) ds,$$

where $\omega_n = (\omega_{n,1}, \omega_{n,2}, \omega_{n,3}) \in \mathcal{H}_t$. By Stokes theorem:

$$X_n = \iint_{S_n} \mathbf{B} \cdot \mathbf{n} \, dS - \int_{\Lambda_n} \mathbf{a} \cdot d\mathbf{r} \qquad (6.145)$$

where Λ_n is the (oriented) segment joining $\omega_n(t)$ with 0, while $\int_{\Lambda_n} \mathbf{a} \cdot d\mathbf{r}$ is the line integral of $\mathbf{a}$ along Λ_n. The symbol S_n denotes any regular oriented surface with oriented boundary given by the close path union of ω_n and Λ_n, **n** denotes the normal unit vector and $\iint_{S_n} \mathbf{B} \cdot \mathbf{n}\, dS$ is the surface integral of **B** on S_n. Our study can be restricted to

$$\iint_{S_n} \mathbf{B} \cdot \mathbf{n}\, dS,$$

as we can immediately see that the second term converges in $L^2(C_t, \mathbb{P})$ independently of $\{P_n\}_n$. Indeed:

$$\int_{\Lambda_n} \mathbf{a} \cdot d\mathbf{r} = \int_0^1 \mathbf{a}(u\omega_n(t))du \cdot \omega_n(t),$$

and for any sequence of finite dimensional projection operators $\{P_n\}_n$ such that $P_n \to \mathbb{I}$ we have

$$\omega_n(t) \to \omega(t), \qquad \int_0^1 \mathbf{a}(u\,\omega_n(t))du \to \int_0^1 \mathbf{a}(u\omega(t))du, \qquad \forall t \geq 0.$$

Let us consider now the surface integral $\iint_{S_n} \mathbf{B} \cdot \mathbf{n}\, dS$. Since by the assumption on **a** the magnetic field **B** is constant, by the Gauss-Green formula we get

$$\iint_{S_n} \mathbf{B} \cdot \mathbf{n}\, dS = \mathbf{B} \iint_{S_n} \mathbf{n}\, dS = \mathbf{B} \cdot \frac{1}{2} \int_0^t \omega_n(s) \wedge \dot{\omega}_n(s)ds.$$

Let us define for any $i = 1, 2, 3$ the sequence of random variables $h_n^i : C_t \to \mathbb{R}$ by

$$h_n^i(\omega) := \hat{e}_i \cdot \int_0^t \omega_n(s) \wedge \dot{\omega}_n(s)\, ds = \langle H^i(\omega_n), \omega_n \rangle, \qquad \omega \in C_t,$$

where $\hat{e}_i$, $i = 1, 2, 3$, are the vectors of the canonical basis of $\mathbb{R}^3$ and the linear operators $H^i : C_t \to \mathcal{H}_t$ are defined by

$$(H^i(\omega)(s))^T := \int_0^s J^i\omega(u)^T du,$$

with ^{T} denoting the transpose and J^i, $i = 1, 2, 3$, are the matrices:

$$J^1 = \begin{pmatrix} 0 & 0 & 0 \\ 0 & 0 & -1 \\ 0 & 1 & 0 \end{pmatrix}, \qquad J^2 = \begin{pmatrix} 0 & 0 & 1 \\ 0 & 0 & 0 \\ -1 & 0 & 0 \end{pmatrix}, \qquad J^3 = \begin{pmatrix} 0 & -1 & 0 \\ 1 & 0 & 0 \\ 0 & 0 & 0 \end{pmatrix}.$$

Actually the operators H^i, $i = 1, \ldots, 3$ have the form (6.132) and by Lemma 6.12 are Hilbert-Schmidt. Further, by Lemma 6.14 and Theorem 6.12, the renormalized random variables

$$h_n^i(\omega) - r_n^i = \langle H^i(\omega_n), \omega_n \rangle - \text{Tr}[P_n H^i]$$

$$= \int_0^t (\omega_n(s) \wedge \dot{\omega}_n(s))_i\, ds - \sum_{k=1}^n (e_k(s) \wedge \dot{e}_k(s))_i\, ds$$

converge in $L^2(C_t, \mathbb{P})$ and the limit does not depend on the sequence $\{P_n\}_n$. By combining these results we obtain the convergence of the sequence

$$\int_0^t \mathbf{a}(\omega_n(s)) \cdot \dot{\omega}_n(s)ds - \mathbf{B} \cdot \frac{1}{2} \sum_{k=1}^n \int_0^t e_k(s) \wedge \dot{e}_k(s)ds$$

and the limit is independent of the sequence $\{P_n\}_n$. Eventually, by choosing the sequence $\{P_n\}_n$ of piecewise linear approximations (6.121), where the elements e_n of the corresponding basis $\{e_n\}$ satisfy $\int_0^t e_n(s) \wedge \dot{e}_n(s)ds = 0$, and by applying Lemma 6.10 we complete the proof. $\qquad\square$

It is interesting to remark that he renormalization term given in Theorem 6.13 contains, besides the magnetic field $\mathbf{B}$, the area integrals of the elements $\{e_n\}$ of the orthonormal basis spanning the finite dimensional Hilbert space $P_n\mathcal{H}_t$. This term is gauge independent. However, for a general orthonormal basis in $\mathcal{H}_t$, it does not converge to a well defined limit as the following example shows.

Let us fix $t \equiv 1$ and let us consider two sequences of real valued functions $\{u_n\}_{n\geq 0}$, $\{v_n\}_{n\geq 1}$ defined on the interval $[0,1]$ by $u_0(s) = s$ and for $n \geq 1$

$$u_n(s) = \frac{\cos(2\pi n s)}{2\pi n}, \quad v_n(s) = \frac{\sin(2\pi n s)}{2\pi n}, \qquad s \in [0,1]$$

and the sequences of vectors in $\mathcal{H}_t$ defined by: $e_{n,1} := (u_n, v_n, 0)$, $e_{n,2} := (u_n, -v_n, 0)$, $e_{n,3} := (v_n, u_n, 0)$, $e_{n,4} := (v_n, -u_n, 0)$, $e_{n,5} := (0, 0, u_n)$, $e_{n,6} := (0, 0, v_n)$, that together with the vectors $e_{0,1} := (u_0, 0, 0)$, $e_{0,2} := (0, u_0, 0)$ and $e_{n,3} := (0.0, u_0)$ provide an orthonormal basis of $\mathcal{H}_t$. Given the linear the vector field $\mathbf{a} : \mathbb{R}^3 \to \mathbb{R}^3$

$$\mathbf{a}(x, y, z) = \left(\frac{z-y}{2}, \frac{x-z}{2}, \frac{y-x}{2} \right).$$

with $\mathbf{B} = \mathrm{rot}\,\mathbf{a} = (1, 1, 1)$, and taking the vectors $e_{k,1}$ and $e_{k,4}$, $k = 1, \ldots, n$, we get

$$\mathbf{B} \cdot \frac{1}{2} \sum_{k=1}^n \int_0^t e_{k,1} \wedge \dot{e}_{k,1} = \mathbf{B} \cdot \frac{1}{2} \sum_{k=1}^n \int_0^t e_{k,4} \wedge \dot{e}_{k,4} = \sum_{k=1}^n \frac{1}{4\pi k}.$$

On the other hand, considering the vectors $e_{k,2}$ and $e_{k,3}$, $k = 1, \ldots, n$, we have

$$\mathbf{B} \cdot \frac{1}{2} \sum_{k=1}^n \int_0^t e_{k,2} \wedge \dot{e}_{k,2} = \mathbf{B} \cdot \frac{1}{2} \sum_{k=1}^n \int_0^t e_{k,3} \wedge \dot{e}_{k,3} = -\sum_{k=1}^n \frac{1}{4\pi k},$$

while the other vectors of the orthonormal basis give vanishing area integrals. Hence, the renormalization term r_n given by (6.144) is not absolutely convergent as $n \to \infty$.

A direct consequence of Lemma 6.13 and Theorem 6.13 is the following result.

Corollary 6.1. *Under the assumptions of Lemma 6.13, the sequence of finite dimensional renormalized oscillatory integrals*

$$(2\pi i\hbar)^{-n/2} \int_{P_n\mathcal{H}_t}^{o} e^{\frac{i}{2\hbar}\|\gamma_n\|^2} e^{-\frac{i}{\hbar}\left(\int_0^t \mathbf{a}(\gamma_n(s)\cdot\dot{\gamma}_n(s)ds - r_n\right)} \psi_0(\gamma_n(t) + x)d\gamma_n,$$

with the renormalization term r_n given by (6.144), converges as $n \to \infty$ to the Wiener integral

$$\mathbb{E}\left[\psi_0(\sqrt{i\hbar}\omega(t) + x)e^{-\frac{i}{\hbar}\sqrt{i\hbar}\int_0^t \mathbf{a}(\sqrt{i\hbar}\omega(s)+x)\circ d\omega(s)}\right] \tag{6.146}$$

and the limit is independent of the sequence $\{P_n\}_n$ of finite dimensional approximations. In addition, it provides the solution of the Schrödinger equation with magnetic field

$$\begin{cases} i\hbar\partial_t\psi(t,x) = \frac{1}{2}\left(-i\hbar\nabla - \mathbf{a}\right)^2 \psi(t,x) \\ \psi(0,x) = \psi_0(x), \end{cases} \quad t \in \mathbb{R}^+, \quad x \in \mathbb{R}^3. \tag{6.147}$$

Proof. The first part of the theorem follows from Lemma 6.13 and Theorem 6.13. The second part can be proved by using the analyticity properties of the semigroup generated by the quantum Hamiltonian operator $H = \frac{1}{2}(-i\hbar\nabla - \mathbf{a})^2$. More precisely for $t \in \mathbb{R}^+$ the action of the heat semigroup on the vector ψ_0 is given by the Feynman-Kac-Itō formula:

$$e^{-\frac{t}{\hbar}H}\psi_0(x) = \mathbb{E}\left[\psi_0(\sqrt{\hbar}\omega(t) + x)e^{-\frac{i}{\sqrt{\hbar}}\int_0^t \mathbf{a}(\sqrt{\hbar}\omega(s)+x)\circ d\omega(s)}\right].$$

For any $\phi \in L^2(\mathbb{R}^3)$ the inner product $\langle\phi, e^{-z\frac{t}{\hbar}H}\psi_0\rangle$ is an analytic function of $z \in D$, $D = \{z \in \mathbb{C}, Re(z) \geq 0\}$, continuous on $\bar{D}$, giving for $z = i$ the inner product between $\phi \in L^2(\mathbb{R}^d)$ and the solution of the Schrödinger Eq. (6.147). For $z \in \mathbb{R}^+$, by the change of variables formula we have

$$\langle\phi, e^{-z\frac{t}{\hbar}H}\psi_0\rangle = \int_{\mathbb{R}^3} \bar{\phi}(x)\mathbb{E}\left[\psi_0(\sqrt{z\hbar}\omega(t) + x)e^{-\frac{i\sqrt{z}}{\sqrt{\hbar}}\int_0^t \mathbf{a}(\sqrt{z\hbar}\omega(s)+x)\circ d\omega(s)}\right] dx.$$

By the assumptions on t, $\mathbf{a}$, and ψ_0, both sides of the equality above are analytic for $z \in D$, continuous in $\bar{D}$ and coincide on $\mathbb{R}^+$. Hence, for $z = i$ we obtain that the solution in $L^2(\mathbb{R}^3)$ of (6.147) is given by (6.146). $\quad\square$

Chapter 7

The stationary phase method and the semiclassical limit of quantum mechanics

The present chapter is devoted to the description of an infinite dimensional version of the classical stationary phase method in the framework of infinite dimensional oscillatory integrals. In particular, Section 7.4 presents the application of these techniques to the study of the semiclassical asymptotic behaviour of the solution of Schrödinger equation in the limit $\hbar \downarrow 0$.

7.1 Asymptotic expansions

In the study of several problems, such as the evaluation of integrals or the solution of differential equations, it is very difficult to find an exact analytical solution, given in terms of known functions. On the other hand, in the applications exact solutions are not always of practical use, both from a computational and an analytical point of view, and it is preferable the knowledge of an approximated solution when a parameter or some variable of the problem can be considered either large or small.

The asymptotic analysis studies the techniques that allow to obtain approximated solutions of a large class of problems, when a parameter approaches a value in which the solution is not analytic. It was introduced at the end of the nineteenth century by Poincaré, who gave a precise definition of an *asymptotic expansion*, and was further developed in the twentieth century in connection with the study of equations arising in mathematical physics, in particular in fluid mechanics.

We recall here the definition an the main properties of asymptotic sequence and expansions, that will be used in the next sections. For a detailed treatment see for instance [53, 130, 135, 180, 269, 279].

Let $V \subset \mathbb{C}$ a domain[1] in the complex plane $\mathbb{C}$ (or, more generally, on the Riemann surface $\tilde{\mathbb{C}}$ of the logarithm) such that $0 \in \partial V$ and

$$z \in V \Rightarrow \forall t \in (0, 1], \ tz \in V.$$

Let us denote $\hat{V} := V \cup \{0\}$. Both V and $\hat{V}$ will be called *angular neighborhoods* of zero.

A set $U \subset \mathbb{C}$, which is the closure of an angular neighborhood of zero, will be called *closed angular neighborhood*.

Definition 7.1. Let V be an angular neighborhood of zero. An asymptotic sequence of functions $(\phi_i)_{i \in \mathbb{N}}$ for $z \to 0$ in V is a sequence of functions $\phi_i : \hat{V} \to \mathbb{C}$, which do not vanish in V and such that for every $i \in \mathbb{N}$:

$$\lim_{z \to 0} \frac{\phi_{i+1}}{\phi_i}(z) = 0.$$

In the following we shall focus on the asymptotic sequence (in any angular neighborhood of zero) $\phi_n(z) = z^{n/k}$, $n \in \mathbb{N}$, for fixed $k > 0$ and denote by $\mathbb{C}[z^{1/k}]$ the space of formal power series with complex coefficients

$$\hat{f}(z) = \sum_{n=0}^{\infty} a_n z^{n/k}, \quad \{a_n\} \subset \mathbb{C}, \ k > 0. \tag{7.1}$$

Definition 7.2. A formal power series $\hat{f}$ is called a ($z^{1/k}-$) asymptotic expansion for a function $f : V \to \mathbb{C}$ as $z \to 0$ in an angular neighborhood V if for each closed angular neighborhood U with $U \subsetneq V$ and any $N \in \mathbb{N}$, there exists a number $C(N) > 0$ such that

$$\forall z \in U : \quad |f(z) - \sum_{n=0}^{N-1} a_n z^{n/k}| \leq C(N) |z^{N/k}|. \tag{7.2}$$

In this case we write $f \sim \hat{f}$, $z \to 0$ in V.

For more details see [180].

Remark 7.1. It is important to recall that the domain V in Definition 7.2 plays a crucial role, indeed the existence of an expansion depends strongly on V.

Remark 7.2. An asymptotic expansions is not necessarily convergent (and usually this is the case!). Indeed condition (7.2) means that for fixed N the function f is approximated by the sum $\sum_{n=0}^{N} a_n z^{n/k}$ for z sufficiently

[1]With the word *domain* we mean a connected open nonempty set.

small, while if the formal power series (7.1) is convergent at $z = 0$ in some domain V to a function f then the following holds:

$$\forall z \in V : \quad \lim_{N \to \infty} \left| f(z) - \sum_{n=0}^{N} a_n z^{n/k} \right| = 0, \tag{7.3}$$

which means that for fixed $z \in V$ the value of the function f is approximated by the sum $\sum_{n=0}^{N} a_n z^{n/k}$ for N sufficiently large.

It is easy to see that if a function f admits an $(z^{1/k}-)$ asymptotic expansion in a given domain V, i.e. $f \sim \hat{f} = \sum_{n=0}^{\infty} a_n z^{n/k}$ for $z \to 0$ in V, then the coefficients a_n, $n \in \mathbb{N}$, are uniquely determined by

$$a_0 = \lim_{z \to 0} f(z)$$

$$a_1 = \lim_{z \to 0} \frac{f(z) - a_0}{z^{1/k}}$$

$$\cdots$$

$$a_n = \lim_{z \to 0} \frac{f(z) - \sum_0^{n-1} a_j z^{j/k}}{z^{n/k}}.$$

On the other hand different functions can have the same asymptotic expansion, for instance the function $f(z) = 0$ and $g(z) = e^{-1/z}$ have both a zero asymptotic expansion in the domain $\{z \in \mathbb{C},\ Re(z) > 0\}$. In other words, if an asymptotic expansion is not convergent (and this is often the case) it does not characterize uniquely a function f asymptotically equivalent to it.

In order to associate in a unique way to formal power series $\sum a_n z^{n/k}$ a function f which is asymptotically equivalent to it, one can apply, under suitable assumptions, a powerful summation tool: Borel summability (see for instance [243, 244, 180, 279, 307, 285]). It works as follows:

(1) transform the given power series $\hat{f}$ into another convergent power series $\hat{B}$;

(2) compute the analytic function B which has $\hat{B}$ as a convergent power series expansion;

(3) apply an integral transform mapping the analytic function B to an analytic function f;

(4) the function f (the so-called "sum of $\hat{f}$") obtained in this way has the power series $\hat{f}$ we started from as asymptotic expansion.

For the applicability of Borel summability method it is necessary to impose suitable growth conditions on the coefficients a_n [279].

Definition 7.3. Given $s > 0$, a formal power series $\hat{f}(z) = \sum_{n=0}^{\infty} a_n z^{n/k} \in$ $\mathbb{C}[z^{1/k}]$ belongs to the s-Gevrey class $\mathbb{C}[z^{1/k}]_s$ if there exist two constants $C, M > 0$, such that

$$|a_n| \leq CM^n (\Gamma(1 + n/k))^s, \qquad \forall n \in \mathbb{N}, \tag{7.4}$$

where Γ is the Euler Γ function.

Remark 7.3. By Stirling formula conditions (7.4) can be replaced by

$$|a_n| \leq CM^n \left(\frac{n}{k}\right)^{ns/k}, \qquad \forall n \in \mathbb{N}, \tag{7.5}$$

or by

$$|a_n| \leq CM^n \Gamma(1 + sn/k), \quad \forall n \in \mathbb{N}. \tag{7.6}$$

The Gevrey classes are connected via the following transform acting on formal series:

Definition 7.4. The map $B_{p,k} : \mathbb{C}[z^{1/k}]_s \to \mathbb{C}[z^{p/k}]_{s-p}$ defined by

$$B_{p,k}\left[\sum_{n=0}^{\infty} a_n z^{n/k}\right](t) := \sum_{n=0}^{\infty} \frac{a_n}{\Gamma(1 + np/k)} t^{np/k}$$

is called the (formal) (p, k)-Borel transform.

It is important to note that the (s, k)-Borel transform maps $\mathbb{C}[z^{1/k}]_s$ to convergent series.

We can now define the concept of μ-*Borel summability*:

Definition 7.5. Let $k, s > 0$, $\mu = 1/s$. A formal power series $\hat{f}(z) = \sum_{n=0}^{\infty} a_n z^{n/k}$ is called μ−Borel summable to the sum f if f is an holomorphic function on V for some angular neighborhood of zero V, $f \sim \hat{f}$ as $z \to 0$ in V and the following procedure is possible:

(1) The (s, k)-Borel transform $B_{s,k}[\hat{f}](t)$ has non-zero radius of convergence and thus converges in a neighborhood of zero to some function $B(\cdot)$.
(2) This holomorphic function B admits an analytic continuation (denoted again by the symbol $B(\cdot)$) onto some open neighborhood of $\mathbb{R}^+$
(3) The Borel-Laplace transform $\mathcal{L}(B)$ of B gives a representation of f on a subset of V: $f(z) = \frac{1}{z^\mu}\mathcal{L}(B)(\frac{1}{z^\mu})$, i.e.[2]:

$$f(z) = \frac{1}{z^\mu} \int_0^\infty B(t) e^{-t/z^\mu} dt \tag{7.7}$$

---

[2]The integral in (7.7) is the conventional Laplace transform in the variable $w = 1/z^\mu$. Note that if the integral converges for some $z_0 \neq 0$ them it converges for all $z \in \mathbb{C}$ such that $Re(1/z^\mu) > Re(1/z_0^\mu)$.

If $\mu = 1$, then $\hat{f}$ is simply said *Borel summable*.

In other words if an asymptotic series is Borel summable to a function f, it characterizes uniquely f, even if it is not convergent.

The following criterion for Borel summability is due to F. Nevanlinna [266] and is a improvement of a result by Watson [328], see also [307, 285] for a modern proof and discussion of Nevanlinna's result:

Theorem 7.1. *Let $k > 0$, $R \in (0, +\infty]$ and define $D_R := \{z \in \mathbb{C} : Re(1/z) > 1/R\}$ if $R \neq \infty$ and $D_R := \{z \in \mathbb{C} : Re(z) > 0\}$ else.*

Let f be an holomorphic function on D_R admitting an asymptotic expansion with respect to the asymptotic sequence $z^{n/k}$ in the domain D_R

$$f(z) \sim \sum_{n=0}^{\infty} a_n z^{n/k} =: \hat{f}.$$

Let us assume that $\exists A > 0, \rho > 0$, such that $\forall \epsilon > 0, z \in \{Re(1/z) \geq \epsilon + 1/R\}$, $\hat{\rho} > \rho, n \in \mathbb{N}$, the following holds:

$$\left| f(z) - \sum_{i=0}^{n-1} a_i z^{i/k} \right| \leq A\Gamma(1 + n/k)\hat{\rho}^n |z|^{n/k}. \tag{7.8}$$

Then the asymptotic power series $\hat{f}$ is Borel summable to the function f.

Remark 7.4. By Stirling formula condition (7.8) can be replaced by

$$\left| f(z) - \sum_{i=0}^{n-1} a_i z^{i/k} \right| \leq A\hat{\rho}^n \left(\frac{n}{k} \right)^{n/k} |z|^{n/k}. \tag{7.9}$$

Proof. The present proof is taken from [307], and it is a generalization of Hardy's proof of Watson's theorem [180], see also [266]. Without loss of generality, we can restrict ourselves to consider the case where $k = 1$. The general case can be handled in a completely similar way.

By estimate (7.8), it is possible to see that the integrals

$$b_m(t) = a_m + \frac{1}{2\pi i} \oint_{Rez^{-1}=r^{-1}} e^{t/z} z^{-(m+1)} \left(f(z) - \sum_{j=0}^{m} a_j z^j \right) dz, \tag{7.10}$$

are absolutely convergent for $t \geq 0$ and independent of r for $0 < r < R$. Moreover b_0 is a C^∞ function whose mth derivative is b_m and the following estimate holds:

$$|b_m(t)| \leq K_1 \rho^{m+1} (m+1)! e^{t/R},$$

with K_1 independent of t and m. By performing a contour integral, one finds

$$b_0(t) = \sum_{j=0}^{N-1} \frac{1}{j!} a_j t^j + \frac{1}{2\pi i} \oint_{Re z^{-1} = r^{-1}} e^{t/z} z^{-1} \left(f(z) - \sum_{j=0}^{N-1} a_j z^j \right) dz.$$

(7.11)

By Eq. (7.8) and by taking $r = t/N$ (with $N > t/R$), the second term in the sum at the right-hand side of Eq. (7.11) goes to 0 as $N \to \infty$ and it follows that

$$b_0(t) = B(t) = \sum_{j=0}^{\infty} \frac{1}{j!} a_j t^j.$$

(7.12)

By condition (7.8) the series on the right hand side of Eq. (7.12) converges in the circle $|t| < 1/\rho$. Moreover it is possible to see that each series

$$B_{t_0}(t) = \sum_{m=0}^{\infty} \frac{1}{m!} b_m(t_0)(t - t_0)^m, \qquad t_0 \geq 0,$$

converges in the circle $|t - t_0| < 1/\rho$ and satisfies there the bound

$$|B_{t_0}(t)| \leq K_1 e^{t_0/R} \frac{1}{1 - \rho(t - t_0)}.$$

It is not difficult to prove that $B_{t_0}(t) = B_{t_1}(t)$ when both functions are defined and that the union of these functions defines an analytic function $B(t)$ defined in the region $S_\rho \subset \mathbb{C}$ given by

$$S_\rho := \{z \in \mathbb{C}, : dist(z, \mathbb{R}^+) < 1/\rho\},$$

and satisfying there the bound

$$|B(t)| \leq K e^{|t|/R}.$$

By inserting (7.10) with $m = 0$ into the expression for the Laplace transform and by interchanging the order of integration, we obtain:

$$\frac{1}{z} \int_0^\infty e^{-t/z} B(t) dt = \frac{1}{z} \int_0^\infty e^{-t/z} a_0 dt$$
$$+ \frac{1}{z} \int_0^\infty e^{-t/z} \frac{1}{2\pi i} \oint_{Re z^{-1} = r^{-1}} e^{t/z'} \frac{f(z') - a_0}{z'} dz' dt = f(z).$$

$\square$

7.2 The stationary phase method: Finite dimensional case

The present and the following sections concern the study of the asymptotic behaviour as $\hbar \downarrow 0$ of function of the variable $\hbar \in \mathbb{R}^+$ defined in terms of oscillatory integrals of the form

$$I(\hbar) := \widetilde{\int_{\mathcal{H}}} e^{\frac{i}{\hbar}\Phi(x)} f(x)dx, \qquad \hbar \in \mathbb{R}^+, \qquad (7.13)$$

with $\Phi : \mathcal{H} \to \mathbb{R}$ and $f : \mathcal{H} \to \mathbb{C}$.

In the finite dimensional case, i.e. when we can identify the Hilbert space $\mathcal{H}$ with $\mathbb{R}^n$, the fundamental tool for the study of the asymptotics of the integral (7.13) is the *stationary phase method* [130]. It was originally developed by Stokes [309] and Kelvin [225] in the 19th century. The physical relevance of this topic is connected with the fact that integrals of the form (7.13) play a fundamental role in the description of wave phenomena. More recent investigations can be found in the work of Maslov [248], in connection with the study of the semiclassical limit of quantum mechanics, and in the work by Hörmander [186, 187], in connection with the theory of Fourier integral operators and the study of partial differential equations. When the phase function Φ has some degenerate critical points, the theory of unfoldings of singularities plays a crucial role in the description of the asymptotics of the integral (7.13) (see for instance Arnold's and Duistermaat's work [44, 125]) and brings the stationary phase method to an high level of mathematical rigour and elegance.

Despite the technical difficulties in the rigorous study of the asymptotic behaviour of oscillatory integrals, the mathematical and physical ideas behind the stationary phase method are rather simple and intuitive. Indeed, when $\hbar$ can be considered very small, the function

$$x \mapsto e^{\frac{i}{\hbar}\Phi(x)}, \qquad x \in \mathbb{R}^n,$$

oscillates very fast, in such a way that the contributions to the integral coming from the positive and the negative parts of the oscillations annul each other. The key point is the fact that the only regions of $\mathbb{R}^n$ giving a non-vanishing contribution to the value of the integral (7.13) are the neighbourhoods of the stationary points of the phase function Φ, i.e. the points x_c satisfying the equation

$$\nabla \Phi(x) = 0.$$

As an illustrative example, let us consider the integral $I(\hbar) := \int_{-\pi/4}^{\pi/4} cos(x^2/\hbar)dx$. As a comparison between Figure 7.1(a) and Figure 7.1(b)

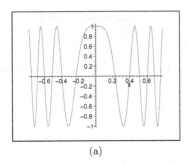   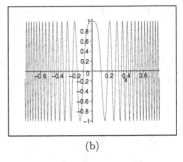

(a)   (b)

Fig. 7.1   The function $\cos(x^2/\hbar)$ for different values of the parameter $\hbar$. (a) $\hbar = 1/30$.
(b) $\hbar = 1/150$.

shows, when the frequency $\hbar^{-1}$ increases, the region giving a significant contribution to the integral is a neighborhood of the stationary point $x = 0$.

These reasoning can be made completely rigorous under rather general assumptions. Following [7], we describe here in some detail the study of the asymptotics of finite dimensional oscillatory integrals of the form

$$I(\hbar) := \int_{\mathbb{R}^n} e^{\frac{i}{2\hbar}\langle x, Tx\rangle} e^{-\frac{i}{\hbar}V(x)} g(x) dx, \qquad (7.14)$$

where $T = (I - L) : \mathbb{R}^n \to \mathbb{R}^n$ is a self-adjoint bijection of $\mathbb{R}^n$, $V : \mathbb{R}^n \to \mathbb{R}$ and $g : \mathbb{R}^n \to \mathbb{C}$ satisfy suitable assumptions. These results will be generalized in the next section to the infinite dimensional case.

Let us consider the phase function $\Phi : \mathbb{R}^n \to \mathbb{R}$,

$$\Phi(x) = \frac{1}{2}\langle x, Tx\rangle - V(x), \qquad x \in \mathbb{R}^n,$$

and the set of critical points of Φ:

$$\mathcal{C}(\Phi) := \{x \in \mathbb{R}^n \, . \, \Phi'(x) = 0\}.$$

Theorem 7.2. *Let* $V : \mathbb{R}^n \to \mathbb{R}$ *be a* C^∞ *function with bounded first derivatives and all higher order derivatives of at most linear growth, i.e.*

$$|V'(x)| \leq M, \qquad D^\alpha V(x) \leq m(1 + |x|), \qquad x \in \mathbb{R}^n,$$

for suitable constants m, M, *where* $V'(x) \equiv \nabla V(x)$ *and*

$$D^\alpha = D_1^{\alpha_1} \dots D_n^{\alpha_n} = \frac{\partial^{\alpha_1}}{\partial x_1^{\alpha_1}} \dots \frac{\partial^{\alpha_n}}{\partial x_n^{\alpha_n}}, \qquad |\alpha| = \alpha_1 + \dots + \alpha_n.$$

Let $g : \mathbb{R}^n \to \mathbb{C}$ *a* C^∞ *function such that for some* $p \geq 0$

$$|D^\alpha g(x)| \leq C_\alpha (1 + |x|^2)^{p/2}, \qquad \forall \alpha \in \mathbb{N}^n, \forall x \in \mathbb{R}^n. \qquad (7.15)$$

Let us assume that the critical set $C(\Phi)$ contains a finite number of points $\{c_1, \ldots, c_s\}$, that are not degenerate, i.e.

$$\det \Phi''(c_i) \neq 0, \qquad i = 1, \ldots, s,$$

where $\Phi'' \equiv D^2 \Phi$.

Let χ_i, with $i = 0, 1, \ldots s$, denote the $C^\infty(\mathbb{R}^n, \mathbb{R})$ functions such that

(1) $0 \leq \chi_i \leq 1$ and $\sum_{i=0}^{s} \chi_i = 1$;
(2) $C(\Phi) \cap supp(\chi_i) = \{c_i\}$ for $i = 1, \ldots s$;
(3) $supp(1 - \chi_0) \subset B(0, 3/2r)$ and $\chi_0^{-1}(\{0\}) = \bar{B}(0, r)$;
(4) $\chi_i(x) = 1$ for $x \in B(c_i, r_i)$ and $i = 1, \ldots s$, for some $r, r_i > 0$;

where $B(a, r)$ denotes the open ball with center a and radius r and $\bar{B}(a, r)$ its closure.

Then the integral $I(\hbar)$ in Eq. (7.14) is well defined and it is given by:

$$I(\hbar) = \sum_{j=0}^{s} \widetilde{\int_{\mathbb{R}^n}} e^{\frac{i}{2\hbar}\langle x, Tx \rangle} e^{-\frac{i}{\hbar}V(x)} g(x) \chi_j(x) dx$$

$$=: \sum_{j=0}^{s} I_j(\hbar). \tag{7.16}$$

Moreover, by defining $I_j^*(\hbar)$ as

$$I_j(\hbar) = e^{\frac{i}{\hbar}\Phi(c_j)} I_j^*(\hbar), \qquad j = 1, \ldots, s,$$

and $I_0^*(\hbar) = I_0(\hbar)$, then each I_j^* is a C^∞ function on $\mathbb{R}$ and in particular

$$I_j^*(0) = (\det(T - D^2 V(c_j)))^{-1/2} g(c_j), \qquad j = 1, \ldots, s,$$

$$I_0^{(k)}(0) = 0, \qquad \forall k \geq 0.$$

Proof. We follow [7], see also [186, 187].

Since both the phase function Φ and the function g satisfy the assumptions of Theorem 5.1, the oscillatory integral $I(\hbar)$ is well defined.

Let us assume that $s = 1$ (the proof in the general case is completely analogous).

Let $\phi \in \mathcal{S}(\mathbb{R}^n)$ such that $\phi(0) = 1$ and let $\epsilon > 0$. The regularized approximations of the oscillatory integral $I(\hbar)$ are given by:

$$
\begin{aligned}
I_\epsilon(\hbar, \phi) &= (2\pi i\hbar)^{-n/2} \int_{\mathbb{R}^n} e^{\frac{i}{2\hbar}\langle x, Tx \rangle} e^{-\frac{i}{\hbar}V(x)} g(x)\phi(\epsilon x)dx \\
&= (2\pi i\hbar)^{-n/2} \int_{\mathbb{R}^n} e^{\frac{i}{2\hbar}\langle x, Tx \rangle} e^{-\frac{i}{\hbar}V(x)} g(x)\chi_1(x)\phi(\epsilon x)dx \\
&\quad + (2\pi i\hbar)^{-n/2} \int_{\mathbb{R}^n} e^{\frac{i}{2\hbar}\langle x, Tx \rangle} e^{-\frac{i}{\hbar}V(x)} g(x)(1 - \chi_1(x))\phi(\epsilon x)dx \\
&= I_\epsilon^1(\hbar, \phi) + I_\epsilon^2(\hbar, \phi).
\end{aligned}
\tag{7.17}
$$

The first integral is equal to

$$
I_\epsilon^1(\hbar, \phi) = (2\pi i\hbar)^{-n/2} e^{\frac{i}{\hbar}\Phi(c_1)} \int_{\mathbb{R}^n} e^{\frac{i}{\hbar}(\Phi(x) - \Phi(c_1))} g(x)\chi_1(x)\phi(\epsilon x)dx,
$$

and, by dominated convergence, it converges as $\epsilon \downarrow 0$ to

$$
(2\pi i\hbar)^{-n/2} e^{\frac{i}{\hbar}\Phi(c_1)} \int_{\mathbb{R}^n} e^{\frac{i}{\hbar}(\Phi(x) - \Phi(c_1))} g(x)\chi_1(x)dx,
$$

which is a C^∞ function of $\hbar \in \mathbb{R}$ (see [187], Theorem 7.7.5). By making in a neighbourhood of c_1, i.e. in the support of χ_1, the change of variable

$$
\frac{\Phi(x) - \Phi(c_1)}{2\hbar} = \frac{yD^2\Phi(c_1)y}{2}
$$

one gets easily

$$
I_1^*(0) = (\det D^2\Phi(c_1))^{-1/2} g(c_1) = (\det(T - D^2V(c_1)))^{-1/2} g(c_1).
$$

By reasoning as in the proof of Theorem 5.1, it is possible to see that also the limit $\lim_{\epsilon \to 0} I_\epsilon^2(\hbar, \phi)$ exists and it is independent of ϕ. Moreover, by denoting this limit with $I^2(\hbar)$, one has

$$
|I^2(\hbar)| \leq C_k |\hbar|^k, \quad |\hbar| \leq 1, \forall k \geq 1.
$$

Indeed let us consider the C^∞ vector field $a(x) = (a_1(x), \ldots, a_n(x))$ with components given by

$$
a_j(x) = \frac{D_j\Phi(x)}{|\nabla\Phi(x)|^2}(1 - \chi_1(x)), \qquad j = 1, \ldots, n,
$$

and the first order differential operator

$$
L_\hbar := (-i\hbar) \sum_{j=1}^n a_j(x)\frac{\partial}{\partial x_j}.
$$

It is simple to see that $L_\hbar^+$, the adjoint of $L_\hbar$, is given by

$$L_\hbar^+ := i\hbar \left(\sum_{j=1}^{n} a_j(x) \frac{\partial}{\partial x_j} + \text{div } a \right) \equiv i\hbar A.$$

By the assumptions on the function V, it is simple to verify that for any $\alpha \in \mathbb{N}^n$, there exists $C_\alpha > 0$ such that

$$|D^\alpha a_j(x)| \leq \frac{C_\alpha}{1 + |x|}, \qquad x \in \mathbb{R}^d.$$

Moreover, since

$$L_\hbar e^{\frac{i}{\hbar}\Phi(x)} = (1 - \chi_1(x)) e^{\frac{i}{\hbar}\Phi(x)},$$

by Stokes formula we have:

$$I_\epsilon^2(\hbar, \phi) = (2\pi i\hbar)^{-n/2} \int_{\mathbb{R}^n} e^{\frac{i}{\hbar}\Phi(x)} L_\hbar^+(f_\epsilon) dx$$

$$= (2\pi i\hbar)^{-n/2} i\hbar \int_{\mathbb{R}^n} e^{\frac{i}{\hbar}\Phi(x)} A(f_\epsilon) dx, \qquad (7.18)$$

where $f_\epsilon(x) = \phi(\epsilon x) g(x)$. By iterating the procedure k times, we get

$$I_\epsilon^2(\hbar, \phi) = (2\pi i\hbar)^{-n/2} (i\hbar)^k \int_{\mathbb{R}^n} e^{\frac{i}{\hbar}\Phi(x)} f_{\epsilon,k}(x) dx,$$

where

$$f_{\epsilon,k}(x) = \sum_{j=0}^{k} \beta_j^k(x) D^j f_\epsilon(x),$$

where $\beta_j^k(x)$ is a system of coefficients with the property that each $\beta_j^k(x)$ is a sum of finite number of terms, and each term is a multiplication of k factors. Each factor is either one of the functions a_j or its derivative, the total number of derivatives in each term being equal to $k - j$.

By taking $k > n + p$, where p is constant in Eq. (7.15), we have that $\beta_j^k(x) D^j g(x) \in L^1(\mathbb{R}^n)$ and, by dominated convergence theorem, the limit

$$I^2(\hbar) := \lim_{\epsilon \to 0} I_\epsilon^2(\hbar, \phi)$$

exists and satisfies the condition

$$|I^2(\hbar)| \leq C_k |\hbar|^k, \qquad |\hbar| \leq 1, \forall k \geq 1.$$

$\square$

Remark 7.5. Under additional assumptions (see for instance [17], Corollaries 2.4 and 2.5) it is also possible to compute all term of the asymptotic expansion of each integral $I_j^*(\hbar)$, which is given by:

$$
I_j^*(\hbar) = (\det T)^{-1/2} \sum_{m=0}^{\infty} \hbar^m \left(\frac{i}{2}\right)^m \sum_{k=0}^{\infty} \left(\frac{1}{2}\right)^k \frac{1}{k!(m+k)!}
$$

$$
\left[\left(\left(\sum_{l=1}^{k} \nabla_{x_l} + \nabla_y\right) T^{-1} \left(\sum_{l=1}^{k} \nabla_{x_l} + \nabla_y\right)\right)_2^{m+k} V(x_1) \dots V(x_k) g(y) \chi_j(y)\right],
$$

$$
\tag{7.19}
$$

where the value of $[\ ]$ is to be taken at the critical point c_j and $\left(\left(\sum_{l=1}^{k} \nabla_{x_l} + \nabla_y\right) T^{-1} \left(\sum_{l=1}^{k} \nabla_{x_l} + \nabla_y\right)\right)_2^{m+k}$ is the sum of all terms in the expansion of $\left(\left(\sum_{l=1}^{k} \nabla_{x_l} + \nabla_y\right) T^{-1} \left(\sum_{l=1}^{k} \nabla_{x_l} + \nabla_y\right)\right)^{m+k}$ which are of at least second degree with respect to each ∇_{x_l}, $l = 1, \dots, k$.

In the case where the phase function Φ in the oscillatory integral (7.13) presents some degenerate critical point c, that is

$$
\nabla\Phi(c) = 0, \qquad \det \Phi''(c) = 0,
$$

the situation is more involved. In this case the asymptotic behaviour as $\hbar \downarrow 0$ of

$$
I(\hbar) := \int e^{\frac{i}{\hbar}\Phi(x)} f(x) dx
$$

is determined by taking into account the higher derivatives of Φ and the classification of different types of degeneracies [48, 47].

In the case of the generalized Fresnel integral studied in Section 5.3, the solution of the problem is simpler and has been described in [31], where the whole asymptotic expansion of an oscillatory integral of the form

$$
\int_{\mathbb{R}^n}^o e^{\frac{i}{\hbar}\Phi(x)} f(x) dx, \tag{7.20}
$$

is studied, where $\Phi : \mathbb{R}^n \to \mathbb{R}$ is an homogeneous polynomial function of even degree $2M$:

$$
\Phi(x) = A_{2M}(x, \dots, x), \qquad x \in \mathbb{R}^d,
$$

and A_{2M} is a completely symmetric $2M$th order covariant tensor on $\mathbb{R}^n$ such that $A_{2M}(x, \dots, x) > 0$ unless $x = 0$.

By the positivity of the phase function it is simple to see that the integral (7.20) is well defined also for $\hbar \in \mathbb{C}$, $Im(\hbar) < 0$, provided that the function f is bounded. For $\hbar \in \mathbb{R}$, sufficient conditions for the definition of the purely oscillatory integral (7.20) are given by Theorem 5.4.

Theorem 7.3. *Let $f \in \mathcal{F}(\mathbb{R}^n)$ be the Fourier transform of a bounded variation measure μ_f admitting moments of all orders.*
Let us suppose μ_f satisfies the following conditions, for all $l \in \mathbb{N}$:

(1)

$$\int_{\mathbb{R}^n} |kx|^l e^{-kx} |d\mu_f|(k) \leq F(l)g(|x|)e^{c|x|^{2M-1}}, \qquad \forall x \in \mathbb{R}^n,$$

where $c \in \mathbb{R}$, $F(l)$ is a positive constant depending on l, $g : \mathbb{R}^+ \to \mathbb{R}$ is a positive function with polynomial growth.
(2)

$$\left| \int_{\mathbb{R}^n} (ku)^l e^{ik\rho u \hbar^{1/2M}} e^{i\pi/4M} \, d\mu_f(k) \right| \leq Ac^l C(l, M, n)$$

for all $u \in S_{n-1}$, $\rho \in \mathbb{R}^+$, $Im(\hbar) \leq 0$ $\hbar \neq 0$, where $A, c, C(l, M, n) \in \mathbb{R}$ (and S_{n-1} is the $(n-1)$-spherical hypersurface of radius 1 and centered at the origin).

Then the oscillatory integral (7.20) admits for $\hbar \in \mathbb{C}$, $Im(\hbar) \leq 0$, the following asymptotic expansion in powers of $\hbar^{1/2M}$:

$$I(\hbar) = \hbar^{n/2M} \frac{e^{in\pi/4M}}{2M} \sum_{l=0}^{N-1} \frac{(i)^l}{l!} (e^{i\pi/4M})^l \hbar^{l/2M} \Gamma\left(\frac{l+n}{2M}\right)$$

$$\int_{\mathbb{R}^n} \int_{S_{n-1}} (ku)^l P(u)^{-\frac{l+n}{2M}} d\Omega_{n-1} d\mu_f(k) + \mathcal{R}_N, \quad (7.21)$$

with $|\mathcal{R}_N| \leq A'|\hbar|^{N/2M}(c')^N \frac{C(N,M,n)}{N!} \Gamma\left(\frac{n+N}{2M}\right)$ where A', $c' \in \mathbb{R}$ are suitable constants and $C(N, M, n)$ is the constant in assumption (2). If $C(N, M, n)$ satisfies the following bound:

$$C(N, M, n) \leq N! \Gamma\left(\frac{n+N}{2M}\right)^{-1} \qquad (7.22)$$

then the series has a positive radius of convergence, while if

$$C(N, M, n) \leq N! \Gamma\left(1 + \frac{N}{2M}\right) \Gamma\left(\frac{n+N}{2M}\right)^{-1} \qquad (7.23)$$

then the expansion is Borel summable in the sense of, e.g. [266, 180] and determines $I(\hbar)$ uniquely.

Remark 7.6. The hypothesis of Theorem 7.3 are slightly different from those presented in [31].

Proof. Let $\tilde{F}(k) \equiv \int_{\mathbb{R}^n} e^{ikx} e^{iA_{2M}(x,\dots,x)} dx$, then by Theorem 5.4 the oscillatory integral (7.20) is given by:

$$\int_{\mathbb{R}^n} e^{\frac{i}{\hbar} A_{2M}(x,\dots,x)} f(x)dx = \hbar^{n/2M} \int_{\mathbb{R}^n} \tilde{F}(\hbar^{1/2M}k)\mu_f(dk). \qquad (7.24)$$

By Lemma 5.1, $\tilde{F}$ is given by

$$\tilde{F}(\hbar^{1/2M}k) = e^{in\pi/4M} \int_{\mathbb{R}^n} e^{i\hbar^{1/2M}kxe^{i\pi/4M}} e^{-A_{2M}(x,\dots,x)} dx, \qquad k \in \mathbb{R}^n,$$

where, if $\hbar = |\hbar|e^{i\phi}$, $\phi \in [-\pi, 0]$, $\hbar^{1/2M} = |\hbar|^{1/2M} e^{i\phi/2M}$. By representing the latter absolutely convergent integral using polar coordinates in $\mathbb{R}^n$ we get:

$$\tilde{F}(\hbar^{1/2M}k) = e^{in\pi/4M} \int_{S_{n-1}} \int_0^\infty e^{i\hbar^{1/2M}e^{i\pi/4M}\rho ku} e^{-\rho^{2M}A_{2M}(u,\dots,u)}$$

$$\rho^{n-1} d\rho d\Omega_{n-1}$$

where $d\Omega_{n-1}$ is the Riemann-Lebesgue measure on the $n-1$-dimensional spherical hypersurface S_{n-1}, $x = \rho u$, $\rho = |x|$, $u \in S_{n-1}$ is a unitary vector. We can expand the latter integral in a power series of $\hbar^{1/2M}$ and apply Fubini theorem:

$$\tilde{F}(\hbar^{1/2M}k) = e^{in\pi/4M} \int_{S_{n-1}} \int_0^\infty \sum_{l=0}^\infty \frac{(i)^l}{l!} (e^{i\pi/4M})^l \hbar^{l/2M} \rho^l (ku)^l$$

$$e^{-\rho^{2M}A_{2M}(u,\dots,u)} \rho^{n-1} d\rho d\Omega_{n-1}$$

$$= e^{in\pi/4M} \sum_{l=0}^\infty \frac{(i)^l}{l!} (e^{i\pi/4M})^l \hbar^{l/2M} \int_{S_{n-1}} (ku)^l \int_0^\infty \rho^{l+n-1}$$

$$e^{-\rho^{2M}A_{2M}(u,\dots,u)} d\rho d\Omega_{n-1}$$

$$= \frac{e^{in\pi/4M}}{2M} \sum_{l=0}^\infty \frac{(i)^l}{l!} (e^{i\pi/4M})^l \hbar^{l/2M} \Gamma\left(\frac{l+n}{2M}\right)$$

$$\times \int_{S_{n-1}} (ku)^l P(u)^{-\frac{l+n}{2M}} d\Omega_{n-1}, \qquad (7.25)$$

where $P(u) \equiv A_{2M}(u,\dots,u)$ is a strictly positive continuous function on the compact set S_{n-1}, so that it admits an absolute minimum denoted by m. This gives

$$\left| \int_{S_{n-1}} (ku)^l P(u)^{-\frac{l+n}{2M}} d\Omega_{n-1} \right| \le |k|^l m^{-\frac{l+n}{2M}} \Omega_{n-1}(S_{n-1})$$

$$= |k|^l m^{-\frac{l+n}{2M}} 2\pi^{n/2} \Gamma\left(\frac{n}{2}\right)^{-1}. \qquad (7.26)$$

The latter inequality and the Stirling formula assure the absolute convergence of the series (7.25). We can now insert this formula into (7.24) and get:

$$
\int_{\mathbb{R}^n} e^{\frac{i}{\hbar} A_{2M}(x,\ldots,x)} f(x)dx =
$$

$$
= \hbar^{n/2M} \frac{e^{in\pi/4M}}{2M} \sum_{l=0}^{N-1} \frac{(i)^l}{l!} (e^{i\pi/4M})^l \hbar^{l/2M} \Gamma\left(\frac{l+n}{2M}\right)
$$

$$
\int_{\mathbb{R}^n} \int_{S_{n-1}} (ku)^l P(u)^{-\frac{l+n}{2M}} d\Omega_{n-1} \mu_f(dk) + \mathcal{R}_N. \quad (7.27)
$$

Equation (7.27) can also be written in the following form:

$$
\int_{\mathbb{R}^n} e^{\frac{i}{\hbar} A_{2M}(x,\ldots,x)} f(x)dx =
$$

$$
= \hbar^{n/2M} \frac{e^{iN\pi/4M}}{2M} \sum_{l=0}^{N-1} \frac{1}{l!} (e^{i\pi/4M})^l \hbar^{l/2M} \Gamma\left(\frac{l+n}{2M}\right)
$$

$$
\int_{S_{n-1}} P(u)^{-\frac{l+n}{2M}} \frac{\partial^l}{\partial u^l} f(0) d\Omega_{n-1} + \mathcal{R}_N, \quad (7.28)
$$

where $\frac{\partial^l}{\partial u^l} f(0)$ denotes the l_{th} partial derivative of f at 0 in the direction u, and

$$
\mathcal{R}_N = \hbar^{n/2M} e^{in\pi/4M} \int_{\mathbb{R}^n} \int_{S_{n-1}} \int_0^\infty \sum_{l=N}^\infty \frac{(i)^l}{l!} (e^{i\pi/4M})^l \hbar^{l/2M} \rho^l (ku)^l
$$

$$
e^{-\rho^{2M} A_{2M}(u,\ldots,u)} \rho^{n-1} d\rho d\Omega_{n-1} \mu_f(dk). \quad (7.29)
$$

In the case where assumptions (1) and (2) are satisfied, we can prove the asymptoticity of the expansion (7.27), indeed

$$
\mathcal{R}_N = \hbar^{n/2M} e^{in\pi/4M} \frac{(i)^N}{N-1!} (e^{i\pi/4M})^N \hbar^{N/2M}
$$

$$
\int_{\mathbb{R}^n} \int_{S_{n-1}} \int_0^\infty \int_0^1 (1-t)^{N-1} e^{iku\rho t \hbar^{1/2M}} e^{i\pi/4M} e^{-\rho^{2M} A_{2M}(u,\ldots,u)}
$$

$$
(ku)^N \rho^{n+N-1} dt d\rho d\Omega_{n-1} \mu_f(dk). \quad (7.30)
$$

By assumptions (1), (2) and Fubini theorem, the latter is bounded by

$$
|\mathcal{R}_N| \leq \frac{A}{M} \pi^{n/2} \Gamma\left(\frac{n}{2}\right)^{-1} |\hbar|^{(n+N)/2M} c^N m^{-\frac{n+N}{2M}} \frac{C(N,M,n)}{N!} \Gamma\left(\frac{n+N}{2M}\right).
$$

If assumption (7.22) is satisfied, then the latter becomes

$$|\mathcal{R}_N| \le \frac{A}{M} \pi^{n/2} \Gamma\left(\frac{n}{2}\right)^{-1} |\hbar|^{(n+N)/2M} c^N m^{-\frac{n+N}{2M}}$$

and the series has a positive radius of convergence, while if assumption (7.23) holds, we get the estimate

$$|\mathcal{R}_N| \le \frac{A}{M} \pi^{n/2} \Gamma\left(\frac{n}{2}\right)^{-1} |\hbar|^{(n+N)/2M} c^N m^{-\frac{n+N}{2M}} \Gamma\left(1 + \frac{N}{2M}\right).$$

This bound, the analyticity of the function $I(\hbar)$ in a sector of the complex plan of amplitude π (i.e. in $\text{Im}(\hbar) < 0$) and Nevanlinna's Theorem 7.1 assure the Borel summability of the power series expansion in $\hbar^{1/2M}$ (7.21). $\qquad\square$

7.3   The stationary phase method: Infinite dimensional case

The implementation of an infinite dimensional version of the stationary phase method allowing the study of the asymptotic behaviour, when $\hbar \,\rightarrow\, 0$, of infinite dimensional oscillatory integrals of form

$$\widetilde{\int_{\mathcal{H}}} e^{\frac{i}{\hbar}\Phi(x)} f(x) dx$$

is not trivial.

The first results can be found in [17] and were further developed in [7] and in [284]. The authors consider infinite dimensional oscillatory integrals of the form

$$I(\hbar) = \widetilde{\int_{\mathcal{H}}} e^{\frac{i}{2\hbar}\langle x, Tx\rangle} e^{-\frac{i}{\hbar}V(x)} g(x) dx, \qquad (7.31)$$

with $V, g \in \mathcal{F}(\mathcal{H})$, $T = I - L$ self-adjoint and invertible operator, $L : \mathcal{H} \to \mathcal{H}$ of trace class. Under additional regularity assumptions on V, g it is possible to prove that the phase function

$$\Phi(x) = \frac{1}{2}\langle x, (I - L)x\rangle - V(x), \qquad x \in \mathcal{H}, \qquad (7.32)$$

has only non-degenerate critical points, that the essential part in $I(\hbar)$ is a C^∞ function of $\hbar$ and its asymptotic expansion at $\hbar = 0$ depends only on the derivatives of V and g at these critical points (see for instance [17, 7]).

The following lemma gives sufficient conditions for the existence and uniqueness of a non-degenerate stationary point of the phase function (7.32)

Lemma 7.1. *Let $V \in \mathcal{F}(\mathcal{H})$, with*

$$V(x) = \int_{\mathcal{H}} e^{i\langle x, y\rangle} d\mu(y).$$

Let us assume that the total variation measure $|\mu|$ of the measure μ satisfies the following inequality

$$\|T^{-1}\| \int_{\mathcal{H}} \|y\|^2 d|\mu|(y) < 1. \tag{7.33}$$

Then there exists a unique $a \in \mathcal{H}$ such that

$$\Phi'(a) = Ta - V'(a) = 0.$$

Moreover $\Phi''(a)$ is invertible.

Proof.

As $V'(x) = i \int_{\mathcal{H}} y e^{i\langle x, y\rangle} d\mu(y)$, from the triangle inequality

$$\|T^{-1}V'(x) - T^{-1}V'(y) \leq \|T^{-1}\| \int_{\mathcal{H}} \|z\| |1 - e^{i\langle x-y,z\rangle}| d|\mu|(z)$$

$$\leq \|T^{-1}\| \|x - y\| \int_{\mathcal{H}} \|z\|^2 \left| \frac{1 - e^{i\langle x-y,z\rangle}}{\|x - y\| \|z\|} \right| d|\mu|(z). \tag{7.34}$$

As

$$\left| \frac{1 - e^{it}}{t} \right| = 2 \left| \frac{\sin \frac{t}{2}}{t} \right| \leq 1,$$

we have

$$\|T^{-1}V'(x) - T^{-1}V'(y) \leq \| \|x - y\| \|T^{-1}\| \int_{\mathcal{H}} \|y\|^2 d|\mu|(y)| < \| \|x - y\|,$$

and by the contraction mapping principle, it follows that there exists a unique $a \in \mathcal{H}$ such that $Ta - V'(a) = 0$.

Concerning the study of the operator $\Phi''(a)$, it is simple to see that

$$\Phi''(a) = T - V''(a),$$

and that

$$V''(a)(x)(z) = i^2 \int_{\mathcal{H}} \langle y, z\rangle \langle y, x\rangle e^{i\langle y,a\rangle} d\mu(y), \qquad x, z \in \mathcal{H}.$$

By condition (7.33), it follows that $\|T^{-1}V''(a)\| < 1$, so $T - V''(a)$ is invertible. $\qquad\square$

The following theorem gives sufficient conditions for the existence of the asymptotic expansion of the integral (7.31) around a unique stationary point of the phase function (7.32). It also states the Borel summability of the asymptotic expansion. We skip here the proof, which is rather technical and involves a large amount of calculations, but we point out to the interested reader the original paper by J. Rezende [284] for more details.

Theorem 7.4. *Let* $V, g \in \mathcal{F}(\mathcal{H})$, *with*

$$V(x) = \int_{\mathcal{H}} e^{i\langle x, y \rangle} d\mu(y),$$

$$g(x) = \int_{\mathcal{H}} e^{i\langle x, y \rangle} d\nu(y). \qquad (7.35)$$

Let us assume that the measures $\mu, \nu \in \mathcal{M}(\mathcal{H})$ *admits finite moments of all orders and there exist constants* $L, M, \epsilon > 0$ *such that the following inequalities hold:*

$$\int_{\mathcal{H}} \|x\|^j d|\mu|(x) \leq L\frac{j!}{\epsilon^j}, \qquad j \in \mathbb{N}, \qquad (7.36)$$

$$\int_{\mathcal{H}} \|x\|^j d|\nu|(x) \leq M\frac{j!}{\epsilon^j}, \qquad j \in \mathbb{N}, \qquad (7.37)$$

where $|\mu|, |\nu|$ *denote the total variation measures of the measures* μ, ν *respectively. Let us assume moreover that the constants* $L, \epsilon > 0$ *satisfy the following inequality*

$$2L\|T^{-1}\|(3 + 2\sqrt{2}) < \epsilon^2 \qquad (7.38)$$

$(\|T^{-1}\|$ *denoting the operator norm of* T^{-1}).
Then the following holds:

(1) There is a unique point $a \in \mathcal{H}$ *such that* $V'(a) = Ta$; $T^{-1}V''(a)$ *is of trace class and its trace-norm* $\| \ \|_1$ *satisfies the inequality*

$$\|T^{-1}V''(a)\|_1 < 1.$$

(2) The infinite dimensional oscillatory integral $I(\hbar)$ *given by Eq. (7.31) is analytic in* $\mathrm{Im}(\hbar) < 0$ *and the function* I^*, *given by*

$$I^*(\hbar) = I(\hbar)e^{\frac{i}{\hbar}V(a) - \frac{i}{2\hbar}\langle a, Ta \rangle}, \qquad \text{for } \mathrm{Im}(\hbar) \leq 0, \hbar \neq 0,$$

is a continuous function of $\hbar$ *in* $\mathrm{Im}(\hbar) \leq 0$, *with*

$$I^*(0) = \det(T - V''(a))^{-1/2}g(a),$$

where $\det(T - V''(a))$ *is the Fredholm determinant of the operator* $(T - V''(a))$.

(3) $I^*(\hbar)$ has the following asymptotic expansion and estimate

$$\left| I^*(\hbar) - \det T^{-1/2} \sum_{m=0}^{l-1} \hbar^m \left(-\frac{i}{2} \right)^m \sum_{n=0}^{\infty} \frac{(-2)^{-n}}{n!(m+n)!} \right.$$

$$\left. \int \cdots \int \left\{ T^{-1/2} \left(\sum_{j=1}^{n} \alpha_j + \beta \right) \right\}_{(\alpha,2)}^{2m+2n} \prod_{j=1}^{n} e^{i\langle a, \alpha_j \rangle} d\mu(\alpha_j) e^{i\langle a, \beta \rangle} d\nu(\beta) \right|$$

$$= \left| I^*(\hbar) - \det(T - V''(a))^{-1/2} \sum_{m=0}^{l-1} \hbar^m \left(-\frac{i}{2} \right)^m \sum_{n=0}^{2m} \frac{(-2)^{-n}}{n!(m+n)!} \right.$$

$$\int \cdots \int \left\{ (T - V''(a))^{-1/2} \left(\sum_{j=1}^{n} \alpha_j + \beta \right) \right\}_{(\alpha,3)}^{2m+2n}$$

$$\left. \prod_{j=1}^{n} e^{i\langle a, \alpha_j \rangle} d\mu(\alpha_j) e^{i\langle a, \beta \rangle} d\nu(\beta) \right|$$

$$\leq \frac{M}{(2-\sqrt{2})\sqrt{\pi}} \left(\frac{2|\hbar| \|T^{-1}\|}{\epsilon^2(6-4\sqrt{2})} \right)^l \left(1 - \frac{2L\|T^{-1}\|(3+2\sqrt{2})}{\epsilon^2} \right)^{-l-1/2} \left(l - \frac{1}{2} \right)!,$$

where

$$(x_1 + \cdots + x_n + y)_{(x,m)}^s = \frac{s!}{(s-mn)!(m-1)!^n} \int_0^t \cdots \int_0^t [(1-t_1)\ldots(1-t_n)]^{m-1}$$

$$(x_1^m, \ldots, x_n^m, (t_1 x_1 + \cdots + t_n x_n + y)^{s-mn}) dt_1 \ldots dt_n,$$

and

$$(x_1, \ldots, x_{2n}) = \frac{1}{(2n)!} \sum_{\sigma} (x_{\sigma(1)}, x_{\sigma(2)}) \ldots (x_{\sigma(2n-1)}, x_{\sigma(2n)}),$$

the summation being over all permutations σ of $\{1, \ldots, 2n\}$.
(4) The asymptotic expansion is Borel summable and determines $I^*(\hbar)$ uniquely.

As we have seen in the previous section in the finite dimensional case, if the phase function Φ has several critical points then the asymptotic expansion of the integral $I(\hbar)$ is just the sum over all the stationary points of the corresponding expansion for each critical point (see Theorem 7.2). In the finite dimensional case this result is obtained by writing the function g as a sum of functions, each one having compact support containing a unique stationary point. In the infinite dimensional case the generalization of this technique is not straightforward, as we do not know whether the integral has an asymptotic expansion at all if condition (7.37) in Theorem 7.4 is not satisfied. Indeed the condition $g(x) = \int_{\mathcal{H}} e^{i\langle x,y \rangle} d\nu(y)$, with

$$\int_{\mathcal{H}} \|x\|^j d|\nu|(x) \le M \frac{j!}{\epsilon^j}, \qquad j \in \mathbb{N},$$

implies that there exists a $\lambda \in \mathbb{R}^+$ such that

$$\int_{\mathcal{H}} e^{\lambda \|x\|} d|\nu|(x) < \infty. \tag{7.39}$$

By condition (7.39) it is simple to see that, if $z = x + iy$ with $x, y \in \mathcal{H}$, the function g defined on the complexification of the real Hilbert space $\mathcal{H}$ by

$$g(z) = \int_{\mathcal{H}} e^{i\langle z,\alpha \rangle} d\nu(\alpha)$$

is analytic in $\|y\| < \lambda$, hence the support of g cannot be compact.

In order to overcome this problem, in [17] an alternative technique has been implemented. Indeed by imposing suitable assumption on the function $V \in \mathcal{F}(\mathcal{H})$, it is possible to prove that there exists a decomposition of the Hilbert space $\mathcal{H}$ into a direct sum $\mathcal{H} = \mathcal{H}_1 \oplus \mathcal{H}_2$, with $\mathcal{H}_2$ finite dimensional and the phase Φ when restricted to $\mathcal{H}_1$ presents a unique stationary point. The oscillatory integral is then studied by means of Fubini theorem (see Theorem 5.7). We present here the results obtained in [17], where the case $T = I$ is handled. Analogous results can be obtained also for $T = I - L$ self-adjoint and invertible operator, $L : \mathcal{H} \to \mathcal{H}$ of trace class (see [7]).

Lemma 7.2. *Let $V \in \mathcal{F}(\mathcal{H})$, with $V(x) = \int_{\mathcal{H}} e^{i\langle x,y \rangle} d\mu(y)$, such that there exists a $\lambda > 0$ with $\|\mu\| < \lambda^2$ and*

$$\int_{\mathcal{H}} e^{\sqrt{2}\lambda \|y\|} d|\mu|(y) < \infty.$$

Then there exists a decomposition $\mathcal{H} = \mathcal{H}_1 \oplus \mathcal{H}_2$, with finite dimensional $\mathcal{H}_2$, such that, with $V(x) = V(y,z)$ for $x = y \oplus z$, $V(y,z)$ satisfies, as a function of y, the following condition uniformly in z:

$$\frac{1}{\lambda^2} \int_{\mathcal{H}_1} e^{\sqrt{2}\lambda \|\beta\|} d|\mu_z|(\beta) < 1,$$

where

$$V(y,z) = \int_{\mathcal{H}_1} e^{i\langle \beta,y \rangle} d\mu_z(\beta).$$

Moreover the equation

$$d_1 V(y,z) = y$$

has a unique solution $y = b(z)$ for all $z \in \mathcal{H}_2$ and the mapping $z \mapsto b(z)$ is real analytic from $\mathcal{H}_2 \to \mathcal{H}_1$.

Proof. Let $\{P_n\}$ be a sequence of orthogonal projections on $\mathcal{H}$ with finite dimensional ranges such that P_n converges strongly to the identity. By Lebesgue dominated convergence theorem

$$\lim_{m \to \infty} \frac{1}{\lambda^2} \int_{\mathcal{H}} e^{\sqrt{2}\lambda\|y - P_n y\|} d|\mu|(y) = \frac{1}{\lambda^2} \int_{\mathcal{H}} d|\mu|(y) < 1,$$

so there exists a finite dimensional projection operator P such that

$$\frac{1}{\lambda^2} \int_{\mathcal{H}} e^{\sqrt{2}\lambda\|y - Py\|} d|\mu|(y) < 1.$$

Let us decompose $\mathcal{H} = \mathcal{H}_1 \oplus \mathcal{H}_2 = (I - P)\mathcal{H} \oplus P\mathcal{H}$ and introduce the notation $x = (y, z)$ for $x = y \oplus z$. The measure $d\mu(\alpha)$ can be considered as a measure $d\mu(\beta, \gamma)$ on the product space $\mathcal{H}_1 \times \mathcal{H}_2$, which is isomorphic as a metric and as a measure space with $\mathcal{H}_1 \oplus \mathcal{H}_2$. We then have

$$V(y, z) = \int_{\mathcal{H}_1 \times \mathcal{H}_2} e^{i\langle \beta, y \rangle} e^{i\langle \gamma, z \rangle} d\mu(\beta, \gamma).$$

Let us denote with μ_z the measure on $\mathcal{H}_1$ given by

$$\int_{\mathcal{H}_1} f(\beta) d\mu_z(\beta) = \int_{\mathcal{H}_1 \times \mathcal{H}_2} f(\beta) e^{i\langle \gamma, z \rangle} d\mu(\beta, \gamma).$$

Clearly we have

$$V(y, z) = \int_{\mathcal{H}_1} e^{i\langle \beta, y \rangle} d\mu_z(\beta),$$

moreover by the Minkowski inequality

$$\frac{1}{\lambda^2} \int_{\mathcal{H}_1} e^{\sqrt{2}\lambda\|\beta\|} d|\mu_z|(\beta) \leq \frac{1}{\lambda^2} \int_{\mathcal{H}} e^{\sqrt{2}\lambda\|\beta\|} d|\mu|(\beta, \gamma) < 1.$$

The last inequality implies that

$$\int_{\mathcal{H}_1} \|\beta\|^2 d|\mu_z|(\beta) < 1$$

and, as in the proof of the first part of Lemma 7.1, this assures that for any $z \in \mathcal{H}_2$, the equation

$$d_1 V(y, z) = y$$

has a unique solution $b(z)$:

$$d_1 V(b(z), z) = b(z) \tag{7.40}$$

(where $d_1 V(y, z)$ denotes the derivative of $V(y, z)$ with respect to y.

In order to prove that $b(z)$ is a smooth function from $\mathcal{H}_2$ to $\mathcal{H}_1$, let us take the derivative of Eq. (7.40):

$$d_1^2 V(b(z), z)db(z) + d_1 d_2 V(b(z), z) = db(z). \tag{7.41}$$

As

$$\|d_1^2 V(y, z)\|_1 \leq \int_{\mathcal{H}_1} \|\beta\|^2 d|\mu_z|(\beta) < 1,$$

($\| \quad \|_1$ denoting the trace norm) we have that $I - d_1^2 V(y, z)$ has a uniformly bounded inverse and from Eq. (7.41)

$$db(z) = (I - d_1^2 V(b(z), z))^{-1} d_1 d_2 V(b(z), z).$$

This proves that $b(z)$ is uniformly continuous and bounded in z, so that $z \mapsto b(z)$ is a smooth mapping.

The assumptions on V implies that $d_1 V(y, z)$ is analytic in $|\mathrm{Im}(y)|^2 + |\mathrm{Im}(z)|^2 < 2\lambda^2$ and since z is a regular solution of

$$d_1 V(b(z), z) = b(z)$$

it is possible to conclude that $b(z)$ is real analytic from $\mathcal{H}_2$ to $\mathcal{H}_1$. $\qquad\square$

Lemma 7.3. *Let $V \in \mathcal{F}(\mathcal{H})$ satisfy the assumptions of Lemma 7.2. Then the equation*

$$V'(x) = x$$

(V' being the Frechet derivative of the function V) has at most a discrete set S of solutions, i.e. S has no limit points in $\mathcal{H}$.

Proof. Under the given assumptions, all the results of Lemma 7.2 hold. Let us consider the partial derivative $d_2 V(y, z)$ of $V(y, z)$ with respect to z, and the equation

$$d_2 V(b(z), z) = z. \tag{7.42}$$

Since $d_2 V(y, z)$ is analytic in y and in z, the function $z \mapsto d_2 V(b(z), z)$ is analytic on the finite dimensional space $\mathcal{H}_2$. It follows that Eq. (7.42) has at most a discrete set of solutions. $\qquad\square$

Theorem 7.5. *Let $\mathcal{H}$ be a real separable Hilbert space, and V and g in $\mathcal{F}(\mathcal{H})$, with*

$$V(x) = \int_{\mathcal{H}} e^{i\langle x, y \rangle} d\mu(y),$$

$$g(x) = \int_{\mathcal{H}} e^{i\langle x, y \rangle} d\nu(y). \tag{7.43}$$

Let us assume V and g are C^∞ functions, i.e. all moments of μ and ν exist. Moreover we assume $\mathcal{H} = \mathcal{H}_1 \oplus \mathcal{H}_2$ where $\dim \mathcal{H}_2 < \infty$, and if $d\mu(\beta,\gamma), d\nu(\beta,\gamma)$ are the measures on $\mathcal{H}_1 \times \mathcal{H}_2$ given by μ and ν, then there is a λ such that $\|\mu\| < \lambda^2$ and

$$\int_{\mathcal{H}} e^{\sqrt{2}\lambda|\beta|} d|\mu|(\beta,\gamma) < \infty, \qquad \int_{\mathcal{H}} e^{\sqrt{2}\lambda|\beta|} d|\nu|(\beta,\gamma) < \infty.$$

If the equation $dV(x) = x$ has only a finite number of solutions $x_1, \ldots, x_n$ on the support of the function g, such that none of the operators $I - d^2V(x_i)$, $i = 1, \ldots, n$, has zero as an eigenvalue, then the function

$$I(\hbar) = \int_{\mathcal{H}} e^{\frac{i}{2\hbar}\|x\|^2} e^{-\frac{i}{\hbar}V(x)} \widetilde{g(x)} dx$$

is of the following form

$$I(\hbar) = \sum_{k=1}^{n} e^{\frac{i}{2\hbar}\|x_k\|^2 - V(x_k)} I_k^*(\hbar),$$

where $I_k^*(\hbar)$ $k = 1, \ldots, n$ are C^∞ functions of $\hbar$ such that

$$I_k^*(0) = e^{\frac{i\pi}{2}n_k} |\det(I - d^2V(x_k))|^{-\frac{1}{2}} g(x_k),$$

where n_k is the number of negative eigenvalues of the operator $d^2V(x_k)$ which are larger than 1.

Moreover if V is gentle, that is there exists a constant $\bar{\lambda} > 0$ with

$$\|\mu\| < \bar{\lambda}^2 \qquad and \qquad \int_{\mathcal{H}} e^{\sqrt{2}\bar{\lambda}\|\alpha\|} d|\mu|(\alpha) < \infty, \tag{7.44}$$

then the solutions of equation $dV(x) = x$ have no limit points.

Proof. By applying Lemma 7.2 to the Hilbert space $\mathcal{H}_1$, we obtain a decomposition $\mathcal{H} = \mathcal{H}_1' \oplus \mathcal{H}_2'$, with $\mathcal{H}_2 \subseteq \mathcal{H}_2'$ and $\mathcal{H}_2'$ finite dimensional such that with $x = y' \oplus z'$, $V(x) = V(y', z')$ satisfies, as a function of $y' \in \mathcal{H}_1'$, the following condition uniformly in $z' \in \mathcal{H}_2'$:

$$\frac{1}{\bar{\lambda}^2} \int_{\mathcal{H}_1'} e^{\bar{\lambda}\sqrt{2}\|\beta'\|} d|\mu|(\beta') \leq \frac{1}{\bar{\lambda}^2} \int_{\mathcal{H}} e^{\bar{\lambda}\sqrt{2}\|\beta'\|} d|\mu|(\beta',\gamma') < 1. \tag{7.45}$$

So, if necessary by using the decomposition $\mathcal{H} = \mathcal{H}_1' \oplus \mathcal{H}_2'$ instead of $\mathcal{H} = \mathcal{H}_1 \oplus \mathcal{H}_2$, we may assume that, with the notation of the theorem,

$$\frac{1}{\bar{\lambda}^2} \int_{\mathcal{H}_1} e^{\bar{\lambda}\sqrt{2}\|\beta\|} d|\mu|(\beta) \leq \frac{1}{\bar{\lambda}^2} \int_{\mathcal{H}} e^{\bar{\lambda}\sqrt{2}\|\beta\|} d|\mu|(\beta,\gamma) < 1. \tag{7.46}$$

Condition (7.46) implies that the equation

$$d_1 V(y, z) = y$$

has a unique solution $y = b(z)$ and $z \mapsto b(z)$ is a smooth mapping of $\mathcal{H}_1$ into $\mathcal{H}_2$. By using the Fubini theorem for oscillatory integrals (Theorem 5.7), $I(\hbar)$ is equal to

$$I(\hbar) = \widetilde{\int_{\mathcal{H}_2}} e^{\frac{i}{2\hbar}\|z\|^2} e^{\frac{i}{2\hbar}b(z)^2 - \frac{i}{\hbar}V(b(z),z)} I_2(\hbar, z) dz, \qquad (7.47)$$

with $I_2(\hbar, z) = e^{-\frac{i}{\hbar}\left(\frac{b(z)^2}{2} - V(b(z),z)\right)} I_1(\hbar, z)$ and

$$I_1(\hbar, z) = \widetilde{\int_{\mathcal{H}_1}} e^{\frac{i}{2\hbar}\|y\|^2} e^{-\frac{i}{\hbar}V(y,z)} g(y, z) dy. \qquad (7.48)$$

It is now possible to prove that the Fresnel integral $I_2(\hbar)$ on the infinite dimensional Hilbert space $\mathcal{H}_1$ is a C^∞ function of $\hbar$ on the real line and it is analytic in $\mathrm{Im}(\hbar) < 0$. Moreover

$$I_2(0) = |1 - d_1^2 V(b(z), z)|^{-1/2} g(b(z), z).$$

The integral $I(\hbar)$ is now given in terms of a finite dimensional oscillatory integral on $\mathcal{H}_2$ and it can be studied by means of the classical method of stationary phase for the asymptotic expansions of finite dimensional oscillatory integrals integrals (see Theorem 7.2).

Since the solutions of the equation

$$d_2 V(b(z), z) = z \qquad (7.49)$$

form a discrete set, there exists a partition of unity

$$1 = \sum_j \phi_j(z), \qquad z \in \mathcal{H}_2$$

by smooth functions $\phi_j : \mathcal{H}_2 \to [0, 1]$ of compact support such that only one solution of Eq. (7.49) is contained in the support of each ϕ_j. One has then to study the asymptotics of integrals of the form

$$I(\hbar) = \widetilde{\int_{\mathcal{H}_2}} e^{\frac{i}{2\hbar}\|z\|^2} e^{\frac{i}{2\hbar}b(z)^2 - \frac{i}{\hbar}V(b(z),z)} \phi(z) I_2(\hbar, z) dz, \qquad (7.50)$$

with ϕ with compact support containing only one solution of Eq. (7.49). As from the assumptions of the theorem it follows that

$$\left| I_2(\hbar, z) - \sum_{m=0}^{N} \frac{\hbar^m}{m!} I_2^{(m)}(0, z) \right| \leq |\hbar|^{n+1} C_N,$$

with C_N independent on z, up to terms of order $|\hbar|^{N+1}$ the integral (7.50) may be written as

$$\sum_{m=0}^{N} \frac{\hbar^m}{m!} \widetilde{\int_{\mathcal{H}_2}} e^{\frac{i}{2\hbar}\|z\|^2} e^{\frac{i}{2\hbar}b(z)^2 - \frac{i}{\hbar}V(b(z),z)} \phi(z) I_2^{(m)}(0, z) dz.$$

It is then enough to study the asymptotic behaviour of the integrals

$$\widetilde{\int_{\mathcal{H}_2}} e^{\frac{i}{2\hbar}\|z\|^2} e^{\frac{i}{2\hbar}b(z)^2 - \frac{i}{\hbar}V(b(z),z)} \phi(z) I_2^{(m)}(0,z)dz,$$

which is determined by the solutions of the equation $d\Phi(z) = 0$, with

$$\Phi(z) = \frac{1}{2}\|z\|^2 + \frac{1}{2}b(z)^2 - V(b(z),z), \qquad z \in \mathcal{H}_2.$$

As

$$d\Phi(z) = z + b(z)db(z) - d_1V(b(z),z)db(z) - d_2V(b(z),z),$$

and $b(z) = d_1(b(z),z)$, we have that

$$d\Phi(z) = z - d_2V(b(z),z),$$

so that the critical points are the solutions of Eq. (7.49) and, by construction, the support of ϕ contains only one of them, which will be denoted by c. By denoting $b = b(c)$ and $a = (b,c)$ it is easy to verify that a is a solution of $dV(x) = x$.

By computing $d^2\Phi(c)$, we have:

$$d^2\Phi(z) = 1 - d_2d_1V(b(z),z)db(z) - d_2^2V(b(z),z).$$

Since $b(z) = d_1V(b(z),z)$, by differentiating we get

$$db(z) = d_1^2V(b(z),z)db(z) + d_1d_2V(b(z),z).$$

By Lemma 7.2, $I - d_1^2V(b(z),z)$ has a bounded inverse, so that

$$db(z) = (I - d_1^2V(b(z),z))^{-1}d_1d_2V(b(z),z),$$

hence

$$d^2\Phi(c) = 1 - d_2d_1V(a)(I - d_1^2V(a))^{-1}d_1d_2V(a) - d_2^2V(a).$$

Let us consider a vector $\xi \in \mathcal{H}_2$, such that $d^2\Phi(c)\xi = 0$. By defining $\eta \in \mathcal{H}_1$ as

$$\eta = -(I - d_1^2V(a))^{-1}d_1d_2V(a)\xi,$$

and $\zeta = (\eta,\xi)$, we have that

$$(I - d^2V(a))\zeta = 0,$$

as this is equivalent to

$$(I - d_1^2V(a))\eta + d_1d_2V(a)\xi = 0,$$

$$d_2d_1V(a)\eta + (I - d_2^2V(a))\xi = 0.$$

If for some $\zeta \in \mathcal{H}$, $(I - d^2V(a))\zeta = 0$, then $d^2\Phi(c)\xi = 0$, where ξ is the projection of ζ on $\mathcal{H}_2$. So the condition of non-degeneracy of $d^2\Phi(c)$ is equivalent to the condition of non-degeneracy of $(I - d^2V(a))$. Moreover

$$\det(d^2\Phi(c)) = \det(I - d_1^2V(a))^{-1} \det(I - d^2V(a)).$$

So, if $I - d^2V(a)$ is not degenerate, the integral (7.50) is a C^∞ function of $\hbar$ and its asymptotic behaviour can be studied by means of the classical method of the stationary phase in the (simplest) non-degenerate case (Theorem 7.2) and obtaining, by summing all contributions given by the stationary points, the final result:

$$I(\hbar) = \sum_{k=1}^{n} e^{\frac{i}{2\hbar}\|x_k\|^2 - V(x_k)} I_k^*(\hbar),$$

$$I_k^*(0) = e^{\frac{i\pi}{2}n_k} |\det(I - d^2V(x_k))|^{-\frac{1}{2}} g(x_k). \qquad \square$$

If some critical point of the phase function is degenerate, the study of the asymptotic behavior of the oscillatory integral $I(\hbar)$ in Eq. (7.31) becomes more complicated. Indeed, as we know from the case of a finite dimensional Hilbert space $\mathcal{H}$, in this situation it is possible that the integral $I(\hbar)$, divided by the above leading term, will not tend to a limit as $\hbar \to 0$. A possible approach to the study of this situation can be found for instance in the work of Duistermaat [125] and has been generalized in [17] to case where the integration is performed in an infinite dimensional setting. The problem is solved by letting the functions V and g depend on an additional parameter $y \in \mathbb{R}^k$, for suitable k, and to study, instead of

$$I(\hbar, y) = \widetilde{\int_{\mathcal{H}}} e^{\frac{i}{2\hbar}\|x\|^2} e^{-\frac{i}{\hbar}V(x,y)} g(x, y) dx$$

an oscillatory integral of a larger Hilbert space of the following form:

$$I(\hbar, \psi) = (2\pi i\hbar)^{k/2} \int_{\mathbb{R}^k} e^{-\frac{i}{\hbar}\psi(y)} \chi(y) \widetilde{\int_{\mathcal{H}}} e^{\frac{i}{2\hbar}\|x\|^2} e^{-\frac{i}{\hbar}V(x,y)} g(x, y) dx dy,$$

$$\tag{7.51}$$

where ψ and χ are C^∞ functions and χ has a compact support. It is also required that the applications

$$y \to V(\,\cdot\,, y), \qquad y \in \mathbb{R}^k,$$

$$y \to g(\,\cdot\,, y), \qquad y \in \mathbb{R}^k,$$

are C^∞ functions from $\mathbb{R}^k$ to $\mathcal{F}(\mathcal{H})$ in the strong topology. By the Fubini Theorem (5.7) the integral (7.51) can be regarded as an oscillatory integral on $\mathcal{H} \oplus \mathbb{R}^k$, with a new phase function Φ given by

$$\Phi(x, y) = \frac{1}{2}\|x\|^2 - V(x, y) - \psi(y), \qquad (x, y) \in \mathcal{H} \oplus \mathbb{R}^k.$$

The key point is the non-degeneracy of the critical point of this new phase function. Indeed, by the Morse theorem, the set of functions $V(x)$ such that $\frac{1}{2}\|x\|^2 - V(x)$ has only non-degenerate critical points form a open and dense set in the space of all C^∞ functions and the complement is in a natural sense of codimension 1. In other words, the case of degenerate critical points is unstable in the sense that degenerate critical points will disappear under arbitrary small perturbations (in the sense of the C^∞ topology). We refer to [125] and to [17] for a detailed discussion of this elegant technique, which involves an amount of differential geometry and allows eventually to prove an expression for the leading term in the asymptotic expansion of the integral (7.51) involving only objects and quantities with an intrinsic geometrical meaning.

A different approach to the study of the degeneracies is described in [7, 6], where some particular examples are handled. In particular in [7], by applying the Fubini theorem for infinite dimensional oscillatory integrals (Theorem 5.7), the authors reduce to the study of the degeneracy on a finite dimensional subspace of the Hilbert space $\mathcal{H}$ and apply the existing theory for finite dimensional oscillatory integrals. In fact they assume that the phase function $\frac{1}{2}\langle x, Tx\rangle - V(x)$ has the point $x_c = 0$ as a unique stationary point, which is degenerate, i.e.

$$Z := \mathrm{Ker}(T - d^2V)(0) \neq \{0\}.$$

Under suitable assumptions on T and V, they prove that Z is finite dimensional. By taking the subspace $Y = T(Z^\perp)$ and applying the Fubini Theorem 5.7 one has

$$I(\hbar) = \widetilde{\int_{\mathcal{H}}} e^{\frac{i}{2\hbar}\langle x, Tx\rangle} e^{-\frac{i}{\hbar}V(x)} g(x)\,dx =$$

$$= C_T \widetilde{\int_Z} e^{\frac{i}{2\hbar}\langle z, T_2 z\rangle} \widetilde{\int_Y} e^{\frac{i}{2\hbar}\langle y, T_1 y\rangle} e^{-\frac{i}{\hbar}V(y+z)} g(y+z)\,dy\,dz, \quad (7.52)$$

where T_1 and T_2 are defined by

$$T_1 y = (\pi_Y \circ T)(y), \qquad y \in Y,$$

$$T_2 z = (\pi_Z \circ T)(z), \qquad z \in Z,$$

and $C_T = (\det T)^{-1/2}(\det T1)^{1/2}(\det T_2)^{1/2}$. By assuming that $V, g \in \mathcal{F}(\mathcal{H})$, $V = \hat{\mu}$ and $g = \hat{\nu}$, and under some growth conditions on μ and ν, one has that the phase function

$$y \mapsto \frac{1}{2}\langle y, T_1 y \rangle - V(y + z), \qquad y \in Y,$$

of the oscillatory integral on Y

$$J(z, \hbar) = \widetilde{\int_Y} e^{\frac{i}{2\hbar}\langle y, T_1 y \rangle} e^{-\frac{i}{\hbar}V(y+z)} g(y + z) dy$$

has only one non-degenerate stationary point $a(z) \in Y$. By applying then the theory developed for the non-degenerate case one has

$$J(z, \hbar) = e^{\frac{i}{2\hbar}\langle a(z), T_1 a(z) \rangle} e^{-\frac{i}{\hbar}V(a(z)+z)} J^*(z, \hbar),$$

$$J^*(z, 0) = \left[\det \left(T_1 - \frac{\partial^2 V}{\partial^2 y}(a(z) + z) \right) \right]^{-1/2} g(a(z) + z).$$

As $I(\hbar) = \widetilde{\int_Z} e^{\frac{i}{\hbar}\Phi(z)} J^*(z, \hbar) dz$, where

$$\Phi(z) = \frac{1}{2}\langle z, T_2 z \rangle + \frac{1}{2}\langle a(z), T_1 a(z) \rangle - V(a(z) + z),$$

the main ingredient for the asymptotic behavior of $I(\hbar)$ comes from $J^*(z, 0)$. The phase function Φ has $z = 0$ as a unique degenerate critical point and, by applying the theory for asymptotic behavior for finite dimensional oscillatory integrals [187], one has to investigate the higher derivatives of Φ at 0. For example if $dim(Z) = 1$ and $\frac{\partial^3 V}{\partial^3 z}(0) \neq 0$ then

$$I(\hbar) \sim C\hbar^{-1/6}, \qquad \text{as} \quad \hbar \to 0.$$

More generally it is possible to handle other cases, taking into account the classification of different types of degeneracies (see, e.g., [48, 47]).

7.4   The semiclassical limit of quantum mechanics

The techniques and the results of the previous sections as well as the infinite dimensional oscillatory integral representation for the solution of the Schrödinger equation $\psi(t, x)$ (Theorem 6.2) allow the rigorous study of the semiclassical limit of $\psi(t, x)$, i.e. the detailed behaviour of the wave function and other quantum mechanical quantities when the Planck constant is regarded as a small parameter converging to 0.

This result is particularly important, as one of the most fascinating features of Feynman path integrals is their power to link, at least heuristically, quantum and classical mechanics. As the formal expression of the Feynman path integral representation for the solution of the Schrödinger equation

$$\psi(t,x) = \int_{\gamma(t)=x} e^{\frac{i}{\hbar}S_t(\gamma)}\psi(0,\gamma(0))d\gamma$$

suggests, according to the stationary phase method as $\hbar \downarrow 0$ the leading contribution to the asymptotic behavior of $\psi(t,x)$ should come from those paths γ that make stationary the action functional $S_t(\gamma)$. These, by Hamilton's least action principle, are exactly the classical orbits of the system.

The problem of the way quantum mechanics, and in particular the solution of the Schrödinger equation, approaches classical mechanics has been studied in different ways by several authors [248, 250, 135, 51, 63, 111, 116, 154, 153, 181, 288, 296, 322, 327] (we only mention some of them without any claim of completeness).

First rigorous results concerning the application of the stationary phase method for infinite dimensional oscillatory (Fresnel) integrals to the study of the semiclassical limit of the solution of the Schrödinger equation can be found in the pioneering paper [17], where it is assumed that the potential belongs to the class $\mathcal{F}(\mathbb{R}^d)$. By applying the general results concerning the infinite dimensional stationary phase method and the theory of Lagrangian manifolds, in the spirit of the geometrical analysis of oscillatory integrals and their asymptotics made in [125], the authors provide an alternative (Feynman path integral) derivation of Maslov's results on the semiclassical asymptotics of the solution of the Schrödinger equation [250]. Part of these results are generalized in [7, 5], by means of slightly different methods, to the case where the potential is the sum of a quadratic function and a smooth function in $\mathcal{F}(\mathbb{R}^d)$, providing not only the leading term, but also the higher order terms of the asymptotic expansion as well as a good control on the remainder. The authors consider a particular but physically relevant form for the initial wave function:

$$\psi_0(x) = e^{\frac{i}{\hbar}S_0(x)}\phi_0(x), \tag{7.53}$$

where S_0 is real and $S_0, \phi_0 \in \mathcal{F}(\mathbb{R}^d)$ are independent of $\hbar$. This initial data corresponds to an initial particle distribution with density $\rho_0(x) = |\phi_0|^2(x)$ and to a limiting value of the density of the probability current $J_{\hbar=0} = S_0'(x)\rho_0(x)/m$, giving an initial particle flux associated to the velocity field $S_0'(x)/m$ (S_0' stands for the gradient of S_0).

In [7, 5] the authors consider the Schrödinger equation

$$i\hbar\frac{\partial}{\partial t}\psi(t,x) = -\frac{\hbar^2}{2m}\Delta\psi(t,x) + \frac{1}{2}x\Omega^2x\psi(t,x) + V_0(x)\psi(t,x), \quad t > 0, x \in \mathbb{R}^d, \tag{7.54}$$

(where Ω^2 is a positive symmetric linear operator in $\mathbb{R}^d$ and $V_0 \in \mathcal{F}(\mathbb{R}^d)$) and provide first of all an infinite dimensional oscillatory integral representation of the solution of the initial value problem, which is similar to the one presented in Theorem 6.2.

Let us assume that $\det(\cos(t\Omega)) \neq 0$ and let $\beta_{t,x}$ be the unique solution of the boundary value problem

$$\ddot{\beta}(s) + \Omega^2 \beta(s) = 0, \qquad 0 \leq s \leq t,$$

$$\dot{\beta}(0) = 0, \quad \beta(t) = x.$$

Let $\mathcal{H}_t$ denote the Cameron-Martin Hilbert space, that is the Sobolev space of absolutely continuous functions $\gamma : [0, t] \to \mathbb{R}^d$, such that $\gamma(t) = 0$, with square integrable weak derivative $\dot{\gamma}$:

$$\int_0^t |\dot{\gamma}(s)|^2 ds < \infty,$$

endowed with the inner product

$$\langle \gamma_1, \gamma_2 \rangle = \int_0^t \dot{\gamma}_1(s) \cdot \dot{\gamma}_2(s) ds.$$

Let $L : \mathcal{H}_t \to \mathcal{H}_t$ be the self-adjoint linear operator defined by Eq. (6.5). Analogously to what has been done in Theorem 6.2, by assuming that the initial datum is of the form (7.53), it is possible to prove [17, 7] that the solution of the Schrödinger equation is given by the infinite dimensional oscillatory integral:

$$\psi^\hbar(t,x) = e^{-\frac{i}{\hbar} x \tan(\Omega t)\Omega x} \widetilde{\int_{\mathcal{H}_t}} e^{\frac{i}{2\hbar}\langle \gamma, (I-L)\gamma \rangle} e^{-\frac{i}{\hbar} \int_0^t V_0(\gamma(s)+\beta_{t,x}(s))ds}$$

$$e^{\frac{i}{\hbar} S_0(\gamma(0)+\beta_{t,x}(0))} \phi_0(\gamma(0) + \beta_{t,x}(0)) d\gamma, \quad (7.55)$$

where the superscript $\hbar$ stresses the dependence on the variable $\hbar \in \mathbb{R}^+$. Equation (7.55) is more convenient then representation (6.16) for the study of the semiclassical asymptotics of $\psi^\hbar(t,x)$ as $\hbar \to 0$, as it points out the role of the classical path $\beta_{t,x}$.

For the study of the asymptotic behaviour of the integral (7.55) one has first of all to determine the stationary points of the phase functional $\Phi : \mathcal{H}_t \to \mathbb{R}$, which is given by

$$\Phi(\gamma) = \frac{1}{2}\|\gamma\| - \frac{1}{2}\int_0^t \gamma(s)\Omega^2 \gamma(s) ds - x\tan(\Omega t)\Omega x - \int_0^t V_0(\gamma(s) + \beta_{t,x}(s)) ds$$

$$+ S_0(\gamma(0) + \beta_{t,x}(0)). \quad (7.56)$$

Under the assumption that $V_0, S_0 \in C^2(\mathbb{R}^d)$, it is possible to compute the first and the second Fréchet derivative of the functional Φ:

$$\Phi'(\gamma)(\delta) = \langle \gamma, \delta \rangle - \int_0^t \gamma(s)\Omega^2\delta(s)ds - \int_0^t V_0'(\gamma(s) + \beta_{t,x}(s))\delta(s)ds$$
$$+ S_0'(\gamma(0) + \beta_{t,x}(0))\delta(0), \qquad \gamma, \delta \in \mathcal{H}_t \quad (7.57)$$

$$\Phi''(\gamma)(\delta_1, \delta_2) = \langle \delta_1, \delta_2 \rangle - \int_0^t \delta_1(s)\Omega^2\delta_2(s)ds + S_0''(\gamma(0) + \beta_{t,x}(0))\delta_1(0)\delta_2(0)$$
$$- \int_0^t V_0''(\gamma(s) + \beta_{t,x}(s))\delta_1(s)\delta_2(s)ds, \quad \gamma, \delta_1, \delta_2 \in \mathcal{H}_t \quad (7.58)$$

By means of Eq. (7.57) it is not difficult to verify the following result (see [5] for details):

Lemma 7.4. *Let $V_0, S_0 \in C^2(\mathbb{R}^d)$. Then the function $\Phi : \mathcal{H}_t \to \mathbb{R}$ defined by Eq. (7.56) is of class C^2. Moreover, for $\gamma \in \mathcal{H}_t$, $\Phi'(\gamma) = 0$ iff γ is a solution of*

$$\ddot{\gamma}(s) + \Omega^2\gamma(s) + V_0'(\gamma(s) + \beta_{t,x}(s)) = 0, \qquad 0 \le s \le t, \quad (7.59)$$

$$\dot{\gamma}(0) = S_0'(\gamma(0) + \beta_{t,x}(0)), \quad \gamma(t) = 0.$$

Equivalently, $\chi(s) := \gamma(s) + \beta_{t,x}(s)$ is a solution of

$$\ddot{\chi}(s) + \Omega^2\chi(s) + V_0'(\chi(s)) = 0, \qquad 0 \le s \le t, \quad (7.60)$$

$$\dot{\chi}(0) = S_0'(\chi(0)), \quad \chi(t) = x.$$

Remark 7.7. One can easily recognize in the function χ the "classical path".

For the study of the degeneracy of the critical points γ_c, one has to determine the spectrum of Hessian of the phase function evaluated in γ_c. Let the operator $A : \mathcal{H}_t \to \mathcal{H}_t$ be defined by

$$I + A = \Phi''(\gamma),$$

i.e. for $\delta_1, \delta_2 \in \mathcal{H}_t$

$$\langle \delta_1, A\delta_2 \rangle = -\int_0^t \delta_1(s)\Omega^2\delta_2(s)ds - \int_0^t V_0''(\chi(s))\delta_1(s)\delta_2(s)ds$$
$$+ S_0''(\chi(0))\delta_1(0)\delta_2(0), \quad (7.61)$$

(χ being the solution of Eq. (7.59)).

The following lemma reduces the computation of the Fredholm determinant of the operator $\Phi''(\gamma)$, to the solution of a finite dimensional Cauchy problem.

Lemma 7.5. *The linear operator $A : \mathcal{H}_t \to \mathcal{H}_t$ defined by Eq. (7.61) is uniquely determined by the following conditions.*

For $\eta \in \mathcal{H}_t$, $A\eta$ is the unique solution to the following boundary value problem:

$$(\ddot{A}\eta)(s) = p(s)\eta(s), \qquad 0 \leq s \leq t,$$

$$(\dot{A}\eta)(0) = -Q\eta(0), \qquad (A\eta)(t) = 0,$$

where

$$p(s) = \Omega^2 + V_0''(\chi(s)),$$

$$Q = S_0''(\chi(0)).$$

Moreover

$$\det(I + A) = \det \Phi''(\gamma) = \det K(t),$$

where $K(s)$ is a (matrix valued) solution to the following second order equation

$$\ddot{K}(s) + p(s)K(s) = 0, \qquad s > 0, \tag{7.62}$$

$$K(0) = I, \qquad \dot{K}(0) = Q,$$

and $\gamma \in \mathcal{H}_t$, $\chi = \gamma + \beta_{t,x}$.

Proof. For a detailed proof see [5], Theorem 2.1, which is based to an idea of [124] and to the Hadamard Factorization Theorem. $\square$

Equation (7.62) is a second order linear equation with non-constant coefficients, hence at a first glance its solution would seem to be an arduous task. On the other hand the following argument provides a simple technique for the construction of the solution.

Let $y(s, x_0)$, for $s \geq 0$, be the unique solution to

$$\ddot{y}(s) + \Omega^2 y(s) + V_0'(y(s)) = 0, \qquad s \geq 0, \qquad (7.63)$$

$$y(0) = x_0, \quad \dot{y}(0) = S_0'(x_0).$$

Let

$$Y(s) = Y(s, x_0) := \frac{\partial y}{\partial x_0}(s, x_0).$$

It is simple to verify that $Y(s)$, for $s \geq 0$, is the unique solution to

$$\frac{d^2}{ds^2}Y(s) + (\Omega^2 + V_0''(y(s, x_0)))Y(s) = 0,$$

$$Y(0) = I, \quad \dot{Y}(0) = S_0''(x_0).$$

This implies the following result.

Lemma 7.6. *If $\gamma \in \mathcal{H}_t$, $\chi = \gamma + \beta_t, x$. Then*

$$\det \Phi''(\gamma) = \det\left(\frac{\partial y}{\partial x_0}(t, x_0)\right).$$

As Lemma 7.6 states, given a critical point γ_c of the phase function Φ, the non-degeneracy condition, i.e. $\det \Phi''(\gamma_c) \neq 0$ is equivalent to the non-vanishing of the quantity $\det\left(\frac{\partial y}{\partial x_0}(t, x_0)\right)$. For the importance of this concept, let us introduce the definition of *focal* and *non-focal point*.

Definition 7.6. A time-space point (t, x) is called *non-focal* if for any $x_0 \in \mathbb{R}^d$ and for any solution $y(\cdot, x_0)$ to Eq. (7.63) such that $y(t, x_0) = x$ the following holds:

$$\det\left(\frac{\partial y}{\partial x_0}(t, x_0)\right) \neq 0.$$

If this condition is not satisfied for some $x_0 \in \mathbb{R}^d$, the time-space point (t, x) is called *focal*.

Moreover, by Morse theorem (see [256], Theorem 15.1), the index of the operator $\Phi''(\gamma)$, with γ solution of Eq. (7.59), is equal to the Morse-Maslov index of the curve χ solution of Eq. (7.60) (see also [45]).

The following result is an application of Theorem 7.4 to the study of the asymptotics of the oscillatory integral (7.55) and handles the case where the phase function has a unique stationary point.

Theorem 7.6. *Let $\psi^\hbar(t,x)$ (given by the oscillatory integral (7.55)) be the unique solution of the Schrödinger Eq. (7.54) and initial condition ψ_0 of the form (7.53). Assume that*

$$V_0 = \hat{\mu}_0, \quad S_0 = \hat{\sigma}_0, \quad \phi_0 = \hat{\nu}_0 \in \mathcal{F}(\mathbb{R}^d),$$

and for some $K_i, \epsilon_i > 0, i = 1, 2, 3$,

$$\int_{\mathbb{R}^d} |y|^j d|\mu_i|(y) \le K_i \frac{j!}{\epsilon_i^j}, \qquad j \in \mathbb{N},$$

with $\mu_1 = \mu_0$, $\mu_2 = \sigma_0$, $\mu_3 = \nu_0$.

Let $L : \mathcal{H}_t \to \mathcal{H}_t$ be the operator defined by Eq. (6.5). Assume that

$$t \ne \left(n + \frac{1}{2}\right)\frac{\pi}{\Omega_j}, \qquad n \in \mathbb{N}, \; j+1, \ldots, d, \tag{7.64}$$

where Ω_j, for $j = 1, \ldots, d$ are the eigenvalues of the operator Ω. Then $(I - L)^{-1}$ exists and it is bounded. Let us put

$$K = K_1 + K_2, \qquad \epsilon = \frac{\min\{\epsilon_1, \epsilon_2\}}{\sqrt{t}}.$$

let us assume that

$$2(3 + 2\sqrt{2})K\|(I - L)^{-1}\|\epsilon^{-2} < 1. \tag{7.65}$$

Then there exists a unique path $\bar\gamma \in \mathcal{H}_t$ making stationary the phase function Φ, i.e.

$$\Phi'(\bar\gamma) = 0,$$

where Φ is given by Eq. (7.56) and its Frechet derivative by Eq. (7.57). Let $x \in \mathbb{R}^d$ and put $\chi := \bar\gamma + \beta_{t,x}$, i.e.

$$\chi(s) = \bar\gamma(s) + \beta_{t,x}(s), \qquad 0 \le s \le t.$$

then χ is a solution of the boundary value problem (7.60). Let $m(\chi)$ denote the Maslov (or Morse) index of the curve χ. then, as $\hbar \downarrow 0$, the following asymptotic formula for the integral (7.55) holds:

$$\psi^\hbar(t,x) = e^{\frac{i}{\hbar}\Phi(\bar\gamma)}e^{-\frac{i\pi}{2}m}\left|\det\frac{\partial y(t,x_0)}{\partial x_0}\Big|_{x_0=\chi(0)}\right|^{-1/2}\phi_0(\chi(0)) + O(\hbar)$$

$$= e^{\frac{i}{\hbar}\left(\int_0^t |\dot\chi(s)|^2 ds/2 - \int_0^t \chi(s)\Omega^2\chi(s)ds/2 - \int_0^t V_0(\chi(s)ds + S_0(\chi(0))\right)}$$

$$e^{-\frac{i\pi}{2}m}\left|\det\frac{\partial y(t,x_0)}{\partial x_0}\Big|_{x_0=\chi(0)}\right|^{-1/2}\phi_0(\chi(0)) + O(\hbar), \tag{7.66}$$

where $y(t,x_0)$ is the unique solution of the Cauchy problem (7.63).

Proof. By Lemma 6.1, if condition (7.64) is satisfied, the operator L given by Eq. (6.5) is such that $(I - L)^{-1}$ exists and it is bounded. In this case we have the matrix $\cos(\Omega t)$ is non-singular and $\tan(\Omega t)$ is well defined.

Let us put

$$W(\gamma) = \int_0^t V_0(\gamma(s) + \beta_{t,x}(s))ds - S_0(\gamma(0) + \beta_{t,x}(0)), \qquad \gamma \in \mathcal{H}_t. \quad (7.67)$$

By reasoning as in the proof of Lemma 6.3, it is easy to verify that $W = \hat{\mu}$, with $\mu \in \mathcal{M}(\mathcal{H}_t)$. Moreover the following holds:

$$\int_{\mathcal{H}_t} \|\gamma\|^j d|\mu|(\gamma) \le t^{j/2} \int_{\mathbb{R}^d} |y|^j d|\sigma_0|(y) + \frac{t^{j/2+1}}{j/2+1} \int_{\mathbb{R}^d} |y|^j d|\mu_0|(y),$$

and

$$\int_{\mathcal{H}_t} \|\gamma\|^j d|\mu|(\gamma) \le K \frac{j!}{\epsilon^j}. \quad (7.68)$$

Since the integral $\psi^\hbar(t, x)$ can be written as:

$$\psi^\hbar(t, x) = e^{-\frac{i}{\hbar} x \tan(\Omega t)\Omega x} \int_{\mathcal{H}_t} e^{\frac{i}{2\hbar}\langle \gamma, (I-L)\gamma \rangle} e^{-\frac{i}{\hbar}W(\gamma)} g(\gamma)d\gamma, \quad (7.69)$$

with

$$g(\gamma) = \phi_0(\gamma(0) + \beta_{t,x}(0)), \quad (7.70)$$

and inequality (7.65) holds by assumption, we can apply Theorem 7.4 and conclude that there exists a unique stationary point $\bar{\gamma} \in \mathcal{H}_t$ of the phase function and

$$\psi^\hbar(t, x) = e^{\frac{i}{\hbar}\Phi(\bar{\gamma})}(\det \Phi''(\bar{\gamma}))^{-1/2} g(a) + O(\hbar) \quad (7.71)$$

where

$$\Phi(\bar{\gamma}) = \int_0^t |\dot{\chi}(s)|^2 ds/2 - \int_0^t \chi(s)\Omega^2\chi(s)ds/2 - \int_0^t V_0(\chi(s))ds + S_0(\chi(0)).$$

The final result follows from Lemma 7.6 $\qquad\qquad\qquad\qquad\qquad \square$

Analogously to what is done in Theorem 7.4, where the complete asymptotic expansion of the oscillatory integral $I(\hbar)$ in powers of $\hbar$ is provided as well as a good control on the remainder, it is possible to compute not only the first term in the expansion of $\psi^\hbar(t, x)$. Indeed it is possible to prove (see [5, 7]), in the case where the phase function has a unique stationary point $\bar{\gamma}$, that

$$\psi^\hbar(t, x) = e^{\frac{i}{\hbar}\Phi(\bar{\gamma})}I^*(\hbar),$$

with $I^*(\hbar)$ being a C^∞ function of $\hbar \in \mathbb{R}$, and

$$I^*(\hbar) = \sum_{j=0}^{n-1} \alpha_j \hbar^j + R_n(\hbar),$$

where the coefficients α_j of the expansions depend on Ω as well as the derivatives of V_0, S_0, ϕ_0, and

$$|R_n(\hbar)| \leq C_n |\hbar|^n$$

for suitable coefficients C_n and $|\hbar|$ sufficiently small.

The assumption (7.65) in Theorem 7.6 assures that the phase function Φ has a unique stationary point. Following [5], it is worthwhile to give also some weaker conditions assuring that the set of critical points is not empty and finite.

Lemma 7.7. *Let us assume that the operator $(I - L)$ is invertible and the functions V_0, S_0 are C^1 with bounded derivative. Then the number of non-degenerate stationary points of the phase function Φ is finite.*

Lemma 7.8. *Let us assume that the functions V_0, S_0 are bounded C^1 functions with bounded derivatives. Then there exists at least a stationary point of the phase function Φ.*

For a proof of both lemmas we refer to the original paper [5]. The next theorem handles the case where the phase function Φ has a finite number of non-degenerate stationary points and can be seen as an application of Theorem 7.5 (see also Theorems 3.7, 4.1 and 5.4 in [7]).

Theorem 7.7. *Let us assume that*

$$V_0 = \hat{\mu}_0, \quad S_0 = \hat{\sigma}_0, \quad \phi_0 = \hat{\nu}_0 \in \mathcal{F}(\mathbb{R}^d),$$

and for some $K_i, \epsilon_i > 0, i = 1, 2, 3$,

$$\int_{\mathbb{R}^d} |y|^j d|\mu_i|(y) \leq K_i \frac{j!}{\epsilon_i^j}, \qquad j \in \mathbb{N},$$

with $\mu_1 = \mu_0, \mu_2 = \sigma_0, \mu_3 = \nu_0$.

Let $L : \mathcal{H}_t \to \mathcal{H}_t$ be the operator defined by Eq. (6.5). Assume that

$$t \neq \left(n + \frac{1}{2}\right) \frac{\pi}{\Omega_j}, \qquad n \in \mathbb{N}, j + 1, \ldots, d, \qquad (7.72)$$

where Ω_j, for $j = 1, \ldots, d$ are the eigenvalues of the operator Ω. Then $(I - L)^{-1}$ exists and it is bounded.

Let the point (t, x) be non-focal.

Then there exists a finite number of solutions $\chi_1, \ldots, \chi_n$ to the problem (7.60) and the solution $\psi^\hbar(t, x)$ of the Schrödinger Eq. (7.54) has the following asymptotic representation as $\hbar \downarrow 0$:

$$\psi^\hbar(t, x) = \sum_{j=1}^{n} e^{\frac{i}{\hbar}\left(\int_0^t |\dot\chi_j(s)|^2 ds/2 - \int_0^t \chi_j(s)\Omega^2\chi_j(s)ds/2 - \int_0^t V_0(\chi_j(s))ds + S_0(\chi_j(0))\right)}$$

$$e^{-\frac{i\pi}{2}m_j}\left|\det\frac{\partial y_j(t, x_0)}{\partial x_0}\right|^{-1/2}\phi_0(\chi_j(0)) + O(\hbar) \quad (7.73)$$

If some critical point of the phase function is degenerate, or, in other words, if the time-space point (t, x) is focal, then the problem can be solved by reducing the study of the degeneracy to a finite dimensional subspace of $\mathcal{H}_t$, in the way described at the end of Section 7.3.

Particular examples are handled in [5, 6]. In the next section we shall describe the application of this technique to the study of the trace of the Schrödinger group. We present briefly here some results concerning the solution of the Schrödinger equation.

Let us assume that the space dimension d is less or equal then 3 and that the following inequalities are satisfied

$$\det \sin(\Omega t) \neq 0 \qquad \det \cos(\Omega t) \neq 0. \quad (7.74)$$

By Lemma 6.1, the second condition assures that the operator $(I - L)$ is invertible. Let us consider the solution of the Schrödinger equation given in terms of the infinite dimensional oscillatory integral $\psi^\hbar(t, x)$ (Eq. (7.55)).

Let us introduce the subspaces of $\mathcal{H}_t$:

$$Y = \{\gamma \in \mathcal{H}_t : \gamma(0) = 0\},$$

$$Z = \{\gamma \in \mathcal{H}_t : \gamma(s) = \sin\Omega(s - t)z \text{ for some } z \in \mathbb{R}^d\}. \quad (7.75)$$

It is easy to verify that $(I - L)Y \subset Z^\perp$. Let us denote by Π_Y and Π_Z the orthogonal projections of $\mathcal{H}_t$ onto Y and Z and by $T_1 : Y \to Y$ and $T_2 : Z \to Z$ the linear operators defined by

$$T_1 := \Pi_Y \circ T_{|Y}, \qquad T_2 := \Pi_Z \circ T_{|Z},$$

where $T : \mathcal{H}_t \to \mathcal{H}_t$ is given by $T = I - L$.

By applying the Fubini Theorem 5.7 to the oscillatory integral (7.55), or equivalently (7.69), we have

$$\widetilde{\int_{\mathcal{H}_t}} e^{\frac{i}{2\hbar}\langle\gamma, T\gamma\rangle} e^{-\frac{i}{\hbar}W(\gamma)} g(\gamma)d\gamma$$

$$= C_T \widetilde{\int_Z} e^{\frac{i}{2\hbar}\langle\zeta, T_2\zeta\rangle} \widetilde{\int_Y} e^{\frac{i}{2\hbar}\langle\eta, T_1\eta\rangle} e^{-\frac{i}{\hbar}W(\zeta+\eta)} g(\zeta + \eta)d\eta d\zeta, \quad (7.76)$$

where $W : \mathcal{H}_t \to \mathbb{R}$ is given by Eq. (7.67) and $g : \mathcal{H}_t \to \mathbb{C}$ by Eq. (7.70), while the constant C_T is given by

$$C_T = (\det T)^{-1/2}(\det T_1)^{1/2}(\det T_2)^{1/2}$$

and it is equal to

$$C_T = 2\big(\sin^2 \Omega t(\Omega t)^{-1}(2\Omega t + \sin 2\Omega t)^{-1}\big)^{1/2}.$$

As $\eta(0) = 0$, Eq. (7.76) becomes

$$\widetilde{\int}_{\mathcal{H}_t} e^{\frac{i}{2\hbar}\langle \gamma, T\gamma\rangle} e^{-\frac{i}{\hbar}W(\gamma)} g(\gamma) d\gamma$$

$$= C_T \widetilde{\int}_Z e^{\frac{i}{2\hbar}\langle \zeta, T_2\zeta\rangle} e^{\frac{i}{\hbar}S(\zeta)} \bigg(\widetilde{\int}_Y e^{\frac{i}{2\hbar}\langle \eta, T_1\eta\rangle} e^{-\frac{i}{\hbar}V(\zeta+\eta)} d\eta\bigg) g(\zeta) d\zeta, \quad (7.77)$$

where $S(\zeta) = S_0(\zeta(0) + \beta_{t,x}(0))$, and $V(\zeta + \eta) = \int_0^t V_0(\zeta(s) + \eta(s) + \beta_{t,x}(s)) ds$. The key point is the fact that the integral over Z is a finite dimensional oscillatory integral and the classical results of the stationary phase method for finite dimensional oscillatory integrals apply [186].

By assuming that the functions $V_\zeta : Y \to \mathbb{R}$, with $\zeta \in Z$, given by

$$V_\zeta(\eta) := \int_0^t V_0(\zeta(s) + \eta(s) + \beta_{t,x}(s)) ds, \qquad \eta \in Y,$$

satisfy assumptions analogous to Eqs. (7.36) and (7.38), i.e.

$$\int_Y \|\eta\|^j d|\mu_\zeta|(\eta) \leq L\frac{j!}{\epsilon^j},$$

$$\frac{L\|T_1^{-1}\|}{\epsilon^2} 2(3 + 2\sqrt{2}) < 1,$$

uniformly in $\zeta \in Z$, one has that the phase function $\Phi_\zeta : Y \to \mathbb{R}$, given by

$$\Phi_\zeta(\eta) = \frac{1}{2}\langle \eta, T_1\eta\rangle - V(\eta + \zeta), \qquad \eta \in Y,$$

has a unique non-degenerate stationary point $a(\zeta) \in Y$. Moreover

$$\widetilde{\int}_Y e^{\frac{i}{2\hbar}\langle \eta, T_1\eta\rangle} e^{-\frac{i}{\hbar}V(\zeta+\eta)} d\eta = e^{\frac{i}{\hbar}\Phi_\zeta(a(\zeta))} J^*(\hbar, \zeta),$$

with J^* being a C^∞ function of $\hbar \in \mathbb{R}$ such that

$$J^*(0, \zeta) = (\det \psi_\zeta''(a(\zeta)))^{-1/2}.$$

The solution of the Schrödinger equation is then given in terms of the following oscillatory integral on a d dimensional space Z:

$$\psi^\hbar(t,x) = C_T e^{-\frac{i}{\hbar}x\tan(\Omega t)\Omega x} \int_Z e^{\frac{i}{2\hbar}\langle \zeta, T_2\zeta\rangle} e^{\frac{i}{\hbar}S(\zeta)} e^{\frac{i}{\hbar}\Phi_\zeta(a(\zeta))} \widetilde{J^*}(\hbar,\zeta)g(\zeta)d\zeta$$

$$(7.78)$$

with a phase function $\Phi : Z \to \mathbb{R}$ equal to

$$\Phi(\zeta) = \frac{1}{2}\langle \zeta, T_2\zeta\rangle + S(\zeta) + \Phi_\zeta(a(\zeta)), \qquad \zeta \in Z \qquad (7.79)$$

and the theory of asymptotic expansions of finite dimensional oscillatory integrals (with degenerate critical points) [125, 44] can be applied. In particular, if Φ has a degenerate critical point, i.e. if the time-space point (t,x) is focal, and if the space dimension d is less or equal then 3, then there exists a complete classification of the types of the possibles degeneracies [164] and the asymptotics of the integral (7.78) is of the form.

$$\psi^\hbar(t,x) \sim \hbar^{-\beta} \qquad \hbar \downarrow 0,$$

where $\beta \in \mathbb{Q}$, $\beta > 0$, is called Coxeter number and depends on the type of degeneracy [46].

7.5   The trace formula

The study of the connections between quantum and classical quantities is very old and goes back to the very origin of quantum theory; it is sufficient to think for instance to the Bohr quantization rules, relating the spectrum of the energy operator to the volume in the phase space enclosed by the classical periodic orbits of the system. The interest in this kind of relations has been renewed in recent years since, according to the theory of quantum chaos, the study of the distribution of the energy eigenvalues of a given quantum mechanical system should reflect that the underlying classical system is integrable (resp. chaotic) [172]. Particularly interesting is an (heuristic) trace formula connecting the semiclassical asymptotics for $\hbar \downarrow 0$ of trace of the Schrödinger group $\mathrm{Tr}(e^{-\frac{i}{\hbar}Ht})$ to the classical periodic orbits of the system. One can look at this as a quantum analogue of Selberg trace formula, relating the trace of the heat kernel on manifolds of constant negative curvature with the periodic geodesics [90, 189].

The first rigorous Feynman path integral derivation of the trace formula for the Schrödinger group and the study of its singularities as a function of the time variable can be found in [4], see also [2, 3]. Part of those results are generalized in [8, 6], where it is shown that in this particular problem the degenerate stationary points of the phase function of the Feynman integral play a fundamental role. Recently in [34, 35] some interesting results on the degenerate case have been applied to the study of a trace formula for the heat semigroup with a polynomial potential. In the present section we give some details of the results of [8, 6].

Let V_0 be a bounded real function belonging to $\mathcal{F}(\mathbb{R}^d)$ and $\Omega^2 > 0$ a positive symmetric $d \times d$ matrix and let us consider an anharmonic oscillator Hamiltonian of the following form

$$H = -\frac{\hbar^2}{2}\Delta + \frac{1}{2}x\Omega^2 x + V_0(x),$$

with domain

$$D(H) = \{\psi \in W^{2,2}(\mathbb{R}^d), \int_{\mathbb{R}^d} |x|^4|\psi(x)|^2 dx < \infty\}.$$

It is well known [281] that H is a self-adjoint operator on $L^2(\mathbb{R}^d)$, it has a pure point spectrum and, by applying Theorem 10.5 in [302] and Tauberian theorem, it follows that for some $\beta > 0$ its eigenvalues λ_n satisfy the following inequality:

$$\liminf_{n\to\infty} \frac{\lambda_n}{n^\beta} > 0.$$

This relation implies that the trace of the unitary group generated by iH, i.e. $\text{Tr}(e^{-\frac{i}{\hbar}Ht})$ is, as a function of t, a well defined distribution over $\mathbb{R}$.

Let $\mathcal{H}_{p,t}$ be the Hilbert space of periodic functions $\gamma \in H^1(0, t; \mathbb{R}^d)$ such that $\gamma(0) = \gamma(t)$, with norm

$$\|\gamma\|^2 = \int_0^t \dot\gamma(s)^2 ds + \int_0^t \gamma(s)^2 ds.$$

Let $\Phi : \mathcal{H}_{p,t} \to \mathbb{R}$ denote the phase function

$$\Phi(\gamma) = \frac{1}{2}\int_0^t \dot\gamma(s)^2 ds - \int_0^t V_1(\gamma(s))ds, \qquad \gamma \in \mathcal{H}_{p,t}, \tag{7.80}$$

where

$$V_1(x) = \frac{1}{2}x\Omega^2 x + V_0(x), \qquad x \in \mathbb{R}^d.$$

Let us consider the infinite dimensional oscillatory integral on $\mathcal{H}_{p,t}$

$$I(t, \hbar) = \widetilde{\int_{\mathcal{H}_{p,t}}} e^{\frac{i}{\hbar}\Phi(\gamma)}d\gamma. \tag{7.81}$$

The integral (7.81) can also be written in the following form

$$I(t, \hbar) = \widetilde{\int_{\mathcal{H}_{p,t}}} e^{\frac{i}{2\hbar}\langle \gamma,(I-L)\gamma\rangle} e^{-\frac{i}{\hbar}V(\gamma)} d\gamma, \tag{7.82}$$

where $L : \mathcal{H}_{p,t} \to \mathcal{H}_{p,t}$ is a self-adjoint trace class operator, given by

$$\langle \gamma, L\gamma \rangle = \int_0^t \gamma(s)\Omega^2\gamma(s)ds + \int_0^t |\gamma(s)|^2 ds, \qquad \gamma \in \mathcal{H}_{p,t} \tag{7.83}$$

and $V : \mathcal{H}_{p,t} \to \mathbb{R}$ is equal to

$$V(\gamma) = \int_0^t V_0(\gamma(s))ds, \qquad \gamma \in \mathcal{H}_{p,t}.$$

As $V_0 \in \mathcal{F}(\mathbb{R}^d)$, it is not difficult to verify that $I(t, \hbar)$ is well defined, provided that $(I - L) : \mathcal{H}_{p,t} \to \mathcal{H}_{p,t}$ is an invertible operator on $\mathcal{H}_{p,t}$. Moreover the following holds [6].

Theorem 7.8. *Let us assume that the time variable t satisfies the following condition*

$$\det \sin \left(\frac{t}{2}\Omega\right) \neq 0.$$

Then the infinite dimensional oscillatory integral (7.81) is well defined and the function $t \mapsto I(t, \hbar)$ is of class C^∞ on the subset D_Ω of the real line defined by

$$D_\Omega = \{t : \det \sin \left(\frac{t}{2}\Omega\right) \neq 0\}.$$

Moreover the trace of the Schrödinger group $\mathrm{Tr}(e^{-\frac{i}{\hbar}Ht})$ is a well defined distribution over $(0, \infty)$, a C^∞ function on D_Ω and it is equal to

$$\mathrm{Tr}(e^{-\frac{i}{\hbar}Ht}) = \left(2(\cosh t - 1)\right)^{-d/2} I(t, \hbar). \tag{7.84}$$

Proof. For a detailed proof we refer to [6]. The definition and the regularity properties of the oscillatory integral $I(t, \hbar)$ can be proved by means of the general theory (see Chapter 5, in particular Theorem 5.5). The proof of Eq. (7.84) is based on an analytic continuation technique and on the proof of an analogous equality for the trace of the corresponding heat semigroup. □

According to Eq. (7.84), the study of the semiclassical limit of the trace of the Schrödinger group $\mathrm{Tr}(e^{-\frac{i}{\hbar}Ht})$ is reduced to the study of the asymptotic behaviour of the integral in the limit $\hbar \downarrow 0$.

If $V_1 : \mathbb{R}^d \to \mathbb{R}$ is of class C^2, then one proves that the functional Φ is of class C^2 and a path $\gamma \in \mathcal{H}_{p,t}$ is a stationary point for Φ if and only if γ is a solution of the Newton equation

$$\ddot{\gamma}(s) + V_1'(\gamma(s)) = 0, \qquad s \in [0, t] \tag{7.85}$$

satisfying the periodic conditions

$$\gamma(0) = \gamma(t), \qquad \dot{\gamma}(0) = \dot{\gamma}(t). \tag{7.86}$$

Concerning the second Frechet derivative of the phase function Φ, it is possible to prove that, given a vector $\gamma \in \mathcal{H}_{p,t}$, the operator $\Phi''(\gamma) : \mathcal{H}_{p,t} \to \mathcal{H}_{p,t}$ is defined by $\Phi''(\gamma) = I + L_\gamma$, where $u = L_\gamma \phi$ if and only if u satisfies Eq. (7.86) and

$$\ddot{u}(s) - u(s) = \big(V_1''(\gamma(s)) + 1\big)\phi(s), \qquad 0 < s < t. \tag{7.87}$$

The symmetry of L_γ can be easily verified, moreover the regularity of the solutions of Eq. (7.87) implies that the range of L_γ is contained in $H^3(0, t; \mathbb{R}^d)$ and L_γ is a trace-class operator.

A first example of stationary path of the phase Φ can easily be found if the potential V_1 admits critical points c_j, $j = 1, \ldots, m$. Indeed in this case the functions γ_{c_j}, given by

$$\gamma_{c_j}(s) = c_j, \qquad s \in [0, t],$$

are periodic solutions of Eq. (7.85). Moreover the Fredholm determinant of $\Phi''(\gamma_{c_j})$ is given by (see [6], Proposition 2.11)

$$\det \Phi''(\gamma_{c_j}) = \det \Big((\cosh t - 1)^{-1}\big(\cos(t\sqrt{V_1''(c_j)}) - 1\big)\Big).$$

From this equality we can easily see that the stationary point γ_{c_j} is non-degenerate if and only if $\cos(t\sqrt{V_1''(c_j)}) - 1 \neq 0$.

Besides the "trivial" solution of Eqs. (7.85) and (7.86), there can be also other stationary points of Φ, that are degenerate. Indeed the following holds [126, 6].

Theorem 7.9. Let $\gamma \in \mathcal{H}_{p,t}$ be a non-constant solution of Eq. (7.85) and Eq. (7.86). Then $\mathrm{Ker}(\Phi''(\gamma)) \neq \{0\}$. In particular:

$$\Phi''(\gamma)(\dot{\gamma}) = 0.$$

Proof. Let us introduce the C_0-group of linearly unitary transformations in $\mathcal{H}_{p,t}$:

$$G_\tau \gamma(s) = \tilde{\gamma}(s + \tau), \qquad s \in [0, t], \ \tau \in \mathbb{R}, \ \gamma \in \mathcal{H}_{p,t},$$

where $\tilde{\gamma}$ is the unique extension of γ to a $t-$ periodic continuous function on $\mathbb{R}$. One can see that Φ is invariant under the action of G_τ, i.e. $\Phi \circ G_\tau = \Phi$. Let us define $\gamma_\tau := G_\tau \gamma$ and by the chain rule we have:

$$\Phi'(\gamma_\tau) \circ G_\tau = d_\gamma(\Phi \circ G_\tau) = d_\gamma(\Phi) = 0.$$

From this we can conclude that $\Phi'(\gamma_\tau) = 0$. Hence

$$\lim_{s \to 0} \frac{1}{s} \Phi'(\gamma_s) - \Phi'(\gamma) = 0,$$

but the last limit is equal to

$$\Phi''(\gamma)\left(\frac{d}{ds}\gamma(s)_{|s=0}\right) = \Phi''(\gamma)(\dot{\gamma}).\square$$

The proof of Theorem 7.9 shows that if $\gamma \in \mathcal{H}_{p,t}$ is a non-constant solution of Eqs. (7.85) and (7.86), then γ is not only a degenerate stationary point, but it is also non-isolated as for each $\tau \in [0, t]$ the path γ_τ is also a degenerate stationary point of Φ. In particular, the set of degenerate stationary points

$$\{\gamma_\tau, \ \tau \in [0, t]\}$$

is a compact manifold diffeomorphic to the circle S^1.

The following result is an extension to this setting of the classical Morse theorem about non-degenerate critical points [6].

Theorem 7.10. *Let $\gamma \in \mathcal{H}_{p,t}$ be a degenerate stationary point of Φ satisfying the following condition*

$$\dim \operatorname{Ker} \Phi''(\gamma) = 1. \tag{7.88}$$

then the set

$$\{\gamma_\tau, \ \tau \in [0, t]\}$$

is isolated within the set of all stationary points of Φ, i.e., there is $\epsilon > 0$ such that if $\phi \in \mathcal{H}_{p,t}$ satisfies $\|\phi - \gamma_\tau\| < \epsilon$ for some $\tau \in \mathbb{R}$, and $\Phi'(\phi) = 0$, then $\phi = \gamma_s$ for some $s \in \mathbb{R}$.

Proof. Let

$$Z = \operatorname{Ker} \Phi''(\gamma) = \{\alpha\dot{\gamma}, \ \alpha \in \mathbb{R}\} \subset \mathcal{H}_{p,t}$$

and let $Y := Z^\perp$ be the orthogonal complement of Z in $\mathcal{H}_{p,t}$. By Eq. (7.87), the operator $\Phi''(\gamma)$ is compact and self-adjoint. By the Fredholm alternative theorem it follows that $\Phi''(\gamma)$ maps injectively Y onto Y.

Let $F : Y \to \mathbb{R}$ be defined by

$$F(\psi) = \Phi(\gamma + \psi), \qquad \psi \in Y.$$

Denoted by $j : Y \to \mathcal{H}_{p,t}$ the natural embedding map and by $j^* : \mathcal{H}_{p,t} \to Y$ its dual operator, we have that $F''(0) = j^* \circ \Phi'' \circ j$ is a linear isomorphism of Y. By the implicit function theorem, there is an $\epsilon_0 > 0$ such that for any $\eta \in Y$, with $\|\eta\| < \epsilon_0$, we have $F'(\eta) \neq 0$. By the tubular neighbourhood theorem [257], the set

$$\tilde{W}_{\epsilon_0} := \cup_{s \in \mathbb{R}} \{ \gamma_s + G_s \eta : \eta \in Y, \|\eta\| < \epsilon_0 \}$$

is an open neighbourhood of the manifold $\hat{\gamma} := \{ \gamma_\tau, \ \tau \in [0, t] \}$, and there exists an $\epsilon \in (0, \epsilon_0)$ such that the open neighbourhood $W_\epsilon(\hat{\gamma})$

$$W_\epsilon(\hat{\gamma}) := \{ \xi \in \mathcal{H}_{p,t} : dist(\xi, \hat{\gamma}) < \epsilon \}$$

is included in $\tilde{W}_{\epsilon_0}$. It is then possible to see that $\Phi' \neq 0$ on $W_\epsilon(\hat{\gamma})$. Indeed if $\Phi(\psi) = 0$ for $\psi = \gamma_s + G_s \eta$ for some $s \in \mathbb{R}$ and some $\eta \in Y$ such that $\|\eta\| < \epsilon$, then, since $\Phi \circ G_{-s} = \Phi$, we have that $\Phi'(G_{-s}(\gamma_s + G_s \eta)) = 0$. On the other hand, $G_{-s}(\gamma_s + G_s \eta) = \gamma + \eta$, $G_{-s}(\eta) \in Y$. Therefore, $F'(\eta) = 0$ and this gives $\eta = 0$. $\qquad\square$

The results stated so far concerning the stationary points of the phase function $\Phi : \mathcal{H}_{p,t} \to \mathbb{R}$ are valid for a general Φ of the form (7.80). Let us now consider the particular case where the potential is of the following form

$$V_1(x) = \frac{1}{2} x \Omega^2 x + V_0(x), \qquad x \in \mathbb{R}^d, \tag{7.89}$$

with V_0 real bounded and $V_0 \in \mathcal{F}(\mathbb{R}^d)$.

In [6] V_1 is also assumed to satisfy the following conditions:

(1) V_1 has a finite number critical points $c_1, \ldots, c_m$, and each of them is non-degenerate, i.e.

$$\det V_1''(c_j) \neq 0, \qquad j = 1, \ldots m.$$

(2) $t > 0$ is such that the function γ_{c_j}, given by $\gamma_{c_j}(s) = c_j$, $s \in [0, t]$, is a non-degenerate stationary point for Φ;

(3) any non-constant $t-$periodic solution γ of Eqs. (7.85) and (7.86) is a "non-degenerate periodic solution", in the sense of [126], i.e. $\dim \operatorname{Ker} \Phi''(\gamma) = 1$.

Then by Theorems 7.9 and 7.10, the set M of stationary points of the phase function Φ is a disjoint union of the following form:

$$M = \{x_{c_1}, \ldots, x_{c_m}\} \cup \bigcup_{k=1}^{r} M_k,$$

where x_{c_i}, $i = 1, \ldots m$, are non-degenerate and M_k are manifolds (diffeomorphic to S^1) of degenerate stationary points, on which the phase function is constant. Under some growth conditions on V_0 it is also possible to compute the asymptotic behaviour as $\hbar \to 0$ of the trace of the Schrödinger group, or, equivalently (see Eq. (7.84)), of the infinite dimensional oscillatory integral $I(t, \hbar)$.

Theorem 7.11. *Let us assume that the potential V_1 is of the form (7.89), with $V_0 \in \mathcal{F}(\mathbb{R}^d)$, $V_0 = \hat{\mu}_0$, satisfying*

$$\int_{\mathbb{R}^d} |y|^j d|\mu_0|(y) \leq K \frac{j!}{\epsilon^j}, \qquad j \in \mathbb{N},$$

for some $K, \epsilon > 0$. Then, as $\hbar \downarrow 0$, the trace of the Schrödinger group has the following asymptotic behaviour

$$\mathrm{Tr}(e^{-\frac{i}{\hbar}Ht}) = \sum_{j=1}^{m} e^{\frac{i}{\hbar}tV_1(c_j)} I_j^*(\hbar) + (2\pi i\hbar)^{-1/2} \left[\sum_{k=1}^{r} e^{\frac{i}{\hbar}\Phi(b_k)} |M_k| I_k^{**}(\hbar) + O(\hbar) \right],$$

where c_j are the points in condition 1, $b_k \in M_k$ are all non-constant t-periodic solutions of (7.85) and (7.86) as in condition (3), $|M_k|$ is the Riemannian volume of M_k, I_j^* and I_k^{**} are C^∞ functions of $\hbar \in \mathbb{R}$ such that, in particular,

$$I_j^*(0) = \left(\det \left[2 \left[\cos \left(t\sqrt{V''(c_j)} \right) - 1 \right] \right] \right)^{-1/2},$$

$$I_k^{**}(0) = \left(\frac{d}{d\epsilon} \det(R_\epsilon^k(t) - I)|_{\epsilon=1} \right)^{-1/2},$$

where $R_\epsilon^k(t)$ denotes the fundamental solution of

$$\begin{cases} \ddot{x}(s) = -\epsilon V''(b_k(s))x(s), & s > 0, \\ x(0) = x_0, & \dot{x}(0) = y_0 \end{cases}$$

written as a first order system of $2d$ equations for real valued functions.

Proof. For a detailed proof we refer to the paper [6], we give here only some hints. By Eq. (7.84), one has to study the asymptotic behaviour of the oscillatory integral $I(t, \hbar)$. By the condition on the potential V_1, the set of stationary points of the phase function is completely determined.

Moreover it is possible to reduce the study of the degeneracy to a finite dimensional subspace of the Hilbert space $\mathcal{H}_{p,t}$ and to apply the technique described at the end of Section 7.3 (see also [7]).

Indeed it is possible to find a finite dimensional subspace $Z \subset \mathcal{H}_{p,t}$ such that, given $Y := T(Z)^{\perp}$ with $T = (I - L)$ and L is given by Eq. (7.83), one has $Z + Y = \mathcal{H}_{p,t}$ and the phase function when restricted to y

$$\Phi_z : Y \to \mathbb{R}, \qquad \Phi_z(y) = \Phi(z + y),$$

has a unique stationary point, denoted by $a(z)$. Moreover the function $a : Z \to Y$, $z \mapsto a(z)$, is of class C^{∞} and bounded with all its derivatives. By the Fubini Theorem 5.7 for oscillatory integrals, $I(t, \hbar)$ is given by

$$I(t, \hbar) = C_T \widetilde{\int_Z} \left(\widetilde{\int_Y} e^{\frac{i}{\hbar} \Phi_z(y)} dy \right) dz,$$

with $C_T = (\det T)^{-1/2} (\det T_{|Y})^{1/2} (\det T_{|Z})^{1/2}$.

By Theorem 7.4, the integral over Y is given by

$$\widetilde{\int_Y} e^{\frac{i}{\hbar} \Phi_z(y)} dy = e^{\frac{i}{\hbar} \Phi_z(a(z))} J^*(\hbar, z),$$

with J^* being a C^{∞} function of both $\hbar \in (-1, -1)$ and $z \in Z$, such that

$$J^*(0, z) = \det \left(\frac{\partial^2}{\partial y^2} \Phi_z(a(z)) \right)^{-1/2}.$$

Therefore, as $\hbar \to 0$,

$$I(t, \hbar) = C_T \widetilde{\int_Z} \det \left(\frac{\partial^2}{\partial y^2} \Phi_z(a(z)) \right)^{-1/2} e^{\frac{i}{\hbar} \Phi_z(a(z))} dz + O(\hbar),$$

and the asymptotic behaviour of $I(t, \hbar)$ is determined by the reduced phase function $\Psi : Z \to \mathbb{R}$, given by

$$\Psi(z) = \Phi(z + a(z)), \qquad z \in Z.$$

One can prove (see [7], Lemma 5.3) that $\Psi'(z) = 0$ iff $\Phi'(z + a(z)) = 0$ and

$$\ker(\Phi''(z + a(z))) = (I_Z, a'(z))(\ker \Psi''(z)).$$

Denoting by Q the projection from $\mathcal{H}_{p,t}$ onto Z along y, i.e. $Q(z + y) = z$, the set N of stationary points of the phase Ψ is the projection $Q(M)$ of the stationary points of Φ:

$$N = \{d_1, \ldots, d_m\} \cup \bigcup_{k=1}^{r} Q(M_k),$$

with $d_1, \ldots, d_m$ are non-degenerate and $Q(M_k)$ for $j = k, \ldots, r$ are compact and connected one dimensional manifolds such that $\dim \ker \Psi''(a) = 1$ for each $a \in \cup_{k=1}^{r} Q(M_k)$. The final results follows by applying the stationary phase method on finite dimensional oscillatory integrals (see [6]).   □

Chapter 8

Open quantum systems

8.1 Feynman path integrals and open quantum systems

Our treatment of the Feynman path integral formulation of (non-relativistic) quantum mechanics has started with the description of the dynamics of a pointwise particle, described by the Schrödinger Eq. (1.4). The generalization of the formalism developed so far to more complicated and more realistic quantum systems can be realized, on the one hand, by increasing the number of degrees of freedom and by considering more complicated Hamiltonian H. On the other hand, in many situations occurring in quantum mechanics, one is not interested in the detailed description of the whole system (that, when the number of degrees of freedom increases, becomes a very difficult task), but only of a part of it. Let us think for instance to the interaction of a radiating atom with the electromagnetic field when one is interested only in the state of the atom and not in the emitted radiation. Another fundamental problem of this kind is the description of the process of quantum measurement. Indeed when a quantum system is submitted to the measurement of one of its observables, it interacts with a (macroscopic) measuring apparatus which is not of primary interest but, on the other hand, whose influence on the system, by the Heisenberg uncertainty principle, cannot be neglected.

The development of a formalism allowing the quantum description of the dynamic of a system interacting with an external "environment" is the task of the quantum theory of open systems [106, 132]. A possible approach is the description of the compound "system plus environment" as a whole, by means of a Schrödinger equation involving an interaction Hamiltonian. As a second step, the environment's degrees of freedom have to be traced out.

An alternative solution to the problem is provided by the path integral formalism. This chapter concerns a few instances of the descriptive power of Feynman path integrals when applied to the theory of open quantum systems. Let us briefly introduce some heuristic considerations which motivate a further detailed treatment.

The simplest example of a system subject to an external influence is a classical one-dimensional linearly forced harmonic oscillator, whose equation of motion is

$$m\ddot{x}(t) + kx(t) = f(t), \qquad t \in \mathbb{R}, \tag{8.1}$$

where $m, k \in \mathbb{R}^+$ and f is a "fluctuating force". The quantization of the classical system (8.1) in terms of Feynman path integrals provides for the Green function of the corresponding Schrödinger equation an heuristic formula of the form

$$G(0, y, t, x) = \int_{\substack{\gamma(t)=x \\ \gamma(0)=y}} e^{\frac{i}{\hbar}\left(\frac{m}{2}\int_0^t \dot{\gamma}(s)^2 ds - \frac{k}{2}\int_0^t \gamma(s)^2 ds - \int_0^t \gamma(s)f(s)ds\right)} D\gamma. \tag{8.2}$$

As a further step one can consider the external force f to be random, and be interested to the averaged value over f of the quantum dynamics given by Eq. (8.2). In order to obtain a physically meaningful expression, the average has to be done on quantum observables, as for instance the transition amplitudes, rather than on mathematical objects, as the wave functions. Therefore let us consider two state vectors $\psi, \phi \in L^2(\mathbb{R})$ and compute the probability of a transition from ψ to ϕ, i.e. $|\langle \phi, \psi(t) \rangle|^2$, where $\psi(t, x) = \int G_f(0, y, t, x)\psi(y)dy$,

$$|\langle \phi, \psi(t) \rangle_f|^2 = \int \int \int \int \phi(x')\bar{\phi}(x)G_f(0, y, t, x)\overline{G_f(0, y', t, x')}\psi(y)$$
$$\bar{\psi}(y')dydxdy'dx', \tag{8.3}$$

where we have introduced the subscript in order to stress the f-dependence of the Green function. By inserting in (8.3) the Feynman path integral representation (8.2), we obtain an heuristic formula which represents the transition amplitude in terms of a double path integral:

$$|\langle \phi, \psi(t) \rangle_f|^2 = \int \int \int \int \phi(x')\bar{\phi}(x)\psi(y)\bar{\psi}(y')$$
$$\int \int e^{\frac{i}{\hbar}\left(S(\gamma) - S(\gamma')\right)} e^{-\frac{i}{\hbar}\int_0^t \left(\gamma(s) - \gamma'(s)\right)f(s)ds} d\gamma d\gamma' dydxdy'dx', \tag{8.4}$$

where $\gamma(0) = y, \gamma(t) = x, \gamma'(0) = y', \gamma'(t) = x'$ and

$$S(\gamma) = \frac{m}{2}\int_0^t \dot{\gamma}(s)^2 ds - \frac{k}{2}\int_0^t \gamma(s)^2 ds.$$

By averaging the above quantity over the random force f we obtain:

$$\mathbb{E}_f|\langle \phi, \psi(t)\rangle_f|^2 = \int\int\int\int \phi(x')\bar{\phi}(x)\psi(y)\bar{\psi}(y')$$

$$\int\int e^{\frac{i}{\hbar}\left(S(\gamma)-S(\gamma')\right)} F(\gamma,\gamma')d\gamma d\gamma' dy dx dy' dx', \quad (8.5)$$

where the influence of the random fluctuating force f is modelled by the functional

$$F(\gamma,\gamma') = \mathbb{E}_f[e^{-\frac{i}{\hbar}\int_0^t \left(\gamma(s)-\gamma'(s)\right)f(s)ds}].$$

This, still heuristic, considerations show some key features of the Feynman path integral description of the quantum open systems, such as the introduction of a *double path integral* of the form (8.5) and the modellization of an interacting environment in terms of an *influence functional* $F(\gamma,\gamma')$ which couples the paths γ,γ'. This particular formalism was introduced by Feynman and Vernon in [137], where a formula similar to Eq. (8.5) was heuristically derived by modelling the environment in terms of a many body quantum system and by tracing out its degrees of freedom. This technique has been applied to the modellization of several complex quantum phenomena, including the quantum Brownian motion [76], i.e. the description of the dynamics of a quantum particle interacting with an macroscopic environment.

Let us consider for instance the time evolution of a quantum system made of two linearly interacting subsystems A and B. Let us assume that the state space of the system A is $L^2(\mathbb{R}^d)$ while the state space of the system B is $L^2(\mathbb{R}^N)$. Let the total Hamiltonian of the compound systems be of the form

$$H_{AB} = H_A + H_B + H_{INT} \quad (8.6)$$

$$H_A = -\frac{\Delta_{\mathbb{R}^d}}{2M} + \frac{1}{2}x\Omega_A^2 x + v_A(x), \quad x \in \mathbb{R}^d,$$

$$H_B = -\frac{\Delta_{\mathbb{R}^N}}{2m} + \frac{1}{2}R\Omega_B^2 R + v_B(R), \quad R \in \mathbb{R}^N,$$

$$H_{INT} = xCR, \quad x \in \mathbb{R}^d, R \in \mathbb{R}^N,$$

with $C : \mathbb{R}^N \to \mathbb{R}^d$ is a linear operator, Ω_A, resp. Ω_B, a symmetric positive $d \times d$ (resp. $N \times N$) matrix, v_A, resp. v_B, real bounded functions. Let us assume that the quadratic part of the total potential, i.e. the function $x, R \mapsto \frac{1}{2}x\Omega_A^2 x + \frac{1}{2}R\Omega_B^2 R + xCR$ is positive definite (so that the total

Hamiltonian is bounded from below). Let us assume moreover that the density matrix of the compound system factorizes $\rho_{AB} = \rho_A \rho_B$ and has a smooth kernel $\rho_{AB}(x, y, R, Q) = \rho_A(x, y)\rho_B(R, Q)$. By writing the Feynman path integral representation for the evolution of the density matrix, we have

$$\rho_t(x, y, R, Q) = \left(e^{-\frac{i}{\hbar}H_{AB}t}\rho_0 e^{\frac{i}{\hbar}H_{AB}t}\right)(x, y, R, Q)$$

$$= \int\int\int\int \rho_A(\gamma(0), \gamma'(0))\rho_B(\Gamma(0), \Gamma'(0))e^{\frac{i}{\hbar}\left(S_A(\gamma) - S_A(\gamma')\right)}e^{\frac{i}{\hbar}\left(S_B(\Gamma) - S_B(\Gamma')\right)}$$

$$e^{\frac{i}{\hbar}\left(S_{AB}(\gamma, \Gamma) - S_{AB}(\gamma', \Gamma')\right)}d\gamma d\gamma' d\Gamma d\Gamma' \quad (8.7)$$

where the integral is taken over the path $\gamma, \gamma' : [0, t] \to \mathbb{R}^d$ and $\Gamma, \Gamma' : [0, t] \to \mathbb{R}^N$ such that $\gamma(t) = x$, $\gamma'(t) = y$, $\Gamma(t) = R$, $\Gamma'(t) = Q$. S_A, S_B and S_{AB} are given by:

$$S_A(\gamma) = \int_0^t \left(\frac{M}{2}\dot{\gamma}(s)^2 - \frac{1}{2}\gamma(s)\Omega_A^2\gamma(s) - v_A((\gamma(s)))\right)ds,$$

$$S_B(\Gamma) = \int_0^t \left(\frac{m}{2}\dot{\Gamma}(s)^2 - \frac{1}{2}\Gamma(s)\Omega_B^2\Gamma(s) - v_B((\Gamma(s)))\right)ds,$$

$$S_{AB}(\gamma, \Gamma) = -\int_0^t \gamma(s)C\Gamma(s)ds$$

By tracing over the coordinates of the system B, one obtains the following (heuristic) Feynman path integral representation for the reduced density matrix $\rho_A^r(t)$ of the system A:

$$\rho_A^r(t)(x, y) = \int \rho_t(x, y, R, R)dR$$

$$= \int\int \rho_A(\gamma(0), \gamma'(0))e^{\frac{i}{\hbar}\left(S_A(\gamma) - S_A(\gamma')\right)}F(\gamma, \gamma')d\gamma d\gamma', \quad (8.8)$$

where

$$F(\gamma, \gamma') = \int\int_{\Gamma(t)=R\ \Gamma'(t)=R}\int \rho_B(\Gamma(0), \Gamma'(0))e^{\frac{i}{\hbar}\left(S_B(\Gamma) - S_B(\Gamma')\right)}$$

$$e^{\frac{i}{\hbar}\left(S_{AB}(\gamma, \Gamma) - S_{AB}(\gamma', \Gamma')\right)}d\Gamma d\Gamma' dR. \quad (8.9)$$

In Section 8.2 we shall give a rigorous mathematical meaning to the double Feynman path integral (8.8) in terms of a well defined infinite dimensional oscillatory integral and analyze the Caldeira-Leggett model [76] for the description of the quantum Brownian motion.

Another fundamental problem of quantum theory is the description of the process of quantum measurement, i.e. the interaction of a physical system with a (macroscopic) measuring apparatus allowing an observer to obtain the result of the measurement of a physical quantity. In the traditional formulation of quantum mechanics, the continuous time evolution described by the Schrödinger Eq. (1.4) is valid if the quantum system is "undisturbed". On the other hand we should not forget that all the information we can obtain about the state of a quantum particle are the result of some measurement process. When the particle interacts with the measuring apparatus, its time evolution is no longer continuous: the state of the system after the measurement is the result of a random and discontinuous change, the so-called "collapse of the wave function", which cannot be described by the ordinary Schrödinger equation. Quoting Dirac [121], after the introduction of the Planck constant $\hbar$ the concept of "large" and "small" are no longer relative: it is "microscopic" [1] one object such that the influence on the measuring apparatus on it cannot be neglected.

Let us recall the main features of the traditional quantum description of the measurement of an observables $\mathcal{A}$, represented by a self-adjoint operator $A : D(A) \subset \mathcal{H} \to \mathcal{H}$ on a complex separable Hilbert space $\mathcal{H}$, whose unitary vectors represents the states of the system. Let us consider for simplicity the case where the operator A is bounded and its spectrum is discrete. Let $\{a_i\}_{i \in \mathbb{N}} \subset \mathbb{R}$ and $\{\psi_i\}_{i \in \mathbb{N}} \subset \mathcal{H}$ be the corresponding eigenvalues and eigenvectors. According to the traditional mathematical formulation by Von Neumann the consequences of the measurement are:

(1) The *decoherence* of the state of the quantum system

Because of the interaction with the measuring apparatus, the initial pure state ψ of the system becomes a mixed state, described by the density operator $\rho^{prior}(t) = \sum_i w_i P_{\psi_i}$, where P_{ψ_i} denotes the projector operator onto the eigenspace which is spanned by the vector ψ_i and $w_i = |\langle \psi_i, \psi \rangle|^2$. By considering another observable $\mathcal{B}$ (represented by a bounded self-adjoint operator $B : \mathcal{H} \to \mathcal{H}$), its expectation value at time t, after the measurement of the observable $\mathcal{A}$ (but without the information of the result of the measurement of $\mathcal{A}$), is given by

$$\mathbb{E}(B)_t^{prior} = \mathrm{Tr}[\rho^{prior}(t)B].$$

The existence of the trace is assumed. The transformation mapping ψ to the so-called "prior state" $\rho^{prior}(t)$ is named "prior dynamics" or non-selective dynamics.

---

[1] It would be more correct the word "quantum" as there exist also macroscopic quantum systems, but they were unknown at Dirac's time.

(2) The so-called "collapse of the wave function"

After the reading of the result of the measurement (i.e. the real number a_i) the state of the system is the corresponding eigenstate of the measured observable:

$$\rho(t)_{a_i}^{post} = P_{\psi_i}.$$

The expectation value of another observable $\mathcal{B}$ of the system at time t (taking into account the information about the value of the measurement of $\mathcal{A}$) is given by

$$\mathbb{E}^{post}(B|A = a_i)_t = \mathrm{Tr}[\rho_{a_i}^{post}(t)B] = \langle \psi_i, B\psi_i \rangle.$$

The transformation mapping the initial state ψ to one of the so-called "posterior states" $\rho_{a_i}^{post}(t)$ is called "posterior dynamics" or selective dynamics and depends on the result a_i of the measurement of $\mathcal{A}$.

As it is suggested by the collapse of the wave function, the non-selective dynamics maps pure states into mixed states, while the selective one maps pure states into pure states. The relation between the posterior state and the prior state is given by

$$\rho^{prior}(t) = \sum_i P(A = a_i)\rho_{a_i}^{post}(t),$$

where $P(A = a_i)$ denotes the probability that the outcome of the measurement of $\mathcal{A}$ is the eigenvalue a_i and it is given by

$$P(A = a_i) = |\langle \psi_i, \psi \rangle|^2.$$

We remark that

$$\mathbb{E}(B)_t^{prior} = \sum_i \mathbb{E}^{post}(B|A = a_i)P(A = a_i). \qquad (8.10)$$

There are several efforts to include the process of measurement into the traditional quantum theory and to deduce from its laws, instead of postulating, both the process of decoherence (see point 1) and the collapse of the wave function (point 2). In particular the aim of the *quantum theory of measurement* is a description of the process of measurement taking into account the properties of the measuring apparatus, which is handled as a quantum system, and its interaction with the system submitted to the measurement [106, 75]. Even if also this approach is not completely satisfactory (also in this case one has to postulate the collapse of the state of the compound system "measuring apparatus plus observed system"), it is able to give a better description of the process of measurement.

An interesting result of the quantum theory of measurement is the so-called "Zeno effect", which seems to forbid a satisfactory description of continuous measurements. Indeed if a sequence of "ideal"[2] measurements of an observable A with discrete spectrum is performed and the time interval between two measurements is sufficiently small, then the observed system does not evolve. In other words a particle whose position is continuously monitored cannot move. This result is in apparent contrast with the experience: indeed in a bubble chamber repeated measurements of the position of microscopical particles are performed without "freezing" their state. For a detailed description of the quantum Zeno paradox see for instance [258, 93, 273].

An heuristic Feynman path integral description of the process of quantum measurement has been proposed by several authors, in particular by M.B. Mensky [254, 255], who considers the selective dynamics of a particle whose position is continuously observed. According to Mensky the state of the particle at time t if the observed trajectory is the path $\omega(s)_{s \in [0,t]}$ is given by the "restricted path integrals"

$$\psi(t, x, \omega) = \int_{\{\gamma(t) = x\}} e^{\frac{i}{\hbar} S_t(\gamma)} e^{-\lambda \int_0^t (\gamma(s) - \omega(s))^2 ds} \phi(\gamma(0)) D\gamma, \qquad (8.11)$$

where $\lambda \in \mathbb{R}^+$ is a real positive parameter which is proportional to the accuracy of the measurement. Heuristically formula (8.11) suggests that, as an effect of the correction term $e^{-\lambda \int_0^t (\gamma(s) - \omega(s))^2 ds}$ due to the measurement, the paths γ giving the main contribution to the integral (8.11) are those closer to the observed trajectory ω.

As one looks to the observed trajectory ω in Eq. (8.11) as a random variable and to the state $\psi(t, x, \omega)$, $t \geq 0$ as a stochastic process, it is natural to think at Eq. (8.11) as the Feynman path integral implementation of a quantum stochastic dynamics. Indeed in the physical and in the mathematical literature a class of stochastic Schrödinger equations giving a phenomenological description of quantum measurements has been proposed by several authors, see for instance [65, 54, 55, 119, 255, 156]. Let us consider for instance Belavkin equation, a stochastic Schrödinger equation describing the selective dynamics of a d-dimensional particle submitted to the measurement of one of its (possible M-dimensional vector) observables, described

---

[2]A measurement is called *ideal* if the correlation between the state of the measuring apparatus and the state of the system after the measurement is maximal.

by the self-adjoint operator R on $L^2(\mathbb{R}^d)$:

$$\begin{cases} d\psi(t,x) = -\frac{i}{\hbar}H\psi(t,x)dt - \frac{\lambda}{2}R^2\psi(t,x)dt + \sqrt{\lambda}R\psi(t,x)dB(t) \\ \psi(0,x) = \psi_0(x) \qquad (t,x) \in [0,T] \times \mathbb{R}^d. \end{cases} \tag{8.12}$$

H is the quantum mechanical Hamiltonian, B is an M-dimensional Brownian motion on a probability space $(\Omega, \mathcal{F}, \mathbb{P})$ (see Section 3.5), $dB(t)$ is the Ito differential and $\lambda > 0$ is a coupling constant, which is proportional to the accuracy of the measurement. In the particular case of the description of the continuous measurement of position one has $R = x$ (the multiplication operator), so that Eq. (8.12) assumes the following form:

$$\begin{cases} d\psi(t,x) = -\frac{i}{\hbar}H\psi(t,x)dt - \frac{\lambda}{2}x^2\psi(t,x)dt + \sqrt{\lambda}x\psi(t,x)dB(t) \\ \psi(0,x) = \psi_0(x) \qquad (t,x) \in [0,T] \times \mathbb{R}^d. \end{cases} \tag{8.13}$$

Belavkin derives Eq. (8.12) by modeling the measuring apparatus (but it is better to say "the informational environment") by means of a one-dimensional bosonic field and by assuming a particular form for the interaction Hamiltonian between the field and the system on which the measurement is performed. The resulting dynamics is such that there exists a family of mutually commuting Heisenberg operators of the compound system, denoted by $X(t)_{t\in[0,T]}$, such that:

$$[X(t), X(s)] = 0, \qquad s,t \in [0,T],$$

(on a dense domain in $L^2(\mathbb{R}^d)$). In this description the concept of trajectory of X is meaningful, even from a quantum mechanical point of view. Furthermore the "non-demolition principle" is fulfilled: the measurement of any future Heisenberg operator $Z(t)$ of the system is compatible with the measurement of the trajectory of X up to time t, that is

$$[Z(t), X(s)] = 0, \qquad s < t,$$

(on a dense domain in $L^2(\mathbb{R}^d)$). The measured observable R is connected to the operator X by the following relation

$$X(t) = R(t) + \lambda(B_t + B_t^+), \tag{8.14}$$

(where $(B_t + B_t^+)$ is a quantum Brownian motion [188]). Equation (8.14) shows how the measurement of $X(t)$ gives some (indirect and not precise) information on the value of R, overcoming the problems of quantum Zeno paradox. Indeed we are dealing with "unsharp" in spite of "ideal" measurements.

For an alternative derivation of Eq. (8.13), we point out to the reader the appendix A of [232], where a physically intuitive model of unsharp continuous quantum measurement is proposed, as well as a Feynman path integral representation of the state ψ analogous to Eq. (8.11).

In Section 8.3 we shall realize, in terms of infinite dimensional oscillatory integrals, a Feynman path integral representation of the solution of Belavkin Eq. (8.13):

$$\psi(t,x) = \widetilde{\int} e^{\frac{i}{2\hbar} \int_0^t |\dot\gamma(s)|^2 ds - \lambda \int_0^t |\gamma(s)+x|^2 ds} e^{-\frac{i}{\hbar} \int_0^t V(\gamma(s)+x) ds}$$

$$e^{\int_0^t \sqrt{\lambda}(\gamma(s)+x)\cdot dB(s)} \psi_o(\gamma(0)+x) d\gamma. \qquad (8.15)$$

and therefore we shall give a rigorous mathematical meaning to Mensky's heuristic formula (8.11).

8.2   The Feynman-Vernon influence functional

Let us introduce a particular type of infinite dimensional oscillatory integral on a real separable Hilbert space $\mathcal{H}$ which will be used in the mathematical definition of double Feynman path integrals of the form (8.8).

Definition 8.1. A Borel measurable function $f : \mathcal{H} \times \mathcal{H} \to \mathbb{C}$ is called *Fresnel integrable* if for any sequence $\{P_n\}$ of projectors onto n-dimensional subspaces of $\mathcal{H}$, such that $P_n \le P_{n+1}$ and $P_n \to 1$ strongly as $n \to \infty$ (1 being the identity operator in $\mathcal{H}$), the finite dimensional oscillatory integrals (suitably normalized)

$$(2\pi\hbar)^{-n} \int_{P_n\mathcal{H}}^{\circ} \int_{P_n\mathcal{H}}^{\circ} e^{\frac{i}{2\hbar}\langle P_n x, P_n x\rangle} e^{-\frac{i}{2\hbar}\langle P_n y, P_n y\rangle} f(P_n x, P_n y) d(P_n x) d(P_n y)$$

are well defined (in the sense of Definition 5.1) and the limit

$$\lim_{n\to\infty} (2\pi\hbar)^{-n} \int_{P_n\mathcal{H}}^{\circ} \int_{P_n\mathcal{H}}^{\circ} e^{\frac{i}{2\hbar}\langle P_n x, P_n x\rangle} e^{-\frac{i}{2\hbar}\langle P_n y, P_n y\rangle} f(P_n x, P_n y) d(P_n x) d(P_n y)$$

exists and is independent of the sequence $\{P_n\}$. In this case the limit is denoted by

$$\widetilde{\int}\widetilde{\int} e^{\frac{i}{2\hbar}\langle x, x\rangle} e^{-\frac{i}{2\hbar}\langle y, y\rangle} f(x,y) dx dy.$$

For the integrals of Definition 8.1 it is possible to prove a result analogous to Theorem 5.5.

Theorem 8.1. *Let $L : \mathcal{H} \to \mathcal{H}$ be a trace-class operator, such that $I - L$ is invertible, and let $f : \mathcal{H} \times \mathcal{H} \to \mathbb{C}$ be the Fourier transform of a complex bounded variation measure μ_f on $\mathcal{H} \times \mathcal{H}$. Then the integral*

$$\widetilde{\int\int} e^{\frac{i}{2\hbar}\langle x,x \rangle} e^{-\frac{i}{2\hbar}\langle y,y \rangle} e^{-\frac{i}{2\hbar}\langle x-y, L(x+y) \rangle} f(x,y) dx\, dy$$

is well defined and is equal to

$$\frac{1}{\det(I-L)} \int_{\mathcal{H}} \int_{\mathcal{H}} e^{-\frac{i\hbar}{2}\langle \alpha+\beta, (I-L)^{-1}(\alpha-\beta) \rangle} d\mu_f(\alpha, \beta)\,,$$

where $\det(I - L)$ is the Fredholm determinant of $I - L$.

Proof. Let us consider a sequence $\{P_n\}$ of projectors onto n-dimensional subspaces of $\mathcal{H}$, such that $P_n \leq P_{n+1}$ and $P_n \to 1$ strongly as $n \to \infty$. The finite dimensional approximations of the oscillatory integral

$$\widetilde{\int\int} e^{\frac{i}{2\hbar}\langle x,x \rangle} e^{-\frac{i}{2\hbar}\langle y,y \rangle} e^{-\frac{i}{2\hbar}\langle x-y, L(x+y) \rangle} f(x,y) dx\, dy$$

are given by

$$\frac{1}{(2\pi\hbar)^n} \int_{P_n\mathcal{H}}^{\circ} \int_{P_n\mathcal{H}}^{\circ} e^{\frac{i}{2\hbar}\langle x_n-y_n, (I_n-L_n)(x_n+y_n) \rangle} f(x_n, y_n) dx_n\, dy_n\,,$$

where $x_n = P_n x$, $x \in \mathcal{H}$, and $I_n - L_n = I|_{P_n\mathcal{H}} - P_n L P_n$. The finite dimensional approximations are defined by the following sequence of regularized integrals:

$$\lim_{\epsilon \to 0} \frac{1}{(2\pi\hbar)^n} \int_{P_n\mathcal{H}} \int_{P_n\mathcal{H}} e^{\frac{i}{2\hbar}\langle x_n-y_n, (I_n-L_n)(x_n+y_n) \rangle} \phi(\epsilon x_n, \epsilon y_n) f(x_n, y_n) dx_n\, dy_n,$$

with $\phi \in \mathcal{S}(\mathbb{R}^n \times \mathbb{R}^n)$, $\phi(0) = 1$.

Since $I - L$ is invertible, for any sequence $\{P_n\}$ of projectors there exists an $\bar{n}$ such that for any $n \geq \bar{n}$ the operator $P_n(I - L)P_n$ is invertible and thus $\det(I_n - L_n) \neq 0$. Hence, for $n \geq \bar{n}$, by introducing the new variables $z_n = x_n - y_n$ and $w_n = x_n + y_n$, and by Fubini theorem, the integral can be written as

$$\lim_{\epsilon \to 0} \frac{1}{(4\pi\hbar)^n} \int_{P_n\mathcal{H}} \int_{P_n\mathcal{H}} \left(\int_{P_n\mathcal{H}} \int_{P_n\mathcal{H}} e^{i\langle \alpha, \frac{z_n+w_n}{2} \rangle + i\langle \beta, \frac{w_n-z_n}{2} \rangle} \right.$$

$$\left. e^{\frac{i}{2\hbar}\langle z_n, (I_n-L_n)w_n \rangle} \phi_T(\epsilon z_n, \epsilon w_n) \, dz_n\, dw_n \right) d\mu_n(\alpha, \beta)\,,$$

where $\mu_n \in \mathcal{F}(P_n\mathcal{H} \times P_n\mathcal{H})$ is defined as $\mu_n = \mu \circ P_n$ and $\phi_T \in \mathcal{S}(P_n\mathcal{H} \times P_n\mathcal{H})$ as

$$\phi_T(z_n, w_n) = \phi\left(\frac{z_n + w_n}{2}, \frac{w_n - z_n}{2}\right).$$

If we write the function ϕ_T in terms of its Fourier transform and apply Fubini again, the integrals over the variables z_n and w_n become

$$\int_{P_n\mathcal{H}} \int_{P_n\mathcal{H}} e^{i\langle \alpha, \frac{z_n+w_n}{2}\rangle + i\langle \beta, \frac{w_n-z_n}{2}\rangle} e^{\frac{i}{2\hbar}\langle z_n, (I_n - L_n)w_n\rangle} \phi_T(\epsilon z_n, \epsilon w_n) \, dz_n dw_n$$

$$= \left(\frac{\hbar}{\pi}\right)^n \det(I_n - L_n)^{-1} \int_{P_n\mathcal{H}} \int_{P_n\mathcal{H}} e^{\frac{-i\hbar}{2}\langle \alpha + \beta - 2\epsilon\gamma_n, (I_n - L_n)^{-1}(\alpha - \beta - 2\epsilon\delta_n)\rangle}$$

$$\widehat{\phi}_T(\gamma_n, \delta_n) d\gamma_n d\delta_n.$$

Lebesgue's dominated convergence theorem allows us to exchange limit and integrals. Thus, by taking into account that

$$\int_{P_n\mathcal{H}} \int_{P_n\mathcal{H}} \widehat{\phi}_T(\gamma_n, \delta_n) d\gamma_n d\delta_n = (2\pi)^{2n}\phi_T(0, 0),$$

we obtain

$$\det(I_n - L_n)^{-1} \int_{P_n\mathcal{H}} \int_{P_n\mathcal{H}} e^{-\frac{i\hbar}{2}\langle \alpha + \beta, (I_n - L_n)^{-1}(\alpha - \beta)\rangle} d\mu_n(\alpha, \beta).$$

The statement follows by taking the limit $n \to \infty$, since $\det(I_n - L_n)$ converges to $\det(I - L)$ (see Section 5.4). $\qquad\square$

The next result is a straightforward consequence of Theorem 8.1.

Corollary 8.1. *Under the assumptions of Theorem 8.1, the functional*

$$f \in \mathcal{F}(\mathcal{H} \times \mathcal{H}) \mapsto \widetilde{\iint} e^{\frac{i}{2\hbar}\langle x, x\rangle} e^{-\frac{i}{2\hbar}\langle y, y\rangle} e^{-i\langle x-y, L(x+y)\rangle} f(x, y) dx dy$$

is continuous in the $\mathcal{F}(\mathcal{H} \times \mathcal{H})$-norm.

For the applications that will follow, it is convenient to introduce the following Fubini-type theorem on the change of order of integration between oscillatory integrals and Lebesgue integrals.

Let $\{\mu_\alpha | \alpha \in \mathbb{R}^d\}$ be a family in $\mathcal{M}(\mathcal{H})$. We denote by $\int_{\mathbb{R}^d} \mu_\alpha d\alpha$ the measure defined by

$$f \mapsto \int_{\mathbb{R}^d} \int_{\mathcal{H}} f(x) d\mu_\alpha(x) d\alpha, \qquad f \in C_0(\mathcal{H}),$$

whenever it exists.

Theorem 8.2. *Let $L : \mathcal{H} \to \mathcal{H}$ be as in the assumptions of Theorem 8.1 and let $\mu : \mathbb{R}^d \to \mathcal{M}(\mathcal{H} \times \mathcal{H})$, $\alpha \mapsto \mu_\alpha$, be a continuous map such that*

$$\int_{\mathbb{R}^d} |\mu_\alpha| d\alpha < \infty.$$

Further, for any $\alpha \in \mathbb{R}^d$, let $f_\alpha \in \mathcal{F}(\mathcal{H} \times \mathcal{H})$ be given by

$$f_\alpha(x,y) = \hat{\mu}_\alpha(x,y), \qquad (x,y) \in \mathcal{H} \times \mathcal{H}.$$

Then $\int_{\mathbb{R}^d} f_\alpha d\alpha \in \mathcal{F}(\mathcal{H} \times \mathcal{H})$ and

$$\int_{\mathbb{R}^d} \widetilde{\int_{\mathcal{H}}} \widetilde{\int_{\mathcal{H}}} e^{\frac{i}{2\hbar}\langle x,x \rangle} e^{-\frac{i}{2\hbar}\langle y,y \rangle} e^{-\frac{i}{2\hbar}\langle x-y, L(x+y) \rangle} f_\alpha(x,y) dx dy d\alpha$$

$$= \widetilde{\int_{\mathcal{H}}} \widetilde{\int_{\mathcal{H}}} e^{\frac{i}{2\hbar}\langle x,x \rangle} e^{-\frac{i}{2\hbar}\langle y,y \rangle} e^{-\frac{i}{2\hbar}\langle x-y, L(x+y) \rangle} \int_{\mathbb{R}^d} f_\alpha(x,y) d\alpha dx dy. \quad (8.16)$$

Proof. Since f_α is assumed to be the Fourier transform of μ_α, by Fubini's theorem

$$\int_{\mathbb{R}^d} f_\alpha d\alpha = \int_{\mathbb{R}^d} \int_{\mathcal{H} \times \mathcal{H}} e^{i\langle k,x \rangle + i\langle h,y \rangle} d\mu_\alpha(k,h) d\alpha$$

$$= \int_{\mathcal{H} \times \mathcal{H}} e^{i\langle k,x \rangle + i\langle h,y \rangle} \int_{\mathbb{R}^d} d\mu_\alpha(k,h) d\alpha,$$

so that $\int_{\mathbb{R}^d} f_\alpha d\alpha \in \mathcal{F}(\mathcal{H} \times \mathcal{H})$.

By applying Theorem 8.1 to the left-hand side of (8.16), we obtain

$$\int_{\mathbb{R}^d} \widetilde{\int_{\mathcal{H}}} \widetilde{\int_{\mathcal{H}}} e^{\frac{i}{2\hbar}\langle x,x \rangle} e^{-\frac{i}{2\hbar}\langle y,y \rangle} e^{-\frac{i}{2\hbar}\langle x-y, L(x+y) \rangle} f_\alpha(x,y) dx dy d\alpha$$

$$= \det(I - L)^{-1} \int_{\mathbb{R}^d} \int_{\mathcal{H}} \int_{\mathcal{H}} e^{-\frac{i\hbar}{2}\langle k+h, (I-L)^{-1}(k-h) \rangle} d\mu_\alpha(k,h) d\alpha.$$

By the usual Fubini theorem the latter is equal to

$$\det(I - L)^{-1} \int_{\mathcal{H}} \int_{\mathcal{H}} e^{-\frac{i\hbar}{2}\langle k+h, (I-L)^{-1}(k-h) \rangle} \int_{\mathbb{R}^d} d\mu_\alpha(k,h) d\alpha.$$

But this expression, by Theorem 8.1, is equal to the r.h.s. of (8.16). □

We can now apply these results to the mathematical realization of the Feynman-Vernon influence functional in the Caldeira-Leggett model [136, 76, 12].

Let us consider the Cameron-Martin space $\mathcal{H}_t$, i.e. the Hilbert space of absolutely continuous paths $\gamma : [0,t] \to \mathbb{R}$, such that $\gamma(t) = 0$ and $\int_0^t |\dot{\gamma}(s)|^2 ds < \infty$, endowed with the inner product

$$\langle \gamma_1, \gamma_2 \rangle = \int_0^t \dot{\gamma}_1(s) \dot{\gamma}_2(s) ds.$$

Let $L : \mathcal{H}_t \to \mathcal{H}_t$ be the trace-class symmetric operator on $\mathcal{H}_t$ given by

$$(L\gamma)(s) = \int_s^t ds' \int_0^{s'} ds'' \gamma(s''), \qquad \gamma \in \mathcal{H}_t.$$

Further let $\mathcal{H}_t^d = \oplus_{i=1}^d \mathcal{H}_t$ and let $L^d : \mathcal{H}_t^d \to \mathcal{H}_t^d$ denote the operator defined by

$$L^d = L^{(1)} \otimes L^{(2)} \otimes \cdots \otimes L^{(d)},$$

where $L^{(j)} = I \otimes \cdots \otimes I \otimes L \otimes I \cdots \otimes I$ with L acting on the jth space. Given a positive symmetric $d \times d$ matrix Ω, let $L_\Omega : \mathcal{H}_t^d \to \mathcal{H}_t^d$ be the trace-class symmetric operator on $\mathcal{H}_t^d$ defined by

$$(L_\Omega \gamma)(s) = \int_s^t ds' \int_0^{s'} (\Omega^2 \gamma)(s'') ds'', \qquad \gamma \in \mathcal{H}_t^d. \tag{8.17}$$

Clearly, $L_\Omega \gamma = L^d \Omega^2 \gamma$, for all $\gamma \in \mathcal{H}_t^d$. One can easily verify that

$$\langle \gamma_1, L_\Omega \gamma_2 \rangle = \int_0^t \gamma_1(s) \Omega^2 \gamma_2(s) ds.$$

By Lemmas 6.1 and 6.2, if $t \neq [(n + 1/2)\pi]/\Omega_j$, for any $n \in \mathbb{N}$ and any eigenvalue Ω_j of Ω, one has that the operator $I - L_\Omega$ is invertible with

$$(I - L_\Omega)^{-1} \gamma(s) = \gamma(s) - \Omega \int_s^t \sin(\Omega(s' - s)) \gamma(s') ds'$$

$$+ \sin(\Omega(t - s)) \int_0^t (\cos \Omega t)^{-1} \Omega \cos(\Omega s') \gamma(s') ds' \tag{8.18}$$

and

$$\det(I - L_\Omega) = \det(\cos(\Omega t))$$

(see Lemmas 6.1 and 6.2).

Let $v \in \mathcal{F}(\mathbb{R}^d)$ be a real bounded function and let H be the quantum Hamiltonian, given on smooth vectors $\psi \in \mathcal{S}(\mathbb{R}^d)$ by

$$H\psi(x) = -\frac{\Delta}{2}\psi(x) + \frac{1}{2}x\Omega^2 x\psi(x) + v(x)\psi(x), \qquad x \in \mathbb{R}^d.$$

Let $U_t = e^{-\frac{i}{\hbar}Ht}$ be the unitary group generated by H. As we have seen in Chapter 6, given an initial datum $\psi_0 \in \mathcal{F}(\mathbb{R}^d)$ and assuming that $t \neq [(n + 1/2)\pi]/\Omega_j, \forall n \in \mathbb{N}$, the solution of the Schrödinger equation $\psi(t) = U_t \psi_0$ is given by an infinite dimensional oscillatory integral on the Cameron Martin space $\mathcal{H}_t$:

$$\psi(t, x) = e^{-\frac{i}{2\hbar}x\Omega^2 xt} \widetilde{\int_{\mathcal{H}_t^d}} e^{\frac{i}{2\hbar}\langle \gamma, (I - L_\Omega)\gamma \rangle} e^{-\frac{i}{\hbar}\int_0^t x\Omega^2 \gamma(s)ds} e^{-\frac{i}{\hbar}\int_0^t v(\gamma(s)+x)ds}$$

$$\psi_0(\gamma(0) + x) d\gamma.$$

(see Theorem 6.2). This result can be generalized to the Feynman path integral representation of the time evolution for a mixed state, represented by a density matrix.

Theorem 8.3. *Let ρ_0 be a density matrix operator on $L^2(\mathbb{R}^d)$, such that ρ_0 admits a regular kernel $\rho_0(x,y)$, $x,y \in \mathbb{R}^d$. Let us assume moreover that ρ_0 admits a decomposition into pure states of the form $\rho_0(x,y) = \sum_i \lambda_i e_i(x)\bar{e}_i(y)$, with $\lambda_i > 0$, $\sum_i \lambda_i = 1$, $\langle e_i, e_j \rangle_{L^2(\mathbb{R}^d)} = \delta_{ij}$, and $e_i(x) = \hat{\mu}_i(x)$, satisfying*

$$\sum_i \lambda_i |\mu_i|^2 < \infty. \tag{8.19}$$

Let $t \neq [(n+1/2)\pi]/\Omega_j$, $\forall n \in \mathbb{N}$. Then the density matrix operator ρ_t at time t, $\rho_t = U_t \rho_0 U_t^\dagger$, admits a smooth kernel $\rho_t(x,y)$, which is given by the infinite dimensional oscillatory integral

$$e^{-\frac{i}{2\hbar}(x\Omega^2 x - y\Omega^2 y)t} \widetilde{\int_{\mathcal{H}_t^d}} \widetilde{\int_{\mathcal{H}_t^d}} e^{\frac{i}{2\hbar}\langle \gamma, (I-L_\Omega)\gamma \rangle} e^{-\frac{i}{2\hbar}\langle \gamma', (I-L_\Omega)\gamma' \rangle} \tag{8.20}$$

$$e^{-\frac{i}{\hbar}\int_0^t (x\Omega^2 \gamma(s) - y\Omega^2 \gamma'(s))ds} e^{-\frac{i}{\hbar}\int_0^t (v(\gamma(s)+x) - v(\gamma'(s)+y))ds}$$

$$\rho_0(\gamma(0) + x, \gamma'(0) + y)d\gamma d\gamma'.$$

Proof. By decomposing ρ into pure states, by Corollary 8.1, and by condition (8.19), the integral (8.20) is equal to

$$\sum_i \lambda_i \left(e^{-\frac{i}{2\hbar} x\Omega^2 xt} \widetilde{\int_{\mathcal{H}_t^d}} e^{\frac{i}{2\hbar}\langle \gamma, (I-L_\Omega)\gamma \rangle} e^{-\frac{i}{\hbar}\int_0^t x\Omega^2 \gamma(s)ds} e^{-\frac{i}{\hbar}\int_0^t v(\gamma(s)+x)ds} e_i(\gamma(0) + x)d\gamma \right)$$

$$\times \left(e^{\frac{i}{2\hbar} y\Omega^2 yt} \widetilde{\int_{\mathcal{H}_t^d}} e^{-\frac{i}{2\hbar}\langle \gamma', (I-L_\Omega)\gamma' \rangle} e^{\frac{i}{\hbar}\int_0^t y\Omega^2 \gamma'(s)ds} e^{\frac{i}{\hbar}\int_0^t v(\gamma'(s)+y)ds} e_i^*(\gamma(0) + y)d\gamma \right).$$

This is equal to

$$\sum_i \lambda_i U_t e_i(x) \overline{(U_t e_i)}(y) = \rho_t(x,y),$$

(where for $z \in \mathbb{C}$, $\bar{z}$ denotes the conjugate of the complex number z). $\qquad\square$

Heuristically, expression (8.20) can be written as a double Feynman path integral:

$$\int e^{\frac{i}{\hbar}(S_t(\gamma+x)-S_t(\gamma'+y))}\rho_0(\gamma(0)+x,\gamma'(0)+y)d\gamma d\gamma'.$$

Let us consider now the time evolution of a quantum system made of two linearly interacting subsystems A and B. Let us assume that A is d-dimensional, B is N-dimensional and the quantum mechanical Hamiltonian of the compound system is given by Eq. (8.6). Let us assume that the density matrix of the compound system factorizes as $\rho_{AB} = \rho_A\rho_B$ and has a regular kernel $\rho_{AB}(x,y,R,Q) = \rho_A(x,y)\rho_B(R,Q)$. We are going to see how it is possible to construct an infinite dimensional oscillatory integral realization for the Feynman path integral (8.8) representing the reduced density operator at time t, namely

$$\int \left(e^{-\frac{i}{\hbar}H_{AB}t}\rho_{AB}e^{\frac{i}{\hbar}H_{AB}t}\right)(x,y,R,R)dR.$$

Heuristically:

$$\int \int_{\substack{\gamma(t)=x \\ \Gamma(t)=R}}\int_{\substack{\gamma'(t)=y \\ \Gamma'(t)=R}} e^{\frac{i}{\hbar}(S_A(\gamma)+S_B(\Gamma)+S_{INT}(\gamma,\Gamma)-S_A(\gamma')-S_B(\Gamma')-S_{INT}(\gamma',\Gamma'))}$$

$$\times\rho_A(\gamma(0),\gamma'(0))\rho_B(\Gamma(0),\Gamma'(0))D\gamma D\gamma' D\Gamma D\Gamma' dR, \qquad (8.21)$$

where γ, resp. Γ, represents a generic path in the configuration space of the system, resp. of the reservoir, and

$$S_A(\gamma) + S_B(\Gamma) + S_{INT}(\gamma,\Gamma)$$
$$-\int_0^t \left(\frac{M}{2}\dot{\gamma}^2(s) - \frac{M}{2}\gamma(s)\Omega_A^2\gamma(s) - v_A(\gamma(s))\right) ds$$
$$+\int_0^t \left(\frac{m}{2}\dot{\Gamma}^2(s) - \frac{m}{2}\Gamma(s)\Omega_B^2\Gamma(s) - v_B(\Gamma(s))\right) ds - \int_0^t \gamma(s)C\Gamma(s)ds.$$

If we rescale γ via $\gamma \to \gamma/\sqrt{M}$ and Γ via $\Gamma \to \Gamma/\sqrt{m}$, formula (8.21) becomes

$$\int \int_{\substack{\gamma(t)=\sqrt{M}x \\ \Gamma(t)=\sqrt{m}R}}\int_{\substack{\gamma'(t)=\sqrt{M}y \\ \Gamma'(t)=\sqrt{m}R}} e^{\frac{i}{2\hbar}\int_0^t \left(\dot{\gamma}^2(s)-\gamma(s)\Omega_A^2\gamma(s)-2v_A\left(\frac{\gamma(s)}{\sqrt{M}}\right)\right)ds}$$

$$e^{\frac{i}{2\hbar}\int_0^t \left(\dot{\Gamma}^2(s)-\Gamma(s)\Omega_B^2\Gamma(s)-2v_B\left(\frac{\Gamma(s)}{\sqrt{m}}\right)\right)ds}$$

$$e^{-\frac{i}{\hbar}\int_0^t \gamma(s)\frac{C}{\sqrt{mM}}\Gamma(s)ds}e^{-\frac{i}{2\hbar}\int_0^t \left(|\dot{\gamma}'|^2(s)-\gamma'(s)\Omega_A^2\gamma'(s)-2v_A\left(\frac{\gamma'(s)}{\sqrt{M}}\right)\right)ds}$$

$$e^{-\frac{i}{2\hbar}\int_0^t \left((\dot{\Gamma}')^2(s)-\Gamma'(s)\Omega_B^2\Gamma'(s)-2v_B\left(\frac{\Gamma'(s)}{\sqrt{m}}\right)\right)ds}e^{\frac{i}{\hbar}\int_0^t \gamma'(s)\frac{C}{\sqrt{mM}}\Gamma'(s)ds}$$

$$\rho_A\left(\frac{\gamma(0)}{\sqrt{M}},\frac{\gamma'(0)}{\sqrt{M}}\right)\rho_B\left(\frac{\Gamma(0)}{\sqrt{m}},\frac{\Gamma'(0)}{\sqrt{m}}\right)D\gamma D\gamma' D\Gamma D\Gamma' dR.$$

Let

$$\Omega_{AB}^2 = \begin{pmatrix} \Omega_A^2 & C' \\ C'^T & \Omega_B^2 \end{pmatrix},$$ (8.22)

where $C' = C/\sqrt{Mm}$, and define for simplicity $L_A\gamma = L_{\Omega_A}\gamma$, $L_B\Gamma = L_{\Omega_B}\Gamma$, and $L_{AB}(\gamma, \Gamma) = L_{\Omega_{AB}}(\gamma, \Gamma)$.

Formula (8.21) can be made completely rigorous under suitable assumptions. In the following we shall assume without loss of generality that $m = M = 1$. The result in the general case can be obtained by replacing C, v_A, v_B, ρ_0^A, and ρ_0^B by $C' = C/\sqrt{mM}$, $v_A'(\cdot) = v_A(\cdot/\sqrt{M})$, $v_B'(\cdot) = v_B(\cdot/\sqrt{m})$, $\rho_A'(\cdot) = \rho_A(\cdot/\sqrt{M})$ and $\rho_B'(\cdot) = \rho_B(\cdot/\sqrt{m})$ respectively.

Theorem 8.4. *Let ρ_A and ρ_B be two density matrix operators on $L^2(\mathbb{R}^d)$ and $L^2(\mathbb{R}^N)$, respectively, with regular kernels $\rho_A(x, x')$ and $\rho_B(R, R')$ and such that they decompose into sums of pure states:*

$$\rho_A = \sum_i w_i^A P_{\psi_i^A}, \qquad \rho_B = \sum_j w_j^B P_{\psi_j^B},$$ (8.23)

with $\psi_i^A \in \mathcal{F}(\mathbb{R}^d)$, $\psi_j^B \in \mathcal{F}(\mathbb{R}^N)$, and

$$\sum_{i,j} w_i^A w_j^B |\mu_i^A|^2 |\mu_j^B|^2 < \infty.$$ (8.24)

Further let $\rho_B \in \mathcal{S}(\mathbb{R}^N \times \mathbb{R}^N)$. Let t satisfy the following assumptions

$$t \neq [(n + 1/2)\pi]/\Omega_j^A, \qquad n \in \mathbb{N}, \quad j = 1 \ldots d,$$ (8.25)

$$t \neq [(n + 1/2)\pi]/\Omega_j^B, \qquad n \in \mathbb{N}, \quad j = 1 \ldots N,$$ (8.26)

$$t \neq [(n + 1/2)\pi]/\lambda_j, \qquad n \in \mathbb{N}, \quad j = 1 \ldots d + N,$$ (8.27)

with $\Omega_j^A, \Omega_j^B, \lambda_j$ being respectively the eigenvalues of $\Omega_A, \Omega_B, \Omega_{AB}$. Let us assume moreover that the determinant of the $d \times d$ left upper block of the matrix $\cos(\Omega_{AB}t)$ is non-vanishing.

Then the kernel $\rho_R(t, x, y)$ of the reduced density operator of the system A evaluated at time t is given by

$\rho_R(t, x, y)$

$= e^{-\frac{it}{2\hbar}x\Omega_A^2 x} e^{\frac{it}{2\hbar}y\Omega_A^2 y} \widetilde{\int_{\mathcal{H}_t^d}} \widetilde{\int_{\mathcal{H}_t^d}} e^{\frac{i}{2\hbar}\langle\gamma,(I_d - L_A)\gamma\rangle} e^{-\frac{i}{2\hbar}\langle\gamma',(I_d - L_A)\gamma'\rangle}$

$e^{-\frac{i}{\hbar}\int_0^t (x\Omega_A^2\gamma(s)ds - y\Omega_A^2\gamma'(s))ds} e^{-\frac{i}{\hbar}\int_0^t (v_A(\gamma(s)+x) - v_A(\gamma(s)+y))ds}$

$F(\gamma, \gamma', x, y)\rho_A(\gamma(0) + x, \gamma'(0) + y)d\gamma d\gamma',$ (8.28) where $F(\gamma, \gamma', x, y)$ is the influence functional

$$F(\gamma, \gamma', x, y)$$
$$= \int_{\mathbb{R}^N} e^{-\frac{it}{\hbar} x C R} e^{+\frac{it}{\hbar} y C R} e^{-\frac{i}{\hbar} \int_0^t (\gamma(s) - \gamma'(s)) C R ds}$$

$$\widetilde{\int_{\mathcal{H}_t^N}} \widetilde{\int_{\mathcal{H}_t^N}} e^{\frac{i}{2\hbar} \langle \Gamma, (I_N - L_B)\Gamma \rangle} e^{-\frac{i}{2\hbar} \langle \Gamma', (I_N - L_B)\Gamma' \rangle} e^{-\frac{i}{\hbar} \langle \Gamma, L^N C^T \gamma \rangle} e^{\frac{i}{\hbar} \langle \Gamma', L^N C^T \gamma' \rangle}$$

$$e^{-\frac{i}{\hbar} \int_0^t R \Omega_B^2 (\Gamma(s) - \Gamma'(s)) ds} e^{-\frac{i}{\hbar} \int_0^t (x C \Gamma(s) - y C \Gamma'(s)) ds}$$

$$e^{-\frac{i}{\hbar} \int_0^t \left(v_B (\Gamma(s) + R) - v_B (\Gamma'(s) + R) \right) ds}$$

$$\rho_B (\Gamma(0) + R, \Gamma'(0) + R) d\Gamma d\Gamma' dR .$$

Proof. The proof of the present theorem involves a large amount of computation. We give here only the main steps and refer to [12] for more details.

First of all one has to prove that the functional $(\gamma, \gamma') \mapsto F(\gamma, \gamma', x, y)$ is well defined for any $\gamma, \gamma' \in \mathcal{H}_t^d$, $x, y \in \mathbb{R}^d$ and is Fresnel integrable in the sense of Definition 8.1. By decomposing the mixed state ρ_B into pure states according to (8.23), the influence functional can be written as

$$\int_{\mathbb{R}^N} \sum_j w_j^B \psi_j^B (x, \gamma; R) \overline{\psi_j^B (y, \gamma'; R)} dR ,$$

where $\psi_j^B (x, \gamma)$ is the solution of the Schrödinger equation on $L^2(\mathbb{R}^N)$ with Hamiltonian

$$H = -\frac{1}{2} \Delta_R + \frac{1}{2} R \Omega_B^2 R + v_B(R) + (x + \gamma(t)) C R = H_B + (x + \gamma(t)) C R$$

and initial state ψ_j^B. In particular,

$$\int_0^t \gamma(s) C \Gamma(s) ds = \langle L^N C^T \gamma, \Gamma \rangle .$$

Because of the unitarity of the evolution operator, $\|\psi_j^B (x, \gamma)\|_{L^2(\mathbb{R}^N)} = 1$ for any $x \in \mathbb{R}^d$, $\gamma \in \mathcal{H}_t^d$ and, by Schwartz inequality,

$$\sum_j w_j^B \int_{\mathbb{R}^N} \psi_j^B (x, \gamma; R) \overline{\psi_j^B (y, \gamma'; R)} dR$$

$$\leqslant \sum_j w_j^B \|\psi_j^B (x, \gamma)\|_{L^2(\mathbb{R}^N)} \|\psi_j^B (y, \gamma')\|_{L^2(\mathbb{R}^N)} = \sum_j w_j^B = 1 .$$

Thus we can conclude that $F(\gamma, \gamma', x, y)$ is well defined for any $x, y \in \mathbb{R}^d$ and $\gamma, \gamma' \in \mathcal{H}_t^d$.

By exploiting the assumptions on $v_B \in \mathcal{F}(\mathbb{R}^n)$ and on ρ_B and by a large amount of explicit computation, it is possible to see that $F(\gamma, \gamma', x, y)$ has the following form:

$$F(\gamma, \gamma', x, y) = e^{-\frac{i}{2\hbar}\langle(\gamma-\gamma'), A(\gamma+\gamma')\rangle} f(\gamma, \gamma')$$

with $f \in \mathcal{F}(\mathcal{H}_t^d \oplus \mathcal{H}_t^d)$ and the operator $A : \mathcal{H}_t^d \to \mathcal{H}_t^d$ is given by

$$\langle \gamma, A\gamma' \rangle = -\int_0^t C^T \gamma(s) \Omega_B^{-1} \int_0^s \sin(\Omega_B(s-r)) C^T \gamma'(r) dr ds.$$

Hence one has to verify that the operator $(I - L_A - A) : \mathcal{H}_t^d \to \mathcal{H}_t^d$ is invertible. In fact a vector $\gamma \in \mathcal{H}_t^d$ belongs to the kernel of the operator $I - L_A - A$, if it satisfies the following equation for all $s \in [0, t]$:

$$
\gamma(s) - \int_s^t ds' \int_0^{s'} \Omega_A^2 \gamma(s'') ds''
$$
$$
+ \int_s^t ds' \int_0^{s'} ds'' \int_0^{s''} C\Omega_B^{-1} \sin(\Omega_B(s'' - r)) C^T \gamma(r) dr = 0, \quad (8.29)
$$

with $\gamma(t) = 0$. By differentiating twice, (8.29) becomes

$$
\begin{cases}
\ddot{\gamma}(s) + \Omega_A^2 \gamma(s) - \int_0^s C\Omega_B^{-1} \sin(\Omega_B(s-r)) C^T \gamma(r) dr = 0, \\
\gamma(t) = 0 = \dot{\gamma}(0).
\end{cases}
$$

By further differentiating (8.29), one can see that its solution, if it exists, is a C^∞-function and its odd derivatives, evaluated for $s = 0$, vanish, while the even derivatives satisfy the following relation:

$$\gamma^{2(M+2)}(0) + \Omega_A^2 \gamma^{2(M+1)}(0) - \sum_{k=0}^{M}(-1)^k C\Omega_B^{2k} C^T \gamma^{2(M-k)}(0) = 0.$$

By induction it is possible to prove that $\gamma^{2M}(0) = (-1)^M [\Omega_{AB}^{2M}]_{d \times d} \gamma(0)$, where $[\Omega_{AB}^{2M}]_{d \times d}$ denotes the $d \times d$ left upper block of the M-th power of the matrix Ω_{AB}^2. One can conclude that the solution of Eq. (8.29) is of the form $\gamma(s) = [\cos(\Omega_{AB}s)]_{d \times d} \gamma(0)$. By imposing the condition $\gamma(t) = 0$, one concludes that if $\det([\cos(\Omega_{AB}t)]_{d \times d}) \neq 0$ then Eq. (8.29) cannot admit non-trivial solutions and the operator $I - L_A - A$ is invertible. Hence, by Theorem 8.1, we can finally conclude that the influence functional is a Fresnel integrable function.

The second step is the proof that the reduced density operator $\rho_R(t, x, y)$ of the system A is given by the infinite dimensional oscillatory integral (8.28). this results follows from a regularization procedure and by Theorems 8.3, 8.2 and Corollary 8.1 (see [12] for further details). $\square$

This result can now be applied to the Caldeira-Leggett model [76], describing the influence of a heat bath on a quantum Brownian particle. The heat bath is described by a finite number of oscillators and thus $v_B = 0$. Further the model presumes that the environment is initially in equilibrium at temperature T, i.e. that its initial density $\rho_B(R, Q)$ is a product of Gaussian functions: $\rho_B(R, Q) = \prod_{j=1}^{N} \rho_B^{(j)}(R_j, Q_j, 0)$, where

$$\rho_B^{(j)}(R_j, Q_j, 0) = \sqrt{\frac{m\Omega_j^B}{\pi\hbar \coth(\hbar\Omega_j^B/2kT)}}$$
$$e^{-\left(\frac{m\Omega_j^B}{2\hbar \sinh(\hbar\Omega_j^B/kT)}\left((R_j^2+Q_j^2)\cosh\frac{\hbar\Omega_j^B}{kT}-2R_jQ_j\right)\right)}$$

with Ω_j^B, $j = 1 \ldots N$, the eigenvalues of the matrix Ω_B.

By a direct computation and by exploiting the results of Theorem 8.4 the influence functional becomes

$$F(\gamma, \gamma', x, y) \tag{8.30}$$
$$= e^{\frac{i}{2\hbar}\int_0^t C^T(\gamma(s)+x-\gamma'(s)-y)\Omega_B^{-1}\int_0^s \sin(\Omega_B(s-r))C^T(\gamma(r)+x+\gamma'(r)+y)drds}$$
$$e^{-\frac{1}{2\hbar}\int_0^t C^T(\gamma(s)+x-\gamma'(s)-y)\Omega_B^{-1}\coth\left(\frac{\hbar\Omega_B}{2kT}\right)\int_0^s \cos(\Omega_B(s-r))C^T(\gamma(r)+x-\gamma'(r)-y)drds},$$

which yields the result heuristically derived in [137, 76].

In Theorem 8.4 the allowed values of the time variable t are restricted by conditions (8.25), (8.26), and (8.27), as well as by $\det[\cos(\Omega_{AB}t)]_{d\times d} \neq 0$. Since the influence functional (8.30) is well defined also for the excluded values of t, we can extend the formula "by continuity" to all times.

8.3   The stochastic Schrödinger equation

In order to realize the heuristic Feynman path integrals (8.11) and (8.15), it is necessary to generalize Definition 5.4 and Theorem (5.5) to complex-valued phase functions.

Let $\mathcal{H}$ be a real separable Hilbert space and let us denote by $\mathcal{H}^{\mathbb{C}}$ its complexification. An element $x \in \mathcal{H}^{\mathbb{C}}$ is a couple of vectors $x = (x_1, x_2)$, with $x_1, x_2 \in \mathcal{H}$, or with a different notation $x = x_1 + ix_2$. The multiplication of the vector $x \in \mathcal{H}^{\mathbb{C}}$ for the pure imaginary scalar $i = \sqrt{-1}$ is given by $ix = (-x_2, x_1)$. A vector $y \in \mathcal{H}$ can be seen as the element $(y, 0) \in \mathcal{H}^{\mathbb{C}}$. With an abuse of notation, let us denote with A the extension to $\mathcal{H}^{\mathbb{C}}$ of a linear operator $A : D(A) \subseteq \mathcal{H} \to \mathcal{H}$:

$$A : D(A) \subseteq H^{\mathbb{C}} \to H^{\mathbb{C}}, \qquad D(A) = D(A) + iD(A),$$

$$Ax = A(x_1, x_2) = (Ax_1, Ax_2).$$

Let $dim(\mathcal{H}) = 1$, i.e. $\mathcal{H} = \mathbb{R}$, $\mathcal{H}^{\mathbb{C}} = \mathbb{C}$. Then, for any $f \in \mathcal{F}(\mathbb{R})$, $f = \hat{\mu}_f$, and any complex constant $\alpha \in \mathbb{C}$, $\alpha \neq 0$, $\mathrm{Im}(\alpha) \geq 0$, one can easily prove the following equality

$$\widetilde{\int_{\mathbb{R}}} e^{\frac{i\alpha}{2\hbar}x^2} f(x)dx = \alpha^{-1/2} \int_{\mathbb{R}} e^{\frac{-i\hbar}{2\alpha}x^2} d\mu_f(x) \tag{8.31}$$

The proof is completely similar to the proof of Theorem 5.2.
More generally, given $\alpha \in \mathbb{C}$, $\alpha \neq 0$, $\mathrm{Im}(\alpha) > 0$ and $\beta \in \mathbb{R}$

$$\widetilde{\int_{\mathbb{R}}} e^{\frac{i\alpha}{2\hbar}x^2} e^{\beta x} f(x)dx = \alpha^{-1/2} \int_{\mathbb{R}} e^{\frac{-i\hbar}{2\alpha}(x-i\beta)^2} d\mu_f(x) \tag{8.32}$$

Such a result can be generalized to the infinite dimensional case [23, 24, 14]:

Theorem 8.5. *Let $\mathcal{H}$ be a real separable Hilbert space, let $y \in \mathcal{H}$ be a vector in $\mathcal{H}$ and let L_1 and L_2 be two self-adjoint, trace class commuting operators on $\mathcal{H}$ such that $I + L_1$ is invertible and L_2 is non-negative. Let moreover $f : \mathcal{H} \to \mathbb{C}$ be the Fourier transform of a complex bounded variation measure μ_f on $\mathcal{H}$:*

$$f(x) = \hat{\mu}_f(x), \qquad f(x) = \int_{\mathcal{H}} e^{i\langle x,k \rangle} d\mu_f(k).$$

Then the function $g : \mathcal{H} \to \mathbb{C}$ given by

$$g(x) = e^{\frac{i}{2\hbar}\langle x, Lx \rangle} e^{\langle y,x \rangle} f(x)$$

(L being the operator on the complexification $\mathcal{H}^{\mathbb{C}}$ of the real Hilbert space $\mathcal{H}$ given by $L = L_1 + iL_2$) is Fresnel integrable (in the sense of Definition 5.4) and its Fresnel integral

$$\widetilde{\int_{\mathcal{H}}} e^{\frac{i}{2\hbar}\langle x,(I+L)x \rangle} e^{\langle y,x \rangle} f(x)dx$$

can be explicitly computed by means of the following Parseval type equality:

$$\widetilde{\int_{\mathcal{H}}} e^{\frac{i}{2\hbar}\langle x,(I+L)x \rangle} e^{\langle y,x \rangle} f(x)dx$$

$$= \det(I+L)^{-1/2} \int_{\mathcal{H}} e^{\frac{-i\hbar}{2}\langle k-iy,(I+L)^{-1}(k-iy) \rangle} d\mu_f(k). \tag{8.33}$$

Proof. First of all one can notice that both sides of Eq. (8.33) are well defined. Indeed one can easily prove that $(I + L) : H_C \to H_C$ is invertible, if $(I + L_1)$ is invertible and that $\det(I + L)$ exists as L is trace class.

On the other hand the function $f_1 : \mathcal{H} \to \mathbb{C}$

$$f_1(x) = e^{-\frac{1}{2\hbar}\langle x, L_2 x\rangle} e^{\langle y, x\rangle} f(x)$$

where $y \in \mathcal{H}$ and $f \in \mathcal{F}(\mathcal{H})$, $f(x) = \hat{\mu}_f(x)$, $\mu_f \in \mathcal{M}(\mathcal{H})$ is the Fourier transform of a complex bounded variation measure μ on $\mathcal{H}$. In fact μ is the convolution of μ_f and the measure ν, with

$$d\nu(x) = e^{\frac{\hbar}{2}\langle y, L_2^{-1} y\rangle - i\hbar\langle y, L_2^{-1} x\rangle} d\mu_{L_2}(x),$$

where μ_{L_2} is the Gaussian measure on $\mathcal{H}$ with covariance operator $L_2/\hbar$.

By Theorem 5.5 the Fresnel integral of the function g is well defined and can be explicitly computed:

$$\widetilde{\int_{\mathcal{H}}} e^{\frac{i}{2\hbar}\langle x,(I+L)x\rangle} e^{\langle y,x\rangle} f(x) dx = \widetilde{\int_{\mathcal{H}}} e^{\frac{i}{2\hbar}\langle x,(I+L_1)x\rangle} e^{-\frac{1}{2\hbar}\langle x, L_2 x\rangle} e^{\langle y,x\rangle} f(x) dx$$

$$= \det(I + L_1)^{-1/2} \int_{\mathcal{H}} e^{\frac{-i\hbar}{2}\langle x,(I+L_1)^{-1}x\rangle} d(\mu_f * \nu)(x)$$

$$= \det(I + L_1)^{-1/2} \int_{\mathcal{H}} \int_{\mathcal{H}} e^{\frac{-i\hbar}{2}\langle x+z,(I+L_1)^{-1}(x+z)\rangle} d\mu_f(z) d\nu(x)$$

$$= \det(I + L_1)^{-1/2} \int_{\mathcal{H}} \int_{\mathcal{H}} e^{\frac{-i\hbar}{2}\langle x+z,(I+L_1)^{-1}(x+z)\rangle}$$

$$e^{\frac{\hbar}{2}\langle y, L_2^{-1} y\rangle - i\hbar\langle y, L_2^{-1} x\rangle} d\mu_{L_2}(x) d\mu_f(z). \tag{8.34}$$

Equation (8.33) can be proved by taking the finite dimensional approximation of the last line of Eq. (8.34) and of the r.h.s. of (8.33) and showing they coincide (see [14] for more details). □

Let us consider now Belavkin Eq. (8.13) describing the posterior dynamics of a quantum particle, whose position is continuously observed. Equation (8.13) can also be written in the Stratonovich equivalent form:

$$\begin{cases} d\psi = -\frac{i}{\hbar}H\psi dt - \lambda|x|^2\psi dt + \sqrt{\lambda}x\psi \circ dB(t) \\ \psi(0, x) = \psi_0(x) \end{cases} \qquad t \geq 0, \ x \in \mathbb{R}^d. \tag{8.35}$$

The existence and uniqueness of a strong solution of Eqs. (8.13) and $(8.35)^3$ is proved in [156]. We shall prove that it can be represented by an infinite dimensional oscillatory integral on a suitable Hilbert space.

---

[3] A strong solution for the stochastic Eq. (8.35) is a predictable process with values in $\mathcal{H} = L^2(\mathbb{R}^d)$, such that:

Let us consider the Cameron Martin space $\mathcal{H}_t$ and let $\mathcal{H}_t^{\mathbb{C}}$ be its complexification. Let $L : \mathcal{H}_t^{\mathbb{C}} \to \mathcal{H}_t^{\mathbb{C}}$ be the linear operator on $\mathcal{H}_t^{\mathbb{C}}$ defined by

$$\langle \gamma_1, L\gamma_2 \rangle = -a^2 \int_0^t \gamma_1(s) \cdot \gamma_2(s) ds,$$

where $a^2 = -2i\lambda\hbar$. The j-th component of $L\gamma$, $L\gamma = (L\gamma_1, \ldots, L\gamma_d)$, is given by

$$(L\gamma)_j(s) = 2i\lambda\hbar \int_s^t ds' \int_0^{s'} \gamma_j(s'')ds'' \qquad j = 1, \ldots, d \qquad (8.37)$$

As we have seen in Section 6.1, the operator $iL : \mathcal{H}_t \to \mathcal{H}_t$ is self-adjoint with respect to the $\mathcal{H}_t$-inner product, it is trace-class and the Fredholm determinant of $(I + L)$ is given by:

$$\det(I + L) = \cos(at).$$

Moreover $(I + L)$ is invertible and its inverse is given by

$$[(I + L)^{-1}\gamma]_j(s) = \gamma_j(s) - a \int_s^t \sin[a(s' - s)]\gamma_j(s')ds'$$

$$+ \sin[a(t - s)] \int_0^t [\cos at]^{-1} a \cos(as')\gamma_j(s')ds', \qquad j = 1, \ldots, d.$$

For any $\omega \in C_t$, let us introduce the vector $l \in \mathcal{H}_t$ defined by

$$\langle l, \gamma \rangle = -\sqrt{\lambda} \int_0^t \omega(s) \cdot \dot{\gamma}(s)ds = \sqrt{\lambda} \int_0^t \gamma(s) \cdot dB(s), \qquad (8.38)$$

$$l(s) = \sqrt{\lambda} \int_s^t \omega(\tau)d\tau.$$

With this notation, we can apply Theorem 8.5 and prove that, under suitable assumptions on the potential V and the initial wave function ψ_0,

$\psi(t) \in D(-i/\hbar H - \lambda|x|^2)$ P-a.s.

$\mathbf{P}\left(\int_0^T (\|\psi(t)\|^2 + \|(-i/\hbar H - \lambda|x|^2)\psi\|^2) \, dt < \infty \right) = 1$

$\mathbf{P}\left(\int_0^T \||x|\psi(t) \, dt\|^2 < \infty \right) = 1$ and

$\mathbf{P}$ a.s. for all $t \in [0, T]$:

$$\begin{cases} d\psi = -\frac{i}{\hbar}H\psi dt - \lambda|x|^2\psi dt + \sqrt{\lambda}x \cdot \psi \circ dB(t) & t \geq 0, \ x \in \mathbb{R}^d \\ \psi(0, x) = \psi_0(x). \end{cases} \qquad (8.36)$$

the heuristic expression (8.15) can be realized as the infinite dimensional oscillatory integral with complex phase on the Cameron-Martin space $\mathcal{H}_t$:

$$C(t,x,\omega)\overbrace{\int}_{\mathcal{H}_t} e^{\frac{i}{2\hbar}\langle\gamma,(I+L)\gamma\rangle}e^{\langle l,\gamma\rangle}e^{-2\lambda x\cdot\int_0^t\gamma(s)ds}e^{-\frac{i}{\hbar}\int_0^t V(\gamma(s)+x)ds}\psi_0(\gamma(0)+x)d\gamma,$$

(8.39)

where $C(t,x,\omega) = e^{-\lambda|x|^2+\sqrt{\lambda}x\cdot\omega(t)}$ is a constant depending on t, $x \in \mathbb{R}^d$, $\omega \in C_t$. Indeed the integrand $\exp(\frac{i}{2\hbar}\Phi)$ in (8.15), where

$$\Phi(\gamma) \equiv \int_0^t |\dot\gamma(s)|^2 ds + 2i\hbar\lambda \int_0^t |\gamma(s)+x|^2 ds - 2i\hbar \int_0^t \sqrt{\lambda}(\gamma(s)+x)\cdot dB(s),$$

can be rigorously defined as the functional on the Cameron Martin space $\mathcal{H}_t$ given by

$$\Phi(\gamma) = \langle\gamma,(I+L)\gamma\rangle - 2i\hbar\langle l,\gamma\rangle - 2\hbar\int_0^t a^2 x\cdot\gamma(s)ds - a^2|x|^2 t - 2i\hbar\sqrt{\lambda}x\cdot\omega(t),$$

where L is the operator (8.37) and l is the vector (8.38).

By means of Theorem 8.5 one can compute the integral (8.39) in terms of an absolutely convergent integral on $\mathcal{H}_t$. Moreover it is possible to prove it represents the solution of Belavkin Eq. (8.35) (see [14]).

Theorem 8.6. *Let V and ψ_0 be Fourier transforms of complex bounded variation measures on $\mathbb{R}^d$. Then there exist a (strong) solution to the Stratonovich stochastic differential Eq. (8.35) and it is given by the infinite dimensional oscillatory integral with complex phase (8.39).*

Remark 8.1. The result can be extended to general initial vectors $\psi_0 \in L^2(\mathbb{R}^d)$, using the fact that $\mathcal{F}(\mathbb{R}^d)$ is dense in $L^2(\mathbb{R}^d)$.

Proof. The proof in divided into 3 steps: in the first two we consider the case $V \equiv 0$. First of all we deal with an approximated problem and we find a representation for its solution via a infinite dimensional oscillatory integral, then we show that the sequence of approximated solutions converges in a suitable sense to the solution of problem (8.35). In the final step we introduce the potential V and show that the right-hand side of (8.39) is in fact the solution of Eq. (8.35).

1. The solution of the approximated problem
We approximate the trajectory $t \to \omega(t)$ of the Wiener process by a sequence of smooth curves. More precisely we consider the sequence of functions[4]

$$n\int_{t-\frac{1}{n}}^t \omega(s)d\,s \equiv \omega_n(t), \qquad n \in \mathbb{N}.$$

---

[4]Here we denote, as usual, the trajectory of the Wiener process $B(t)$ as $\omega(t)$.

We have $\omega_n \to \omega$ uniformly on $[0, T]$, more precisely

$$\sup_{s \in [0,T]} |w_n(s) - w(s)| \to 0 \quad \text{as } n \to \infty \qquad \mathbb{P} \text{ a.s.}$$

Let us consider the sequence of approximated problems:

$$\begin{cases} d\psi_n = -\frac{i}{\hbar} H \psi_n dt - \lambda |x|^2 \psi_n dt + \sqrt{\lambda} x \cdot \psi_n dB_n(t) \\ \psi_n(0, x) = \psi_0(x), \end{cases} \tag{8.40}$$

where $dB_n(t)$ is an ordinary differential, i.e. $dB_n(t) = \dot{\omega}_n(t)dt$, and we can also write:

$$\begin{cases} \dot{\psi}_n = -\frac{i}{\hbar} H \psi_n - \lambda |x|^2 \psi_n + \sqrt{\lambda} x \cdot \psi_n \dot{\omega}_n(t) \\ \psi_n(0, x) = \psi_0(x), \end{cases} \tag{8.41}$$

which can be recognized as a family of Schrödinger equations, with a complex potential, labeled by the random parameter $\omega \in \Omega$.

Now we compute a representation of the solution of (8.41) by means of an infinite dimensional oscillatory integral with complex phase, under suitable assumptions on the (real) potential V and on the initial datum $\psi_n(0, x, \omega) = \psi_0(x)$.

We can write Eq. (8.41) in the following form:

$$\begin{cases} \dot{\psi}_n = -\frac{i}{\hbar} \left(\frac{-\hbar^2 \Delta}{2m} - i\lambda\hbar |x|^2 \right) \psi_n - \frac{i}{\hbar} V \psi_n + \sqrt{\lambda} x \cdot \psi_n \dot{\omega}_n(t) \\ \psi_n(0, x) = \psi_0(x), \end{cases} \tag{8.42}$$

so that we can recognize in it the Schrödinger equation for an anharmonic oscillator with a complex potential, i.e.

$$\begin{cases} \dot{\psi}_n = -\frac{i}{\hbar} \left(\frac{-\hbar^2 \Delta}{2m} + \frac{a^2}{2} |x|^2 \right) \psi_n - \frac{i}{\hbar} U \psi_n \\ \psi_n(0, x) = \psi_0(x), \end{cases} \tag{8.43}$$

where $a^2 = -2i\lambda\hbar$ and $U = U(t, x, \omega) = V(x) + i\hbar\sqrt{\lambda} x \cdot \dot{\omega}_n(t)$.

We introduce the sequence of vectors $l_n \in \mathcal{H}_t$ defined by

$$\langle l_n, \gamma \rangle = \sqrt{\lambda} \int_0^t \gamma(s) \cdot \dot{\omega}_n(s) ds = -\sqrt{\lambda} \int_o^t \omega_n(s) \cdot \dot{\gamma}(s) ds,$$

which is given by

$$l_n(s) = \sqrt{\lambda} \int_s^t \omega_n(\tau) d\tau. \tag{8.44}$$

First of all let us consider Eq. (8.35) with H replaced by the free Hamiltonian $H = -\hbar^2 \Delta / 2$. The following result holds:

Lemma 8.1. *Let $\psi_0 \in \mathcal{S}(\mathbb{R}^d)$. Then the solution of the Cauchy problem:*

$$\begin{cases} \dot{\psi}_n(t, x) = \frac{i\hbar}{2}\Delta\psi_n(t, x) - \lambda|x|^2\psi_n(t, x) + \sqrt{\lambda}x \cdot \dot{\omega}_n(t)\psi_n(t, x) \\ \psi_n(0, x) = \psi_0(x), \qquad x \in \mathbb{R}^d \end{cases} \tag{8.45}$$

is given by

$$\psi_n(t, x) = \widetilde{\int_{\mathcal{H}_t}} e^{\frac{i}{2\hbar}\int_0^t |\dot{\gamma}(s)|^2 ds - \lambda\int_0^t |\gamma(s)+x|^2 ds} e^{\sqrt{\lambda}\int_0^t (\gamma(s)+x)\cdot\dot{\omega}_n(s) ds}\psi_0(\gamma(0) + x)d\gamma,$$
$$\tag{8.46}$$

(where the right-hand side is interpreted as the infinite dimensional oscillatory integral of $\psi_0(\gamma(0) + x)e^{\langle l_n, \gamma\rangle}$ with complex quadratic phase function $\langle\gamma, (I + L)\gamma\rangle/\hbar$, with $\mathcal{H}_t$ the Cameron-Martin space, l_n the vector defined by (8.44) and L the operator defined by (8.37).)

Proof. Formula (8.46) can be realized as

$$\widetilde{\int_{\mathcal{H}_t}} e^{\frac{i}{2\hbar}\int_0^t |\dot{\gamma}(s)|^2 ds - \lambda\int_0^t |\gamma(s)+x|^2 ds} e^{\sqrt{\lambda}\int_0^t (\gamma(s)+x)\cdot\dot{\omega}_n(s) ds}\psi_0(\gamma(0) + x)d\gamma$$

$$= e^{\frac{-ia^2|x|^2 t}{2\hbar} + \sqrt{\lambda}x\cdot\omega_n(t)}\widetilde{\int_{\mathcal{H}_t}} e^{\frac{i}{2\hbar}\langle\gamma, (I+L)\gamma\rangle} e^{\langle l_n, \gamma\rangle}\int_{\mathbb{R}^d} e^{i\alpha\cdot x} e^{i\langle b(\alpha,x),\gamma\rangle} \tilde{\psi}_0(\alpha)d\alpha d\gamma,$$

where $b(\alpha, x) \in \mathcal{H}_t$, precisely:

$$b(\alpha, x)(s) = \alpha(t - s) - \frac{xa^2}{2\hbar}(t^2 - s^2),$$

One can directly verify that the function

$$f(\gamma) \equiv \int_{\mathbb{R}^d} e^{i\alpha\cdot x} e^{i\langle b(\alpha,x),\gamma\rangle} \tilde{\psi}_0(\alpha)d\alpha, \qquad \gamma \in \mathcal{H}_t$$

is the Fourier transform of a measure $\mu \in \mathcal{M}(\mathcal{H}_t)$, that is:

$$\mu(d\gamma) = \int_{\mathbb{R}^d} e^{i\alpha\cdot x} \tilde{\psi}_0(\alpha)\delta_{b(\alpha,x)}(d\gamma)d\alpha$$

so we can apply Theorem 8.5 and the integral (8.46) is equal to:

$$e^{\frac{-ia^2|x|^2 t}{2\hbar} + \sqrt{\lambda}\cdot\omega_n(t)}\int_{\mathbb{R}^d} e^{i\alpha\cdot x} \det(I + L)^{-1/2} e^{\frac{-i\hbar}{2}\langle b(\alpha,x) - il_n, (I+L)^{-1}(b(\alpha,x) - il_n)\rangle}$$
$$\tilde{\psi}_0(\alpha)d\alpha.$$

By simple calculations we get the final result:

$$\psi_n(t, x) = \int_{\mathbb{R}^d} G_n(t, x, y)\psi_0(y)dy,$$

where $G_n(t, x, y)$ is given by

$$G_n(t, x, y) \equiv \frac{1}{\sqrt{2\pi i\hbar}} \sqrt{\frac{a}{\sin(at)}} e^{\sqrt{\lambda}x \cdot \omega_n(t) - \frac{\sqrt{\lambda}ax}{\sin(at)} \cdot \int_0^t \omega_n(s) \cos(as) ds}$$

$$e^{\frac{i\hbar\lambda}{2} \int_0^t |\omega_n(s)|^2 ds} e^{\frac{i\hbar\lambda}{2} (-a \int_0^t \omega_n(s) \cdot \int_s^t \omega_n(s') \sin[a(s'-s)] ds' ds)}$$

$$\cdot e^{\frac{i\hbar\lambda}{2} (-a \int_0^t \sin(as)\omega_n(s) ds \cdot \int_0^t \cos(as)\omega_n(s) ds - a \cot(at)| \int_0^t \cos(as)\omega_n(s) ds|^2)}$$

$$e^{\frac{i}{2\hbar} (\cot(at)(|x|^2 + |y|^2) - \frac{2x \cdot y}{\sin(at)})} \cdot e^{a\sqrt{\lambda}y \cdot (\cot(at) \int_0^t \cos(as)\omega_n(s) ds + \int_0^t \sin(as)\omega_n(s) ds)},$$

$$(8.47)$$

which is, as one can easily directly verify, the fundamental solution to the approximate Cauchy problem (8.40). $\qquad\square$

2. The convergence of the sequence of approximated solutions
We will prove the following result:

Lemma 8.2. *The following equation*
$$\begin{cases} d\psi = -\frac{i}{\hbar}H\psi dt - \lambda|x|^2\psi dt + \sqrt{\lambda}x \cdot \psi \circ dB(t) & t > 0 \\ \psi(0, x) = \psi_0(x), \quad \psi_0 \in S(\mathbb{R}^d), \end{cases} \qquad (8.48)$$
with $H = -\hbar^2\Delta/2$, has a unique strong solution given by the Feynman path integral

$$\psi(t, x) = \int \widetilde{e^{\frac{i}{2\hbar} \int_0^t |\dot{\gamma}(s)|^2 ds - \lambda \int_0^t |\gamma(s) + x|^2 ds}} e^{\sqrt{\lambda} \int_0^t (\gamma(s) + x) \cdot dB(s)}$$

$$\psi_0(\gamma(0) + x) d\gamma$$

rigorously realized as the infinite dimensional oscillatory integral with complex phase on the Hilbert space $\mathcal{H}_t$

$$e^{-\lambda|x|^2 + \sqrt{\lambda}x \cdot \omega(t)} \int_{\mathcal{H}_t} e^{\frac{i}{2\hbar} \langle \gamma, (I+L)\gamma \rangle} e^{\langle l, \gamma \rangle} e^{-2\lambda x \cdot \int_0^t \gamma(s) ds} \psi_0(\gamma(0) + x) d\gamma.$$

Moreover it can be represented by the process
$$\psi(t, x) = \int_{\mathbb{R}^d} G(t, x, y)\psi_0(y) dy,$$

where

$$G(t, x, y) = \frac{1}{\sqrt{2\pi i\hbar}} \sqrt{\frac{a}{\sin(at)}} e^{\sqrt{\lambda}x \cdot \omega(t) - \frac{\sqrt{\lambda}ax}{\sin(at)} \cdot \int_0^t \cos(as)\omega(s) ds}$$

$$e^{\frac{i\hbar\lambda}{2} (-a \int_0^t \omega(s) \cdot \int_s^t \omega(s') \sin[a(s'-s)] ds' ds)}$$

$$\cdot e^{\frac{i\hbar\lambda}{2} (-a \int_0^t \sin(as)\omega(s) ds \cdot \int_0^t \cos(as)\omega(s) ds - a \cot(at)| \int_0^t \cos(as)\omega(s) ds|^2)}$$

$$e^{\frac{i}{2\hbar} \left(\cot(at)(|x|^2 + |y|^2) - \frac{2x \cdot y}{\sin(at)} \right)} e^{a\sqrt{\lambda}y \cdot \frac{1}{\sin(at)} (\int_0^t \cos[a(s-t)]\omega(s) ds)}.$$

Proof. Let us consider the sequence of approximated solutions

$$\psi_n(t,x) = \int_{\mathbb{R}^d} G_n(t,x,y)\psi_0(y)dy.$$

Using the dominated convergence theorem we have that

$$\mathbf{P}\left(\lim_{n\to\infty}\int_{\mathbb{R}^d}|\psi_n(t,x)-\tilde{\psi}(t,x)|^2dx \to 0\right) = 1 \qquad (8.49)$$

with $\tilde{\psi}(t,x) = \int_{\mathbb{R}} G(t,x,y)\psi_0(y)dy$, as

$$\lim_{n\to\infty}|G_n(t,x,y) - G(t,x,y)| \to 0$$

for all $t \in [0,T]$ and $x,y \in \mathbb{R}^d$. Moreover, one can see by a direct computation that $a = \sqrt{-2i\hbar\lambda}$ can be chosen is such a way that:

$$\left|\int_{\mathbb{R}^d} G_n(t,x,y)\psi_0(y)dy\right|^2 \le C(t)e^{P(t,x)}\|\psi_0(y)\|^2, \qquad (8.50)$$

where $P(t,x)$ is a second order polynomial with negative leading coefficient and $C(t)$ and $P(t,x)$ are continuous functions of the variable $t \in [0,T]$. Applying the Itô formula to the limit process $\tilde{\psi}(t)$ we see that it verifies Eq. (8.48) for every (t,x,y). Since the kernel $G(t,x,y)$ is adapted to the filtration of the Brownian motion by construction, it follows that the solution is predictable. By direct computation and using estimates analogous to (8.50) one can verify that $\tilde{\psi}$ is a strong solution. On the other hand every $\psi_n(t,x)$ is equal to

$$\int_{\mathcal{H}_t} e^{\frac{i}{2\hbar}\int_0^t |\dot{\gamma}(s)|^2 ds - \lambda \int_0^t |\gamma(s)+x|^2 ds} e^{\sqrt{\lambda}\int_0^t (\gamma(s)+x)\cdot\dot{\omega}_n(s)ds} \psi_0(\gamma(0)+x)d\gamma$$

$$= e^{\frac{-ia^2|x|^2 t}{2\hbar}+\sqrt{\lambda}x\cdot\omega_n(t)} \int_{\mathcal{H}_t} e^{\frac{i}{2\hbar}\langle\gamma,(I+L)\gamma\rangle} e^{\langle l_n,\gamma\rangle} e^{-i\int_0^t a^2 x\cdot\gamma(s)ds} \psi_0(\gamma(0)+x)d\gamma$$

$$= e^{\frac{-ia^2|x|^2 t}{2\hbar}+\sqrt{\lambda}x\cdot\omega_n(t)} \det(I+L)^{-1/2} \int_{\mathcal{H}_t} e^{\frac{-i\hbar}{2}\langle\gamma-il_n,(I+L)^{-1}(\gamma-il_n)\rangle} d\mu(\gamma),$$

where $\mu(d\gamma)$ is the measure on $\mathcal{H}_t$ whose Fourier transform is the function $\gamma \to e^{-i\int_0^t a^2 x\cdot\gamma(s)ds}\psi_0(\gamma(0)+x)$.

We have $\|l_n - l\|_H^2 \to 0$ as $n \to \infty$, where $l(s) = \sqrt{\lambda}\int_s^t \omega(r)dr$. Therefore, by the Lebesgue's dominated convergence theorem, we have that, for every $x \in \mathbb{R}^d$:

$$\lim_{n\to\infty} e^{\frac{-ia^2|x|^2 t}{2\hbar}+\sqrt{\lambda}x\cdot\omega_n(t)} \det(I+L)^{-1/2} \int_{\mathcal{H}_t} e^{\frac{-i\hbar}{2}\langle\gamma-il_n,(I+L)^{-1}(\gamma-il_n)\rangle} d\mu(\gamma)$$

$$= e^{\frac{-ia^2|x|^2 t}{2\hbar}+\sqrt{\lambda}x\omega(t)} \det(I+L)^{-1/2} \int_{\mathcal{H}_t} e^{\frac{-i\hbar}{2}\langle\gamma-il,(I+L)^{-1}(\gamma-il)\rangle} d\mu(\gamma).$$

Therefore, taking into account the uniqueness of the pointwise limit, we have shown that:

$$\psi(t,x) = \int_{\mathbb{R}} G(t,x,y)\psi_0(y)dy$$

$$= \widetilde{\int_{\mathcal{H}_t}} e^{\frac{i}{2\hbar}\int_0^t |\dot\gamma(s)|^2 ds - \lambda \int_0^t |\gamma(s)+x|^2 ds} e^{\int_0^t (\gamma(s)+x)\cdot dB(s)} \psi_0(\gamma(0)+x)d\gamma. \quad \square$$

Remark 8.2. The result can be extended by continuity to all $\psi_0 \in L^2(\mathbb{R}^d)$, using the density of $S(\mathbb{R}^d)$ in $L^2(\mathbb{R}^d)$.

3. The proof of Feynman-Kac-Ito formula by means of Dyson expansion

Let us consider now the general case where $H = -\hbar^2\Delta/2 + V$ and complete the proof of Theorem 8.6. We follow here the technique by Elworthy and Truman [129].

We set for $t > 0$, $x \in \mathbb{R}^d$:

$$\Theta(t,0)\psi_0(x) = \widetilde{\int_{\mathcal{H}_t}} e^{\frac{i}{2\hbar}\int_0^t |\dot\gamma(s)|^2 ds - \lambda \int_0^t |\gamma(s)+x|^2 ds} e^{-\frac{i}{\hbar}\int_0^t V(\gamma(s)+x)ds}$$

$$\cdot e^{\sqrt{\lambda}\int_0^t (\gamma(s)+x)\cdot dB(s)} \psi_0(\gamma(0)+x)d\gamma$$

and

$$\Theta_0(t,0)\psi_0(x) = \widetilde{\int_{\mathcal{H}_t}} e^{\frac{i}{2\hbar}\int_0^t |\dot\gamma(s)|^2 ds - \lambda \int_0^t |\gamma(s)+x|^2 ds} e^{\sqrt{\lambda}\int_0^t (\gamma(s)+x)\cdot dB(s)}$$

$$\psi_0(\gamma(0)+x)d\gamma.$$

Then we have:

$$\Theta(t,0)\psi_0(x) = e^{\frac{-ia^2|x|^2 t}{2\hbar} + \sqrt{\lambda}x\cdot\omega(t)} \widetilde{\int_{\mathcal{H}_t}} e^{\frac{i}{2\hbar}\langle\gamma,(I+L)\gamma\rangle} e^{\langle l,\gamma\rangle} e^{-i\int_0^t a^2 x\cdot\gamma(s)ds}$$

$$\cdot e^{-\frac{i}{\hbar}\int_0^t V(x+\gamma(s))ds} \psi_0(\gamma(0)+x)d\gamma$$

Let $\mu_0(\psi)$ be the measure on $\mathcal{H}_t$ such that its Fourier transform evaluated in $\gamma \in \mathcal{H}_t$ is $\psi_0(\gamma(0)+x)$.

For $0 \leq u \leq t$ let $\mu_u(V,x)$, $\nu_u^t(V,x)$ and $\eta_u^t(x)$ be the measures on $\mathcal{H}_t$, whose Fourier transforms when evaluated at $\gamma \in \mathcal{H}_t$ are respectively $V(x+\gamma(u))$, $\exp\left(-\frac{i}{\hbar}\int_u^t V(x+\gamma(s))ds\right)$, and $\exp\left(\frac{i}{\hbar}\int_u^t a^2 x\gamma(s)ds\right)$. We shall often write $\mu_u \equiv \mu_u(V,x)$, $\nu_u^t \equiv \nu_u^t(V,x)$ and $\eta_u^t \equiv \eta_u^t(x)$. If $\{\mu_u : a \leq u \leq b\}$ is a family in $\mathcal{M}(\mathcal{H}_t)$, we shall let $\int_a^b \mu_u du$ denote the measure on $\mathcal{H}_t$ given by:

$$f \to \int_a^b \int_{\mathcal{H}_t} f(\gamma)d\mu_u(\gamma)du,$$

whenever it exists.

Then, since for any continuous path γ

$$\exp\left(-\frac{i}{\hbar}\int_0^t V(\gamma(s)+x)ds\right)$$
$$= 1 - \frac{i}{\hbar}\int_0^t V(\gamma(u)+x)\exp\left(-\frac{i}{\hbar}\int_u^t V(\gamma(s)+x)ds\right)du,$$

we have

$$\nu_0^t = \delta_0 - \frac{i}{\hbar}\int_0^t (\mu_u * \nu_u^t)du \tag{8.51}$$

where δ_0 is the Dirac measure at $0 \in \mathcal{H}_t$.

By the Parseval-type equality:

$$\Theta(t,0)\psi_0(x) = e^{\frac{-ia^2|x|^2 t}{2\hbar}+\sqrt{\lambda}x\cdot\omega(t)}\det(I+L)^{-1/2}$$
$$\cdot \int_{\mathcal{H}_t} e^{\frac{-i\hbar}{2}\langle\alpha-il,(I+L)^{-1}(\alpha-il)\rangle}d(\eta_0^t * \nu_0^t * \mu_0(\psi))(\alpha)$$

Applying to this equality (8.51) we obtain:

$$\Theta(t,0)\psi_0(x) = e^{\frac{-ia^2|x|^2 t}{2\hbar}+\sqrt{\lambda}x\cdot\omega(t)}\det(I+L)^{-1/2}$$
$$\int_{\mathcal{H}_t} e^{\frac{-i\hbar}{2}\langle\alpha-il,(I+L)^{-1}(\alpha-il)\rangle}(\eta_0^t * \mu_0(\psi))(d\alpha)$$
$$- \frac{i}{\hbar}\int_0^t e^{\frac{-ia^2|x|^2 t}{2\hbar}+\sqrt{\lambda}x\cdot\omega(t)}\det(I+L)^{-1/2}$$
$$\cdot \int_{\mathcal{H}_t} e^{\frac{-i\hbar}{2}\langle\alpha-il,(I+L)^{-1}(\alpha-il)\rangle}(\eta_0^t * \mu_u(V,x) * \nu_u^t * \mu_0(\psi))(d\alpha)du$$
$$= \Theta_0(t,0)\psi_0(x) - \frac{i}{\hbar}\int_0^t \widetilde{\int_{\mathcal{H}_t}} e^{\frac{i}{2\hbar}\int_0^t |\dot\gamma(s)|^2 ds - \lambda \int_0^t |\gamma(s)+x|^2 ds}$$
$$e^{-\frac{i}{\hbar}\int_u^t V(\gamma(s)+x)ds}$$
$$e^{\sqrt{\lambda}\int_0^t (\gamma(s)+x)\cdot dB(s)}V(\gamma(u)+x)\psi_0(\gamma(0)+x)d\gamma du.$$

By the Fubini theorem for oscillatory integrals (see Theorem 5.7), we get

$$\widetilde{\int_{\mathcal{H}_t}} e^{\frac{i}{2\hbar}\int_0^t |\dot\gamma(s)|^2 ds - \lambda\int_0^t |\gamma(s)+x|^2 ds}e^{-\frac{i}{\hbar}\int_u^t V(\gamma(s)+x)ds}e^{\sqrt{\lambda}\int_0^t (\gamma(s)+x)\cdot dB(s)}$$
$$V(\gamma(u)+x)\cdot\psi_0(\gamma(0)+x)d\gamma$$
$$= \widetilde{\int_{\mathcal{H}_{u,t}}} e^{\frac{i}{2\hbar}\int_u^t |\dot\gamma_2(s)|^2 ds - \lambda\int_u^t |\gamma_2(s)+x|^2 ds}e^{-\frac{i}{\hbar}\int_u^{tu} V(\gamma_2(s)+x)ds}.$$

$$e^{\sqrt{\lambda}\int_u^t (\gamma_2(s)+x)\cdot dB(s)} V(\gamma_2(u)+x) \overbrace{\int_{\mathcal{H}_{0,u}}} e^{\frac{i}{2\hbar}\int_0^u |\dot{\gamma}_1(s)|^2 ds - \lambda \int_0^u |\gamma_1(s)+\gamma_2(u)+x|^2 ds}.$$

$$e^{\sqrt{\lambda}\int_0^u (\gamma_1(s)+\gamma_2(u)+x)\cdot dB(s)} \psi_0(\gamma_1(0)+\gamma_2(u)+x) d\gamma_1 d\gamma_2.$$

Here $\gamma_1 \in \mathcal{H}_{0,u}$ and $\gamma_2 \in \mathcal{H}_{u,t}$ are the integration variables, where we have denoted by $\mathcal{H}_{r,s}$ the Cameron-Martin space of paths $\gamma : [r,s] \to \mathbb{R}^d$.

Finally we have:

$$\Theta(t,0)\psi_0(x) = \Theta_0(t,0)\psi_0(x) - i\int_0^t \Theta(t,u)(V\Theta_0(u,0)\psi_0)(x)du. \quad (8.52)$$

Now the iterative solution of the latter integral equation is the Dyson series for $\Theta(t,0)$, which coincides with the corresponding power series expansion of the solution of the stochastic Schrödinger equation, which converges strongly in $L^2(\mathbb{R}^d)$. The equality holds pointwise. On the other hand, following [156], it is possible to prove that the problem (8.48) has a strong solution that verifies (8.52) in the L^2 sense, therefore $\Theta(t,0)\psi_0$ coincides with the solution $\psi(t)$. This concludes the proof of Theorem 8.6. $\qquad\square$

It is also possible to consider Belavkin's equation describing the continuous measurement of the momentum p of a d-dimensional quantum particle:

$$\begin{cases} d\psi(t,x) = -\frac{i}{\hbar}H\psi(t,x)dt + \frac{\lambda\hbar^2}{2}\Delta\psi(t,x)dt - i\sqrt{\lambda}\hbar\nabla\psi(t,x)dB(t) \\ \psi(0,x) = \psi_0(x) \qquad (t,x) \in [0,T] \times \mathbb{R}^d. \end{cases}$$
$$(8.53)$$

The stochastic term plays the role of a complex random potential depending on the momentum of the particle. In this case one has to use the phase space Feynman path integrals described in Section 6.3. More precisely, by means of a infinite dimensional oscillatory integral with complex phase on the space of paths in phase space one can give a rigorous mathematical meaning to the following heuristic expression:

$$\psi(t,x) = \overbrace{\int} e^{\frac{i}{\hbar}(\int_0^t (\dot{q}(s)p(s) - \frac{1}{2m}p(s)^2)ds - \lambda\int_0^t p(s)^2 ds} e^{-\frac{i}{\hbar}\int_0^t V(q(s)+x)ds}$$

$$\cdot e^{\sqrt{\lambda}\int_0^t p(s)\cdot dB(s)} \psi_0(\gamma(0)+x)dqdp. \quad (8.54)$$

See [15] for a detailed study of this problem.

Chapter 9

Alternative approaches to Feynman path integration

Since the first introduction of Feynman path integrals [138], several approaches for their mathematical definition and several applications have been proposed, both in the mathematical and in the physical literature. The present chapter is a brief survey of this topic, without any claim of completeness.

9.1 Analytic continuation of Wiener integrals

One of the first attempts to the rigorous mathematical realization of Feynman path integrals involves analytic continuation of Gaussian Wiener integrals. The first rigorous results can be found in Cameron's paper in 1960 [77] and where further developed in a series of papers [78, 79, 81, 80, 82–84, 204, 205, 207, 206, 208, 209, 198]. As Cameron proved, it is not possible to define a Wiener measure W_λ on the space of continuous paths $C([0,t])$ with a complex covariance $\lambda \in \mathbb{C}$, as, unless $\lambda \in \mathbb{R}^+$, it would have infinite total variation. Therefore the expression

$$\int_{C([0,t])} f(\omega) dW_\lambda(\omega),$$

is meaningless. On the other hand, for $\lambda \in \mathbb{R}^+$ and for a Borel function f on $C_t \equiv C([0,t])$ satisfying suitable conditions, the following formula holds [77]:

$$\int_{C_t} f(\omega) dW_\lambda(\omega) = \int_{C_t} f(\sqrt{\lambda}\omega) dW(\omega). \tag{9.1}$$

If λ is complex, the left-hand side or Eq. (9.1) is not well defined, but the right-hand side can still be meaningfull, provided that the function f has suitable analyticity and measurability properties. In particular, for

$\lambda = i$, the right-hand side of Eq. (9.1) is the natural candidate for an "analytic Wiener integral" or "analytic Feynman integral". The class of functions which are "integrable" in this sense does not substantially differ from the Fresnel class $\mathcal{F}(\mathcal{H}_t)$ which can be handled by means of the infinite dimensional oscillatory integral approach (see [198]).

Concerning the application of analytically continued Wiener Gaussian integrals to the Feynman path integral representation for the solution of the Schrödinger equation, the fundamental idea is the extension to the complex case of the Wiener integral representation for the solution of the heat equation

$$\begin{cases} -\frac{\partial}{\partial t}u = -\frac{1}{2}\Delta u + V(x)u \\ u(0,x) = u_0(x), \qquad x \in \mathbb{R}^d, \end{cases} \tag{9.2}$$

i.e. the Feynman-Kac formula:

$$u(t,x) = \int_{C_t} e^{-\int_0^t V(\omega(s)+x))ds} u_0(\omega(t)+x) dW(\omega). \tag{9.3}$$

Indeed, by introducing in Eq. (9.2) a real parameter λ, related to the time t

$$-\lambda\hbar\frac{\partial}{\partial t}u = -\frac{1}{2m}\hbar^2\Delta u + V(x)u$$
$$u(t,x) = \int_{C_t} e^{-\frac{1}{\lambda\hbar}\int_0^t V(\sqrt{\hbar/(m\lambda)}\omega(s)+x)ds} u_0(\sqrt{\hbar/(m\lambda)}\omega(t)+x) dW(\omega),$$

or the Planck constant $\hbar$ [123]

$$\lambda\frac{\partial}{\partial t}u = \frac{1}{2m}\lambda^2\Delta u + V(x)u$$
$$u(t,x) = \int_{C_t} e^{\frac{1}{\lambda}\int_0^t V(\sqrt{\lambda/m}\,\omega(s)+x)ds} u_0(\sqrt{\lambda/m}\,\omega(t)+x) dW(\omega),$$

or the mass m [265, 215, 10]

$$\frac{\partial}{\partial t}u = \frac{1}{2\lambda}\Delta u - iV(x)u,$$
$$u(t,x) = \int_{C_t} e^{-i\int_0^t V(\sqrt{1/\lambda}\omega(s)+x)ds} u_0(\sqrt{1/\lambda}\omega(t)+x) dW(\omega),$$

and by substituting respectively $\lambda = -i$, $\lambda = i\hbar$, or $\lambda = -im$ (when the resulting Wiener integral is well defined), one gets, at least heuristically, Schrödinger equation (with $\hbar = 1$ in the latter case) and its solution. These procedures can be made completely rigorous under suitable conditions on the potential V and the initial datum u_0. For further results and details, see for instance [94, 95, 100, 199, 200, 203, 214, 217, 245, 260, 264, 303, 311, 312] and [314–318, 322, 323, 336].

The class of classical potentials V which can be handled by means of these methods does not substantially differ from those of the type "quadratic plus Fourier transform of measure". However, it is worthwhile to point out that Nelson [265] handles potentials which are singular at the origin, Doss [123] can deal with some polynomially growing potentials, while Albeverio, Brzeźniak and Haba [10] study the case of potentials that are Laplace transform of measures and can have an exponential growth at infinity (for analogous results in the framework of the white noise calculus see [236]).

Furthermore, this particular approach allows the study of the semi-classical limit of the solution of the Schrödinger equation [50, 51, 63] by means of (a modified version of) the Laplace method for the study of the asymptotic of the Wiener integrals [295,62,64,127, 128,216,275,276,288,107,108,109,110,43,25].

It is worthwhile to recall the alternative way to define Feynman integrals by means of convergent Wiener integrals proposed by Daubechies and Klauder [104, 105, 102, 103, 228, 229]. The authors study the matrix elements of the unitary evolution operator $U(t) = e^{-\frac{i}{\hbar}Ht}$ between two coherent states $\phi_{q',p'}, \phi_{q'',p''}$, defined by

$$\phi_{q,p} = e^{i(pQ-qP)}\phi_0,$$

where Q, P are respectively the quantum position and momentum operators and ϕ_0 is the ground state of an harmonic oscillator. Formally the matrix elements $\langle \phi_{q'',p''}, U(t)\phi_{q',p'} \rangle$, in the case $\hbar = 1$ and $d = 1$, should be given by the phase space Feynman path integral:

$$\langle \phi_{q'',p''}, U(t)\phi_{q',p'} \rangle = \mathcal{N} \int e^{\frac{i}{2}\int_0^t (p(s)\dot{q}(s)-q(s)\dot{p}(s))ds-i\int_0^t H(p(s),q(s))ds} dpdq,$$
(9.4)

where $(q(s), p(s))_{s\in[0,t]}$ represents a generic path in the phase space, while $H : \mathbb{R}^2 \to \mathbb{R}$ is defined as the matrix element of the quantum Hamiltonian operator, i.e.

$$H(q,p) = \langle \phi_{q,p}, H\phi_{q,p} \rangle, \qquad (q,p) \in \mathbb{R}^2,$$

and $\mathcal{N}$ represents a normalization constant. The heuristic expression (9.4) is defined by inserting into the integrand an extra factor

$$e^{-\frac{1}{2\nu}\int_0^t (\dot{p}(s)^2+\dot{q}(s)^2)ds},$$

representing formally the density of a Wiener measure (on the space of paths $(q(s), p(s))_{s\in[0,t]}$ in the phase space) with diffusion constant $\nu > 0$. The exponent $\int_0^t (p(s)\dot{q}(s) - q(s)\dot{p}(s))ds$ is replaced by $\int_0^t (p(s)dq(s) - q(s)dp(s))$ and interpreted as a (Ito or Stratonovich) stochastic integral. In this way, for $\nu > 0$, the resulting expression

$$\int e^{\frac{i}{2}\int_0^t (p(s)dq(s)-q(s)dp(s))-i\int_0^t H(p(s),q(s))ds} dW_\nu(p,q)$$
(9.5)

is a well defined Wiener integral. The main result is the proof that, by replacing in Eq. (9.5) the function $H : \mathbb{R}^2 \to \mathbb{R}$ with $h : \mathbb{R}^2 \to \mathbb{R}$, given by

$$h(p,q) = \exp\left(-\frac{1}{2}(\partial_p^2 + \partial_q^2)\right) H(p,q), \qquad (p,q) \in \mathbb{R}^2, \qquad (9.6)$$

the matrix elements $\langle \phi_{q'',p''}, U(t)\phi_{q',p'} \rangle$ are given by the following limit

$$\langle \phi_{q'',p''}, U(t)\phi_{q',p'} \rangle$$
$$= \lim_{\nu \to \infty} 2\pi e^{\nu t/2} \int e^{\frac{i}{2} \int_0^t (p(s)dq(s) - q(s)dp(s)) - i \int_0^t h(p(s),q(s))ds} dW_\nu(p,q),$$

where W_ν is the product of two independent Wiener measures (one in p and one in q) with diffusion constant ν pinned at p', q' for $s = 0$ and at p'', q'' for $s = t$. This formula is valid for all self-adjoint Hamiltonian operators H on $L^2(\mathbb{R})$ for which the linear span of the harmonic oscillator eigenstates is a core and such that

$$H = \int h(p,q) P_{p,q} \frac{dpdq}{2\pi},$$

where $h : \mathbb{R}^2 \to \mathbb{R}$ is given by Eq. (9.6) and $P_{p,q} : L^2(\mathbb{R}) \to L^2(\mathbb{R})$ for $(p,q) \in \mathbb{R}^2$ is the projection operator

$$P_{p,q}(\psi) = \phi_{q',p'} \langle \phi_{q',p'}, \psi \rangle, \qquad \psi \in L^2(\mathbb{R}).$$

One has also to impose that for all $\alpha > 0$ the bound

$$\int_{\mathbb{R}^2} |h(p,q)|^2 e^{-\alpha(p^2+q^2)} dpdq < \infty$$

is satisfied. The class of Hamiltonian operator satisfying these condition includes all Hamiltonians that are polynomial in P and Q. The same technique has been also applied to systems with spin.

The key point of the whole procedure is the "duplication" of the path integral by working on the phase space rather than in the configuration space. Indeed if one tries to apply the same regularization technique on a configuration space path integral and to define an heuristic integral of the form $\int e^{\frac{i}{2} \int_0^t \dot{q}(s)^2 ds} dq$ by inserting the term $e^{-\frac{1}{2\nu} \int_0^t \dot{q}(s)^2 ds}$, $\nu > 0$, one has to face with Cameron's result [77] on the non-existence of a Wiener measure with complex covariance (in this case $i + 1/\nu$), or equivalently with the non-integrability of the function $e^{\frac{i}{2} \int_0^t \dot{q}(s)^2 ds}$.

9.2 The sequential approach

An alternative approach to the rigorous mathematical definition of Feynman path integrals which is very close to Feynman's original derivation of formula (1.6) is the "sequential approach", which has already been briefly discussed in the introduction (see Eqs. (1.7)–(1.9)).

The starting point is the Lie-Kato-Trotter product formula [320, 91, 92, 265], that, under suitable assumptions on the potential V, allows one to write the unitary evolution operator $e^{-\frac{i}{\hbar}tH}$, whose generator is the Hamiltonian $H = H_0 + V$, $H_0 = -\frac{\hbar^2}{2m}\Delta$, in terms of the following strong operator limit:

$$e^{-\frac{i}{\hbar}tH} = s - \lim_{n\to\infty} \left(e^{-\frac{i}{\hbar}\frac{t}{n}V} e^{-\frac{i}{\hbar}\frac{t}{n}H_0} \right)^n. \tag{9.7}$$

Let us consider a vector ϕ in Schwartz space $S(\mathbb{R}^d)$. For $V = 0$, the vector $e^{-\frac{i}{\hbar}tH_0}\phi$ can be expressed in terms of the Green function of $e^{-\frac{i}{\hbar}tH_0}$:

$$e^{-\frac{i}{\hbar}tH_0}\phi(x) = \left(2\pi i \frac{\hbar}{m}t \right)^{-\frac{d}{2}} \int e^{im\frac{(x-y)^2}{2\hbar t}}\phi(y)dy. \tag{9.8}$$

On the other hand, if we drop the term H_0, the Hamiltonian operator H is a multiplication operator and $e^{-\frac{i}{\hbar}tH}\phi$ is simply given by:

$$e^{-\frac{i}{\hbar}tH}\phi(x) = e^{-\frac{i}{\hbar}tV(x)}\phi(x). \tag{9.9}$$

By substituting Eqs. (9.8) and (9.9) into (9.7), one gets the following expression:

$$e^{-\frac{i}{\hbar}tH}\phi(x)$$

$$= \lim_{n\to\infty} \left(2\pi i \frac{\hbar}{m}\frac{t}{n} \right)^{-\frac{dn}{2}} \int_{\mathbb{R}^{nd}} e^{-\frac{i}{\hbar}\sum_{j=1}^{n}\left[\frac{m}{2}\frac{(x_j-x_{j-1})^2}{\left(\frac{t}{n}\right)^2} - V(x_j) \right]\frac{t}{n}}\phi(x_0)$$

$$dx_0\ldots dx_{n-1} \tag{9.10}$$

where $x_n = x$. The right-hand side of Eq. (9.10) can be interpreted as the finite dimensional approximation of a path integral. Indeed, if γ is a continuous trajectory from $[0,t]$ to $\mathbb{R}^d$, with $\gamma(t) = x$, let us set $x_j := \gamma(jt/n)$, for $j = 0,\ldots n$. The exponent in the integrand can be interpreted in terms of the Riemann sum of the classical action functional evaluated along the path γ:

$$S_t(\gamma) = \int_0^t \left(\frac{m}{2}\dot{\gamma}^2(s) - V(\gamma(s)) \right)ds$$

$$= \lim_{n\to\infty} \sum_{j=1}^{n} \left[\frac{m}{2}\frac{(x_j - x_{j-1})^2}{\left(\frac{t}{n}\right)^2} - V(x_j) \right]\frac{t}{n}.$$

The sequential approach for the mathematical definition of the Feynman path integrals has in fact two versions. In the first one, the attention is focused on the definition of the evolution operator of the quantum system in terms of a strong operator limit by applying Chernoff-formula, i.e. the semigroup product formula

$$s - \lim_{n \to \infty} (F(t/n))^n = \exp(tF'(0)), \qquad (9.11)$$

where $t \mapsto F(t)$ is a strongly continuous function from $\mathbb{R}$ (or $\mathbb{R}^+$) into the space of bounded linear operators on an Hilbert space $\mathcal{H}$, while $F'(0)$ has to be interpreted as some operator extension of the strong limit $s - \lim_{t \to 0} t^{-1}(F(t) - I)$. In particular if A, B are self-adjoint operators in $\mathcal{H}$ and $F(t) = e^{itA}e^{itB}$, one gets formally the Trotter product formula (see Lemma 6.5):

$$s - \lim_{n \to \infty} (e^{itA/n}e^{itB/n})^n = e^{it(A+B)} \qquad (9.12)$$

(where the sum $A+B$ has to be suitably interpreted). Trotter and Chernoff approximations, as well as their application to the problem of Feynman integration, have been extensively studied (see e.g. [304] and references therein as well as recent results in [282]).

Nelson in 1964 [265] proved Eq. (9.12) in connection with the rigorous mathematical definition of Feynman path integrals, under the assumption that the potential V belongs to the class considered by Kato [220]. Some time later Friedman [144] studied formula (9.11) in connection with continuous quantum observation. Feynman himself in [138] considered particular "ideal" quantum measurements of position, made to determine whether or not the trajectory of a particle lies in a certain space-time region. By substituting in formula (9.11) for $F(t)$ the operator $EP(t)E$, where $P(t)$ is a contraction semigroup in the Hilbert space $\mathcal{H}$ and E is an orthogonal projection, and by letting $P(t) = e^{-itH_0}$ and $E : L^2(\mathbb{R}^d) \to L^2(\mathbb{R}^d)$ be the orthogonal projection given by multiplication by the characteristic function of a suitable region $\mathcal{R}$ of $\mathbb{R}^d$, the limit

$$\lim_{n \to \infty} \|(Ee^{-itH_0/n}E)^n \phi\|^2$$

(if it exists) should give the probability that a continual observation during the time interval $[0, t]$ yields the result that the particle, whose initial state is the vector $\phi \in L^2(\mathbb{R}^d)$, lies constantly in the region $\mathcal{R}$.

It is worthwhile to mention here recent rigorous results by F. Nicola and S. Trapasso [268] on the pointwise convergence of Lie-Trotter approximations.

The second version of the sequential approach has been extensively studied by Fujiwara and Kumano-Go [146–150, 152–155, 235, 151] (see also [191, 192]) and has the right-hand side of Eq. (9.10) as a starting point. A formal path integral on a space of path $\gamma : [0, t] \to \mathbb{R}^d$ of the form

$$\int e^{\frac{i}{\hbar} S(\gamma)} F(\gamma) d\gamma \tag{9.13}$$

is realized in terms of a "time slicing approximation". More precisely for any $n \in \mathbb{N}$ one considers a partition of the interval $[0, t]$ into n subintervals $t_0 = 0 < t_1 < \cdots < t_j < \cdots < t_n = t$ and for each $j = 0, ..., n$ a point $x_j \in \mathbb{R}^d$. The path γ in expression (1.6) is then approximated by a broken line path, passing, for each $j = 0, ..., n$, from the point x_j at time t_j. There are two possible approaches to this "time slicing approximation": one can connect the point x_j at time t_j with the point x_{j+1} at time t_{j+1} by means of a straight line path [324, 235, 152], i.e.

$$\gamma(\tau) = x_j + \frac{x_{j+1} - x_j}{t_{j+1} - t_j} (\tau - t_j), \qquad \tau \in [t_j, t_{j+1}],$$

or by means of a classical path [146], i.e. the (unique for suitable V and if $|t_{j+1} - t_j|$ is sufficiently small) solution of the classical equation of motion

$$\begin{cases} m\ddot{\gamma}(\tau) = -\nabla V(\tau, \gamma(\tau)) \\ \gamma(t_j) = x_j, \\ \gamma(t_{j+1}) = x_{j+1} \end{cases}$$

In particular the term $\frac{(x_j - x_{j-1})}{\left(\frac{t}{n}\right)}$ is the (constant) velocity of the path connecting the points x_{j-1} and x_j in the time interval $\frac{t}{n}$.

An heuristic expression like Eq. (9.13) is then realized as the limit of the time slicing approximation for suitable functional F on the path space. Indeed by denoting with γ_n the broken line path (straight resp. piecewise classical) associated to the partition $t_0 = 0 < t_1 < \cdots < t_j < \cdots < t_n = t$, and by Δ_j the amplitude of each time subinterval, i.e. $\Delta_j = |t_{j+1} - t_j|$, one defines

$$\int e^{\frac{i}{\hbar} S_t(\gamma)} F(\gamma) d\gamma := \lim_{|\Delta| \to 0} \prod_{j=1}^{n} \left(\frac{1}{2\pi i \hbar \Delta_j} \right)^{d/2} \int_{\mathbb{R}^{nd}} e^{\frac{i}{\hbar} S_t(\gamma_n)} F(\gamma_n) \prod_{j=1}^{n} dx_j,$$

$$\tag{9.14}$$

(where $|\Delta| := \sup_j \Delta_j$) whenever the limit exists. The integrals on the right-hand side do not converge absolutely and are meant as (finite dimensional) oscillatory integrals.

D. Fujiwara in the case of approximation with piecewise classical paths and D. Fujiwara and N. Kumano-go in the case of broken line paths prove the existence of the limit (9.14) for a suitable class of functionals F. They assume that the potential $V(t,x)$ is a real valued function of $(t,x) \in \mathbb{R} \times \mathbb{R}^d$ and for any multi-index α, $\partial_x^\alpha V(t,x)$ is continuous in $\mathbb{R} \times \mathbb{R}^d$. Moreover they assume that for any integer $k \geq 2$ there exists a positive constant A_k such that

$$|\partial_x^\alpha V(t,x)| \leq A_k, \qquad |\alpha| = k,$$

(this excludes polynomial behaviour at infinity, except for at most quadratic degree).

The so defined functional has some important properties. Integration by parts and Taylor expansion formula with respect to functional differentiation hold. The functional is invariant under orthogonal transformations and transforms naturally under translations. The fundamental theorem of calculus holds and it is possible to interchange the order of integration with Riemann-Stieltjes integrals as well as to interchange the order with a limit [152].

The semiclassical approximation as $\hbar \to 0$ of the integral has been detailed studied, providing not only the leading term, but also the other ones. We refer in particular to the review paper [154] for a complete treatment of the topic which provides also an intuitive explanation of the main ideas.

We point out to the reader that a phase space Feynman path integral version of this approach has recently been implemented [155] by approximating a generic path $(q,p) : [0,t] \to \mathbb{R}^{2d}$ in the phase space of the system by means of piecewise bicharacteristic paths, i.e the solutions of the Hamilton equation of motion with particular boundary condition (i.e. the initial velocity and the final position are fixed).

Finally we recall that the time slicing approximation, in particular with piecewise polygonal paths, is extensively used in the physical literature not only as a tool for the definition of the Feynman integral, but also as a practical method of computation for particular solvable models [165–167, 173, 230, 294, 296, 157, 331].

9.3 White noise calculus

Another idea which has been largely implemented is the definition of the Feynman integral as an "infinite dimensional distribution". In other words, the heuristic expression $\int e^{\frac{i}{\hbar} S(\gamma)} f(\gamma) d\gamma$, which cannot be defined in terms of a Lebesgue integral, is realized as the distributional pairing between $e^{\frac{i}{\hbar}S}$ and a suitable function f.

The first proposal of the definition of the infinite dimensional oscillatory integrals in terms of a duality relation can be found in two papers by Ito [195, 196] and further developed in Albeverio and Høegh-Krohn's work [18, 17], where an infinite dimensional Fresnel integral on a real separable Hilbert space $\mathcal{H}$

$$\widetilde{\int_{\mathcal{H}}} e^{\frac{i}{2\hbar}\langle\gamma,\gamma\rangle} f(\gamma)d\gamma$$

is defined for $f \in \mathcal{F}(\mathcal{H})$ in term of the Parseval type equality:

$$\widetilde{\int_{\mathcal{H}}} e^{\frac{i}{2\hbar}\langle\gamma,\gamma\rangle} f(\gamma)d\gamma = \int_{\mathcal{H}} e^{-\frac{i\hbar}{2}\langle\gamma,\gamma\rangle} d\mu_f(\gamma), \qquad f = \hat{\mu}_f. \tag{9.15}$$

The main difference with the infinite dimensional oscillatory integral approach described in Chapter 5 is the fact that here Parseval equality (9.15) plays the role of the *definition* of the Fresnel integral, while the same equality in Chapter 5 is a theorem.

The definition of the Feynman integrand $e^{\frac{i}{\hbar}S}$ as a special kind of infinite dimensional distribution was also realized in C. DeWitt-Morette's work [114, 115, 117, 116, 87, 86, 85, 88, 118]. This idea has been recently implemented in a mathematical rigorous setting by means of white noise calculus [182, 183, 134, 133, 170, 236, 240, 300, 310]. In white noise calculus the integral $\int e^{\frac{i}{2\hbar}\langle\gamma,\gamma\rangle} f(\gamma)d\gamma$ is realized as a distributional pairing on a well defined measure space. The idea can be simply explained in a finite dimensional setting. Indeed for $\mathcal{H} = \mathbb{R}^d$ and $\hbar = 1$, one has

$$(2\pi i)^{-d/2} \int_{\mathbb{R}^d} e^{\frac{i}{2}(x,x)} f(x)dx$$
$$= (2\pi i)^{-d/2} \int_{\mathbb{R}^d} e^{\frac{i}{2}(x,x)+\frac{1}{2}(x,x)} f(x) e^{-\frac{1}{2}(x,x)} dx$$
$$= \frac{1}{i^{d/2}} \int_{\mathbb{R}^d} e^{\frac{i}{2}(x,x)+\frac{1}{2}(x,x)} f(x) d\mu_G(x),$$

where the latter line can be interpreted as the distributional pairing of $i^{-d/2}e^{\frac{i}{2}(x,x)+\frac{1}{2}(x,x)}$ and f not with respect to Lebesgue measure dx on $\mathbb{R}^d$, but rather with respect to the centered Gaussian measure μ_G on $\mathbb{R}^d$ with covariance the identity.

This idea can be generalized to the infinite dimensional case by exploiting the fact that, even if Lebesgue measure does not exist, Gaussian measures are still well defined in this setting.

The first step is the construction of the underlying measure space, the infinite dimensional analogous of $(\mathbb{R}^d, \mu_G)$.

Let E be a real separable Hilbert space, with inner product $\langle\,,\,\rangle$, and let $\mathcal{E}_1 \supset \mathcal{E}_2 \supset \ldots$ be vector subspaces of E, each $\mathcal{E}_p$ being a Hilbert space with inner product $\langle\,,\,\rangle_p$ such that:

(1) $\mathcal{E} := \cap_p \mathcal{E}_p$ is dense in E and in each $\mathcal{E}_p$,
(2) $|u|_p \leq |u|_q$ for every $q \geq p$ and $u \in \mathcal{E}_q$,
(3) for every p, the Hilbert-Schmidt norm $\|i_{qp}\|_{HS}$ of the inclusion $i_{qp} :$ $\mathcal{E}_q \to \mathcal{E}_p$ is finite for some $q \geq p$ and $\lim_{q \to \infty} \|i_{qp}\|_{HS} = 0$.

By identifying $E := \mathcal{E}_0$ with its dual $\mathcal{E}_0^*$, it is possible to construct the chain of spaces

$$\mathcal{E} := \cap_p \mathcal{E}_p \subset \cdots \subset \mathcal{E}_2 \subset \mathcal{E}_1 \subset \mathcal{E}_0 = E \simeq \mathcal{E}_0^* \subset \mathcal{E}_{-1} \subset \mathcal{E}_{-2} \cdots \subset \mathcal{E}^* := \cup_p \mathcal{E}_{-p},$$

where $\mathcal{E}_{-p} = \mathcal{E}_p^*$, $p \in \mathbb{Z}$. In a typical example one has an Hilbert-Schmidt operator K, with Hilbert-Schmidt norm $\|K\|_{HS} < 1$, $\mathcal{E}_p$ is taken as the range $Im(K^p)$ and

$$\langle u, v \rangle_p = \langle K^{-p}u, K^{-p}v \rangle. \tag{9.16}$$

According to Milnos' theorem there is a unique probability measure μ on the Borel σ-algebra (of the weak topology) on $\mathcal{E}^*$ such that for every $x \in \mathcal{E}$ the function on $\mathcal{E}^*$ given by

$$\phi \mapsto (\phi, x) := \phi(x), \qquad \phi \in \mathcal{E}^*$$

is a mean zero Gaussian variable of variance $|x|_0^2$. By unitary extension, for every $x \in E$ there is a μ-almost-everywhere defined Gaussian random variable $(\cdot, x)$ on $\mathcal{E}^*$ with mean 0 and variance $|x|_0^2$. This extends by complex linearity to complex Gaussian random variables corresponding to elements z of the complexification $E_{\mathbb{C}}$ of E.

The space $(\mathcal{E}^*, \mu)$ is then taken as the underlying measure space for the realization of the (at this level still heuristic) expression

$$\int_{\mathcal{E}^*} e^{\frac{i}{2}(x,x) + \frac{1}{2}(x,x)} f(x) d\mu(x), \tag{9.17}$$

i.e. $(\mathcal{E}^*, \mu)$ plays the role of the infinite dimensional analogue of $(\mathbb{R}^d, \mu_G)$.

Let us consider the symmetric Fock space $\mathcal{F}_s(E_{\mathbb{C}})$ on $E_{\mathbb{C}}$, i.e. the Hilbert space obtained by completing the symmetric tensor algebra over $E_{\mathbb{C}}$ with respect to the inner product given by

$$\langle\langle \sum_n u_n, \sum_m v_m \rangle\rangle_0 = \sum_n n! \langle u_n, v_n \rangle_0,$$

where u_n and v_n are n-tensors and $\langle \cdot, \cdot \rangle_0$ denotes the inner product on n-tensors induced by the inner product on $E_{\mathbb{C}}$. Analogously the inner products $\langle \cdot, \cdot \rangle_p$ produce inner products $\langle\langle \cdot, \cdot \rangle\rangle_p$ on $\mathcal{F}_s(E_{\mathbb{C}})$.

$\mathcal{F}_s(E_{\mathbb{C}})$ is unitary equivalent to $L^2(\mathcal{E}^*, \mu)$ by the Hermite-Ito-Segal isomorphism $\mathcal{I} : L^2(\mathcal{E}^*, \mu) \to \mathcal{F}_s(E_{\mathbb{C}})$, which is defined by

$$\mathcal{I}(e^{(\cdot, z) - (z, z)_0/2}) = \mathrm{Exp}(z)$$

for every $z \in E_{\mathbb{C}}$, where

$$(\cdot, z) : \mathcal{E}^* \to \mathbb{C} : \phi \mapsto \phi(z)$$

and

$$\mathrm{Exp}(z) = 1 + z + \frac{z^{\otimes 2}}{2!} + \frac{z^{\otimes 3}}{3!} + \cdots \in \mathcal{F}_s(E_{\mathbb{C}}).$$

It allows to transfer the inner products $\langle\langle \cdot, \cdot \rangle\rangle_p$ from $\mathcal{F}_s(E_{\mathbb{C}})$ to $L^2(\mathcal{E}^*, \mu)$, denoted again with $\langle\langle \cdot, \cdot \rangle\rangle_p$.

Let us call *white noise distribution* over $\mathcal{E}^*$ any element of the completion $[\mathcal{E}_{-p}]$ of $L^2(\mathcal{E}^*, \mu)$ with respect to the dual norm $\langle\langle \cdot, \cdot \rangle\rangle_{-p}$, for any integer $p \geq 0$.

It is possible to construct the following chain of Hilbert spaces

$$[\mathcal{E}] := \cap_p [\mathcal{E}_p] \subset \cdots \subset [\mathcal{E}_2] \subset [\mathcal{E}_1] \subset [\mathcal{E}_0]$$
$$= L^2(\mathcal{E}^*, \mu) \simeq [\mathcal{E}_0^*] \subset [\mathcal{E}_{-1}] \subset [\mathcal{E}_{-2}] \cdots \subset [\mathcal{E}^*] := \cup_p [\mathcal{E}_{-p}].$$

The elements of $[\mathcal{E}]$ are taken to be test functions over $[\mathcal{E}^*]$, which is the corresponding space of distributions.

The aim of the white noise approach is the proof that the expression $e^{\frac{i}{2}(x, x) + \frac{1}{2}(x, x)}$ can be rigorously defined in terms of a white noise distribution. In particular in this framework it is possible to realize the Feynman path integral representation for the fundamental solution $G(t, x; 0, y)$ of the Schrödinger equation over $\mathbb{R}^d$

$$G(t, x; 0, y) = \int_{\substack{\gamma(t) = x \\ \gamma(0) = y}} e^{\frac{i}{\hbar} \int_0^t (\dot{\gamma}^2(s)/2) - V(\gamma(s))) ds} d\gamma, \qquad t > 0, \ x, y \in \mathbb{R}^d,$$

in terms of a white noise distributional pairing on a suitable "path space" $\mathcal{E}^*$ [183, 134, 236, 240]. We present here some results given in [236].

Let us consider the Hilbert space $E = L_d := L^2(\mathbb{R}) \otimes \mathbb{R}^d$, the nuclear space $\mathcal{E} = S_d := S(\mathbb{R}) \otimes \mathbb{R}^d$, i.e. the space of d-dimensional Schwartz test functions, and the corresponding dual space $\mathcal{E}^* = S_d' := S'(\mathbb{R}) \otimes \mathbb{R}^d$. Let

μ be the Gaussian measure on the Borel σ-algebra of S_d' identified by its characteristic function

$$\int_{S_d'} e^{i \int X(s)f(s)ds} d\mu(X) = e^{-\frac{1}{2} \int f^2(s)ds}, \qquad f \in S_d.$$

Heuristically the "paths" $X \in S_d'$ can be interpreted as the velocities of Brownian paths, as the d-dimensional Brownian motion is given by

$$B(t) = \left(\int_0^t X_1(s)ds, \ldots, \int_0^t X_d(s)ds \right),$$

X_i being the i_{th} component of the path X. One then considers the triple

$$[S_d] \subset L^2(S_d') \subset [S_d^*]$$

and realizes the Feynman integrand as an element of $[S_d^*]$. More precisely the paths are modeled by

$$\gamma(s) = x - \sqrt{\hbar} \int_s^t X(\sigma)d\sigma := x - \sqrt{\hbar}(X, 1_{(s,t]}),$$

and the Feynman integrand for the free particle is realized as the distribution

$$I_0(x,t;x_0,t_0) = N e^{\frac{i+1}{2} \int_0^t X(s)^2 ds} \delta(\gamma(0) - y),$$

(where N stands for normalization and $\delta(\gamma(0) - y)$ fixes the initial point of the path). The same technique allows one to handle more general potentials, such as those which are Laplace transform of bounded measures (see [236] for a detailed exposition).

Particularly interesting is the application of this formalism [42, 16, 242, 174–176, 178, 177, 179, 40, 37] to the study of the Chern-Simons topological field theory [49, 297, 298, 333] (for similar results by means of different methods see also [41, 73, 145, 171, 224, 221–223, 246]).

9.4   Poisson processes

An alternative mathematical definition of Feynman integrals has been proposed by V.P. Maslov and A.M. Chebotarev [249, 233] and further developed by Ph. Combe, R. Høegh-Krohn, R. Rodriguez, M. Sirugue, M. Sirugue-Collin, Ph. Blanchard [67, 99, 98, 97]. Some recent results and new applications have been proposed by V.N. Kolokoltsov [232, 231].

The main idea is the definition of the Feynman path integral for the solution of the Schrödinger equation in momentum representation:

$$\begin{cases} \frac{\partial}{\partial t}\tilde{\psi}(p) = -\frac{i}{2}p^2\tilde{\psi}(p) - iV(-i\nabla_p)\tilde{\psi}(p) \\ \tilde{\psi}(0,p) = \tilde{\phi}(p) \end{cases} \qquad (9.18)$$

in terms of the expectation with respect to a Poisson process.

The potentials V which can be handled by this method are those belonging to the Fresnel class $\mathcal{F}(\mathbb{R}^d)$:

$$V(x) = \int_{\mathbb{R}^d} e^{ikx} d\mu_v(k), \qquad x \in \mathbb{R}^d.$$

In fact for any $\mu_v \in \mathcal{M}(\mathbb{R}^d)$, there exist a positive finite measure ν and a complex-valued measurable function f such that

$$d\mu_v(k) = f(k)d\nu(k).$$

One can assume that $\nu(\{0\}) = 0$ without loss of generality, because otherwise the condition can be fulfilled by a translation of the potential. In this case ν is a finite Lévy measure and one can consider the Poisson process having Lévy measure ν (see, e.g. [277], for these concepts). This process has almost surely piecewise constant paths. More precisely, a typical path P on the time interval $[0, t]$ is defined by a finite number of independent random jumps $\delta_1, ..., \delta_n$, distributed according to the probability measure ν/λ_ν, with $\lambda_\nu = \nu(\mathbb{R}^d)$, occurring at random times $\tau_1, ...\tau_n$, distributed according to a Poisson measure with intensity λ_ν.

Under the assumption that $\tilde{\phi}(p)$ is a bounded continuous function, it is possible to prove that the solution of the Cauchy problem (9.18) can be represented by the following path integral:

$$\tilde{\psi}(t,p) = e^{t\lambda_\nu} \mathbb{E}_p^{[0,t]} [e^{-\frac{i}{2}\sum_{j=0}^n (P_j, P_j)(\tau_{j+1}-\tau_j)} \prod_{j=1}^n (-if(\delta_j)) \tilde{\phi}(P(t))],$$

where the expectation is taken with respect to the measure associated to the Poisson process and the sample path $P(\cdot)$ is given by

$$P(\tau) = \begin{cases} P_0 = p, & 0 \le \tau < \tau_1 \\ P_1 = p + \delta_1, & \tau_1 \le \tau < \tau_2 \\ ... \\ P_n = p + \delta_1 + \delta_2 + ... + \delta_n, & \tau_n \le \tau \le t. \end{cases}$$

The present approach has also been successfully applied to the study of relativistic quantum theory [1], to Klein-Gordon equation [99, 97], to Fermi systems [98] and to the solution of Dirac equation [233]. We refer to Kolokoltsov's book [232] for a detailed exposition of more general applications of this technique.

9.5 Further approaches and results

In recent years alternative mathematical definitions of Feynman path integrals have been proposed.

As an example we recall the work of Belokurov, Smolianov, Solov'ev and Shavgulidze [299, 305, 58–61], who introduce a regularized path integral which can be expressed in terms of a convergent power series in the coupling constant. This procedure allows a perturbative treatment of polynomial potentials.

In [19] an analytic operator-valued Feynman integral is defined for potentials given by a class of generalized signed measures described in terms of additive functions associated with Dirichlet forms.

Another interesting approach which has not been systematically developed yet makes use of non-standard analysis [13]. The main idea is the extension of the expression

$$\left(2\pi i \frac{\hbar}{m} \frac{t}{n}\right)^{-\frac{dn}{2}} \int_{\mathbb{R}^{dn}} e^{\frac{i}{\hbar} S_t(x_0, \ldots, x_n)} \phi(x_0) dx_0 \ldots dx_{n-1},$$

representing the finite $n-$ dimensional approximation of the heuristic Feynman integral, from $n \in \mathbb{N}$ to a non-standard hyperfinite infinite $n \in \ {}^*\mathbb{N}$. The result is just an *internal* quantity. For a suitable class of potentials, its standard part can be shown to exist and to solve the Schrödinger equation. Analogously the semiclassical approximation of the solution of the Schrödinger equation can be obtained by taking the parameter $\hbar$ as an infinitesimal quantity. Some interesting results can be found in [261, 262].

Besides Feynman integration theory, it is worthwhile to mention also Feynman operational calculus, that is an heuristic procedure which allows to define product and functions of non-commuting operators. It was developed by Feynman in 1951 [142] and applied to quantum electrodynamics. The study of this technique from a rigorous mathematical point of view is extensively described in Johnson and Lapidus' book [203] and in the recent monograph [210] (see also [112, 113, 201, 202, 238, 239, 160–162]).

The number of applications of (heuristic) Feynman path integrals to quantum theory is really huge and it is impossible to give a complete description. Let us only mention some rigorous results concerning the solution of Dirac equation [21, 22, 190, 193, 262, 263, 337, 339, 338], the application to hyperbolic systems [313, 166], the mathematical approaches to supersymmetric Feynman path integrals [289].

Finally we should not forget the enormous predictive power of Feynman description of quantum dynamics when this is applied to quantum field theory [89, 197, 227, 271, 279, 329, 330, 340]. In this case however the gap between what can be heuristically calculated and what can be rigorously mathematically defined is particularly deep. Much work has still to be done...

Bibliography

[1] Albeverio, S., Blanchard, P., Combe, P., Høegh-Krohn, R. and Sirugue, M. (1983). Local relativistic invariant flows for quantum fields, *Comm. Math. Phys.* **90**, 3, pp. 329–351.

[2] Albeverio, S., Blanchard, P. and Høegh-Krohn, R. (1980). Stationary phase for the Feynman integral and the trace formula, in *Functional integration* (Plenum, New York), pp. 341–361.

[3] Albeverio, S., Blanchard, P. and Høegh-Krohn, R. (1981). Feynman path integrals, the Poisson formula and the theta function for the Schrödinger operators, in *Trends in Applications of Pure Mathematics to Mechanics, Vol III* (Pitman, Boston), pp. 1–21.

[4] Albeverio, S., Blanchard, P. and Høegh-Krohn, R. (1982). Feynman path integrals and the trace formula for the Schrödinger operators, *Comm. Math. Phys.* **83**, 1, pp. 49–76.

[5] Albeverio, S., Boutet de Monvel-Berthier, A. M. and Brzeźniak, Z. (1995). Stationary phase method in infinite dimensions by finite dimensional approximations: applications to the Schrödinger equation, *Potential Analysis* **4**, pp. 469–502.

[6] Albeverio, S., Boutet de Monvel-Berthier, A. M. and Brzeźniak, Z. (1996). The trace formula for Schrödinger operators from infinite dimensional oscillatory integrals, *Math. Nachr.* **182**, pp. 21–65.

[7] Albeverio, S. and Brzeźniak, Z. (1993). Finite-dimensional approximation approach to oscillatory integrals and stationary phase in infinite dimensions, *J. Funct. Anal.* **113**, 1, pp. 177–244.

[8] Albeverio, S. and Brzeźniak, Z. (1994). Feynman path integrals as infinite-dimensional oscillatory integrals: some new developments, *Acta Appl. Math.* **35**, pp. 5–27.

[9] Albeverio, S. and Brzeźniak, Z. (1995). Oscillatory integrals on Hilbert spaces and Schrödinger equation with magnetic fields, *J. Math. Phys.* **36**, 5, pp. 2135–2156.

[10] Albeverio, S., Brzeźniak, Z. and Haba, Z. (1998). On the Schrödinger equation with potentials which are Laplace transform of measures, *Potential Anal.* **9**, 1, pp. 65–82.

[11] Albeverio, S., Cangiotti, N. and Mazzucchi, S. (2020). A rigorous mathematical construction of Feynman path integrals for the Schrödinger equation with magnetic field, *Communications in Mathematical Physics* **377**, 2, pp. 1461–1503.

[12] Albeverio, S., Cattaneo, L., Di Persio, L. and Mazzucchi, S. (2007). A rigorous approach to the Feynman-Vernon influence functional and its applications. I, *J. Math. Phys.* **48**, 10, p. 102109.

[13] Albeverio, S., Fenstad, J., Høegh-Krohn, R. and Lindstrøm, T. (1986). *Non Standard Methods in Stochastic Analysis and Mathematical Physics* (Pure and Applied Mathematics 122, Academic Press, Orlando, FL).

[14] Albeverio, S., Guatteri, G. and Mazzucchi, S. (2003). Representation of the Belavkin equation via Feynman path integrals, *Probab. Theory Relat. Fields* **125**, pp. 365–380.

[15] Albeverio, S., Guatteri, G. and Mazzucchi, S. (2004). Representation of the Belavkin equation via phase space Feynman path integrals, *Infin. Dimens. Anal. Quantum Probab. Relat. Top.* **7**, 4, pp. 507–526.

[16] Albeverio, S., Hahn, A. and Sengupta, A. (2004). Rigorous Feynman path integrals, with applications to quantum theory, gauge fields, and topological invariants, in *Stochastic analysis and mathematical physics (SAMP/ANESTOC 2002),* (World Sci. Publishing, River Edge, NJ), pp. 1–60.

[17] Albeverio, S. and Høegh-Krohn, R. (1977). Oscillatory integrals and the method of stationary phase in infinitely many dimensions, with applications to the classical limit of quantum mechanics I, *Invent. Math.* **40**, 1, pp. 59–106.

[18] Albeverio, S., Høegh-Krohn, R. and Mazzucchi, S. (2008). *Mathematical theory of Feynman path integrals. An Introduction,* 2nd edn. (Springer, Berlin).

[19] Albeverio, S., Johnson, J. and Ma, Z. (1996). The analytic operator-value Feynman integral via additive functionals of Brownian motion, *Acta Appl. Math.* **42**, 3, pp. 267–295.

[20] Albeverio, S., Khrennikov, A. and Smolyanov, O. (1999). The probabilistic Feynman-Kac formula for an infinite-dimensional Schrödinger equation with exponential and singular potentials, *Potential Anal.* **11**, 2, pp. 157–181.

[21] Albeverio, S., Khrennikov, A. and Smolyanov, O. (2001). Representation of solutions of Liouville equations for generalized Hamiltonian systems by functional integrals. (Russian), *Dokl. Akad. Nauk.* **381**, 2, pp. 155–159.

[22] Albeverio, S., Khrennikov, A. and Smolyanov, O. (2002). A local Liouville theorem for infinite-dimensional Hamilton-Dirac systems, *Russ. J. Math. Phys.* **9**, 2, pp. 123–139.

[23] Albeverio, S., Kolokoltsov, V. N. and Smolyanov, O. G. (1996). Représentation des solutions de l'équation de Belavkin pour la mesure quantique par une version rigoureuse de la formule d'intégration fonctionnelle de Menski, *C. R. Acad. Sci. Paris Sér. I Math.* **323**, 6, pp. 661–664.

[24] Albeverio, S., Kolokoltsov, V. N. and Smolyanov, O. G. (1997). Continuous quantum measurement: local and global approaches, *Rev. Math. Phys.* **9**, 8, pp. 907–920.

[25] Albeverio, S. and Liang, S. (2005). Asymptotic expansions for the Laplace approximations of sums of Banach space-valued random variables, *Ann. Probab.* **33**, 1, pp. 300–336.

[26] Albeverio, S. and Mazzucchi, S. (2004). Generalized infinite-dimensional Fresnel integrals, *C. R. Acad. Sci. Paris* **338**, 3, pp. 255–259.

[27] Albeverio, S. and Mazzucchi, S. (2004). Some New Developments in the Theory of Path Integrals, with Applications to Quantum Theory, *J. Stat. Phys.* **112**, 12, pp. 191–215.

[28] Albeverio, S. and Mazzucchi, S. (2005). Feynman path integrals for polynomially growing potentials, *J. Funct. Anal.* **221**, 1, pp. 83–121.

[29] Albeverio, S. and Mazzucchi, S. (2005). Feynman path integrals for the time dependent quartic oscillator, *C. R. Acad. Sci. Paris* **341**, 10, pp. 647–650.

[30] Albeverio, S. and Mazzucchi, S. (2005). Feynman path integrals for time-dependent potentials, in G. D. Prato and L. Tubaro (eds.), *Stochastic Partial Differential Equations and Applications - VII* (Lecture Notes in Pure and Applied Mathematics, vol. 245, Taylor & Francis), pp. 7–20.

[31] Albeverio, S. and Mazzucchi, S. (2005). Generalized Fresnel Integrals, *Bull. Sci. Math.* **129**, 1, pp. 1–23.

[32] Albeverio, S. and Mazzucchi, S. (2006). The time dependent quartic oscillator - a Feynman path integral approach, *J. Funct. Anal.* **238**, 2, pp. 471–488.

[33] Albeverio, S. and Mazzucchi, S. (2007). Theory and applications of infinite dimensional oscillatory integrals, in *Stochastic Analysis and Applications. Proceedings of the Abel Symposium 2005 in honor of Prof. Kiyosi Ito* (Springer), pp. 73–91.

[34] Albeverio, S. and Mazzucchi, S. (2011). The trace formula for the heat semigroup with polynomial potential. *Seminar on Stochastic Analysis, Random Fields and Applications VI*, 3–21, Progr. Probab., 63, *Birkhäuser/Springer Basel AG, Basel*.

[35] Albeverio, S. and Mazzucchi, S. (2009). Infinite dimensional oscillatory integrals with polynomial phase function and the trace formula for the heat semigroup, *Astérisque* **327**, pp. 17–45.

[36] Albeverio, S. and Mazzucchi, S. (2009). An asymptotic functional-integral solution for the schrödinger equation with polynomial potential, *Journal of Functional Analysis* **257**, 4, pp. 1030–1052.

[37] Albeverio, S. and Mazzucchi, S. (2009). A survey on mathematical Feynman path integrals: construction, asymptotics, applications, in *Quantum field theory* (Springer), pp. 49–66.

[38] Albeverio, S. and Mazzucchi, S. (2015). An introduction to infinite-dimensional oscillatory and probabilistic integrals, in *Stochastic analysis: a series of lectures* (Springer), pp. 1–54.

[39] Albeverio, S. and Mazzucchi, S. (2016). A unified approach to infinite-dimensional integration, *Reviews in Mathematical Physics* **28**, 02, p. 1650005.

[40] Albeverio, S. and Mitoma, I. (2007). Asymptotic Expansion of Perturbative Chern-Simons Theory via Wiener Space, *Bull. Sci. Math.*

[41] Albeverio, S. and Schäfer, J. (1995). Abelian Chern-Simons theory and linking numbers via oscillatory integrals, *J. Math. Phys.* **36**, pp. 2157–2169.

[42] Albeverio, S. and Sengupta, A. (1997). A mathematical construction of the non-Abelian Chern-Simons functional integral, *Commun. Math. Phys.* **186**, pp. 563–579.

[43] Albeverio, S. and Steblovskaya, V. (1999). Asymptotics of infinite-dimensional integrals with respect to smooth measures I, *Infin. Dimens. Anal. Quantum Probab. Relat. Top.* **2**, 4, pp. 529–556.

[44] Arnold, V. (1972). Integrals of quickly oscillating functions and singularities of projections of Lagrange manifolds, *Funct. Analys. and its Appl.* **6**, pp. 222–224.

[45] Arnold, V. (1972). Une classe caractéristique intervenant dans les conditions de quantification, in *Théorie des perturbations et méthodes asymptotiques* (Dunod. Paris), pp. 341–361.

[46] Arnold, V. (1981). Remarks on the stationary phase method and Coxeter numbers, in *Singularity Theory Selected Papers* (Cambridge Univerity Press. Cambridge), pp. 61–90.

[47] Arnold, V., Goryunov, V. V., Lyashko, O. V. and Vassiliev, V. A. (1993). *Singularity theory. I.*, Vol. 6 (Springer, Berlin).

[48] Arnold, V., Gusein-Zade, S. M. and Varchenko, A. (eds.) (1988). *Singularities of differentiable maps Vol II* (Birhäuser, Basel).

[49] Atiyah, M. (1990). *The Geometry and Physics of Knots* (Cambridge University Press, Cambridge).

[50] Azencott, R. (1982). Formule de taylor stochastique et développement asymptotique d'intégrales de Feynmann, in *Séminaire de Probabilités XVI, 1980/81 Supplément: Géométrie Différentielle Stochastique* (Springer), pp. 237–285.

[51] Azencott, R. and Doss, H. (1985). L'equation de Schrödinger quand h tend vers zero; une approche probabiliste, in *Stochastic aspects of classical and quantum systems* (Springer), pp. 1–17.

[52] Baldi, P. (2017). *Stochastic Calculus* (Springer, Cham).

[53] Balser, W. (1994). *From divergent power series to analytic functions* (LNM 1582, Springer Berlin-Heidelberg).

[54] Barchielli, A. (1993). On the quantum theory of measurements continuous in time, *Rep. Math. Phys.* **33**, pp. 21–34.

[55] Barchielli, A. and Belavkin, V. (1991). Measurements continuous in time and a posteriori states in quantum mechanics, *J. Phys. A.* **24**, 7, pp. 1495–1514.

[56] Bauer, H. (1996). Probability theory, translated from the fourth (1991) german edition by Robert B. Burckel and revised by the author, .

[57] Bauer, H. (2011). *Measure and integration theory*, Vol. 26 (Walter de Gruyter).

[58] Belokurov, V. V., Solovev, Y. P. and Shavgulidze, E. T. (1995). New perturbation theory for quantum field theory: convergent series instead of asymptotic expansions, *Mod.Phys. Let. A* **19**, 39, pp. 3033–3041.

[59] Belokurov, V. V., Solovev, Y. P. and Shavgulidze, E. T. (1997). A method for the approximate calculation of path integrals using perturbation theory with convergent series. I. *Theor. and Math. Phys.* **109**, 1, pp. 1287–1293.

[60] Belokurov, V. V., Solovev, Y. P. and Shavgulidze, E. T. (1997). A method for the approximate calculation of path integrals using perturbation theory with convergent series. II. Euclidean quantum field theory, *Theor. and Math. Phys.* **109**, 1, pp. 1294–1301.

[61] Belokurov, V. V., Solovev, Y. P. and Shavgulidze, E. T. (1997). Perturbation theory with convergent series for functional integrals with respect to the Feynman measure. (Russian), *Math. Surveys* **52**, 2, pp. 392–393.

[62] Ben Arous, G. (1988). Méthode de Laplace et de la phase stationnaire sur l'espace de Wiener. (French) [The Laplace and stationary phase methods on Wiener space], *Stochastics* **25**, 3, pp. 125–153.

[63] Ben Arous, G. and Castell, F. (1996). A probabilistic approach to semi-classical approximations, *J. Funct. Anal.* **137**, 1, pp. 243–280.

[64] Ben Arous, G. and Léandre, R. (1991). Décroissance exponentielle du noyau de la chaleur sur la diagonale. II. (French) [Exponential decay of the heat kernel on the diagonal. II], *Probab. Theory Related Fields* **90**, 3, pp. 377–402.

[65] Benatti, F., Ghirardi, G., Rimini, A. and Weber, T. (1987). Quantum mechanics with spontaneous localization and the quantum theory of measurement, *Nuovo Cimento B* **11**, 100, pp. 27–41.

[66] Billingsley, P. (2008). *Probability and measure* (John Wiley & Sons).

[67] Blanchard, P., Combe, P., Sirugue, M. and Sirugue-Collin, M. (1987). Jump processes: an introduction and some applications in quantum theories, *Rend. Circ. Mat. Palermo. II*, 17, pp. 47–104.

[68] Bochner, S. (2020). *Harmonic analysis and the theory of probability* (University of California press).

[69] Bogachev, V. I. (1998). *Gaussian measures*, 62 (American Mathematical Soc.).

[70] Bogachev, V. I. (2007). *Measure theory*, Vol. 1 (Springer Science & Business Media).

[71] Bogachev, V. I. and Smolyanov, O. G. (2017). *Topological vector spaces and their applications* (Springer).

[72] Bonaccorsi, S. and Mazzucchi, S. (2015). High order heat-type equations and random walks on the complex plane, *Stochastic processes and their Applications* **125**, 2, pp. 797–818.

[73] Broda, B. (1994). A path-integral approach to polynomial invariants of links. Topology and physics, *J. Math. Phys.* **35**, 10, pp. 5314–5320.

[74] Broderix, K., Hundertmark, D. and Leschke, H. (2000). Continuity properties of schrödinger semigroups with magnetic fields, *Reviews in Mathematical Physics* **12**, 02, pp. 181–225.

[75] Busch, P., Lahti, P. and Mittelstaedt, P. (1996). *The quantum theory of measurement*, 2nd edn. (Springer-Verlag, Berlin).

[76] Caldeira, A. and Leggett, A. (1983). Path integral approach to quantum Brownian motion, *Physica* **121 A**, pp. 587–616.

[77] Cameron, R. (1960). A family of integrals serving to connect the Wiener and Feynman integrals, *J. Math. and Phys.* **39**, pp. 126–141.

[78] Cameron, R. (1968). Approximation to certain Feynman integrals, *Journal d'Analyse Math.* **21**, pp. 337–371.

[79] Cameron, R. and Storvick, D. (1966). A translation theorem for analytic Feynman integrals, *Trans. Am. Math. Soc.* **125**, pp. 1–6.

[80] Cameron, R. and Storvick, D. (1968). A Lindelöf theorem and analytic continuation for functions of several variables, with an application to the Feynman integral, in *Proc. Symposia Pure Mathem., XI* (Am. Math. Soc., Providence), pp. 149–156.

[81] Cameron, R. and Storvick, D. (1968). An operator valued function space integral and a related integral equation, *J. Math. and Mech.* **18**, pp. 517–552.

[82] Cameron, R. and Storvick, D. (1970). An integral equation related to Schrödinger equation with an application to integration in function space, in C. Gunning (ed.), *Problems in Analysis, A Sympos. In Honor of S. Bochner* (Princeton Univ. Press, Princeton), pp. 175–193.

[83] Cameron, R. and Storvick, D. (1980). Some Banach algebras of analytic Feynman integrable functionals, in *Analytic functions, Kozubnik, 1979* (Lecture Notes in Mathematics, n 798, Springer-Verlag, Berlin), pp. 18–67.

[84] Cameron, R. and Storvick, D. (1983). A simple definition of the Feynman integral with applications, *Memoirs of the American Mathematical Society* **228**, pp. 1–46.

[85] Cartier, P. and DeWitt-Morette, C. (1995). A new perspective on functional integration, *J. Math. Phys.* **36**, 5, pp. 2237–2312.

[86] Cartier, P. and DeWitt-Morette, C. (1997). A rigorous mathematical foundation of functional integration, in *Functional integration (Cargése, 1996). NATO Adv. Sci. Inst. Ser. B Phys., 361* (Plenum, New York), pp. 1–50.

[87] Cartier, P. and DeWitt-Morette, C. (2000). Functional integration, *J. Math. Phys.* **41**, 6, pp. 4154–4187.

[88] Cartier, P. and DeWitt-Morette, C. (2006). *Functional integration: action and symmetries* (Cambridge University Press, Cambridge).

[89] Chang, S. J. (1990). *Introduction to quantum field theory* (World Scientific, Singapore).

[90] Chazarain, J. (1974). Formula de Poisson pour les Variétés Riemmaniennes, *Invent. Math.* **24**, pp. 65–82.

[91] Chernoff, P. (1968). Note on product formulas for operator semigroups, *J. Functional Analysis* **2** pp. 238–242.

[92] Chernoff, P. (1974). Product formulas, nonlinear semigroups and addition of unbounded operators, *Mem. Am. Math. Soc.* **140**.

[93] Chiu, C. B., Misra, B. and Sudarshan, E. C. G. (1977). Time evolution of unstable quantum states and a resolution of Zeno's paradox, *Phys. Rev. D* **16**, 2, pp. 520–529.

[94] Chung, D. (1991). Conditional analytic Feynman integrals on Wiener spaces, *Proc. AMS* **122**, pp. 479–488.

[95] Chung, L. and Zambrini, J. (2001). *Introduction to random time and quantum randomness* (Monographs of the Portuguese Mathematical Society. McGraw-Hill, Lisbon).

[96] Cohn, D. (ed.) (1980). *Measure Theory* (Birkhäuser, Boston).

[97] Combe, P., Høegh-Krohn, R., Rodriguez, R., Sirugue, M. and Sirugue-Collin, M. (1978). Feynman path integrals and Poisson processes with piecewise classical paths, *J. Math. Phys.* **23**, 2, pp. 405–411.

[98] Combe, P., Høegh-Krohn, R., Rodriguez, R., Sirugue, M. and Sirugue-Collin, M. (1980). Poisson Processes on Groups and Feynman Path Integrals, *Commun. Math. Phys.* **77**, pp. 269–288.

[99] Combe, P., Høegh-Krohn, R., Rodriguez, R., Sirugue, M. and Sirugue-Collin, M. (1981). Generalized Poisson processes in quantum mechanics and field theory, *Phys. Rep.* **77**, pp. 221–233.

[100] Cruzeiro, A. and Zambrini, J. (1991). Feynman's functional calculus and stochastic calculus of variations, in *Stochastic analysis and applications (Lisbon, 1989)* (Progr. Probab., 26, Birkhuser Boston, Boston, MA), pp. 82–95.

[101] Cycon, H. L., Froese, R. G., Kirsch, W. and Simon, B. (2009). *Schrödinger operators: With application to quantum mechanics and global geometry* (Springer).

[102] Daubechies, I. and Klauder, J. R. (1982). Constructing measures for path integrals, *J . Math. Phys.* **23**, 10, pp. 1806–1822.

[103] Daubechies, I. and Klauder, J. R. (1983). Measures for more quadratic path integrals. *Lett. Math. Phys.* **7**, 3, pp. 229–234.

[104] Daubechies, I. and Klauder, J. R. (1984). Quantum-mechanical path integrals with Wiener measure for all polynomial Hamiltonians. *Phys. Rev. Lett.* **52**, 14, pp. 1161–1164.

[105] Daubechies, I. and Klauder, J. R. (1985). Quantum-mechanical path integrals with Wiener measure for all polynomial Hamiltonians II. *J. Math. Phys.* **26**, pp. 2239–2256.

[106] Davies, E. (1976). *Quantum Theory of Open Systems* (Academic Press. London).

[107] Davies, I. M. (1998). Laplace asymptotic expansions for Gaussian functional integrals, *Electron. J. Probab.* **3**, 13, pp. 1–10.

[108] Davies, I. M. and Truman, A. (1982). Laplace asymptotic expansions of conditional Wiener integrals and generalized Mehler kernel formulas, *J. Math. Phys.* **23**, 11, pp. 2059–2070.

[109] Davies, I. M. and Truman, A. (1982). Laplace expansions of conditional Wiener integrals and applications to quantum physics, in *Stochastic processes in quantum theory and statistical physics (Marseille, 1981)* (Lecture Notes in Phys., 173, Springer, Berlin), pp. 40–55.

[110] Davies, I. M. and Truman, A. (1983). On the Laplace asymptotic expansion of conditional Wiener integrals and the Bender-Wu formula for x^{2N}-anharmonic oscillators, *J. Math. Phys.* **24**, 2, pp. 255–266.

[111] de Verdière, Y. C. (2003). Singular Lagrangian manifolds and semiclassical analysis, *Duke Math. J.* **116**, 2, pp. 263–298.

[112] DeFacio, B., Johnson, J. and Lapidus, M. (1992). Feynman's operational calculus as a generalized path integral, in C. S. et al. (ed.), *Stochastic Processes. A Festrischritf in Honour of Gopinath Kallianpur* (Springer-Verlag. New York), pp. 51–60.

[113] DeFacio, B., Johnson, J. and Lapidus, M. (1997). Feynman's operational calculus and evolution equations, *Acta Applicandae Mathematicae* **47**, pp. 155–211.

[114] DeWitt-Morette, C. (1972). Feynman's path integral. Definition without limiting procedure, *Commun. Math. Phys.* **28**, pp. 47–67.

[115] DeWitt-Morette, C. (1974). Feynman path integrals, I. Linear and affine techniques, II. The Feynman – Green function, *Commun. Math. Phys.* **37**, pp. 63–81.

[116] DeWitt-Morette, C. (1984). Feynman path integrals, *Acta Phys. Austr. Suppl.* **XXVI**, pp. 101–170.

[117] DeWitt-Morette, C. (1984). Feynman path integrals: from the prodistribution definition to the calculation of glory scattering, in *Stochastic methods and computer techniques in quantum dynamics (Schladming, 1984)*, Vol. XXVI (Springer, Vienna), pp. 101–170.

[118] DeWitt-Morette, C., Maheshwari, A. and Nelson., B. (1977). Path integration in phase space, *General Relativity and Gravitation* **8**, 8, pp. 581–593.

[119] Diosi, L. (1988). Continuous quantum measurements and Ito formalism, *Phys. Lett. A* **129**, 8, pp. 419–423.

[120] Dirac, P. (1933). The Lagrangian in quantum mechanics, *Phys. Zeitschr. d. Sowjetunion* **3**, 1, pp. 64–72.

[121] Dirac, P. (1958). *The Principles of Quantum Mechanics* (Clarendon Press, Oxford).

[122] Doob, J. L. (1953). *Stochastic processes* (Wiley, New York).

[123] Doss, H. (1980). Sur une Résolution Stochastique de l'Equation de Schrödinger à Coefficients Analytiques, *Commun. Math. Phys.* **73**, pp. 247–264.

[124] Dreyfus, T. and Dym, H. (1978). Product formulas for the Eigenvalues of a Class of Boundary Value Problems, *Duke Math. J.* **45**, 1, pp. 15–37.

[125] Duistermaat, J. (1974). Oscillatory integrals, Lagrange immersions and unfoldings of singularities, *Comm. Pure Appl. Math.* **27**, pp. 207–281.

[126] Ekeland, I. (1990). *Convexity Methods in Hamiltonian Mechanics* (Springer Verlag, Berlin).

[127] Ellis, A. and Rosen, J. (1981). Asymptotic analysis of Gaussian integrals II. Manifold of mimimum points, *Commun. Math. Phys.* **82**, pp. 153–181.

[128] Ellis, A. and Rosen, J. (1982). Asymptotic analysis of Gaussian integrals I: Isolated minimum points, *Trans. Amer. Math. Soc.* **273**, pp. 447–481.

[129] Elworthy, D. and Truman, A. (1984). Feynman maps, Cameron-Martin formulae and anharmonic oscillators, *Ann. Inst. H. Poincaré Phys. Théor.* **41**, 2, pp. 115–142.

[130] Erdélyi, A. (1956). *Asymptotic expansions* (Dover Publications, Inc., New York).

[131] Ethier, S. N. and Kurtz, T. G. (2009). *Markov processes: characterization and convergence*, Vol. 282 (John Wiley & Sons).

[132] Exner, P. (1985). *Open quantum systems and Feynman integrals* (Reidel, Dortrecht).

[133] Faria, M. D., Oliveira, M. J. and Streit, L. (2005). Feynman integrals for nonsmooth and rapidly growing potentials, *J. Math. Phys.* **46**, 6, pp. 063505, 14 pp.

[134] Faria, M. D., Potthoff, J. and Streit, L. (1991). The Feynman integrand as a Hida distribution, *J. Math. Phys.* **32**, pp. 2123–2127.

[135] Fedoryuk, M. (1993). *Asymptotic analysis* (Springer-Verlag, Berlin).

[136] Feynman, R. and Hibbs, A. (1965). *Quantum mechanics and path integrals* (McGraw-Hill, New York).

[137] Feynman, R. and Vernon, F. (1963). The theory of a general quantum system interacting with linear dissipative system, *Ann. Phys.* **24**, pp. 118–173.

[138] Feynman, R. P. (1948). Space-time approach to non-relativistic quantum mechanics, *Rev. Mod. Phys.* **20**, pp. 367–387.

[139] Feynman, R. P. (1949). Space-time approach to quantum electro-dynamics, *Phys. Rev.* **76**, pp. 769–789.

[140] Feynman, R. P. (1949). The theory of positrons, *Phys. Rev.* **76**, pp. 749–759.

[141] Feynman, R. P. (1950). Mathematical formulation of the quantum theory of electromagnetic interaction, *Phys. Rev.* **80**, pp. 440–457.

[142] Feynman, R. P. (1951). An operator calculus having applications in quantum electrodynamics, *Phys. Rev.* **84**, pp. 108–128.

[143] Freidlin, M. I. (2016). *Functional Integration and Partial Differential Equations.(AM-109), Volume 109* (Princeton university press).

[144] Friedman, C. (1972). Semigroup product formulas, compression and continuous observation in quantum mechanics, *Indiana Univ. Math. J.* **21**.

[145] Fröhlich, J. and King, C. (1989). The Chern-Simons Theory and Knot Polynomials, *Comm. Math. Phys.* **126**, pp. 167–199.

[146] Fujiwara, D. (1980). Remarks on convergence of Feynman path integrals, *Duke Math. J.* **47**, pp. 559–600.

[147] Fujiwara, D. (1990). The Feynman path integral as an improper integral over the Sobolev space, in *Journées Équations aux Dérivées Partielles (Saint Jean de Monts, 1990), Exp. No. XIV* (École Polytech. Palaiseau), pp. 1–15.

[148] Fujiwara, D. (1991). The stationary phase method with an estimate of the remainder term on a space of large dimension, *Nagoya Math. J.* **124**, pp. 51–67.

[149] Fujiwara, D. (1993). Some Feynman path integrals as oscillatory integrals over a Sobolev manifold, in *Functional analysis and related topics,1991 (Kyoto)* (Lecture Notes in Math., 1540, Springer, Berlin), pp. 39–53.

[150] Fujiwara, D. (1999). *Mathematical Methods for Feynman Path Integrals.* (*Japanese*) (Springer, Tokyo,).

[151] Fujiwara, D. (2017). *Rigorous time slicing approach to Feynman path integrals* (Springer).

[152] Fujiwara, D. and Kumano-go, N. (2005). Smooth functional derivatives in Feynman path integrals by time slicing approximation, *Bull. Sci. math* **129**, pp. 57–79.

[153] Fujiwara, D. and Kumano-go, N. (2006). The second term of the semiclassical asymptotic expansion for Feynman path integrals with integrand of polynomial growth, *J. Math. Soc. Japan* **58**, 3, pp. 837–867.

[154] Fujiwara, D. and Kumano-go, N. (2008). Feynman Path Integrals and Semiclassical Approximation, *RIMS KoKyuroku Bessatsu* **B5**, pp. 241–263.

[155] Fujiwara, D. and Kumano-go, N. (2008). Phase space Feynman path integrals via piecewise bicharacteristic paths and their semiclassical approximations, *Bull. Sci. math.* **132**, pp. 313–357.

[156] Gatarek, D. and Gisin, N. (1991). Continuous quantum jumps and infinite-dimensional stochastic equation, *J. Math. Phys* **32**, 8, pp. 2152–2157.

[157] Gaveau, B. and Schulman, L. S. (1986). Explicit time-dependent Schrödinger propagators, *J. Phys. A* **19**, 10, pp. 1833–1846.

[158] Gelfand, I. and Vilenkin, N. (1964). *Generalized Functions. IV Applications of Harmonic Analysis* (Academic Press, New York).

[159] Ghandour, N. J. (2009). A note on the inverse limit of finite dimensional vector spaces, *Int. J. Algebra* **3**, 13-16, pp. 619–628.

[160] Gill, T. (1981). Time-ordered operators, I. Foundations for an alternate view of reality, *Trans. Amer. Math. Soc.* **266**, pp. 161–181.

[161] Gill, T. (1983). Time-ordered operators, II. *Trans. Amer. Math. Soc.* **279**, pp. 617–634.

[162] Gill, T. and Zachary, W. (1987). Time-ordered operators and Feynman-Dyson algebras, *J. Math. Phys.* **28**, pp. 1459–1470.

[163] Gill, T. L. and Zachary, W. W. (2016). *Functional analysis and the Feynman operator calculus* (Springer).

[164] Golubitsky, M. (1978). An Introduction to Catastrophe Theory and Applications, *SIAM Mathematical Review* **20**, pp. 352–387.

[165] Grabert, H., Inomata, A., Schulman, L. S. and Weiss, U. (eds.) (1993). *Path integrals from meV to MeV: Tutzing 1992* (World Scientific Publishing Co., Inc., River Edge, NJ).

[166] Grosche, C. (1996). *Path integrals, hyperbolic spaces, and Selberg trace formulae* (World Scientific Publishing Co., Inc, River Edge, NJ).

[167] Grosche, C. and Steiner, F. (1998). *Handbook of Feynman path integrals* (Springer Tracts in Modern Physics, 145. Springer-Verlag, , Berlin).

[168] Gross, L. (1962). Measurable functions on Hilbert spaces, *Trans. Amer. Mat. Soc.* **105**, pp. 375–390.

[169] Gross, L. (1965). Abstract Wiener Spaces, in 5th *Berkeley Symp. Math. Stat. Prob.*, pp. 31–42.

[170] Grothaus, M., Khandekar, D., da Silva, J. and Streit, L. (1997). The Feynman integral for time-dependent anharmonic oscillators, *J. Math. Phys.* **38**, 6, pp. 3278–3299.

[171] Guadagnini, E. (1993). *The link invariants of the Chern-Simons field theory*, de Gruyter Expositions in Mathematics, Vol. 10 (Walter de Gruyter & Co., Berlin).

[172] Gutzwiller, M. C. (1990). *Chaos in classical and quantum mechanics* (Springer-Verlag, New York).

[173] Gutzwiller, M. C., Inomata, A., Klauder, J. R. and Streit, L. (eds.) (1986). *Path integrals from meV to MeV* (Bielefeld Encounters in Physics and Mathematics, VII. World Scientific Publishing Co., Singapore).

[174] Hahn, A. (2001). *Chern-Simons theory on* $\mathbb{R}^3$ *in axial gauge*, Ph.D. thesis, Bonn University.

[175] Hahn, A. (2004). Chern-Simons theory on $\mathbb{R}^3$ in axial gauge: a rigorous approach, *J. Funct. Anal.* **211**, 2, pp. 483–507.

[176] Hahn, A. (2004). The Wilson loop observables of Chern-Simons theory on $\mathbb{R}^3$ in axial gauge, *Commun. Math. Phys.* **248**, 3, pp. 467–499.

[177] Hahn, A. (2005). An analytic approach to Turaevs shadow invariants, *SFB 611 preprint 233, Bonn* .

[178] Hahn, A. (2005). Chern-Simons models on $S^2 \times S^1$, torus gauge fixing, and link invariants I, *J. Geom. Phys.* **53**, 3, pp. 275–314.

[179] Hahn, A. (2005). White noise analysis, quantum field theory, and topology, in *Stochastic analysis: classical and quantum* (World Sci. Publ. Hackensack, NJ), pp. 13–30.

[180] Hardy, G. H. (1963). *Divergent series* (Oxford University Press, London).

[181] Helffer, B. (1988). *Semi-classical analysis for the Schrödinger operator and applications* (Lecture Notes in Mathematics, 1336. Springer-Verlag, Berlin).

[182] Hida, T. (2001). White noise approach to Feynman integrals, *J. Korean Math. Soc.* **38**, 2, pp. 275–281.

[183] Hida, T., Kuo, H., Potthoff, J. and Streit, L. (1993). *White Noise. An infinite-dimensional calculus* (Mathematics and its Applications, 253. Kluwer Academic Publishers Group, Dordrecht).

[184] Hille, E. and Phillips, R. S. (1996). *Functional analysis and semi-groups*, Vol. 31 (American Mathematical Soc.).

[185] Hochberg, K. J. (1978). A signed measure on path space related to wiener measure, *The Annals of Probability* , pp. 433–458.

[186] Hörmander, L. (1971). Fourier integral operators I, *Acta Math.* **127**, 1, pp. 79–183.

[187] Hörmander, L. (1983). *The Analysis of Linear Partial Differential Operators. I. Distribution Theory and Fourier Analysis* (Springer-Verlag, Berlin/Heidelberg/New York/Tokyo).

[188] Hudson, R. L. and Parthasarathy, K. R. (1984). Quantum Ito's Formula and Stochastic Evolutions, *Commun. Math. Phys.* **93**, pp. 301–323.

[189] Hurd, N. (1983). *Geometric quantization in action* (D. Reidel, Dortrecht).

[190] Ichinose, T. (1982). Path integral for the Dirac equation in two space-time dimensions, *Proc. Japan Acad. Ser. A Math. Sci.* **58**, 7, pp. 290–293.

[191] Ichinose, T. (1997). On the formulation of the Feynman path integral through broken line paths, *Comm. Math. Phys.* **189**, 1, pp. 17–33.

[192] Ichinose, T. (2003). Convergence of the Feynman path integral in the weighted Sobolev spaces and the representation of correlation functions, *J. Math. Soc. Japan* **55**, 4, pp. 957–983.

[193] Ichinose, T. and Tamura, H. (1984). Propagation of a Dirac particle. A path integral approach, *J. Math. Phys.* **25**, 6, pp. 1810–1819.

[194] Ikeda, N. and Watanabe, S. (2014). *Stochastic differential equations and diffusion processes* (Elsevier).

[195] Ito, K. (1961). Wiener integral and Feynman integral, in *Proc. Fourth Berkeley Symp. Math. Stat. and Prob.*, Vol. II (Univ. California Press, Berkeley), pp. 227–238.

[196] Ito, K. (1967). Generalized uniform complex measures in the Hilbertian metric space with their application to the Feynman path integral, in *Proc. Fifth Berkeley Symposium on Mathematical Statistics and Probability*, Vol. II.1 (Univ. California Press, Berkeley), pp. 145–161.

[197] Itzykson, C. and Zuber, J. (1980). *Quantum field theory* (McGraw-Hill International Book Co., New York).

[198] Johnson, G. (1982). The equivalence of two approaches to Feynman integral, *J. Math. Phys.* **23**, pp. 2090–2096.

[199] Johnson, G. and Kallianpur, G. (1993). Homogeneous chaos, p-forms, scaling and the Feynman integral, *Trans. Amer. Math. Soc.* **340**, 2, pp. 503–548.

[200] Johnson, G. and Kim, J. G. (2000). A dominated-type convergence theorem for the Feynman integral, *J. Math. Phys.* **41**, 5, pp. 3104–3112.

[201] Johnson, G. and Lapidus, M. (1986). Generalized Dyson series, generalized Feynman diagrams, the Feynman integral and Feynman's operational calculus, *Mem. Amer. Math. Soc.* **62**, 351, pp. 1–78.

[202] Johnson, G. and Lapidus, M. (1988). Noncommutative operations on Wiener functionals and Feynman's operational calculus, *J. Funct. Anal.* **81**, 1, pp. 74–99.

[203] Johnson, G. and Lapidus, M. (2000). *The Feynman integral and Feynman's operational calculus* (Oxford University Press, New York).

[204] Johnson, G. and Skoug, D. (1970). Operator-valued Feynman integrals of certain finite-dimensional functionals, *Proc. Am. Math. Soc.* **24**, pp. 774–780.

[205] Johnson, G. and Skoug, D. (1970). Operator-valued Feynman integrals of finite-dimensional functionals, *Pac. Journ. of Math.* **34**, pp. 415–425.

[206] Johnson, G. and Skoug, D. (1971). Feynman integrals of non-factorable finite-dimensional functionals, *Pac. Journ. of Math.* **37**, pp. 303–317.

[207] Johnson, G. and Skoug, D. (1971). An operator valued function space integral: A sequel to Cameron and Storvick's paper, *Proc. Am. Math. Soc.* **27**, pp. 514–518.

[208] Johnson, G. and Skoug, D. (1973). A Banach algebra of Feynman integrable functionals with application to an integral equation formally equivalent to Schrödinger's equation, *J. Funct. Analysis* **12**, pp. 129–152.

[209] Johnson, G. and Skoug, D. (1974). A function space integral for a Banach space of functionals on Wiener space, *Proc. Am. Math. Soc.* **43**, pp. 141–148.

[210] Johnson, G. W., Lapidus, M. L. and Nielsen, L. (2015). *Feynman's operational calculus and beyond: noncommutativity and time-ordering* (Oxford Mathematical Monographs).

[211] Kac, M. (1949). On distributions of certain Wiener functionals, *Trans. Amer. Math. Soc.* **65**, pp. 1–13.

[212] Kac, M. (1951). On some connections between probability theory and differential and integral equations, in *Proc. Second Berkeley Symp.* (Univ. California Press, Berkeley), pp. 189–215.

[213] Kac, M. (1980). *Integration in function spaces and some of its applications* (Accademia Nazionale dei Lincei, Pisa).

[214] Kallianpur, G. and Bromley, C. (1984). Generalized Feynman integrals using analytic continuation in several complex variables, in *Stochastic analysis and applications* (Adv. Probab. Related Topics, 7, Dekker, New York), pp. 217–267.

[215] Kallianpur, G., Kannan, D. and Karandikar, R. (1985). Analytic and sequential Feynman integrals on abstract Wiener and Hilbert spaces, and a Cameron Martin formula, *Ann. Inst. H. Poincaré, Prob. Th.* **21**, pp. 323–361.

[216] Kallianpur, G. and Oodaira, H. (1978). Freĭdlin-Wentzell type estimates for abstract Wiener spaces, *Sankhyā Ser. A* **40**, 2, pp. 116–137.

[217] Kallianpur, G. and Üstünel, A. S. (1992). Distributions, Feynman integrals and measures on abstract Wiener spaces, in *Stochastic analysis and related topics (Silivri, 1990)* (Birkhuser. Boston (MA)), pp. 237–284.

[218] Kanasugi, H. and Okada, H. (1975). Systematic Treatment of General Time-Dependent Harmonic Oscillator in Classical and Quantum Mechanics, *Progr. Theoret. Phys.* **16**, 2, pp. 384–388.

[219] Karatzas, I. and Shreve, S. (1988). *Brownian motion and stochastic calculus* (Springer-Verlag).

[220] Kato, T. (1966). *Perturbation Theory for Linear Operators* (Springer).

[221] Kauffman, L. (1995). Functional integration and the theory of knots, *J. Math. Phys.* **36**, 5, pp. 2402–2429.

[222] Kauffman, L. (1997). An introduction to knot theory and functional integrals, in *Functional integration (Cargèse, 1996)* (NATO Adv. Sci. Inst. Ser. B Phys., 361, Plenum, New York), pp. 247–308.

[223] Kauffman, L. (1998). Witten's integral and the Kontsevich integral, in *Particles, fields, and gravitation (Łódź, 1998)* (AIP Conf. Proc., 453, Amer. Inst. Phys., Woodbury, NY), pp. 368–381.

[224] Kauffman, L. (2004). Vassiliev invariants and functional integration without integration, in *Stochastic analysis and mathematical physics (SAMP/ANESTOC 2002)* (World Sci. Publishing, River Edge, NJ), pp. 91–114.

[225] Kelvin, L. (1887). *Phil. Mag.* **23**, pp. 252–255.

[226] Khandekar, D. C. and Lawande, S. V. (1975). Exact propagator for a time-dependent harmonic oscillator with and without a singular perturbation, *J. Math. Phys.* **16**, 2, pp. 384–388.

[227] Kinoshita, T. (1990). *Quantum Electrodynamics* (World Scientific, Singapore).

[228] Klauder, J. (1974). *Functional techniques and their application in quantum field theory* (Colorado Assoc. Univ. Press), pp. 329–421.

[229] Klauder, J. R. (2010). *A modern approach to functional integration* (Springer Science & Business Media).

[230] Kleinert, H. (2004). *Path integrals in quantum mechanics, statistics, polymer physics, and financial markets*, 3rd edn. (World Scientific Publishing Co., Inc., River Edge, NJ).

[231] Kolokoltsov, V. N. (1999). Complex measures on path space: an introduction to the Feynman integral applied to the Schrödinger equation, *Methodol. Comput. Appl. Probab.* **1**, 3, pp. 349–365.

[232] Kolokoltsov, V. N. (2000). *Semiclassical analysis for diffusions and stochastic processes* (Lecture Notes in Mathematics, 1724. Springer-Verlag, Berlin).

[233] Kostantinov, A., Maslov, V. and Chebotarev, A. (1990). Probability representations of solutions of the Cauchy problem for quantum mechanical equations, *Russian Math. Surveys* **45**, 6, pp. 1–26.

[234] Krylov, V. Y. (1960). Some properties of the distribution corresponding to equation $\partial u/\partial t = (-1)^{q+1}\partial^{2q}u/\partial x^{2q}$, *Doklady Akademii Nauk* **132**, 6, pp. 1254–1257.

[235] Kumano-go, N. (2004). Feynman path integrals as analysis on path space by time slicing approximation, *Bull. Sci. Math.* **128**, 3, pp. 197–251.

[236] Kuna, T., Streit, L. and Westerkamp, W. (1998). Feynman integrals for a class of exponentially growing potentials, *J. Math. Phys.* **39**, 9, pp. 4476–4491.

[237] Kuo, H. (1975). *Gaussian Measures in Banach Spaces* (Lecture Notes in Math., Springer-Verlag Berlin-Heidelberg-New York).

[238] Lapidus, M. (1987). The Feynman-Kac formula with a Lebesgue-Stieltjes measure and Feynman's operational calculus, *Stud. Appl. Math.* **76**, 2, pp. 93–132.

[239] Lapidus, M. (1996). The Feynman integral and Feynman's operational calculus: a heuristic and mathematical introduction, *Ann. Math. Blaise Pascal* **3**, 1, pp. 89–102.

[240] Lascheck, A., Leukert, P., Streit, L. and Westerkamp, W. (1993). Quantum Mechanical Propagators in terms of Hida Distribution, *Rep. Math. Phys.* **33**, pp. 221–232.

[241] Lefschetz, S. (1942). *Algebraic topology*, Vol. 27 (American Mathematical Soc.).

[242] Leukert, S. and Schäfer, J. (1996). A Rigorous Construction of Abelian Chern-Simons Path Integral Using White Noise Analysis, *Reviews in Math. Phys.* **8**, pp. 445–456.

[243] Malgrange, B. (1995). Sommations des séries divergentes, *Expo. Math.* **13**, pp. 163–222.

[244] Malgrange, B. and Ramis, J. P. (1992). Fonctions Multisommables, *Ann. Inst. Fourier* **42**, 1-2, pp. 353–368.

[245] Mandrekar, V. (1983). Some remarks on various definitions of Feynman integrals, in K. J. Beck (ed.), *Lectures Notes Math.* (Springer. Berlin), pp. 170–177.

[246] Manoliu, M. (1998). Abelian Chern-Simons theory. II. A functional integral approach, *J. Math. Phys.* **39**, 1, pp. 207–217.

[247] Maslov, V. P. (1971). Characteristics of pseudo-differential operators, in *Proc. Internat. Congress Math. (Nice 1970) vol 2* (Gautier-Villars, Paris), pp. 290–294.

[248] Maslov, V. P. (1972). *Théorie des perturbations et méthodes asymptotiques* (Dunod, Paris).

[249] Maslov, V. P. and Chebotarev, A. M. (1979). Processus à sauts et leur application dans la mécanique quantique, in S. A. et al. (ed.), *Feynman path integrals* (Springer Lecture Notes in Physics 106), pp. 58–72.

[250] Maslov, V. P. and Fedoryuk, M. V. (1981). *Semiclassical approximation in quantum mechanics* (Reidel Publishing Co., Dordrecht-Boston, Mass.).

[251] Mazzucchi, S. (2008). Feynman path integrals for the inverse quartic oscillator, *J. Math. Phys.* **49**, 9, p. 093502 (15 pages).

[252] Mazzucchi, S. (2013). Probabilistic representations for the solution of higher order differential equations, *International Journal of Partial Differential Equations* **2013**.

[253] Mazzucchi, S. (2018). Infinite dimensional oscillatory integrals with polynomial phase and applications to higher-order heat-type equations, *Potential Analysis* **49**, 2, pp. 209–223.

[254] Mensky, M. (1993). *Continuous Quantum Measurements and Path Integrals* (Taylor & Francis, Bristol and Philadelphia).

[255] Mensky, M. (2000). *Quantum measurements and decoherence. Models and phenomenology* (Fundamental Theories of Physics, 110. Kluwer Academic Publishers, Dordrecht).

[256] Milnor, J. (1963). *Morse theory* (Princeton University Press).

[257] Milnor, J. (1965). *Topology from the differentiable viewpoint* (The University Press of Virginia, Charlottesville).

[258] Misra, B. and Sudarshan, E. C. G. (1975). The Zeno's paradox in quantum theory, *J. Math. Phys.* **10**, 4, pp. 756–763.

[259] Moretti, V. (2017). *Spectral theory and quantum mechanics. Second edition* (Springer).

[260] Muldowney, P. (1987). *A general theory of integration in function spaces, including Wiener and Feynman integration* (Pitman Research Notes in Mathematics Series, 153. Longman Scientific & Technical, Harlow; John Wiley & Sons, Inc., New York).

[261] Nakamura, T. (1991). A nonstandard representation of Feynman's path integrals, *J. Math. Phys.* **32**, 2, pp. 457–463.

[262] Nakamura, T. (1997). Path space measures for Dirac and Schrödinger equations: nonstandard analytical approach, *J. Math. Phys.* **38**, 8, pp. 4052–4072.

[263] Nakamura, T. (2000). Path space measure for the $3 + 1$-dimensional Dirac equation in momentum space, *J. Math. Phys.* **41**, 8, pp. 5209–5222.

[264] Nelson, B. and Sheeks, B. (1982). Path integration for velocity-dependent potentials, *Comm. Math. Phys.* **84**, 4, pp. 515–530.

[265] Nelson, E. (1964). Feynman integrals and the Schrödinger equation, *J. Math. Phys.* **5**, pp. 332–343.

[266] Nevanlinna, F. (1919). Zur Theorie der asymptotischen Potenzreihen, *Ann. Acad. Sci. Fenn. (A)* **12**, 3, pp. 1–81.

[267] Neveu, J. (1965). *Mathematical Foundations of the Calculus of Probability* (Holden-Day, Inc).

[268] Nicola, F. and Trapasso, S. I. (2020). On the pointwise convergence of the integral kernels in the Feynman-Trotter formula, *Communications in Mathematical Physics* **376**, 3, pp. 2277–2299.

[269] Olver, F. W. J. (1974). *Asymptotics and special functions.* (Computer Science and Applied Mathematics. Academic Press, New York-London).

[270] Osborn, T., Papiez, L. and Corns, R. (1987). Constructive representations of propagators for quantum systems with electromagnetic fields, *Journal of mathematical physics* **28**, 1, pp. 103–123.

[271] Parisi, G. (1988). *Statistical Field Theory* (Addison-Wesley, Redwood City, Cal.).

[272] Pechukas, P. and Light, J. C. (1966). On the Exponential Form of Time-Displacement Operators in Quantum Mechanics, *J. Chem. Phys.* **44**, 10, pp. 3897–3912.

[273] Peres, A. (1980). Zeno paradox in quantum theory, *Am. J. Phys.* **48**, 11, pp. 931–932.

[274] Pesin, I. N. (2014). *Classical and modern integration theories* (Academic Press).

[275] Piterbarg, V. I. (1996). *Asymptotic methods in the theory of Gaussian processes and fields* (Translations of Mathematical Monographs, 148. American Mathematical Society, Providence, RI).

[276] Piterbarg, V. I. and Fatalov, V. R. (1995). The Laplace method for probability measures in Banach spaces, *Russian Math. Surveys* **50**, 6, pp. 1151–1239.

[277] Protter, P. (1990). *Stochastic Integration and Differential Equations* (Applications of Mathematics 21, Springer-Verlag).

[278] Ramer, R. (1974). On nonlinear transformations of gaussian measures, *Journal of Functional Analysis* **15**, 2, pp. 166–187.

[279] Ramis, J. P. (1993). Divergent series and asymptotic theories, *Bull. Soc. Math. France, Panoramas et Synthèses* .

[280] Reed, M. and Simon, B. (1973). *Methods of Modern Mathematical Physics I. Functional Analysis* (Academic Press, New York).

[281] Reed, M. and Simon, B. (1975). *Methods of Modern Mathematical Physics II. Fourier Analysis, Self-Adjointness* (Academic Press, New York).

[282] Remizov, I. D. (2016). Quasi-Feynman formulas–a method of obtaining the evolution operator for the Schrödinger equation, *Journal of Functional Analysis* **270**, 12, pp. 4540–4557.

[283] Rezende, J. (1984). Quantum systems with time dependent harmonic part and the Morse index, *J. Math. Phys.* **25**, 11, pp. 3264–3269.

[284] Rezende, J. (1985). The method of stationary phase for oscillatory integrals on Hilbert spaces, *Comm. Math. Phys.* **101**, pp. 187–206.

[285] Rezende, J. (1993). A note on Borel summability, *J. Math. Phys.* **34**, 9, pp. 4330–4339.

[286] Rezende, J. (1994). Feynman integrals and Fredholm determinants, *J. Math. Phys.* **35**, 8, pp. 4357–4371.

[287] Rezende, J. (1996). Time-Dependent Linear Hamiltonian Systems and Quantum Mechanics, *Lett. Math. Phys.* **38**, pp. 117–127.

[288] Robert, D. (1987). *Autour de l'approximation semi-classique. (French) [On semiclassical approximation]* (Progress in Mathematics, 68. Birkhauser Boston, Inc., Boston, MA).

[289] Rogers, A. (1987). Supersymmetric path integration, *Phys. Lett. B* **193**, 1, pp. 48–54.

[290] Rotman, J. J. (2008). *An introduction to homological algebra* (Springer Science & Business Media).

[291] Rudin, W. (1962). *Fourier Analysis on Groups* (Interscience publishers, New York).

[292] Rudin, W. (1973). *Functional Analysis* (McGraw-Hill, New York).

[293] Rudin, W. (1974). *Real and Complex Analysis* (McGraw-Hill, New York).

[294] Sa-yakanit, V., Sritrakool, W. and Berananda, J.-O. (eds.) (1989). *Path integrals from meV to MeV* (World Scientific Publishing Co., Inc., Teaneck, NJ).

[295] Schilder, M. (1966). Some asymptotic formulas for Wiener integrals, *Trans. Amer. Math. Soc.* **125**, pp. 63–85.

[296] Schulman, L. S. (1981). *Techniques and applications of path integration* (John Wiley & Sons, Inc., New York).

[297] Schwartz, A. S. (1978). The partition function of a degenerate quadratic functional and Ray-Singer invariants, *Lett. Math. Phys* **2**, pp. 247–252.

[298] Schwartz, A. S. (1979). The partition function of a degenerate functional, *Comm. Math. Phys.* **67**, 1, pp. 1–16.

[299] Shavgulidze, E. T. (1996). Solutions of Schrödinger equations with polynomial potential by path integrals, in *Path integrals: Dubna '96* (Joint Inst. Nuclear Res.), pp. 269–272.

[300] Silva, J. L. and Streit, L. (2004). Feynman integrals and white noise analysis, in *Stochastic analysis and mathematical physics (SAMP/ANESTOC 2002)* (World Sci. Publ., River Edge, NJ), pp. 285–303.

[301] Simon, B. (1979). *Trace ideals and their applications* (London Mathematical Society Lecture Note Series 35. Cambridge U.P., Cambridge).

[302] Simon, B. (2005). *Functional integration and quantum physics*, 2nd edn. (AMS Chelsea Publishing, Providence, RI).

[303] Skoug, D. (1974). Partial differential Systems of generalized Wiener and Feynman integrals, *Portug. Math.* **33**, pp. 27–33.

[304] Smolyanov, O. and Shavgulidze, E. (1990). *Path integrals. (Russian)* (Moskov. Gos. Univ., Moscow).

[305] Smolyanov, O. and Shavgulidze, E. (2003). Feynman formulas for solutions of infinite-dimensional Schrödinger equations with polynomial potentials. (Russian), *Dokl. Akad. Nauk.* **390**, 3, pp. 321–324.

[306] Smorodina, N. V. and Faddeev, M. M. (2010). Convergence of independent random variable sum distributions to signed measures and applications to the large deviations problem, *Theory of Stochastic Processes* **16**, 1, pp. 94–102.

[307] Sokal, A. D. (1980). An improvement of Watson's theorem on Borel summability, *J. Math. Phys.* **21**, 2, pp. 261–263.

[308] Sommerfeld, A. (1954). *Optics. Lectures on theoretical physics, Vol. IV* (Academic Press Inc., New York).

[309] Stokes, G. G. (1857). *Camb. Phil. Trans.* **10**, pp. 106–128.

[310] Streit, L. (2003). *Feynman paths, sticky walls, white noise* (World Sci. Publ. River Edge (NJ)), pp. 105–113.

[311] Sugita, H. and Taniguchi, S. (1998). Oscillatory integrals with quadratic phase function on a real abstract Wiener space, *J. Funct. Anal.* **155**, 1, pp. 229–262. '

[312] Sugita, H. and Taniguchi, S. (1999). A remark on stochastic oscillatory integrals with respect to a pinned Wiener measure, *Kyushu J. Math.* **53**, 1, pp. 151–162.

[313] Takeo, F. (1989). Generalized vector measures and path integrals for hyperbolic systems, *Hokkaido Math. J.* **18**, 3, pp. 497–511.

[314] Taniguchi, S. (1998). On the exponential decay of oscillatory integrals on an abstract Wiener space, *J. Funct. Anal.* **154**, 2, pp. 424–443.

[315] Taniguchi, S. (1998). Oscillatory integrals with quadratic phase function on a real abstract Wiener space, *J. Funct. Anal.* **155**, 1, pp. 229–262.

[316] Taniguchi, S. (1998). A remark on stochastic oscillatory integrals with respect to a pinned Wiener measure, *Kyushu J. Math.* **53**, 1, pp. 151–162.

[317] Thaler, H. (2003). Solution of Schrödinger equations on compact Lie groups via probabilistic methods, *Potential Anal.* **18**, 2, pp. 119–140.

[318] Thaler, H. (2005). The Doss Trick on Symmetric Spaces, *Letters in Mathematical Physics* **72**, pp. 115–127.

[319] Thomas, E. G. (2001). Projective limits of complex measures and martingale convergence, *Probab. Theory Related Fields* **119**, 4, pp. 579–588.

[320] Trotter, H. (1959). On the product of semigroups of operators, *Proc. Amer. Math. Soc.* **10**, pp. 545–551.

[321] Truman, A. (1976). Feynman path integrals and quantum mechanics as $\hbar \to 0$, *J. Math. Phys.* **17**, pp. 1852–1862.

[322] Truman, A. (1977). Classical mechanics, the diffusion (heat) equation, and the Schrödinger equation, *J. Math. Phys.* **18**, pp. 2308–2315.

[323] Truman, A. (1978). The Feynman maps and the Wiener integral, *J. Math. Phys.* **19**, pp. 1742–1750.

[324] Truman, A. (1979). The polygonal path formulation of the Feynman path integral, in *Feynman path integrals* (Springer), pp. 73–102.

[325] Truman, A. and Zastawniak, T. (1999). Stochastic PDE's of Schrödinger type and stochastic Mehler kernels—a path integral approach, in *Seminar on Stochastic Analysis, Random Fields and Applications (Ascona, 1996)* (Birkhäuser, Basel), pp. 275–282.

[326] Truman, A. and Zastawniak, T. (2001). Stochastic Mehler kernels via oscillatory path integrals, *J. Korean Math. Soc.* **38**, 2, pp. 469–483.

[327] Watling, K. D. (1992). Formulae for solutions to (possibly degenerate) diffusion equations exhibiting semi-classical asymptotics, in *Stochastics and quantum mechanics (Swansea, 1990)* (World Sci. Publ., River Edge, NJ), pp. 248–271.

[328] Watson, G. N. (1911). A Theory of Asymptotic Series, *Philosophical Transaction of the Royal Society of London, ser A* **CCXI**, pp. 279–31.

[329] Weinberg, S. (2005). *The quantum theory of fields. I. Foundations* (Cambridge University Press, Cambridge).

[330] Weinberg, S. (2005). *The quantum theory of fields. II. Modern applications* (Cambridge University Press, Cambridge).

[331] Wiegel, F. W. (1986). *Introduction to path-integral methods in physics and polymer science* (World Scientific Publishing Co., Singapore).

[332] Wiener, N. (1976). *Norbert Wiener. Collected Works.* (MIT Press, Cambridge).

[333] Witten, E. (1989). Quantum field theory and the Jones polynomial, *Commun. Math. Phys.* **121**, pp. 353–389.

[334] Yajima, K. (1996). Smoothness and Non-Smoothness of the Fundamental Solution of Time Dependent Schrödiger Equations, *Commun. Math. Phys.* **181**, pp. 605–629.

[335] Yamasaki, Y. (1985). *Measures on infinite dimensional spaces*, Vol. 5 (World Scientific).

[336] Zambrini, J. C. (2001). Feynman integrals, diffusion processes and quantum symplectic two-forms, *J. Korean Math. Soc.* **38**, 2, pp. 385–408.

[337] Zastawniak, T. J. (1988). Path integrals for the telegrapher's and Dirac equations; the analytic family of measures and the underlying Poisson process, *Bull. Pol. Acad. Sci. Math.* **36**, 5-6, pp. 341–356.

[338] Zastawniak, T. J. (1989). The nonexistence of the path-space measure for the Dirac equation in four space-time dimensions, *J. Math. Phys.* **30**, 6, pp. 1354–1358.

[339] Zastawniak, T. J. (1989). Path integrals for the Dirac equation—some recent developments in mathematical theory, in *Stochastic analysis, path integration and dynamics (Warwick, 1987)* (Pitman Res. Notes Math. Ser., 200, Longman Sci. Tech., Harlow), pp. 243–263.

[340] Zinn-Justin, J. (1993). *Quantum field theory and critical phenomena*, 2nd edn. (Oxford University Press, New York).

Index

σ-algebra , 19
  Borel, 20, 74
  generated by a collection of sets, 19
  product, 20

abstract Wiener spaces, 81, 162, 200
action functional, 1, 9, 189
Airy integrals, 128
algebra of sets, 26
anharmonic oscillator, 15, 169, 174, 198
asymptotic expansion, 232
asymptotic sequence, 232

Belavkin equation, 284, 295
Bochner theorem, 44
Bochner-Milnos-Sazonov theorem, 78
Borel summability, 234
Borel transform, 234
Brownian motion, 7

Cameron-Martin space, 6, 116, 170
Caratheodory theorem, 29
Carleman determinant, 153
coherent states, 309
consistent family of measures, 52
critical set, 129, 131, 238
cylinder function, 57, 90
cylinder measure, 80
cylinder set, 8, 52, 56, 62, 66, 75

decoherence, 281

degenerate critical points, 242, 256, 273
direct limit, 91
direct system, 91
directed set, 68
Dyson expansion, 124, 177, 212, 306

Euler-Lagrange equation of motion, 1
evolution operator, 2, 3, 170

Feynman measure, 12
Feynman operational calculus, 320
Feynman path integrals
  nonstandard analysis, 320
  as analytic continuation of Wiener integrals, 14, 208, 307
  in terms of complex Poisson measures, 14, 318
  phase space, 188, 314
  sequential approach, 311
  solution of the Schrödinger equation, 170, 174, 175
  white noise approach, 14, 314
Feynman-Kac formula, 8, 67
focal point, 263
Fourier transform of complex measures, 42, 77
Fredholm determinant, 147, 262
  regularized, 153
Fresnel algebra, 15
Fresnel integrals, 12, 14, 99, 128, 133
  definition, 133

with polynomial phase function,
135
Fubini theorem, 115, 149

Gaussian measure, 200
generalized Fresnel integrals, 135
Gevrey class, 234
Green function, 3, 181, 186, 198, 278

Hamilton's least action principle, 1, 3
Hamiltonian operator, 2, 169, 179,
　　182, 187, 199
harmonic oscillator
　　linearly forced, 179, 278
　　time dependent frequency, 182
heat equation, 7, 60, 67, 208
heat semigroup, 10, 208
high-order heat-type equation, 104,
　　121

infinite dimensional Fresnel integrals,
　　15, 110
infinite dimensional Fresnel integrals
　　with polynomial phase, 119
infinite dimensional oscillatory
　　integrals, 14, 145
　　class p normalized, 153
　　definition, 145
　　Fubini theorem, 149
　　normalized with respect to an
　　　　operator B, 152
　　translation formula, 149
　　with complex phase, 295
influence functional, 279, 285

Kolmogorov theorem, 52, 70

Lagrangian, 1

magnetic field, 5, 210
Markov transition function, 59
measurable functions, 30
measurable space, 19
measure, 22
measure
　　Borel, 25, 74

complex, 38, 74
finitely additive, 27
Gaussian, 79
image, 32, 38
Lebesgue, 7, 33, 46
Radon, 25
regular, 25
total variation, 39
measure space, 22
Morse theorem, 257, 273

Nevanlinna's theorem, 235

oscillatory integrals
　　definition, 129
　　on $\mathbb{R}^n$, 127
　　on a real Hilbert space, 145

Parseval-type equality, 15, 133, 143,
　　148, 152, 154, 160, 165, 296, 315
phase function, 238
phase space Feynman path integrals,
　　188, 314
Polish space, 25
polygonal path approximation, 4, 6,
　　178, 313
polynomial phase functions, 135, 154
posterior dynamics, 282
prior dynamics, 281
projective (or inverse) limit, 69
projective family of measures, 52
projective family of sets, 68
projective system of functionals, 89

quantum Brownian motion, 279, 295
quantum measurement, 277, 281, 312
quartic potential, 198

restricted path integrals, 283
Riesz-Markov representation
　　theorem, 42

Schrödinger equation, 2, 169, 174,
　　198, 210
semiclassical limit, 3, 258, 314
semigroup, 207

stationary phase method, 3, 128, 259
  finite dimensional case, 237
  infinite dimensional case, 246
stochastic extension, 84
stochastic Schrödinger equation, 284,
  295
symbols, 130

time dependent potentials, 179
time slicing approximation, 313
total variation of a complex measure,
  11

trace formula
  Gutzwiller, 269
  Selberg, 269
transition amplitude, 278
Trotter product formula, 3, 10, 189,
  311, 312

Wiener measure, 7, 8, 11, 59, 66, 82,
  200, 307, 309
Wiener space, 82

Zeno effect, 283

Printed in the United States
by Baker & Taylor Publisher Services